Encyclopedia of Botany
Volume I

Edited by **Austin Balfour**

R CALLISTO
REFERENCE

New York

Published by Callisto Reference,
106 Park Avenue, Suite 200,
New York, NY 10016, USA
www.callistoreference.com

Encyclopedia of Botany: Volume I
Edited by Austin Balfour

International Standard Book Number: 978-1-63239-217-6 (Hardback)

Printed in the United States of America.

Contents

Preface

Botany as we all know is also known as the science of plants, basically this field of biology deals with the study of plant life. Though fungi and algae have some non-green species their study is also included in botany. It is believed that approx. 4 lac species are studied under this field of biology. Botany is one of the most ancient field of science as it has been studied since the evolution of humans. They started identifying the harmful and useful plants at a very early age. Finally, humans started cultivating edible and medicinal plants in the early 15th century; Padua botanical garden is one of the most ancient gardens to have studied plants with written records and botanical study parameters.

Now in present age the advancement in technology and technical assistance is used to study plants; some of the most advanced techniques present today are electron and optical microscopy, cell imaging, enzymes analysis, genetic engineering, etc. The advancement of botany is even capable to create hybrids of plants these days, and this is because botanists invented the techniques to explore molecular genetics, DNA sequencing to classify species into classes more specifically.

I would especially like to thank the publisher for believing in me and giving me this immense opportunity. I also want to thank my family and friends for their endless support. Finally, I would like to thank all the contributing authors, publishing team members and my family for the support they gave me throughout the project.

Editor

Glomus intraradices Attenuates the Negative Effect of Low Pi Supply on Photosynthesis and Growth of Papaya Maradol Plants

Nava-Gutiérrez Yolanda,[1,2] Ronald Ferrera-Cerrato,[3] and Jorge M. Santamaría[1]

[1] *Laboratorio de Fisiología Vegetal Molecular, Centro de Investigación Científica de Yucatán, Calle 43 No. 130, Colonia Chuburná de Hidalgo, Mérida, YUC, México 97200, Mexico*
[2] *Laboratorio de Micorrizas, Centro de Investigación en Ciencias Biológicas (CICB), Universidad Autónoma de Tlaxcala, Ixtacuixtla, TLAX, México 90000, Mexico*
[3] *Área de Microbiología, Especialidad de Edafología, Instituto de Recursos Naturales, Colegio de Postgraduados Campus, Montecillo, MEX, México 56230, Mexico*

Correspondence should be addressed to Jorge M. Santamaría, jorgesm@cicy.mx

Academic Editor: Kang Chong

Low inorganic phosphorus (Pi) supply limits the photosynthetic process and hence plants growth and development. Contradictory reports exist in the literature on whether mycorrhyzal association can attenuate the negative effects of low Pi supply on photosynthesis and growth. In the present paper, the effect that low Pi supply may have on photosynthesis and growth of papaya Maradol plants was evaluated in intact plants and in those inoculated with two different strains of the arbuscular mycorrhizal fungi *Glomus intraradices*. Plant growth was significantly reduced as the Pi supply decreased. However, inoculation with any strain of *G. intraradices* was able to attenuate such effect. Without Pi in the nutrient solution, the mycorrhizal plants had on average 6.1 times and 7.5 higher photosynthesis than non mycorrhizal plants. The chlorophyll fluorescence values were significantly higher in mycorrhizal than in non-mycorrhizal plants. These results could be associated to an increased ability of mycorhyzal plants to take up Pi from the substrate, as they had higher Pi content than non-mycorrhizal plants. A high correlation was found between internal Pi content and plant biomass. The lower correlation between Pi content and photosynthesis, suggests that some photosynthates could had been used to maintain the symbiosis.

1. Introduction

The scarcity of nutrients is a factor that limits physiological processes in plants, including photosynthesis, their growth, and development [1]. Phosphorus (P) deficiency normally reduces the root system development and plant establishment because it has an important role in cell division, growth, photosynthesis (*Pn*), respiration, energy storage, and transfer [2–4]. So that, in many agricultural systems, it is necessary to supply P in order to have good productivity [5].

When plants grow with nutrient scarcity, some of them are capable of modifying their root architecture, to exude organic acids, or to establish associations with some beneficial organisms as strategies to compensate for the low concentration of nutrient in the substrate. Among the beneficial organisms for plants, the arbuscular mycorrhizal (AM) fungi have a very important role in plant nutrition

[6, 7]. It is through AM association that plants increase their capacity to take up organic phosphorus (Pi) from the soil solution [8–10] that normally translates into better growth than that of non-AM plants [11, 12].

Nevertheless, it has been reported that AM association could be more or less beneficial to the plant growth depending on the plant species and the AM fungi species [8]. In commercial papaya varieties Sunrise and Tainung, the inoculation with *Gigaspora margarita* induced higher production of dry matter than with *Glomus clarum*, growing under similar conditions [13].

Most of the studies on the role of the AM association on plants behavior are undertaken using soil [13] or a mix of inert materials-soil [11, 14, 15] with different chemical properties in every case, which is a factor that can modify the Pi availability for the plants. In order to control the availability of nutrients, the present research was made in a hydroponics system under greenhouse conditions, focusing on the effect of colonisation with two strains of *Gloms intraradices* on the growth and *Pn* of papaya Maradol, growing with low availability of Pi in Hoagland's solution (HS).

2. Materials and Methods

2.1. Plants, AM Fungi, and Growing Conditions. Seeds of papaya Maradol (*Carica papaya* var. Maradol, Carisem) were germinated in wet chambers in darkness at 35°C ± 3°C. After that, they were grown in nurseries for 15–18 days until the plants had three to four leaves. Then, plants were transplanted into single pots (black nursery bags of 35 × 25 centimeters) which were previously filled with agrolite, a sterile and, inert substrate, chemically stable in neutral and acid pH. At all times, the experiment was maintained under greenhouse conditions, with 35°C temperature, 55% relative humidity, and a photosynthetic photon flow density of around 300 μmol m^{-2}s^{-1}.

The inoculants used were two strains of the AM fungi *Gloms intraradices* from different origin: (1) a commercial strain Endo-Rhyza (+Mic1), that consists in spores immersed in vermiculite (2) a noncommercial strain (+Mic2), isolated and propagated in the Microbiology Area, Natural Resources, Postgraduates College, Montecillo, Mexico, consisting of spores and little segments of colonized roots in a soil-tezontle mix used as support.

When the plants were transplanted to single pots, each plant was inoculated with approximately 160 spores of each AM fungi strain, and the nonmycorrhizal and control groups were supplied with the same support from the noncommercial strain inoculant, previously sterilized. After that, in order to maintain all plants with the same pretreatment supply of Pi in solution and to stimulate AM colonization, all plants were supplied with 200 mL of low Pi (0.3 mM) Hoagland solution (HS) [16] twice a week and 250 mL of distilled water once a week for five weeks. At the end of these five weeks the AM colonization was verified taking a sample from the roots of five inoculated plants.

Then, the different Pi concentration treatments were started, by supplying modified HS containing the various Pi concentrations (0, 0.3, 0.6 mM). The pH was adjusted to 6.7, and the necessary amount of ammonium nitrate (0.040 g, 0.028 g y 0.016 g, resp.) was added to HS in order to maintain the N concentration in each case.

The experiment included nine treatments by combining three Pi concentrations (0, 0.3, and 0.6 mM) and three AM conditions (+Mic1, +Mic2, and without inoculation, denoted as −Mic). Every treatment had five plants (as experimental units). The control was a papaya Maradol group without AM fungi supplied with HS without modification (Pi concentration of 1.0 mM).

Data from *Pn* and F_V/F_M were taken after the presence of AM structures was verified (week five) in the roots, at weeks seven, nine, and ten (weeks two, four, and five after the treatments were initiated). Growth parameters, leaves P content and the percentage of AM colonization, were taken only at week ten.

2.2. Growth Measurements. The stem diameter was measured two centimeters from the base to the apex. The plants height was measured from the base to the apex of the stem. Dry weight was determined dividing each plant into shoot and root before they were oven-dried at 70°C to constant weight in order to obtain shoot, root, and total dry weight (DW).

2.3. Photosynthesis and F_V/F_M Measurements. *Pn* was evaluated as CO_2 assimilation, using the third and fourth leaves from the apex to the base of the stem. It was measured between 11 and 13:30 hours, with an infrared gas analyzer (IRGA) LICOR model LI-6200. F_V/F_M data were taken on the same leaves that were used to measure photosynthesis after they were preacclimated to darkness for 30 minutes. A Plant Efficiency Analyzer (PEA, Hansatech Instruments Ltd., Kings Lynn, UK) was used set at 70% for two seconds.

2.4. Phosphorus Measurements. Leaves P content measurements were carried out by digesting the plant material with nitric acid at 80°C [17]. The measurements were carried out using an optical emission spectrometer inductively coupled to plasma (OES-ICP), Perkin Elmer 400, at 253.4 nm.

2.5. Arbuscular Mycorrhizal Colonization. To estimate the AM colonization, the roots were stained with trypan blue [18]. A total of 20 fragments of stained roots (about 1 cm long) for each treatment were mounted on slides in lactic acid and examined with an optical microscope (10x) to ascertain the presence of AM structures. Colonization percentage was calculated by

$$\% \text{ Colonization} = \frac{\text{Number of colonized segments}}{\text{number of total segments viewed}} \times 100. \tag{1}$$

Data are presented as the means of five replicates ± standard deviation (SD) and were analyzed by ANOVA test using the Statgraphics 4.1 Plus program. Differences between the treatments were tested using the Tukey test at 95% of confidence ($P \leq 0.05$).

3. Results

3.1. Plant Growth. Most of the growth parameters measured decreased significantly when the Pi supply in HS decreased. However, this effect was attenuated by the AM association. Stem diameter decreased as the Pi supply was reduced. However, this drop was attenuated by the AM association. When Pi was reduced from 1.0 to 0.6 and 0.3 mM, a similar reduction in stem diameter occurred in both −Mic and +Mic plants. However, when HS had no added Pi (0 mM), stem diameter was reduced 91% in −Mic, while only 74% in +Mic1 and +Mic2 plants compared to control plants (1.0 mM) (Figure 1).

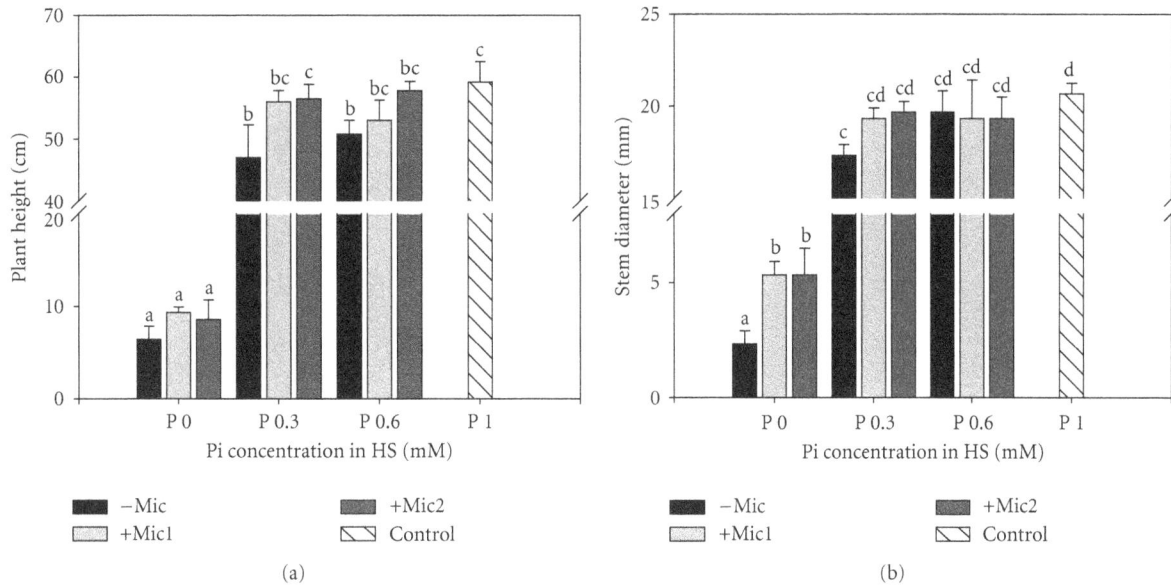

FIGURE 1: (a) Plant height, (b) stem diameter from papaya Maradol plants at week five of the experiment, grown with or without mycorrhiza and treated with Pi 0 mM, Pi 0.3 mM, and Pi 0.6 mM. Control plants were grown without mycorrhiza but treated with Pi 1 mM. Data are means ± SD ($n = 5$). Values with the same letter for treatments within each Pi concentration had no statistical difference (Tukey 95%).

Shoots, roots, and total plant DW also decreased as the Pi supply was reduced and again the drop was attenuated in +Mic plants. With Pi 0.3 mM in HS total DW was reduced by 20% in −Mic, while only 13% and 7% in +Mic1 and +Mic2, respectively (Figure 2(b)). When grown in HS solution with no added Pi, +Mic plants maintained 10 times higher biomass than −MR plants (Figure 2(a)).

3.2. Photosynthesis and F_V/F_M. Pn also decreased drastically as the Pi supply was reduced but MA not only was able to attenuate this effect but in some cases *Pn* in +Mic plants grown at low Pi showed higher *Pn* than control plants grown without Pi limitation. With Pi 0.3 mM −Mic plants reduced their *Pn* by 63.6% while +Mic plants showed *Pn*, values that were in fact higher than those in control plants grown with 1.0 mM (Figure 3(b)). Even when no Pi was added to the HS (0 mM), +Mic plants maintained significantly higher *Pn* than −Mic plants (Figure 3(a)).

F_V/F_M also decreased as the Pi supply was reduced. However at week 5, F_V/F_M values remained significantly higher in +Mic plants than in −Mic plants treated with both 0.3 mM and 0 mM Pi (Figures 4(a) and 4(b)).

3.3. Phosphorus Content. The foliar P content decreased as the Pi supply was reduced and again AM was able to attenuate this drop. In the case of total absence of added Pi through the HS (0 mM), +Mic plants had significantly higher leaf P content than −Mic plants (Figure 5). Also, mycorrhizal plants inoculated with +Mic2 had significantly higher P content than the −Mic plants at P1 0.6 mM, and in fact this treatment was the only one that showed laef P contents similar to the control plants grown with full HS containing 1 mM Pi.

3.4. AM Colonization. No AM colonization was observed in roots of −Mic plants throughout the fertilization gradient. The higher AM colonization occurred in +Mic1 without Pi supply (Table 1). For both strains, AM colonization decreased as the Pi supply in HS increased. Nevertheless, this reduction was lower in AM plants inoculated with +Mic2, whose AM colonization remained two and tree times higher (with Pi 0.3 mM and 0.6 mM, resp.) than those inoculated with +Mic1 (Table 1).

4. Discussion

Our results indicate that both strains of *G. intraradices* were able to attenuate the negative effect caused by low Pi supply on the growth (total DW) of papaya Maradol plants. One possible explanation for that could be the fact that mycorrhizal papaya Maradol plants kept more leaves than non mycorrhizal plants (data not shown) and that surely contributed to the larger shoot DW. With low Pi supply (0.3 mM) the AM association allowed DW and plant height to remain similar to those at plants grown with higher Pi supply (0.6 mM) or even those at control plants. In other study models like strawberry [19] or *Medicago truncatula* [15] it has also been reported that AM association reduces the stress due to the low Pi availability on plant growth. Some commercial papaya varieties like Baixinho, Tainung, Sunrise, and Improved also had greater dry matter production in association with AM fungi *G. clarum* and *Gigaspora margarita* [13]. The fact that the growth response of papaya Maradol to the AM inoculation was greater, particularly at lower concentration of Pi, and that this response decreased at higher Pi concentration in HS, places papaya Maradol in the facultative arbuscular mycorrhizal species group.

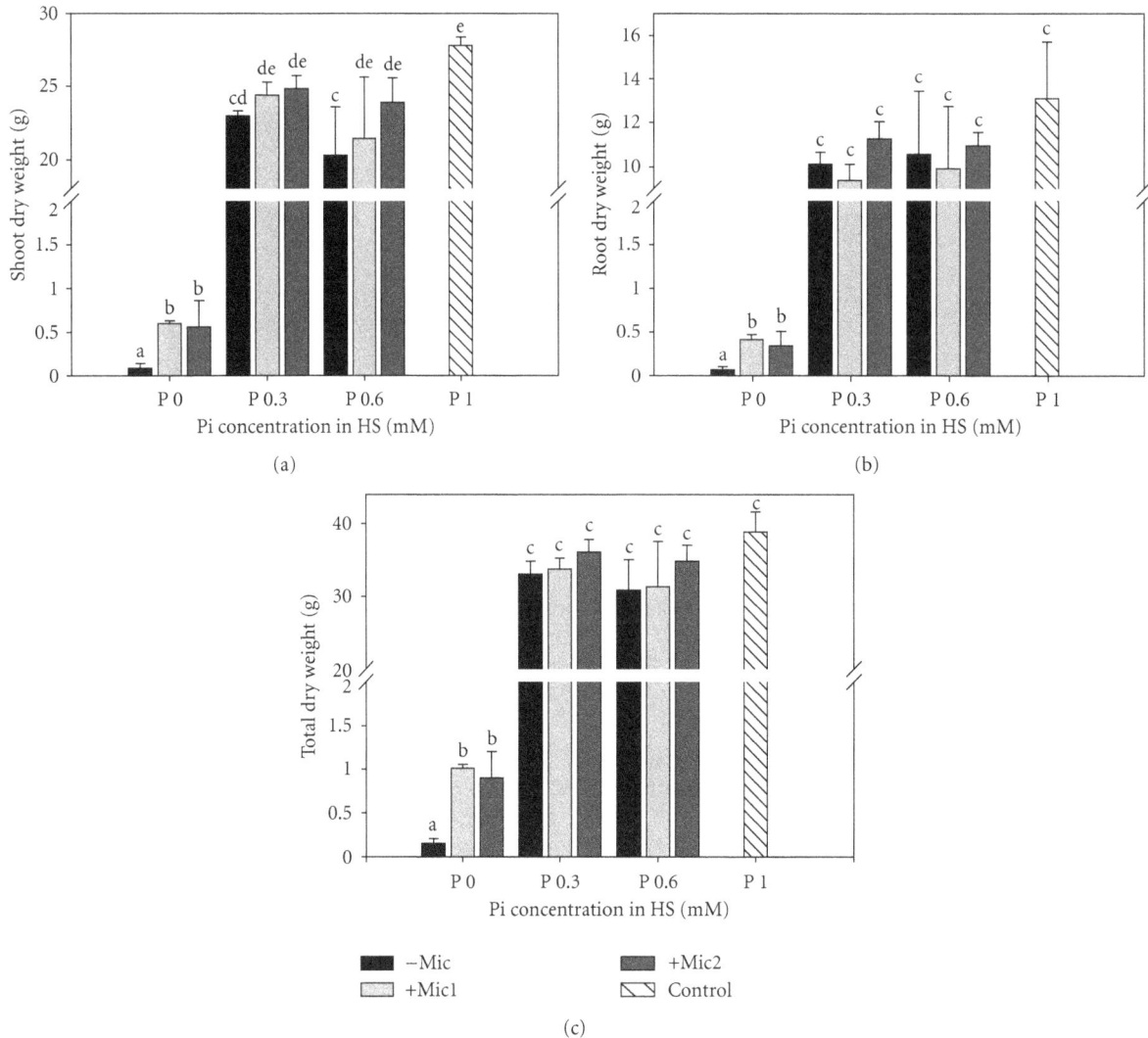

FIGURE 2: Dry weight of (a) shoots, (b) roots, (c) total plant weight from papaya Maradol plants at week five of the experiment, grown with or without mycorrhiza and treated with Pi 0 mM, Pi 0.3 mM, and Pi 0.6 mM. Control plants were grown without mycorrhiza but treated with Pi 1 mM. Data are means ± SD ($n = 5$). Values with the same letter for treatments within each Pi concentration had no statistical difference (Tukey 95%).

TABLE 1: Arbuscular mycorrhizal colonization (%) in papaya Maradol roots, at week five of the experiment. Data are means ($n = 5$). Values with the same letter in each strain across the Pi gradient had no statistical difference (Tukey 95%). Values with the same uppercase letter in the row had no statistical difference.

Mycorrhizal condition	Phosphorus (mM)			
	0.0	0.3	0.6	1.0*
−Mic	0.0	0.0	0.0	0.0
+Mic1	82.0aA	16.0bA	10.0bA	nc
+Mic2	76.0aA	37.0bB	32.0bB	nc

*Total control with Pi 0 mM where arbuscular mycorrhizal colonization was not observed; −Mic: without mycorrhiza, +Mic1: commercial strain, +Mic2: noncommercial strain. nc = not considered.

The beneficial effect of AM association on papaya Maradol CO_2 assimilation is important because it indicates that somehow the mycorrhizal plants are capable of reducing the stress on the photosynthetic apparatus caused by the low availability of Pi.

It has been reported that the negative effect of the nutrient scarcity on the development of many crops is associated with the reduction of Pn, F_V/F_M, and g_s, in corn [20] while in sugar beet, the low Pi reduced CO_2 assimilation by 33% [21]. This negative effect of low Pi supply can be related to many aspects such as reduction of chlorophyll a and b, affecting also, the PSI and PSII efficiency [20]. Furthermore, Pi is necessary to couple photosynthetic reactions in the dark and light phases, because of its structural and regulatory role in energy conservation and CO_2 assimilation [22].

The ratio variable fluorescence over maximum fluorescence (F_V/F_M) has been used as an important and easily measurable parameter of the physiological status of the photosynthetic apparatus in intact plant leaves. When plants

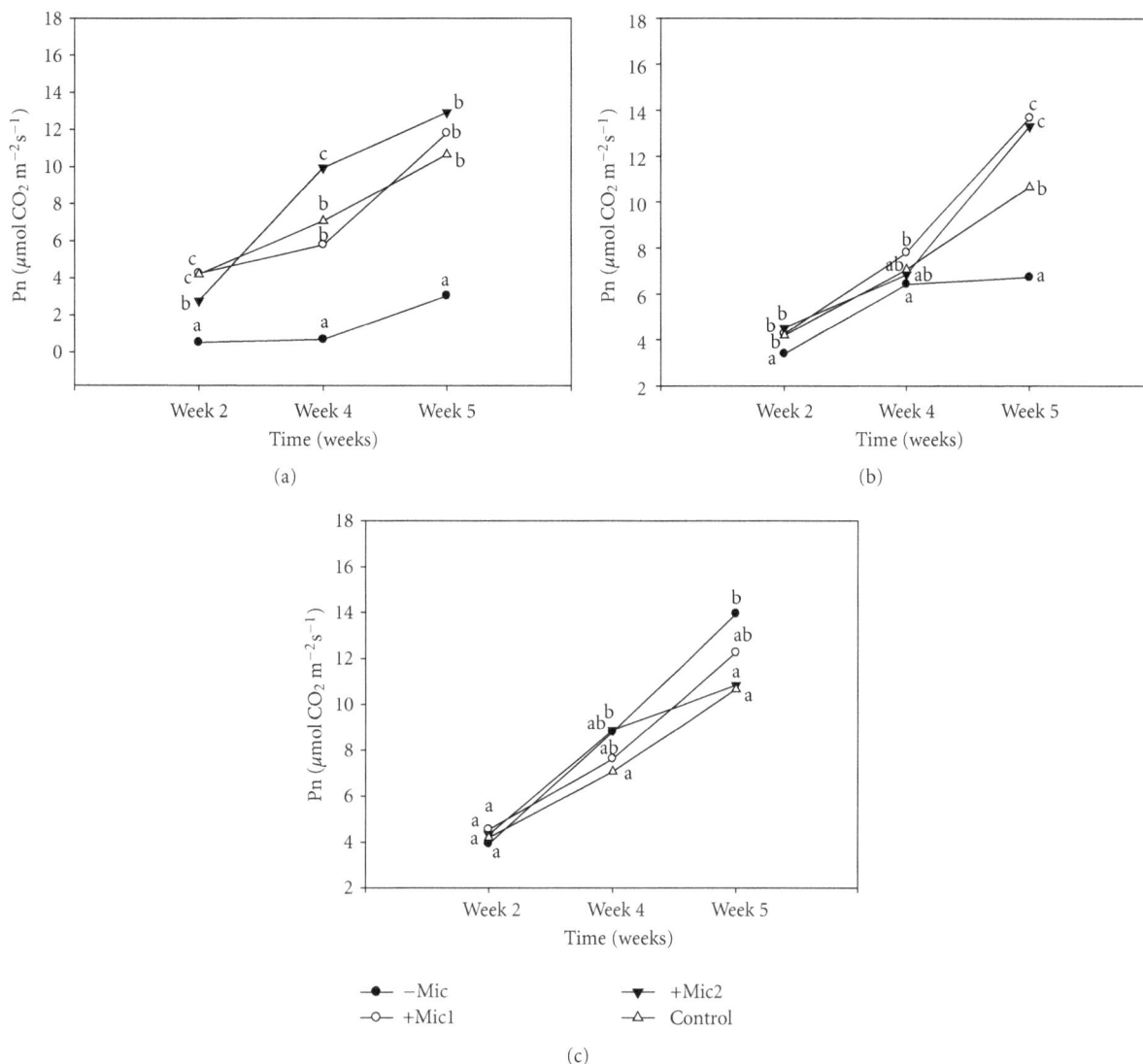

FIGURE 3: *Pn* changes observed at weeks 2, 4, and 5 of the experiment in papaya Maradol plants grown with or without mycorrhiza. (a) Pi 0 mM, (b) Pi 0.3 mM, (c) Pi 0.6 mM. Control group was grown without AM fungi supplied, with Pi 1.0 mM. Data are means ± SD ($n = 5$). Values with the same letter for treatments within each week, had no statistical difference (Tukey 95%).

grow in good conditions, the photosynthetic efficiency has values about 0.83 but when an environmental factor is stressing them, the PSII efficiency is affected and the F_V/F_M values fall [23]. In the present work, the F_V/F_M values suggest that AM association reduced the damage of Pi stress on the maximum efficiency of PSII.

On the other hand, it has been reported that AM colonization improves *Pn* in young leaves but this is not necessarily translated into an increased dry weight [24] as the gained carbon could be used to maintain the symbiosis in addition to growth [25]. In the present work, the AM colonization benefit was reduced when papaya Maradol was fertilized with Pi 0.6 mM. This may be because, as it has been proposed by Olsson et al. [26], under conditions of sufficient P the root reduced the carbon flow to the fungus, therefore the response of AM association to P sufficient conditions would depend on the P status of the colonized root.

Our results show that even without supply of added Pi in HS (0 mM), papaya Maradol plants associated with the AM fungus *G. intraradices* had higher leaf P content than their −Mic counterparts. It has been shown that the Pi taken up by AM fungus *G. intraradices* is delivered almost totally to the plant, contributing significantly to the plants P take-up capacity from the soil [15].

The higher P content reported in the present work in +Mic plants compared to those at −Mic plants in the absence of supplied Pi through the HS (Pi = 0 mM) may be explained by the fact that the mycorrhizal association enabled plants to take up more Pi from that still present in the substrate from the pretreatment Pi supply (see Section 2).

The P foliar concentrations found in our experiments are in agreement with other reports [21]. The total P reported here is on the range of those reported for other plant species [7, 8, 27]. The ranges of P content reported through the

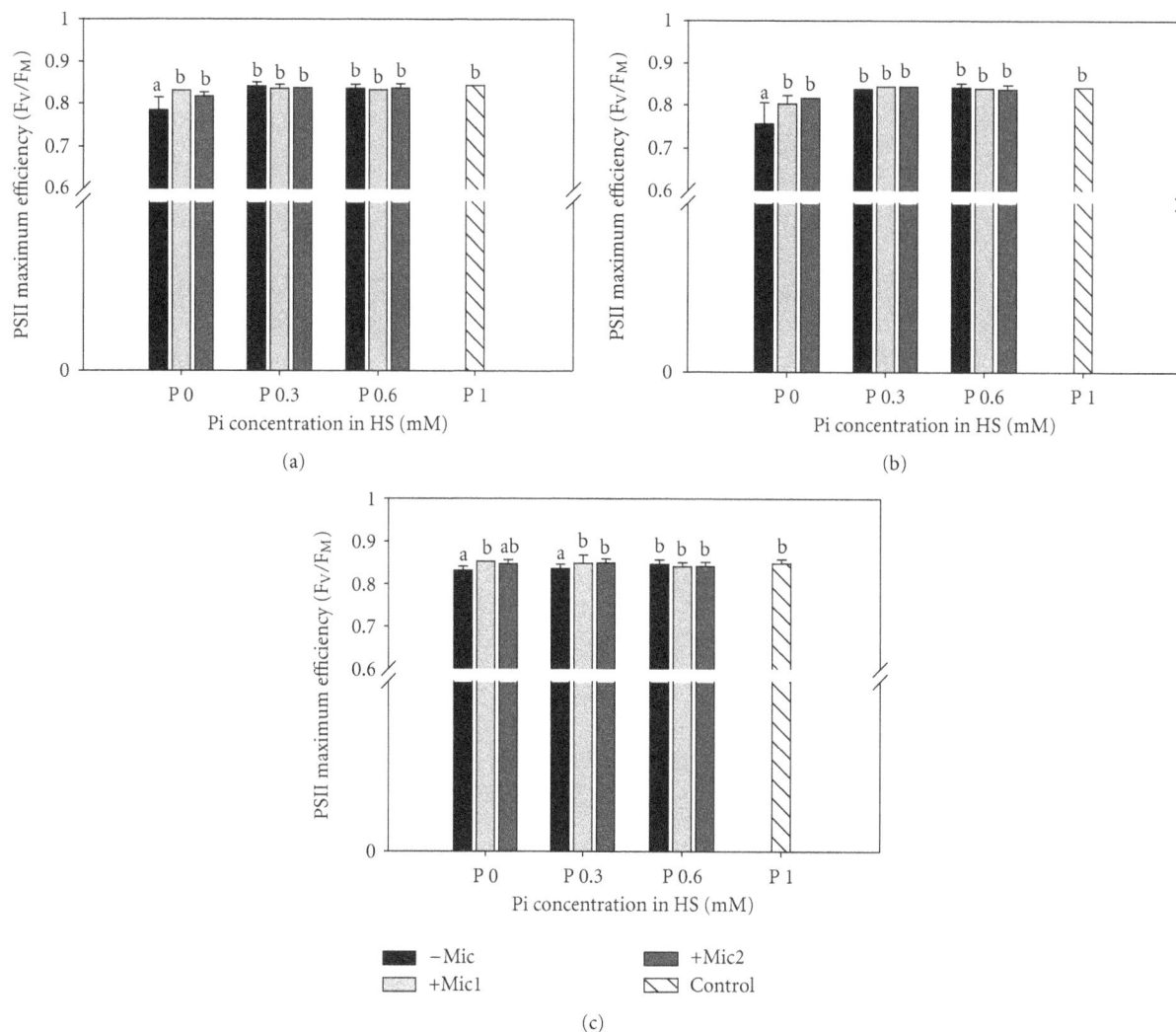

FIGURE 4: Chlorophyll fluorescence (F_V/F_M) found at week five of the experiment in papaya Maradol plants grown with or without mycorrhiza with (a) Pi 0 mM, (b) Pi 0.3 mM, and (c) 0.6 Pi mM. Control group was grown without AM fungi supplied, with Pi 1.0 mM. Data are means $\pm$ SD ($n = 5$). Values with the same letter for treatments within each week had no statistical difference (Tukey 95%).

literature may be the result of the fact that in every case the Pi availability of the substrate is different or because the P uptake is different not as a result of the AM colonization intensity but perhaps because of differences in the expression of genes involved in Pi transport in the different species [7].

The AM colonization intensity in a plant, depends on the AM fungi species to which the plant is associated [15]. Although, it has been reported that sometimes the mineral fertilization is not decreasing the AM colonization [19, 28], in most cases it has been suggested that increased supply of Pi decreases the AM colonization in crops like wheat or pea [29], *Zea mays* and *Cucurbita pepo* [30]. In the present work, colonization decreased as the Pi concentration was reduced but at concentrations as high as 0.6 mM Pi, AM colonization was not totally inhibited, particularly with +Mic2 (Table 1).

It is suggested that under the experimental conditions of this work, the papaya Maradol efficiency to take up Pi was increased when it was associated with *G. intraradices* resulting in higher *Pn* and dry weight than their noninoculated counterparts grown at the same Pi condition. Furthermore, *Pn* and F_V/F_M values found in +MIC plants at low Pi levels were closer to those of control plants grown without Pi limitation. However, the Pearson's coefficient shows low correlation between *Pn* and leaves P content ($r = 0.52$ and 0.6 for +Mic1 and +Mic2, resp.) while the correlation coefficient between dry weight and leaves P content was high ($r = 0.96$) with the two strains of *G. intraradices* suggesting that AM association allowed more plant growth but part of photosynthates resulting from the increased *Pn* were used to maintain the association.

From our results, it can be concluded that growth and *Pn* in papaya Maradol plants decreased strongly under low Pi supply. However, the AM association with *G. intraradices* contributed to reduce the impact of the Pi scarcity on these parameters. On the other hand, it is suggested that the larger dry weight of AM plants, especially when Pi was not added in HS, was related to a better efficiency to take up Pi from their substrate. Finally, and despite

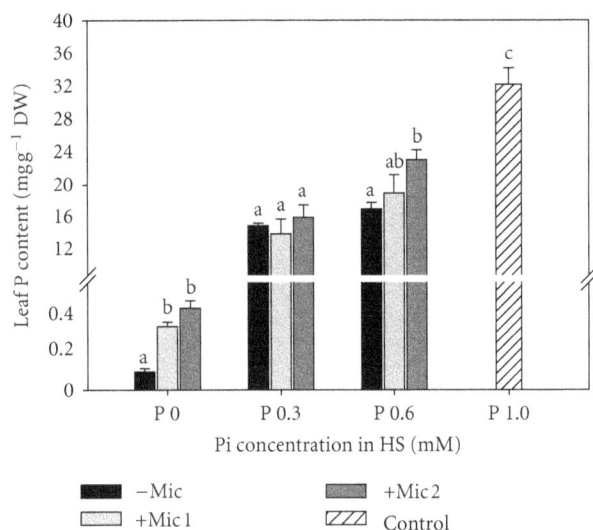

FIGURE 5: Total leaf P content found at week 5 of the experiment, in papaya Maradol plants grown with or without mycorrhiza and treated with Pi 0 mM, Pi 0.3 mM, and Pi 0.6 mM. Control plants were grown without mycorrhiza but treated with Pi 1 mM. Data are means $\pm$ SD ($n = 5$). Values with the same letter for treatments within each Pi concentration had no statistical difference (Tukey 95%).

the fact that AM association enhanced uptake of Pi and increased Pn, the lower correlation between them suggests that some photosynthates could had been used to maintain the symbiosis.

Acknowledgments

YNG was holder of a PROMEP scholarship from the Universidad Autónoma de Tlaxcala. The authors are grateful to Dra Laura Hernández Terrones and Melina Soto (CICY-CEA, Quintana Roo, México and to Dr Enrique Sauri (ITM, Yucatán, México) for performing the leaf P content evaluations and providing expert advice and to Dr. Alejandro Alarcón (Colegio de Postgraduados, Montecillo, México) for revising the paper.

References

[1] B. S. Ripley, S. P. Redfern, and J. Dames, "Quantification of the photosynthetic performance of phosphorus-deficient *Sorghum* by means of chlorophyll-a fluorescence kinetics," *South African Journal of Science*, vol. 100, no. 11-12, pp. 615–618, 2004.

[2] J. Jacob and D. W. Lawlor, "Stomatal and mesophyll limitations of photosynthesis in phosphate deficient sunflower, maize and wheat plants," *Journal of Experimental Botany*, vol. 42, no. 8, pp. 1003–1011, 1991.

[3] H. Marschner, *Mineral Nutrition of Higher Plants*, Academic Press, San Diego, Calif, USA, 1995.

[4] K. G. Raghothama, "Phosphate acquisition," *Annual Review of Plant Biology*, vol. 50, pp. 665–693, 1999.

[5] D. P. Schachtman, R. J. Reid, and S. M. Ayling, "Phosphorus uptake by plants: from soil to cell," *Plant Physiology*, vol. 116, no. 2, pp. 447–453, 1998.

[6] S. E. Smith and D. J. Read, "Uptake, translocation and transfer of nutrients in mycorrhizal symbioses," in *Mycorrhizal Symbiosis*, S. E. Smith and J. D. Read, Eds., pp. 379–308, Academic Press, New York, NY, USA, 1990.

[7] S. Ravnskov and I. Jakobsen, "Functional compatibility in arbuscular mycorrhizas measured as hyphal P transport to the plant," *New Phytologist*, vol. 129, no. 4, pp. 611–618, 1995.

[8] S. H. Burleigh, T. Cavagnaro, and I. Jakobsen, "Functional diversity of arbuscular mycorrhizas extends to the expression of plant genes involved in P nutrition," *Journal of Experimental Botany*, vol. 53, no. 374, pp. 1593–1601, 2002.

[9] G. Al-Karaki, B. McMichael, and J. Zak, "Field response of wheat to arbuscular mycorrhizal fungi and drought stress," *Mycorrhiza*, vol. 14, no. 4, pp. 263–269, 2004.

[10] P. E. Mortimer, E. Archer, and A. J. Valentine, "Mycorrhizal C costs and nutritional benefits in developing grapevines," *Mycorrhiza*, vol. 15, no. 3, pp. 159–165, 2005.

[11] A. Alarcón, F. T. Davies, J. N. Egilla, T. C. Fox, A. A. Estrada-Luna, and R. Ferrera-Cerrato, "Short term effects of *Gloms claroideum* and Azospirillum brasilense on growth and root acid phosphatase activity of *Carica papaya* L. under phosphorus stress," *Revista Latinoamericana de Microbiologia*, vol. 44, no. 1, pp. 31–37, 2002.

[12] G. Bohrer, V. Kagan-Zur, N. Roth-Bejerano, D. Ward, G. Beck, and E. Bonifacio, "Effects of different Kalahari-desert VA mycorrhizal communities on mineral acquisition and depletion from the soil by host plants," *Journal of Arid Environments*, vol. 55, no. 2, pp. 193–208, 2003.

[13] A. V. Trindade, J. O. Siqueira, and P. F. de Almeida, "Dependencia micorrízica de variedades comerciais de mamoeiro," *Pesquisa Agropecuária Brasileira*, vol. 36, pp. 1485–1494, 2001.

[14] C. A. Martin and J. C. Stutz, "Interactive effects of temperature and arbuscular mycorrhizal fungi on growth, P uptake and root respiration of *Capsicum annuum* L," *Mycorrhiza*, vol. 14, no. 4, pp. 241–244, 2004.

[15] S. E. Smith, F. A. Smith, and I. Jakobsen, "Functional diversity in arbuscular mycorrhizal (AM) symbioses: the contribution of the mycorrhizal P uptake pathway is not correlated with mycorrhizal responses in growth or total P uptake," *New Phytologist*, vol. 162, no. 2, pp. 511–524, 2004.

[16] D. R. Hoagland and H. I. Arnon, *The Water-Culture Method for Growing Plants without Soil*, vol. 347, California Experimental Agriculture Station Circular, Berkeley, Claif, USA, 1950.

[17] T. Hoffmann, C. Kutter, and J. M. Santamaría, "Capacity of Salvinia minima baker to tolerate and accumulate As and Pb," *Engineering in Life Sciences*, vol. 4, no. 1, pp. 61–65, 2004.

[18] J. M. Phillips and D. S. Hayman, "Improved procedures for clearing roots and staining parasitic and vesicular-arbuscular mycorrhizal fungi for rapid assessment of infection," *Transactions of the British Mycological Society*, vol. 55, pp. 158–161, 1970.

[19] M. P. Sharma and A. Adholeya, "Effect of arbuscular mycorrhizal fungi and phosphorus fertilization on the post vitro growth and yield of micropropagated strawberry grown in a sandy loam soil," *Canadian Journal of Botany*, vol. 82, no. 3, pp. 322–328, 2004.

[20] T. S. L. Lau, E. Eno, G. Goldstein, C. Smith, and D. A. Christopher, "Ambient levels of UV-B in Hawaii combined with nutrient deficiency decrease photosynthesis in near-isogenic maize lines varying in leaf flavonoids: flavonoids decrease photoinhibition in plants exposed to UV-B," *Photosynthetica*, vol. 44, no. 3, pp. 394–403, 2006.

[21] I. M. Rao and N. Terry, "Leaf phosphate status, photosynthesis, and carbon partitioning in sugar beet. IV. Changes

with time following increased supply of phosphate to low-phosphate plants," *Plant Physiology*, vol. 107, no. 4, pp. 1313–1321, 1995.

[22] E. Epstein and A. J. Bloom, *Mineral Nutrition of Plants: Principles and Perspectives*, Sinauer Associates, Sunderland, Mass, USA, 2004.

[23] G. H. Krause and E. Weis, "Chlorophyll fluorescence and photosynthesis: the basics," *Annual Review of Plant Physiology and Plant Molecular Biology*, vol. 42, no. 1, pp. 313–349, 1991.

[24] D. P. Wright, J. D. Scholes, and D. J. Read, "Effects of VA mycorrhizal colonization on photosynthesis and biomass production of *Trifolium repens* L," *Plant, Cell and Environment*, vol. 21, no. 2, pp. 209–216, 1998.

[25] D. P. Wright, D. J. Read, and J. D. Scholes, "Mycorrhizal sink strength influences whole plant carbon balance of *Trifolium repens* L," *Plant, Cell and Environment*, vol. 21, no. 9, pp. 881–891, 1998.

[26] P. A. Olsson, I. M. Van Aarle, W. G. Allaway, A. E. Ashford, and H. Rouhier, "Phosphorus effects on metabolic processes in monoxenic arbuscular mycorrhiza cultures," *Plant Physiology*, vol. 130, no. 3, pp. 1162–1171, 2002.

[27] G. Cruz-Flores, S. Avilés Marín, and J. C. Cortés Castelán, "Estudio de adaptabilidad del triticale a diferentes dosis de calcio y fósforo en andisoles," *Terra*, vol. 16, no. 1, pp. 63–69, 1998.

[28] M. Gryndler, J. Larsen, H. Hrselová, V. Rezácová, H. Gryndlerová, and J. Kubát, "Organic and mineral fertilization, respectively, increase and decrease the development of external mycelium of arbuscular mycorrhizal fungi in a long-term field experiment," *Mycorrhiza*, vol. 16, no. 3, pp. 159–166, 2006.

[29] M. H. Ryan and J. F. Angus, "Arbuscular mycorrhizae in wheat and field pea crops on a low P soil: increased Zn-uptake but no increase in P-uptake or yield," *Plant and Soil*, vol. 250, no. 2, pp. 225–239, 2003.

[30] M. S. Schroeder and D. P. Janos, "Plant growth, phosphorus nutrition, and root morphological responses to arbuscular mycorrhizas, phosphorus fertilization, and intraspecific density," *Mycorrhiza*, vol. 15, no. 3, pp. 203–216, 2005.

2

The Role of Pathogenesis-Related Proteins in the Tomato-*Rhizoctonia solani* Interaction

Parissa Taheri and Saeed Tarighi

Department of Crop Protection, Faculty of Agriculture, Ferdowsi University of Mashhad, P.O. Box 1163, Mashhad 9177948978, Iran

Correspondence should be addressed to Parissa Taheri, p-taheri@um.ac.ir

Academic Editor: Olivier Honnay

Rhizoctonia solani is one of the most destructive pathogens causing foot rot disease on tomato. In this study, the molecular and cellular changes of a partially resistant (Sunny 6066) and a susceptible (Rio Grande) tomato cultivar after infection with necrotrophic soil-borne fungus *R. solani* were compared. The expression of defense-related genes such as chitinase (*LOC544149*) and peroxidase (*CEVI-1*) in infected tomato cultivars was investigated using semiquantitative reverse transcription-polymerase chain reaction (RT-PCR). This method revealed elevated levels of expression for both genes in the partially resistant cultivar compared to the susceptible cultivar. One of the most prominent facets of basal plant defense responses is the formation of physical barriers at sites of attempted fungal penetration. These structures are produced around the sites of potential pathogen ingress to prevent pathogen progress in plant tissues. We investigated formation of lignin, as one of the most important structural barriers affecting plant resistance, using thioglycolic acid assay. A correlation was found between lignification and higher level of resistance in Sunny 6066 compared to Rio Grande cultivar. These findings suggest the involvement of chitinase, peroxidase, and lignin formation in defense responses of tomato plants against *R. solani* as a destructive pathogen.

1. Introduction

Tomato (*Lycopersicon esculentum*) is one of the most important vegetables in the world which suffers from various fungal diseases [1, 2]. Foot rot disease of tomato plants was found in various greenhouses in Iran. Symptoms were characterized by soft rot of the seedling near the soil line. *Rhizoctonia solani* was consistently isolated from the damaged plant tissues. *Rhizoctonia solani* is a species complex composed of several anastomosis groups (AGs). This pathogen can survive in soil within diseased plant material as mycelia or sclerotia during unfavorable environmental conditions for several years. The pathogen is transported in infested soil or through movement of diseased plant tissues. Potential for seed borne inoculum also exists. In nature, usually *R. solani* has asexual reproduction and exists primarily as vegetative mycelium and/or sclerotia. The teleomorph of *R. solani*, *Thanatephorus cucumeris*, is classified in the phylum Basidiomycota. Formation of basidiospores on diseased host plants in nature is rarely observed. In favorable environmental conditions,

following infection of the host plant by *R. solani*, sexual spores are formed on specialized structures called basidia. Four spores are produced on each basidium. Basidia are formed when enough moisture is available and sufficient growth of the fungus has occurred. Basidiospores are wind-dispersed and germinate in high moisture. Each basidiospore has a single nucleus. The hyphae produced by germinating spores will fuse or anastomose with each other to form new hyphae with a mixture of different types of nuclei [3].

To date, *R. solani* is divided into 14 AGs designated as AG 1 through 13 and bridging isolate (BI) group [4, 5]. Isolates belonging to AG BI have the ability to anastomose not only with isolates of AG BI, but also with other AGs of *R. solani* such as AG 2, AG 3, AG 6, and AG 8 [6, 7]. Several AGs are further subdivided into intraspecific groups (ISGs) based on cultural morphology, nutritional requirements, temperature effect on growth, host specificity, frequency of hyphal anastomosis, and pathogenicity [7]. Isolates of AGs 1, 2, 3, and 4 have been shown to be pathogenic on tomato [8, 9] and cause high yield losses every year worldwide.

The growing concern about negative environmental effects of fungicides and appearance of fungicide-resistant pathogen strains is a motivating research for alternative protection methods. Among such novel strategies, understanding basal defense mechanisms to plan effective disease control methods has emerged as potential supplement in crop protection measures. One of the most important and effective plant defense mechanisms against various pathogens is activation of defense-related genes which are encoding defense proteins known as pathogenesis-related (PR) proteins. On the basis of amino acid sequence data and biochemical functions, PR proteins have been classified into 17 families so far [10].

The PR-3 and -4 families are comprised of chitinases. They hydrolyze the β-1,4 linkages between N-acetylglucosamine residues of chitin, a structural polysaccharide of the cell wall of many fungi, such as *R. solani*. The enzyme is linked with the thinning of the growing hyphal tips of fungi, followed by a balloon-like swelling that eventually leads to a bursting of hyphae. Furthermore, the degradation products of the fungal cell wall, especially the oligomers, could serve as resistance elicitors [8]. A combination of chitinase and -1,3-glucanase is known to be more effective than each enzyme alone against many fungi [7, 10]. Application of an antiserum against a tomato chitinase as a probe to study the subcellular localization of this enzyme in the roots infected with *Fusarium oxysporum* f. sp. *radicis-lycopersici* revealed that chitinase accumulated earlier in the incompatible interaction than in the compatible one. However, in both systems, chitinase deposition was largely correlated with pathogen distribution [11]. The enzyme was found to accumulate in areas where host walls were in close contact with fungal cells. In contrast, the enzyme could not be detected in vacuoles and intracellular spaces. The substantial amount of chitinase found at the fungus cell surface supports the view of an antifungal activity. However, the preferential association of the enzyme with altered fungal wall areas indicates that chitinase activity is either preceded by the hydrolytic action of other enzymes such as β-1,3-glucanases or coincides with these enzymes [11]. More rapidly, induction of various classes of chitinases is also reported during an incompatible *Cladosporium fulvum*-tomato interaction than during a compatible interaction [12]. In the *C. fulvum*-tomato pathosystem, resistance against the fungus correlates with early induction of transcription of genes encoding apoplastic chitinase and 1,3-b-glucanase and accumulation of these proteins in inoculated tomato leaves. For vacuolar, basic isoforms of chitinase and 1,3-b-glucanase, however, early gene transcript accumulation was observed in both incompatible and compatible interactions. Upregulation of the chitinase gene was observed most abundantly in resistant tomato genotypes which correlates well with the differences in gene expression observed in time course studies of compatible and incompatible *C. ful.um*-tomato interactions [13].

The PR-9 or peroxidases are key enzymes in the cell wall-building process, and it has been suggested that extracellular or wall-bound peroxidases would enhance resistance in various plant species against phytopathogens by the construction of a cell wall barrier that may hamper pathogen ingress and spread in plant cells. The PR-9 group contains a specific type of peroxidase that could act in cell wall reinforcement by catalyzing lignification [14] and enhance resistance against multiple pathogens with different life styles. It has been suggested that peroxidases are involved in the resistance of heated tomato fruits to infection with Botrytis cinerea as a destructive necrotrophic phytopathogen [15]. During the heat treatment, peroxidase activity was higher than in tomato fruits held at 20°C, and new peroxidase isoforms appeared. These isoforms disappeared when the tomato fruits were removed to 20°C and the fruits became sensitive to fungal infection. These findings indicate the role of peroxidases in defense responses of tomato fruit against *B. cinerea* and high temperature stress may help in defense against the fungal infection [15].

Basal resistance should be a key component of a disease management strategy for *R. solani* on tomato. Regrettably, only a few moderately resistant tomato cultivars are available with commercially acceptable horticultural traits. More sources of resistance from commercial cultivars or wild species need to be identified. Breeders use wild *Solanum* spp. as sources of genes controlling traits of economic importance such as fruit characteristics, nutritional content, and general disease resistance [16]. Identifying sources of resistant tomato germplasm and the mechanisms involved in basal resistance to the pathogen would aid in the development of cultivars suitable for production in *R. solani*-infested fields.

Foot rot disease of tomato caused by *R. solani* has not been investigated at cellular and molecular levels yet. Therefore, the objective of this study was to examine basal defense and the cytomolecular aspects of tomato-*R. solani* interaction by chitinase and peroxidase gene expression and determining levels of lignin in the plant cells.

2. Materials and Methods

2.1. Plant Material and Growth Conditions. The tomato cultivars Sunny 6066 and Rio Grande, which are partially resistant and susceptible to *R. Solani*, respectively [17], were used in this study. Tomato seedlings grown in 10 cm-diameter plastic pots for 6 weeks were used. The pots were filled with a commercial potting soil (Klassmann-Deilmann, Germany) and the plants were grown in greenhouse (30 ± 4°C; 16/8 h light/dark photoperiod). Six-week-old plants were used for inoculations.

2.2. Pathogen Inoculation, Disease Evaluation, and Microscopic Analysis. *Rhizoctonia solani* isolate M-2.3 belonging to AG 4 HG I, obtained from symptomatic tomato in the Khorasan-Razavi province of Iran, was maintained on potato dextrose agar (PDA). In pathogenicity tests on the stem near the soil line, plants were transplanted into 15 cm diameter plastic pots and soil-inoculum mixture filled 2 cm above the original ground level, which contained 10 g of wheat bran inoculum incubated at 25°C for 3 weeks. Sterilized wheat bran served as the control. Disease evaluation was done 14 days post inoculation (dpi) by measuring the lesion length. Symptoms were observed, and the pathogen was reisolated as described by Misawa and Kuninaga [8]. Six replicate plants were inoculated in a completely randomized design and the experiment was repeated twice. Data were analyzed using the Mann-Whitney comparison test at $P = 0.05$. Microscopic analysis of infected leaves was carried out by staining of fungal hyphae with aniline blue [18].

2.3. Thioglycolic Acid Assay. Lignin was quantified using thioglycoli acid (TGA) assay in which the lignin was bound to TGA to form lignin thioglycolic acid (LTGA) derivatives which can easily be extracted from tissue using sodium hydroxide and measured spectrophotometrically [19]. Each sample, consisting of 0.5 g of fresh plant tissue collected at different time points before and after inoculation, was ground in liquid nitrogen. LTGA derivatives were purified as described by Campbell and Ellis [20]. The purified LTGA derivatives were dissolved in 1 M NaOH, and the absorbance of the samples at 280 nm was recorded using a UVIKON922 spectrophotometer (Kontron, B. R. S., Anderlecht, Belgium).

2.4. RNA Extraction and Semiquantitative Reverse Transcription-Polymerase Chain Reaction (RT-PCR). Total RNA was prepared from tomato leaves by TRIzol reagent (Invitrogen). After treatment with RNase-free DNase (TURBO DNA-free kit, Ambion, USA) to remove contaminating DNA, the RNA was quantified spectrophotometrically. Their qualities were checked by ethidium bromide staining of agarose gels. Afterward, cDNA was synthesized with oligo (dT) 18 primer and SuperScript Reverse Transcriptase (Invitrogen) and used for PCR amplification. The gene-specific PCR primers used in this study were designed for the chitinase (*LOC544149*) and peroxidase (*CEVI-1*) genes of tomato using primer 3 program. Tomato actin gene (*LOC100147713*) was used as a constitutive control in all gene expression experiments. Primer sequences were checked by BLAST analysis against rice genome sequence in order to guarantee gene-specific amplifications by the primers. The reaction included an initial 5 min denaturation at 94°C, followed by 25–35 cycles of PCR (94°C, 30 s; 58°C, 30 s, and 72°C, 1 min), and a final extension period at 72°C for 7 min. Several tests were performed in order to determine the optimal amount of cDNA (RT samples) per reaction and the optimal number of cycles of amplifications within the exponential range. PCR products were sampled after 20, 25, 30, and 35 cycles and were separated on 1% agarose gels and stained with ethidium bromide to determine the linearity of the PCRs. In each RT-PCR experiment, the RNA samples treated with the DNase (-RT samples) were added to test for the presence of genomic DNA contamination in the samples. Each RT-PCR was repeated twice to ensure about reproducibility of the obtained data.

3. Results

3.1. Correlation between Resistance and Lignification. Thioglycolic acid (TGA) extractable cell wall complexes which are commonly thought to be lignin [20] were measured in order to quantitatively investigate the effect of lignin formation in resistance of tomato cultivars to the pathogen.

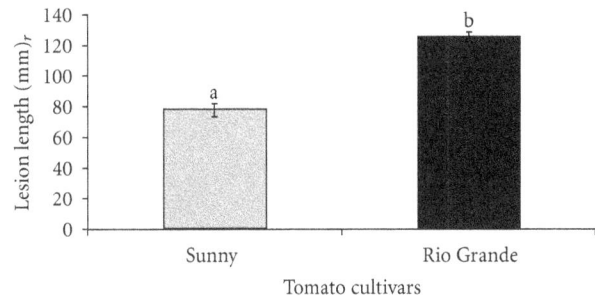

FIGURE 1: Comparison of two different tomato cultivars in response to *Rhizoctonia solani* infection by measuring lesion length at 14 days after inoculation (dpi). Tomato plants (36 plants/cultivar, including 12 replicate plants/cultivar and 3 repetition/cultivar) were used in this experiment. Statistical data analysis was carried out using SPSS 12.0 for Windows. Different letters indicate significant differences between cultivars according to the Mann-Whitney comparison test ($P = 0.05$).

TGA assay revealed that the accumulation of lignin in Sunny cultivar at 24 hpi was significantly higher than in Rio Grande plants and reached to the highest level at 72 hpi and then slightly decreased compared to Rio Grande cultivar. The level of lignin detected in partially resistant sunny plants (Figure 1) using TGA assay was significantly higher than that of susceptible Rio Grande plants at all time points after challenge inoculation with *R. solani* (Figure 2). The progress of fungal infection was investigated by microscopic analysis. After contact with the tomato plant surface, the mycelia grew and produced infection structures which directly penetrate into the plant tissues. Initiation of *R. solani* growth on the epidermis was detected at 1 dpi, followed by formation of infection structures and infrequent penetration at 2 to 3 dpi (data not shown). Compared with disease development in the Sunny cultivar, the onset of disease in the Rio Grande leaves was stronger and faster which might be linked to lower level of lignin formation as a structural barrier in the Rio Grande plants.

3.2. Transcription Analyses. Time course studies of defense-related gene expression were carried out for the inoculated tomato leaves (Figure 3) to determine the time point of up-regulation for each gene investigated in this study. The expression level of peroxidase gene (*CEVI-1*) increased at 12 hpi and reached its maximum at 48 hpi in the Rio Grande plants. However, an increase in PO-C1 expression at 6 hpi and peaking at 18 hpi was observed in Sunny cultivar (Figure 3).

Both Sunny and Rio Grande plants showed the highest level of chitinase (*LOC544149*) expression at 24 hpi. However, higher level of *LOC544149* expression was observed in Sunny compared to Rio Grande cultivar at 12 hpi (Figure 3), indicating priming the chitinase gene expression in the partially resistant cultivar.

4. Discussion

The involvement of peroxidase, chitinase, and lignin formation in basal resistance of tomato to *R. solani* was examined

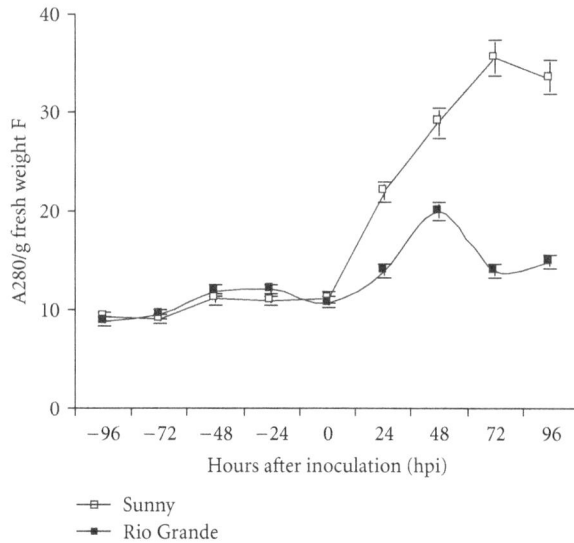

FIGURE 2: Lignin formation in the leaves of tomato cultivars at various time points before and after inoculation with *Rhizoctonia solani*. Data are means (± standard error) of four replicates of a representative experiment. Each replicate consisted of one sample pooled from six individual plants. The experiment was repeated twice with similar results.

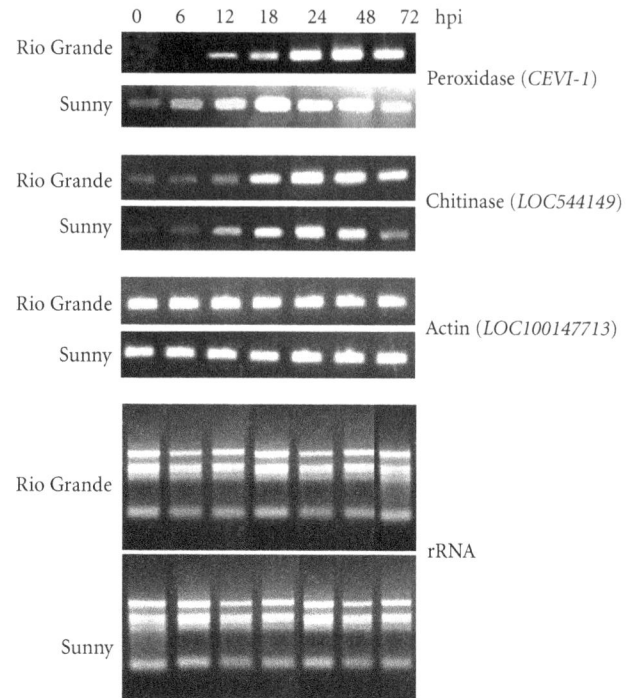

FIGURE 3: Expression patterns of the chitinase (*LOC544149*) and peroxidase (*CEVI-1*) genes in the leaf samples of tomato cultivars at different time points after inoculation with *R. solani* (hpi, hours after inoculation). Each time point represents eight leaves from eight different plants used for RNA extraction. Gene-specific primers for each of the defense genes were used in reverse transcription-polymerase chain reaction (RT-PCR). Control RT-PCR reactions were carried out with specific primers for tomato actin gene. The experiment was repeated three times with similar results.

in the present study. We demonstrated that basal resistance to *R. solani* in Sunny cultivar was correlated with higher levels of lignin formation in this cultivar compared to Rio Grande as a susceptible tomato cultivar (Figures 1 and 2). These observations suggest that enhanced lignin formatin in the partially resistant Sunny plants may be a defense strategy against *R. solani* which is a destructive necrotrophic fungal pathogen on tomato and many other plant species. This is in agreement with our previous findings about defense mechanisms of rice, as a monocot model plant, against this phytopathogen [21]. Therefore, we can conclude that similar defense mechanisms are involved in resistance of monocot and dicot plants against *R. solani*.

Plants exhibit a variety of responses during infection by pathogens, insects, or abiotic stresses, many of which involve the activation of host defense genes. Activation of these genes leads to physical and biochemical changes in plant cells which are not favorable for damage progress in plant. Among the major biochemical changes is biosynthesis and the accumulation of inducible defense related proteins. Most of these proteins correspond to pathogenesis-related proteins. These proteins are mostly of low molecular weight, preferentially extracted at low pH, resistant to proteolysis, and localized predominantly in the intercellular spaces of leaves.

The role of peroxidases, as a group of pathogenesis-related (PR) proteins which are involved in lignin production in phenylpropanoid pathway, in the interaction of tomato with some pathogens has been investigated. Peroxidases, grouped in PR9 family, have been associated with deposition of phenolics into plant cell walls during resistance responses [10]. Sareenal and associates [22] reported that enhancement

of the activity of peroxidase enzymes in response to the challenge inoculation with the pathogen under controlled conditions resulted in reduced symptom development and suppression of sheath blight disease in transgenic Pusa Basmati1 (PB1) rice lines, which are engineered with rice chitinase gene (chi11) for resistance against the rice sheath blight pathogen, *R. solani*, compared to nontransgenic control plants, indicating the importance of the phenylpropanoid pathway and peroxidase in defense responses of rice plants against the pathogen. This is in accordance with our results demonstrating the role of peroxidase in resistance of Sunny cultivar to *R. solani* and its importance in defense mechanisms. A possible function of peroxidases is the formation of structural barriers such as cell wall enhancement and deposition of cell wall appositions, both of which can be involved in the polymerization of lignin or suberin, the cross-linking of wall glycoproteins or polysaccharides, and the dimerization of antimicrobial phenols. Lignin formation was investigated using phloroglucinol/HCl test [23], and lignin was detected in Sunny plants at 24 hpi, which is downstream of peroxidase upregulation (Figure 3). This finding indicates that peroxidase might be involved in lignin deposition in partially resistant Sunny cultivar against *R. solani*. Therefore, one possible mechanism of basal resistance in Sunny plants can be linked to the defense pathways leading to formation of structural barriers such as lignin. In the transcript analyses, increased expression of the peroxidase gene (*CEVI-1*) was recorded in inoculated tomato plants. Together, these findings point to the importance of peroxidase-dependent phenolics production as a defense mechanism involved in basal resistance in our pathosystem. Several reports likewise suggest that priming of peroxidase transcripts or activities in various plants which contributes to phenolics-mediated resistance against necrotrophic phytopathogens [10, 24].

Chitinase gene could activate the expression of pathogenesis-related (PR) genes, subsequently enhancing transgenic plant resistance to necrotrophic pathogens [25, 26]. In the present study, a correlation was observed between higher level of resistance in Sunny plants and earlier upregulation of the chitinase gene (*LOC544149*) in this cultivar at 12 hpi. Therefore, chitinase might be a defense gene involved in basal resistance of Sunny plants to the pathogen. This finding is in agreement with the observations of Sridevi et al. [26] on the role of chitinase in resistance of rice plants to *R. solani*. Similarly in tomato, investigations of Chen and associates [25] revealed the involvement of chitinase in defense of the plants against *Botrytis cinerea* as a destructive phytopathogen with necrotrophic lifestyle.

So far, no detailed analyses on the defense mechanisms and signaling pathways involved in resistance of tomato to *R. solani* had been conducted. Thus, our results contribute to the understanding of basal resistance in the important tomato-*R. solani* pathosystem. Disease management strategies rely on our knowledge concerning cellular and molecular aspects of the tomato-*R. solani* interaction, defense mechanisms and signal transduction pathways involved in resistance against the pathogen [14]. Therefore, our knowledge about defense responses involved in basal resistance of tomato could be useful in breeding programs leading to obtain cultivars with higher levels of resistance. Application of partially resistant cultivars may prevent *Rhizoctonia* disease epidemics and decrease yield losses caused by this destructive soil-borne phytopathogen.

Acknowledgment

The authors thank Ferdowsi University of Mashhad for the financial support of this research.

References

[1] A. Fakhro, D. R. Andrade-Linares, S. von Bargen et al., "Impact of *Piriformospora indica* on tomato growth and on interaction with fungal and viral pathogens," *Mycorrhiza*, vol. 20, no. 3, pp. 191–200, 2010.

[2] B. Pharand, O. Carisse, and N. Benhamou, "Cytological aspects of compost-mediated induced resistance against Fusarium crown and root rot in tomato," *Phytopathology*, vol. 92, no. 4, pp. 424–438, 2002.

[3] J. Webster and R. W. S. Weber, *Inrtoduction to Fungi*, Cambridge University Press, 3rd edition, 2007.

[4] D. E. Carling, "Grouping in *Rhizoctonia solani* by hyphal anastomosis," in *Rhizoctonia Species: Taxonomy, Molecular Biology, Ecology, Pathology and Disease Control*, B. Sneh, S. Jabaji-Hare, S. Neate, and G. Dijst, Eds., pp. 37–47, Kluwer Academic, Dordrecht, The Netherlands, 1996.

[5] D. E. Carling, R. E. Baird, R. D. Gitaitis, K. A. Brainard, and S. Kuninaga, "Characterization of AG-13, a newly reported anastomosis group of *Rhizoctonia solani*," *Phytopathology*, vol. 92, no. 8, pp. 893–899, 2002.

[6] S. Kuninaga, R. Yokosawa, and A. Ogoshi, "Some properties of anastomosis group 6 and BI in *Rhizoctonia solani* Kuhn," *Annals of the Phytopathological Society of Japan*, vol. 45, no. 2, pp. 207–214, 1979.

[7] B. Sneh, L. Burpee, and A. Ogoshi, *Identification of Rhizoctonia Species*, The American Phytopathological Society Press, St. Paul, Minn, USA, 1991.

[8] T. Misawa and S. Kuninaga, "The first report of tomato foot rot caused by *Rhizoctonia solani* AG-3 PT and AG-2-Nt and its host range and molecular characterization," *Journal of General Plant Pathology*, vol. 76, no. 5, pp. 310–319, 2010.

[9] E. E. Kuramae, A. L. Buzeto, M. B. Ciampi, and N. L. Souza, "Identification of *Rhizoctonia solani* AG 1-IB in lettuce, AG 4 HG-I in tomato and melon, and AG 4 HG-III in broccoli and spinach, in Brazil," *European Journal of Plant Pathology*, vol. 109, no. 4, pp. 391–395, 2003.

[10] L. C. Van Loon, M. Rep, and C. M. J. Pieterse, "Significance of inducible defense-related proteins in infected plants," *Annual Review of Phytopathology*, vol. 44, pp. 135–162, 2006.

[11] N. Benhamou, M. H. A. J. Joosten, and P. J. G. M. De Wit, "Subcellular localization of chitinase and of its potential substrate in tomato root tissues infected by *Fusarium oxysporum* f. sp. *Radicis-lycopersici*," *Plant Physiology*, vol. 92, no. 4, pp. 1108–1120, 1990.

[12] N. Danhash, C. A. M. Wagemakers, J. A. L. van Kan, and P. J. G. M. de Wit, "Molecular characterization of four chitinase cDNAs obtained from *Cladosporium fulvum*-infected tomato," *Plant Molecular Biology*, vol. 22, no. 6, pp. 1017–1029, 1993.

[13] J. P. Wubben, C. B. Lawrence, and P. J. G. M. De Wit, "Differential induction of chitinase and 1,3-β-glucanase gene expression in tomato by *Cladosporium fulvum* and its race-specific elicitors," *Physiological and Molecular Plant Pathology*, vol. 48, no. 2, pp. 105–116, 1996.

[14] P. Taheri and S. Tarighi, "Cytomolecular aspects of rice sheath blight caused by *Rhizoctonia solani*," *European Journal of Plant Pathology*, vol. 129, no. 4, pp. 511–528, 2011.

[15] S. Lurie, E. Fallik, A. Handros, and R. Shapira, "The possible involvement of peroxidase in resistance to *Botrytis cinerea* in heat treated tomato fruit," *Physiological and Molecular Plant Pathology*, vol. 50, no. 3, pp. 141–149, 1997.

[16] C. M. Rick and J. I. Yoder, "Classical and molecular genetics of tomato: highlights and perspectives," *Annual Review of Genetics*, vol. 22, pp. 281–300, 1988.

[17] A. Yildiz and M. Timur Döken, "Anastomosis group determination of *Rhizoctonia solani* Kühn (Telemorph: *Thanatephorus cucumeris*) isolates from tomatoes grown in Aydin, Turkey and their disease reaction on various tomato cultivars," *Journal of Phytopathology*, vol. 150, no. 10, pp. 526–528, 2002.

[18] K. Schmidt, B. Heberle, J. Kurrasch, R. Nehls, and D. J. Stahl, "Suppression of phenylalanine ammonia lyase expression in sugar beet by the fungal pathogen *Cercospora beticola* is mediated at the core promoter of the gene," *Plant Molecular Biology*, vol. 55, no. 6, pp. 835–852, 2004.

[19] M. A. Doster and R. M. Bostock, "Effects of low temperature on resistance of almond trees to *Phytophthora* pruning wound

cankers in relation to lignin and suberin formation in wounded bark tissues," *Phytopathology*, vol. 78, no. 4, pp. 478–483, 1988.

[20] M. M. Campbell and B. E. Ellis, "Fungal elicitor-mediated responses in pine cell cultures—I. Induction of phenylpropanoid metabolism," *Planta*, vol. 186, no. 3, pp. 409–417, 1992.

[21] P. Taheri and S. Tarighi, "Riboflavin induces resistance in rice against *Rhizoctonia solani* via jasmonate-mediated priming of phenylpropanoid pathway," *Journal of Plant Physiology*, vol. 167, no. 3, pp. 201–208, 2010.

[22] S. Sareena, K. Poovannan, K. K. Kumar et al., "Biochemical responses in transgenic rice plants expressing a defence gene deployed against the sheath blight pathogen, *Rhizoctonia solani*," *Current Science*, vol. 91, no. 11, pp. 1529–1532, 2006.

[23] B. Mauch-Mani and A. J. Slusarenko, "Production of salicylic acid precursors is a major function of phenylalanine ammonia-lyase in the resistance of arabidopsis to Peronospora parasitica," *Plant Cell*, vol. 8, no. 2, pp. 203–212, 1996.

[24] P. Taheri and S. Tarighi, "A survey on basal resistance and riboflavin-induced defense responses of sugar beet against *Rhizoctonia solani*," *Journal of Plant Physiology*, vol. 168, no. 3, pp. 1114–1122, 2011.

[25] S. C. Chen, A. R. Liu, F. H. Wang, and G. J. Ahammed, "Combined overexpression of chitinase and defensin genesin transgenic tomato enhances resistance to *Botrytis cinerea*," *African Journal of Biotechnology*, vol. 8, no. 20, pp. 5182–5188, 2009.

[26] G. Sridevi, C. Parameswari, N. Sabapathi, V. Raghupathy, and K. Veluthambi, "Combined expression of chitinase and β-1,3-glucanase genes in indica rice (*Oryza sativa* L.) enhances resistance against *Rhizoctonia solani*," *Plant Science*, vol. 175, no. 3, pp. 283–290, 2008.

3

Seed Cryopreservation of Some Medicinal Legumes

Alla B. Kholina and Nina M. Voronkova

Institute of Biology and Soil Science, Far Eastern Branch of Russian Academy of Sciences, 159 Prospect 100-letiya Vladivostoka, Vladivostok 690022, Russia

Correspondence should be addressed to Alla B. Kholina, kholina@biosoil.ru

Academic Editor: Olivier Honnay

Seed survival after storage in liquid nitrogen ($-196°C$) was examined in 12 wild medicinal legume species occurred Far East of Russia. Dry seeds of all species survived cryostorage without loss of viability. Initial germinability varied from 3 to 85%. The stimulatory effect of cryogenic temperature on germination, with or without subsequent chemical scarification, was observed in all species studied with deep physical dormancy or heterogeneous levels of hardseededness. Frozen seeds demonstrated higher germination percentages (the percentage of germinated seeds) and germination rates (time for first seed to germinate (T_0) and time required (in days) to reach 50% of the final germination percentage (T_{50})) than the control ones. The anomalous seedlings were not observed after storage of seeds in liquid nitrogen. This study shows that cryostorage may be successfully applied for conservation of native species without detrimental effects on germination and growth.

1. Introduction

Fabaceae family is the one of the richest families of Russian Far-Eastern flora, and it includes a lot of valuable medicinal plants. The set of Far-Eastern legume species arouse particular interest due to pharmaceutical properties. *Astragalus membranaceus* (Fisch.) Bge., a herbaceous perennial, inhabits the western to northern part of China, Korea, and Japan. This herb is one of the important medicinal plants used as an adaptogenic in China, Korea, and Japan [1]; compounds from *A. membranaceus* roots may be utilized as immunostimulants, tonics, diuretics, antidiabetics, and sedatives [2]. *Hedysarum austrokurilense* (N.S. Pavlova) N.S. Pavlova and *H. sachalinense* B. Fedtsch. are perennial herb species endemic to the Sakhalin Island and South Kuril Islands [3]. The chemical components found in the plants of both related species include flavonoids, xanthones-mangiferin and isomangiferin, and polysaccharides many of them are of interest for traditional and modern medicines [4]. A lot of representatives of the genus *Oxytropis* have attracted attention thanks to the wide use in Tibetan and Mongolian medicine and also in folk medicines of Siberia and central Asia as the effective drugs [5]. The plants of some *Oxytropis* species are the richest in an alkaloid content [6], and alkaloid

from *O. ochrocephala* has antitumor activity [7]; one can suppose that other *Oxytropis* species (*O. chankaensis* Jurtz., *O. kamtschatica* Hult., *O. ochotensis* Bunge, *O. revoluta* Ledeb., and *O. retusa* Matsum.) have also valuable bioactive substances. *Sophora flavescens* Soland., a perennial shrub, has been used in the herbal medicine for centuries in China, Japan, and Korea; the dried roots of *S. flavescens* have well-known antibacterial, antiviral, antiprotozoal, anti-inflammatory, antitumor, and antipyretic effects ([8] and references herein, [9]). A perennial herb *Trifolium lupinaster* L. contains flavonoids and vitamins and has been used in Tibetan medicine against hepatitis [10]. *Vicia amurensis* Oett. is a perennial plant occurred in northern China, Japan, Korea, and Manchuria, contained flavonol glycosides [11], and used against tick-borne encephalitis and respiratory infections [12]. It was shown that alcohol extract of *Vicia subrotunda* seeds has antioxidant properties in experiments [13].

Nowadays, these legume species are attracting increasing attention of the modern pharmaceutical industry; however, natural resources of medicinal plants are limited, so it is reasonable to bring medicinal legumes under cultivation by seed conservation and reproduction. Cryopreservation (storage in liquid nitrogen (LN) at $-196°C$) is the only currently available long-term storage technique that ensures

TABLE 1: Seed collection site and ecological preference for 12 wild legume species studied.

Species	Collection site	Distribution and ecological preference
Astragalus membranaceus	Primorye, Muraviev-Amurskii Peninsula	Widespread (Russia-Siberia, Far East; Mongolia, China, Korea)/Forest meadows, dry meadows, steppe
Hedysarum austro-kurilense	Sakhalin, Makarov region, Magutan mud volcano	Endemic of North Sakhalin, South Kurils/Rocky hillsides, coastal terraces
H. sachalinense	Sakhalin, Makarov region, Tikhaia River, rocky hillsides	Endemic of Sakhalin/Coastal cliffs, rocky hillsides
Oxytropis chankaensis	Primorye, Khanka Lake, Sosnovyi Island	Endemic of Primorye, west coast of Khanka Lake/Sandy lakeshores
O. kamtschatica	Kamchatka Peninsula, Central Kamchatka, Kljuchevskaia Sopka volcano	Endemic of Far East of Russia (South Chukotka, Koryak Coast, Kamchatka Peninsula)/Alpine tundra, rock debris, old lava flows, shingle beach
O. ochotensis	Kamchatka Peninsula, Central Kamchatka, Kljuchevskaia Sopka volcano	Regional (Russia-East Siberia, Far East (West Chukotka, Kolyma, coast of the Okhotsk Sea, Kamchatka Peninsula))/Alpine tundra, cliffs, rocky hillsides
O. revoluta-1	Kamchatka Peninsula, Central Kamchatka, Tolbachik volcano	Regional (Russia-Far East (Koryak Coast, Kamchatka Peninsula, North Kuril Islands); Aleutians)/Alpine tundra, stone, rock debris, old lava flows, tephra, slag fields
O. revoluta-2	Kamchatka Peninsula, Central Kamchatka, Kljuchevskaia Sopka volcano	As before
O. retusa	Kuril Islands, Paramushir Island, Shelihov Bay, coastal rocky hillside	Endemic of Kuril Islands/Coastal cliffs, rocky hillsides
Sophora flavescens	Primorye, Hasan region, Gamov Peninsula	Widespread (Russia-East Siberia, Far East; Mongolia, China, Japan)/Forest meadows, dry meadows, river banks, xerophytic scrub
Trifolium lupinaster	Primorye, Ternei region, Svetlaya Bay, rock outcrops	Widespread (Russia-Siberia, Far East; Mongolia, China, Japan)/Open forest, meadows, shores
Vicia amurensis	Primorye, Muraviev-Amurskii Peninsula, sea shores	Widespread (Russia-East Siberia, Far East; China, Japan/Hillsides, roadsides, old railway, xerophytic scrub, meadows

safe conservation of genetic resources of valuable plant species [14–16]. Orthodox seeds of wild-growing legumes (moisture content from 5 to 12%) are one of the most convenient systems available for prolonged storage of genetic information. Cryogenic storage has been successfully carried out in many orthodox seed species [15–20]; however in some species authors pointed out problems after cryopreservation, such as cotyledon breakage [15, 16, 21], abnormal germination [1], or seed death by internal freezing injury [22]. In these cases, the role of seed characteristics, in particular seed size has been studied. The conventional regime of seed storage in seed banks at 5°C is not reliable, because low above-zero temperatures retard the loss of viability but fail to ensure its long-term preservation. The aim of the present study was to evaluate the effect of cryopreservation on the twelve Far-Eastern wild-growing legume species with different seed characteristics (size and mass of seeds) for their *ex situ* conservation. The germination of wild legumes is often complicated by physical dormancy which may require scarification for germination, and thus combinations of dormancy-breaking procedures and exposure to LN have been tested to examine the interaction of cryopreservation and scarification. The results obtained suggest that seeds of species studied have the ability to tolerate cryostorage.

2. Material and Methods

2.1. Plant Material. The study was performed from 2002 to 2008. Mature seeds of twelve wild legume species were collected from wild plants growing in natural habitats from different regions of Far East of Russia (Table 1). Collected seeds stored in tightly closed bags in a laboratory at ambient temperature for 2–4 months until assays were carried out. Nomenclature, range, and distribution for species are according to the monograph on vascular plants of the Russian Far East [3]. For each species or population, seed mass was determined by weighting three samples of 100 seeds each; seed size, by measuring 25 seeds from each sample (Table 2). A batch of 50 seeds from each species was subjected to moisture determination by oven drying at 105°C for 24 h, when the seeds had reached constant dry mass. Moisture content (MC) was obtained from 3 independent determinations and expressed as mean percentage of fresh weight (Table 2). For 3 species (*Hedysarum austrokurilense, H. sachalinense,*

TABLE 2: Rate of germination and final germination percentage (G) of Far-Eastern legumes after scarification, cryopreservation (LN), or cryopreservation and scarification (scarification after LN). MC: seed moisture content. L: seed length. SM: seed mass, g per 100 seeds.

Species	Control			Scarification			LN			Scarification after LN			$G \pm SE$, %*	MC, %	L, mm	SM, g
	T_0, days	T_{50}, days	$G \pm SE$, %	T_0, days	T_{50}, days	$G \pm SE$, %	T_0, days	T_{50}, days	$G \pm SE$, %	T_0, days	T_{50}, days	$G \pm SE$, %				
Astragalus membranaceus	2	11	3.3 ± 1.2	1	2	99.3 ± 1.2	2	11	14.0 ± 3.5	2	2	75.3 ± 6.1	89.3 ± 4.2	6.90 ± 0.18	2.8 ± 0.03	0.4 ± 0.01
Sophora flavescens	7	10	3.0 ± 1.0	3	7	86.8 ± 7.9	4	13	48.9 ± 1.9	4	4	29.9 ± 4.3	78.8 ± 3.3	8.60 ± 0.04	4.86 ± 0.06	4.0 ± 0.18
Trifolium lupinaster	1	7	8.0 ± 2.0	1	2	94.0 ± 5.3	1	3	84.0 ± 2.0	1	2	12.7 ± 1.2	96.7 ± 2.3	7.98 ± 0.75	2.0 ± 0.04	0.33. ± 0.01
Oxytropis chankaensis	2	6	6.0 ± 2.0	2	2	80.7 ± 8.1	2	6	91.3 ± 4.6	2	2	4.7 ± 1.2	96.0 ± 3.5	7.05 ± 0.34	2.2 ± 0.03	0.24 ± 0.003
O. kamtschatica	2	11	68.0 ± 1.2	2	3	93.6 ± 4.2	2	9	89.3 ± 4.8	—	—	—	—	6.64 ± 0.77	2.26 ± 0.02	0.25 ± 0.005
O. ochotensis	2	12	36.0 ± 1.2	2	2	87.3 ± 5.2	2	9	62.0 ± 3.1	—	—	—	—	7.10 ± 0.18	1.98 ± 0.03	0.2 ± 0.002
O. retusa	6	10	9.0 ± 1.0	2	3	67.0 ± 2.0	4	7	41.3 ± 6.4	—	—	—	—	5.74 ± 0.71	2.68 ± 0.04	0.41 ± 0.003
O. revoluta-1	2	33	42.0 ± 2.0	2	2	79.3 ± 1.9	2	6	84.7 ± 7.1	—	—	—	—	6.68 ± 0.25	1.79 ± 0.03	0.13 ± 0.001
O. revoluta-2	2	25	49.3 ± 4.8	2	2	79.3 ± 3.0	2	7	80.7 ± 2.7	—	—	—	—	6.45 ± 0.40	1.81 ± 0.03	0.13 ± 0.002
Hedysarum austrokurilense	1	8	70.7 ± 1.3	1	2	87.7 ± 4.6	1	7	85.0 ± 1.5	—	—	—	—	—	2.84 ± 0.05	0.47 ± 0.011
H. sachalinense	—	2	85.3 ± 1.8	—	—	—	1	3	73.0 ± 7.5	—	—	—	—	—	4.46 ± 0.10	0.93 ± 0.056
Vicia amurensis	2	11	30.0 ± 4.2	2	3	100	5	11	33.3 ± 2.7	—	—	—	—	—	3.35 ± 0.04	2.47 ± 0.06
V. subrotunda	8	26	5.3 ± 0.7	2	3	74.3 ± 1.8	5	16	53.3 ± 7.1	—	—	—	—	6.97 ± 0.23	3.79 ± 0.07	2.57 ± 0.123

* Summarized germination percentage after cryopreservation (LN) and scarification (scarification after LN).

Vicia amurensis), MC was not determined due to small number of seeds. No desiccation procedures were carried out due to low level of seed moisture. Seeds of species tested contained 5.7–8.6% moisture; seed mass ranged from 0.13 to 4.0 g per 100 seeds (Table 2).

2.2. Cryopreservation. The seeds in tightly closed aluminium foil bags were immersed directly into liquid nitrogen ($-196°C$) and stored for 1 month, then warmed at room temperature ($21 \pm 3°C$) for few hours before planting in germination dishes. Unfrozen seeds remained at room temperature.

2.3. Germination Test. For all treatments, the germination tests were carried out at 22–25°C under a 16 h light/8 h night photoperiod. Three replicates of 50 seeds were placed on moist filter paper in Petri dishes and watered daily with tap water. The tests were monitored every day, and germinated seeds (radicle extension ≥ 2 mm) were counted and removed. Incubation was continued until it was obvious that no further germination could occur. The first test was a control germination without any treatments. The second one presents germination after scarification. The water-impermeable legume seeds before plating in Petri dishes were scarified with a concentrated sulphuric acid: *O. kamtschatica, O. ochotensis, O. revoluta, O. retusa*—20 min; *A. membranaceus, T. lupinaster, H. austrokurilense*—30 min; *S. flavescens*—45 min; *V. amurensis, O. chankaensis*—60 min; *V. subrotunda*—70 min. Then the seeds were washed in tap water and sown as described previously. The third attempt was germination after cryogenic storage. For most species, the major part of not-germinated after-cryopreservation seeds became soft and decayed. In four species—*A. membranaceus, O. chankaensis, S. flavescens, T. lupinaster*, the seeds, which failed to germinate during 2 months after cryopreservation, remained intact and undamaged. Then they were treated with a concentrated sulphuric acid, then washed in a tap water, and sown to evaluate a combined effect of cryopreservation and scarification. For each treatment, time of initial germination (T_0) was determined, and final germination percentage (mean values $\pm$ SE) and a number of days needed to reach 50% of the final germination percentage (T_{50}) were calculated. To analyse the significance of differences between mean values, analysis of variance (ANOVA) was applied, as was the Tukey test for pairwise comparisons. The Tukey test was performed after arc-sine transformation at a significance level of $P < 0.05$.

In two species—*H. austrokurilense, H. sachalinense*, the seedling emergence tests, like the germination tests, were conducted out at 22–25°C under 16 h light/8 h night photoperiod, until the seedlings reached ca. 2-3 cm in height. Seedling vigor was measured by a shoot and root length on 15 d after sowing. Twenty five plantlets were measured in each sample.

In *S. flavescens*, plantlets from the control and cryopreserved seeds were transplanted to 1-L pots with a soil to assess the quantitative traits (plant height, leaf size and number of the leaves, and root diameter) over three months (by measuring 30 plantlets from each sample). The data were analyzed using Student's t-test with significance at $P = 0.05$.

3. Results

In control, the time of initial germination varied from 1 to 8 days, with both *Hedysarum* and *T. lupinaster* seeds germinating most rapidly. The germination of studied species seeds varied considerably, ranging from 3 to 85%; this indicates the different levels of hardseededness. The seeds of six species (*A. membranaceus, O. chankaensis, O. retusa, S. flavescens, T. lupinaster, V. subrotunda*) showed very low germination percentages (less than 10%). The germination rate also showed evident differences among species, with slowest rates in *V. subrotunda* and two populations of *O. revoluta*. The depth of hardseededness is species-specific and does not depend on seed size.

After scarification, germination was substantially higher and much faster as compared with control. It is indicative that seed dormancy was only of the physical type. For the most species T_0 and T_{50} generally occurred at the same time (2-3 days after incubation), regardless of seed size. Seeds of *A. membranaceus, O. chankaensis, S. flavescens,* and *T. lupinaster* after scarification germinated about 30–40 days, though most of them germinated in first 10–14 days (Table 2). For the rest of the species, there were no any marked differences among them with respect to the germination rate; all seeds germinated rapidly for 7–12 days. Seed size and weight did not affect germination of scarified seeds apparently.

No damage was observed in control and scarified seeds. Cryopreservation was successful, and seeds of all species began to germinate after exposure to a cryogenic temperature. All tested species, except *H. sachalinense* and *V. amurensis*, exhibited a significant positive response to the 30-day exposure to LN when compared with control. Seed germination of *H. sachalinense* after cryostorage was high ($\geq 70\%$) and did not differ significantly from control ($P = 0.203$), while for *V. amurensis* one was rather low (Table 2) and also did not differ significantly from control ($P = 0.527$). Both populations of *O. revoluta* respond identically to the treatment though they differ in the initial levels of hardseededness. Cryostorage generally did not influence T_0, with an exception of seeds of *O. retusa, S. flavescens,* and both *Vicia* species, which germinated earlier than in control. For all species T_{50} decreased to 3–16 days.

The germination rate of cryopreserved seeds was generally higher than for the unfrozen seeds, but lower than for scarified ones. No general trends between seed size and the ability of seed to survive to LN storage were observed. Nevertheless, a germination of the large seeds with high and moderate levels of hardseededness (*S. flavescens, V. amurensis, V. subrotunda*) did not exceed 50% after LN exposure, while a germination of the small seeds with high levels of hardseededness (*O. chankaensis, T. lupinaster*) was above 80%. For all species with small seed size (*O. chankaensis, O. kamtschatica, O. ochotensis, O. revoluta, T. lupinaster*) germination after cryostorage was very high (60–90%). For these species and *H. austrokurilense* there was no significant differences between germination after scarification and LN treatment. A germination of *A. membranaceus*, *O. chankaensis*, *S. flavescens*, and *T. lupinaster* after freezing was enhanced by a scarification of the nongerminated seeds (Table 2). *A. membranaceus* and *S. flavescens* seeds reached significantly higher germination percentage than before scarification (from 14 to 89% and from 49 to 79%, resp.).

Despite some variability in the seed germination among species studied, all seeds germinated after storage in LN developed into the normal seedlings with healthy shoot and root formation; no diseases were observed. The initial stages of seedling ontogenesis in the control and experimental samples of *Hedysarum* species went on synchronously. The seedlings coming from frozen and control seeds were morphologically and developmentally similar; that is, we could not detect differences in growth rate of shoot and root and in stem and root shape as well.

For *S. flavescens*, the formation of juvenile plants (an appearance of first two leaves) from the control seeds began earlier than from seeds stored in LN. There were no significant differences in the plant height, size and number of leaves, and root diameter between the plantlets derived from cryopreserved and control seeds during the entire period of culturing [23]. It is indicative of normal development of plantlets derived from the seeds treated by LN. Only slight increase of a root diameter was found in the 70-day-old seedlings from frozen seeds.

4. Discussion

The effect of an exposure to ultralow temperature (liquid nitrogen, LN) on germination of the seeds was investigated in 12 wild medicinal legume species occurring Far East of Russia. The results show that the seeds of all investigated species can survive cryostorage.

There was no trend between seed size and survival of seeds after LN treatment. We observed higher germination percentages after cryostorage in the species with small seeds as compared to species with large seeds, which could be conditioned by presence of physical dormancy (hard-seededness) in the tested species. The only exception was *A. membranaceus* seeds (0.4 ± 0.01 g per 100 seeds) with the lowest germination after cryostorage, which evidently had a highest level of hardseededness, and scarification following LN treatment, allowed to reach high germination values (above 90%).

The seeds of the studied species have a hard seed coat with a layer of macroscleids (palisade-like cells with strongly lignified walls), typical for many wild Fabaceae species [24, 25], and most part of them is characterized by high and moderate levels of hardseededness (Table 2). A physical dormancy often occurs as a means of adaptation to the different habitats with the severe climatic conditions, and such seeds are highly resistant to unfavorable environmental factors [26, 27]. It is low permeability of hard seed coat that may play some role in the cryotolerance of these seeds [19, 28]. However, dormancy poses one of the main problems for work with wild species [15, 29]. For the tested seeds of wild legumes, apparently, immersion in LN effects as a means of overcoming seed dormancy like scarification. It was reported that seed coat cracks, formed in the seeds impermeable to water during freezing, favoured a water uptake, and consequently germination was stimulated [30].

The final germination capacity of cryostored seeds differed significantly from nonstored ones for most species studied except *H. sachalinense* and *V. amurensis*. The stimulatory effects of LN exposure were revealed for other wild legumes with hard seeds [1, 15, 21, 31, 32]. Besides, after LN exposure, the seeds required less time to germinate than controls. In agreement with those results, it could be permitted to correlate them to the beneficial effects of low-temperature exposure, namely, stimulating germination and increasing germination rate. In addition, previous scarification or other dormancy-breaking treatment before germination seems not to be necessary for the most of these seeds. However, for seeds with deep physical dormancy LN is not effective scarifying agent, as it was shown for seeds of some *Rubus* species [33] and for seeds of *A. membranaceus*, *S. flavescens*, and *V. amurensis* in the present study. At the same time, for *A. membranaceus* and *S. flavescens* LN treatment, when combined with chemical scarification, favoured the germination resulting in considerable increase in germination percentage (Table 2). The similar action of combined treatment by LN and sulfuric acid has been observed in some orthodox legume seeds of Brazilian tropical species [32]. Necessity of scarification (mechanical or chemical scarification, heat treatment) combined with cryogenic storage to enhance the final germination percentage was shown for other species with hard seed coats [28, 29, 34]. The intensity of seed hardness varies among species and within individual species (e.g., in populations of *O. revoluta*), possibly due to the differences in their seed coat structure, physical, and chemical properties [25]. These differences might be reflected in the seeds response to LN treatment. As it is shown in previous studies, it is not possible to predict the behaviour of the seeds of a species after cryopreservation by their size or by taxonomic relatedness [19, 35]. Therefore, it is important to continue screening different species for effective cryostorage and to investigate possible interactions with germination enhancement treatment.

In our experiments, cryopreserved seeds of *Hedysarum* species and *S. flavescens* developed into the normal seedlings and plants. Similar results were reported by Tikhonova [36] for 30 wild-growing plant species: plant growth and development was not adversely affected by cryopreservation. The normal growth of plants from the seeds treated by LN was shown for two Spanish endemic species [35] seven species of Orchidaceae [37, 38]. It appeared that LN exposure only or combined treatments did not induce seed physical damage or abnormal seedling development for all species studied.

Thus, the obtained results have demonstrated that the seeds of the Far-Eastern medicinal legume species can be deep-frozen at $-196°C$ without loss of viability. Taking into account the tolerance to liquid nitrogen, cryogenic storage would be a suitable method for a long-term conservation of wild medicinal legumes. Additionally, the very small size of seeds of the most studied species results in an effective cryogenic procedure which is easy to operate.

Acknowledgments

The work was supported by the Program of the Presidium of the Russian Academy of Sciences "Biological Diversity", the Program "Genetic Diversity of Natural Populations of Far East Plants" (Project no. 09-1-P23-06), and the Program of the Presidium of the Russian Academy of Sciences "Molecular and Cellular Biology" (Project no. 09-1-P22-03).

References

[1] T. Shibata, E. Sakai, and K. Shimomura, "Effect of rapid freezing and thawing on hard-seed breaking in *Astragalus mongholicus* Bunge (Leguminosae)," *Journal of Plant Physiology*, vol. 147, no. 1, pp. 127–131, 1995.

[2] S. Sinclair, "Chinese herbs: a clinical review of Astragalus, Ligusticum, and Schizandrae," *Alternative Medicine Review*, vol. 3, no. 5, pp. 338–344, 1998.

[3] N. S. Pavlova, "Family Fabaceae," in *The Vascular Plants of the Soviet Far East*, S. S. Kharkevich, Ed., vol. 4, pp. 191–339, Nauka, Leningrad, Russia, 1989.

[4] O. V. Neretina, A. S. Gromova, V. I. Lutsky, and A. A. Semenov, "Component composition of species of the genus Hedysarum (Fabaceae)," *Rastitel'nye Resursy*, vol. 40, no. 4, pp. 111–138, 2004.

[5] K. F. Blinova and E. I. Sakanyan, "Species of *Oxytropis* used in Tibetan medicine and their flavonoid composition," *Rastitel'nye Resursy*, vol. 22, no. 2, pp. 266–272, 1986.

[6] D. Batsuren, S. Tsetsegmaa, N. Batbayar et al., "Alkaloids of *Oxytropis*. I," *Chemistry of Natural Compounds*, vol. 28, no. 3-4, pp. 340–344, 1992.

[7] L. Long and Q. Li, "The effect of alkaloid from *Oxytropis ochrocephala* on growth inhibition and expression of PCNA and p53 in mice bearing H_{22} hepatocellular carcinoma," *Yakugaku Zasshi*, vol. 125, no. 8, pp. 665–670, 2005.

[8] J. D. Cha, M. R. Jeong, S. I. Jeong, and K. Y. Lee, "Antibacterial activity of sophoraflavanone G isolated from the roots of *Sophora flavescens*," *Journal of Microbiology and Biotechnology*, vol. 17, no. 5, pp. 858–864, 2007.

[9] S. Y. Ryu, S. U. Choi, S. K. Kim et al., "*In vitro* antitumour activity of flavonoids from *Sophora flavescens*," *Phytotherapy Research*, vol. 11, no. 1, pp. 51–53, 1997.

[10] G. A. Denisova, V. I. Dorofeev, and G. I. Kapranova, "Genus Trifolium L.," in *Plant Resources of the USSR: Flowering Plants, Their Chemical Composition and Use*, P. D. Sokolov, Ed., vol. 3 of *Hydrangeaceae—Haloragaceae Families*, pp. 181–191, Nauka, Leningrad, Russia, 1987.

[11] S. S. Kang, Y. S. Chang, and J. S. Kim, "Two new acylated flavonol glycosides from *Vicia amurensis*," *Chemical and Pharmaceutical Bulletin*, vol. 48, no. 8, pp. 1242–1245, 2000.

[12] P. L. Popov, "Plant species, using against virus infections of man and animals: regularities of the distribution in the phylogenetic classification system," *Journal of Stress Physiology and Biochemistry*, vol. 4, no. 3, pp. 17–64, 2008.

[13] O. V. Maksimov, P. G. Gorovoy, and G. N. Chumak, "The content of antioxidants in the seeds of some species of Primorie flora," *Rastitel'nye Resursy*, vol. 26, no. 4, pp. 487–498, 1990.

[14] F. Engelmann, "Plant cryopreservation: progress and prospects," *In Vitro Cellular and Developmental Biology—Plant*, vol. 40, no. 5, pp. 427–433, 2004.

[15] V. C. Pence, "Cryopreservation of seeds of Ohio native plants and related species," *Seed Science and Technology*, vol. 19, no. 2, pp. 235–251, 1991.

[16] P. C. Stanwood, "Cryopreservation of seed germplasm for genetic conservation," in *Cryopreservation of Plant Cells and Organs*, K. K. Kartha, Ed., pp. 199–226, CRC Press, Boca Raton, Fla, USA, 1985.

[17] P. Chmielarz, "Cryopreservation of dormant European ash (*Fraxinus excelsior*) orthodox seeds," *Tree Physiology*, vol. 29, no. 10, pp. 1279–1285, 2009.

[18] P. Chmielarz, "Cryopreservation of the non-dormant orthodox seeds of *Ulmus glabra*," *Acta Biologica Hungarica*, vol. 61, no. 2, pp. 224–233, 2010.

[19] D. H. Touchell and K. W. Dixon, "Cryopreservation of seed of Western Australian native species," *Biodiversity and Conservation*, vol. 2, no. 6, pp. 594–602, 1993.

[20] C. Walters, L. Wheeler, and P. C. Stanwood, "Longevity of cryogenically stored seeds," *Cryobiology*, vol. 48, no. 3, pp. 229–244, 2004.

[21] H. W. Pritchard, K. R. Manger, and F. G. Prendergast, "Changes in *Trifolium arvense* seed quality following alternating temperature treatment using liquid nitrogen," *Annals of Botany*, vol. 62, no. 1, pp. 1–11, 1988.

[22] C. W. Vertucci, "Effect of cooling rate on seeds exposed to liquid nitrogen temperatures," *Plant Physiology*, vol. 90, no. 4, pp. 1478–1485, 1989.

[23] N. M. Voronkova and A. B. Kholina, "An influence of temperature factor and scarification on seed germination and growth of seedlings of *Sophora flavescens* Soland," *Rastitel'nye Resursy*, vol. 39, no. 1, pp. 43–49, 2003.

[24] J. M. Baskin, C. C. Baskin, and X. Li, "Taxonomy, anatomy and evolution of physical dormancy in seeds," *Plant Species Biology*, vol. 15, no. 2, pp. 139–152, 2000.

[25] K. M. Kelly, J. Van Staden, and W. E. Bell, "Seed coat structure and dormancy," *Plant Growth Regulation*, vol. 11, no. 3, pp. 201–209, 1992.

[26] A. B. Kholina and N. M. Voronkova, "Conserving the gene pool of Far Eastern plants by means of seed cryopreservation," *Biology Bulletin*, vol. 35, no. 3, pp. 262–269, 2008.

[27] N. M. Voronkova, A. B. Kholina, and V. P. Verkholat, "Plant biomorphology and seed germination in pioneer species of Kamchatka volcanoes," *Biology Bulletin*, vol. 35, no. 6, pp. 599–605, 2008.

[28] P. Chmielarz, "Sensitivity of *Tilia* cordata seeds to dehydration and temperature of liquid nitrogen," *Dendrobiology*, vol. 47, pp. 71–77, 2002.

[29] S. Gonçalves, L. Fernandes, F. Pérez-García, M. E. González-Benito, and A. Romano, "Germination requirements and cryopreservation tolerance of seeds of the endangered species *Tuberaria major*," *Seed Science and Technology*, vol. 37, no. 2, pp. 480–484, 2009.

[30] A. N. Salomao, "Effects of liquid nitrogen storage on *Zizyphus joazeiro* seeds," *Cryo-Letters*, vol. 16, no. 2, pp. 85–90, 1995.

[31] F. Pérez-García, "Effect of cryopreservation, gibberellic acid and mechanical scarification on the seed germination of eight endemic species from the Canary Islands," *Seed Science and Technology*, vol. 36, no. 1, pp. 237–242, 2008.

[32] A. N. Salomão, "Tropical seed species' responses to liquid nitrogen exposure," *Brasilian Journal of Plant Physiology*, vol. 14, no. 2, pp. 133–138, 2002.

[33] D. N. Peacock and K. E. Hummer, "Pregermination studies with liquid nitrogen and sulfuric acid on several *Rubus* species," *HortScience*, vol. 31, no. 2, pp. 238–239, 1996.

[34] F. Pérez-García and M. E. González-Benito, "Seed cryopreservation of *Halimium* and *Helianthemum* species," *Cryo-Letters*, vol. 29, no. 4, pp. 271–276, 2008.

[35] M. E. González-Benito, F. Fernández-Llorente, and F. Pérez-Garcia, "Interaction between cryopreservation, rewarming rate and seed humidification on the germination of two Spanish endemic species," *Annals of Botany*, vol. 82, no. 5, pp. 683–686, 1998.

[36] V. L. Tikhonova, "Long-term storage of seeds," *Russian Journal of Plant Physiology*, vol. 46, no. 3, pp. 400–408, 1999.

[37] T. V. Nikishina, A. S. Popov, G. L. Kolomeitseva, and B. N. Golovkin, "Effect of cryoconservation on seed germination of rare tropical orchids," *Russian Journal of Plant Physiology*, vol. 48, no. 6, pp. 810–815, 2001.

[38] A. S. Popov, E. V. Popova, T. V. Nikishina, and G. L. Kolomeytseva, "The development of juvenile plants of the hybrid orchid *Bratonia* after seed cryopreservation," *Cryo-Letters*, vol. 25, no. 3, pp. 205–212, 2004.

4

Reactive Oxygen Species, Oxidative Damage, and Antioxidative Defense Mechanism in Plants under Stressful Conditions

Pallavi Sharma,[1] Ambuj Bhushan Jha,[2] Rama Shanker Dubey,[1] and Mohammad Pessarakli[3]

[1] Department of Biochemistry, Faculty of Science, Banaras Hindu University, Varanasi 221005, India
[2] Crop Development Centre, Department of Plant Sciences, College of Agriculture and Bioresources, University of Saskatchewan, 51 Campus Drive, Saskatoon SK, Canada SK S7N 5A8
[3] School of Plant Sciences, The University of Arizona, Forbes Building, Room 303, P.O. Box 210036, Tucson, AZ 85721-0036, USA

Correspondence should be addressed to Mohammad Pessarakli, pessarak@email.arizona.edu

Academic Editor: Andrea Polle

Reactive oxygen species (ROS) are produced as a normal product of plant cellular metabolism. Various environmental stresses lead to excessive production of ROS causing progressive oxidative damage and ultimately cell death. Despite their destructive activity, they are well-described second messengers in a variety of cellular processes, including conferment of tolerance to various environmental stresses. Whether ROS would serve as signaling molecules or could cause oxidative damage to the tissues depends on the delicate equilibrium between ROS production, and their scavenging. Efficient scavenging of ROS produced during various environmental stresses requires the action of several nonenzymatic as well as enzymatic antioxidants present in the tissues. In this paper, we describe the generation, sites of production and role of ROS as messenger molecules as well as inducers of oxidative damage. Further, the antioxidative defense mechanisms operating in the cells for scavenging of ROS overproduced under various stressful conditions of the environment have been discussed in detail.

1. Introduction

An unavoidable consequence of aerobic metabolism is production of reactive oxygen species (ROS). ROS include free radicals such as superoxide anion ($O_2^{\bullet-}$), hydroxyl radical ($^{\bullet}OH$), as well as nonradical molecules like hydrogen peroxide (H_2O_2), singlet oxygen ($^1O_2$), and so forth. Stepwise reduction of molecular oxygen (O_2) by high-energy exposure or electron-transfer reactions leads to production of the highly reactive ROS. In plants, ROS are always formed by the inevitable leakage of electrons onto O_2 from the electron transport activities of chloroplasts, mitochondria, and plasma membranes or as a byproduct of various metabolic pathways localized in different cellular compartments [1–5]. Environmental stresses such as drought, salinity, chilling, metal toxicity, and UV-B radiation as well as pathogens attack lead to enhanced generation of ROS in plants due to disruption of cellular homeostasis [6–15]. All ROS are extremely harmful to organisms at high concentrations. When the level of ROS exceeds the defense mechanisms, a cell is said to be in a state of "oxidative stress." The enhanced production of ROS during environmental stresses can pose a threat to cells by causing peroxidation of lipids, oxidation of proteins, damage to nucleic acids, enzyme inhibition, activation of programmed cell death (PCD) pathway and ultimately leading to death of the cells [6–8, 11, 13, 14, 16, 17].

Despite their destructive activity, ROS are well-described second messengers in a variety of cellular processes including tolerance to environmental stresses [18–20]. Whether ROS will act as damaging or signaling molecule depends on the delicate equilibrium between ROS production and scavenging. Because of the multifunctional roles of ROS, it is necessary for the cells to control the level of ROS tightly to avoid any oxidative injury and not to eliminate them completely. Scavenging or detoxification of excess ROS is achieved by an efficient antioxidative system comprising of the nonenzymic as well as enzymic antioxidants [21]. The enzymic antioxidants include superoxide dismutase (SOD), catalase (CAT), guaiacol peroxidase (GPX), enzymes of ascorbate-glutahione (AsA-GSH) cycle such as ascorbate peroxidase

(APX), monodehydroascorbate reductase (MDHAR), dehydroascorbate reductase (DHAR), and glutathione reductase (GR) [21]. Ascorbate (AsA), glutathione (GSH), carotenoids, tocopherols, and phenolics serve as potent nonenzymic antioxidants within the cell. Various workers have reported increased activities of many enzymes of the antioxidant defense system in plants to combat oxidative stress induced by various environmental stresses. Maintenance of a high antioxidant capacity to scavenge the toxic ROS has been linked to increased tolerance of plants to these environmental stresses [22, 23]. Considerable progress has been made in improving stress-induced oxidative stress tolerance in crop plants by developing transgenic lines with altered levels of antioxidants [24, 25]. Simultaneous expression of multiple antioxidant enzymes has been shown to be more effective than single or double expression for developing transgenic plants with enhanced tolerance to multiple environmental stresses [26]. The present review focuses on types of ROS, their site of production, and their role as messenger and inducer of oxidative stress. Further, role of antioxidative defense system in combating danger posed by overproduced ROS under stresses has been discussed in detail.

2. Reactive Oxygen Species, Sites of Production, and Their Effects

ROS are a group of free radicals, reactive molecules, and ions that are derived from O_2. It has been estimated that about 1% of O_2 consumed by plants is diverted to produce ROS [27] in various subcellular loci such as chloroplasts, mitochondria, peroxisomes. ROS are well recognized for playing a dual role as both deleterious and beneficial species depending on their concentration in plants. At high concentration ROS cause damage to biomolecules, whereas at low/moderate concentration it acts as second messenger in intracellular signaling cascades that mediate several responses in plant cells.

2.1. Types of ROS. The most common ROS include $^1O_2$, $O_2^{\bullet-}$, H_2O_2, $^{\bullet}OH$. O_2 itself is a totally harmless molecule as in its ground state it has two unpaired electrons with parallel spin which makes it paramagnetic and, hence, unlikely to participate in reactions with organic molecules unless it is activated [28]. Activation of O_2 may occur by two different mechanisms: (i) absorption of sufficient energy to reverse the spin on one of the unpaired electrons and (ii) stepwise monovalent reduction (Figure 1). In the former, $^1O_2$ is formed, whereas in latter, O_2 is sequentially reduced to $O_2^{\bullet-}$, H_2O_2, and $^{\bullet}OH$ (Figure 1).

Electrons in the biradical form of oxygen have parallel spin. Absorption of sufficient energy reverses the spin of one of its unpaired electrons leading to formation of singlet state in which the two electrons have opposite spin. This activation overcomes the spin restriction and $^1O_2$ can consequently participate in reactions involving the simultaneous transfer of two electrons (divalent reduction) [28]. In the light, highly reactive $^1O_2$ can be produced *via* triplet chlorophyll (Chl) formation in the antenna system and in the reaction centre of photosystem II [29]. In the antenna, insufficient energy

dissipation during photosynthesis can lead to formation of chlorophyll (Chl) triplet state, whereas in the reaction centre it is formed *via* charge recombination of the light-induced charge pair [29]. The Chl triplet state can react with $^3O_2$ to give up the very highly destructive ROS $^1O_2$:

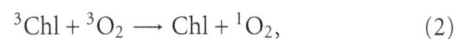

$$Chl \xrightarrow{light} {}^3Chl, \tag{1}$$

$$^3Chl + {}^3O_2 \longrightarrow Chl + {}^1O_2, \tag{2}$$

Further, limited CO_2 availability due to closure of stomata during various environmental stresses such as salinity, drought favors the formation of $^1O_2$. The life time of $^1O_2$ within the cell is probably $3\,\mu s$ or less [30, 31]. A fraction of $^1O_2$ has been shown to be able to diffuse over considerable distances of several hundred nanometers (nm). $^1O_2$ can last for $4\,\mu s$ in water and $100\,\mu s$ in a nonpolar environment [1]. $^1O_2$ reacts with most of the biological molecules at near diffusion-controlled rates [1]. It directly oxidizes protein, unsaturated fatty acids, and DNA [32]. It causes nucleic acid modification through selective reaction with deoxyguanosine [33]. It is thought to be the most important species responsible for light-induced loss of photosystem II (PSII) activity which may trigger cell death [34]. $^1O_2$ can be quenched by β-carotene, α-tocopherol or can react with the D1 protein of photosystem II as target [29].

Due to spin restriction, molecular O_2 cannot accept four electrons at a time to produce H_2O. It accepts one electron at a time and hence during reduction of O_2 stable intermediates are formed in the step-wise fashion [35]. $O_2^{\bullet-}$ is the primary ROS formed in the cell which initiates a cascade of reactions to generate "secondary" ROS, either directly or prevalently through enzyme- or metal-catalysed processes [36] depending on the cell type or cellular compartment. $O_2^{\bullet-}$ is a moderately reactive, short-lived ROS with a half-life of approx. $1\,\mu s$. $O_2^{\bullet-}$ is a nucleophilic reactant with both oxidizing and reducing properties [37]. Anionic charge of $O_2^{\bullet-}$ inhibits its electrophilic activity toward electron-rich molecules. $O_2^{\bullet-}$ has been shown to oxidize enzymes containing the [4Fe-4S] clusters (aconitase or dehydratase as examples) [38] and reduce cytochrome C [39]. $O_2^{\bullet-}$ can accept one electron and two protons to form H_2O_2. It is easily dismutated to H_2O_2 either nonenzymatically or by SOD catalyzed reaction to hydrogen peroxide:

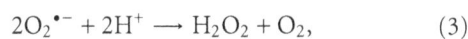

$$2O_2^{\bullet-} + 2H^+ \longrightarrow H_2O_2 + O_2, \tag{3}$$

$$2O_2^{\bullet-} + 2H^+ \xrightarrow{SOD} H_2O_2 + O_2, \tag{4}$$

H_2O_2 is generated in the cells under normal as well as wide range of stressful conditions such as drought, chilling, UV irradiation, exposure to intense light, wounding and intrusion by pathogens. Electron transport chain (ETC) of chloroplast, mitochondria, endoplasmic reticulum and plasma membrane, β-oxidation of fatty acid and photorespiration are major sources of H_2O_2 generation in plant cells. Photooxidation reactions, NADPH oxidase as well as xanthine oxidase (XOD) also contribute to H_2O_2 production in plants. It is also generated in tissues requiring it as a substrate

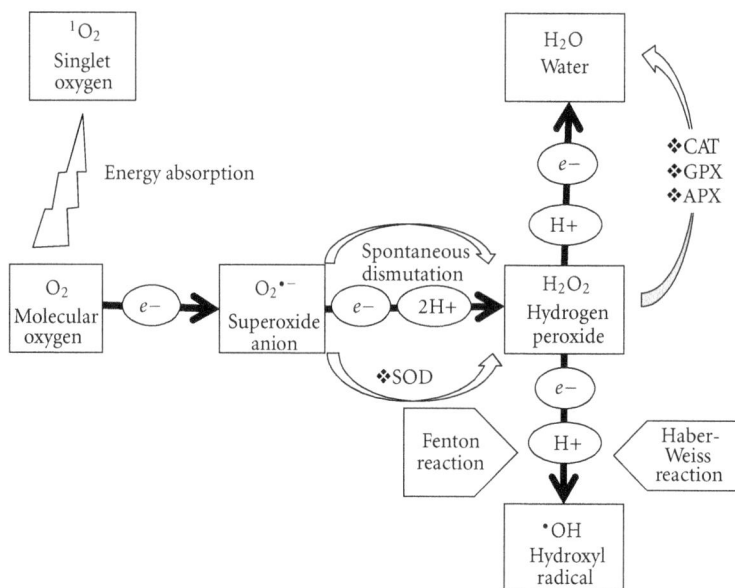

FIGURE 1: Schematic representation of generation of reactive oxygen species (ROS) in plants. Activation of O_2 occurs by two different mechanisms. Stepwise monovalent reduction of O_2 leads to formation of $O_2^{\bullet-}$, H_2O_2, and $^\bullet OH$, whereas energy transfer to O_2 leads to formation of $^1O_2$. $O_2^{\bullet-}$ is easily dismutated to H_2O_2 either nonenzymatically or by superoxide dismutase (SOD) catalyzed reaction to H_2O_2. H_2O_2 is converted to H_2O by catalase (CAT), guaiacol peroxidase (GPX), and ascorbate peroxidase (APX).

for biosynthesis such as for lignification and suberization. H_2O_2 is moderately reactive and is relatively long-lived molecule with a half-life of 1 ms [40]. H_2O_2 has no unpaired electrons, unlike other oxygen radicals, it can readily cross biological membranes and consequently can cause oxidative damage far from the site of its formation. Because H_2O_2 is the only ROS that can diffuse through aquaporins in the membranes and over larger distances within the cell [41] and is relatively stable compared to other ROS, it has received particular attention as a signal molecule involved in the regulation of specific biological processes and triggering tolerance against various environmental stresses such as plant-pathogen interactions at low concentration [19, 20, 42]. At high concentration, H_2O_2 can oxidize the cysteine (−SH) or methionine residues (−SCH_3), and inactivate enzymes by oxidizing their thiol groups, such as enzymes of Calvin cycle, Cu/Zn-SOD, and Fe-SOD [43]. When hydrogen peroxide accumulates at levels of $10\,\mu M$, the enzymes in the Calvin cycle, such as fructose-1,6-bisphosphatase, sedoheptulose-1,7-bisphosphatase, and phosphoribulokinase, lose 50% of their activity [44, 45]. It also oxidizes protein kinases, phosphatases, and transcription factors containing thiolate residues. At high concentrations, it orchestrates programmed cell death [46].

Both $O_2^{\bullet-}$ and H_2O_2 are only moderately reactive. The cellular damage by ROS appears to be due to their conversion into more reactive species. The formation of $^\bullet OH$ is dependent on both H_2O_2 and $O_2^{\bullet-}$ and, thus, its formation is subject to inhibition by both SOD and CAT.

The Haber-Weiss reaction generates $^\bullet OH$ from H_2O_2 and $O_2^{\bullet-}$. It consists of the following two reactions:

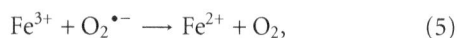

$$Fe^{3+} + O_2^{\bullet-} \longrightarrow Fe^{2+} + O_2, \qquad (5)$$

First, Fe(III) is reduced by $O_2^{\bullet-}$, followed by oxidation by dihydrogen peroxide (Fenton reaction)

$$Fe^{2+} + H_2O_2 \longrightarrow Fe^{3+} + OH^- +^\bullet OH, \qquad (6)$$

and reaction:

$$O_2^{\bullet-} + H_2O_2 \longrightarrow {}^\bullet OH + OH^- + O_2. \qquad (7)$$

Metal catalysis is necessary for this reaction since the rate of uncatalyzed reaction is negligible [47]. $^\bullet OH$ is the most reactive among all ROS. It has a single unpaired electron, thus, it can react with oxygen in triplet ground state. $^\bullet OH$ interacts with all biological molecules and causes subsequent cellular damages such as lipid peroxidation, protein damage, and membrane destruction [48]. Because cells have no enzymatic mechanism to eliminate $^\bullet OH$, its excess production can eventually lead to cell death [49]. Under illumination, formation of $^\bullet OH$ by the Fenton reaction at the active site of the enzyme RbcL leads to its fragmentation in chloroplast lysates [50, 51]. The oxidation of organic substrates by $^\bullet OH$ may proceed by two possible reactions, either by addition of $^\bullet OH$ to organic molecules or due to abstraction of a hydrogen atom from it. Because of short lifetime and the strongly positive redox potential (close to +2 V) of "free" $^\bullet OH$, its sites of reaction are close to its point of formation [52]. In this context, organic oxygen radicals such as alkoxy, peroxy, semiquinones, reduced hydrogen peroxide, and hydrogen peroxide-electron donor complexes (crypto-OH), as well as metallo-oxygen complexes, have been proposed as the ultimately active species besides destructive free $^\bullet OH$ [53].

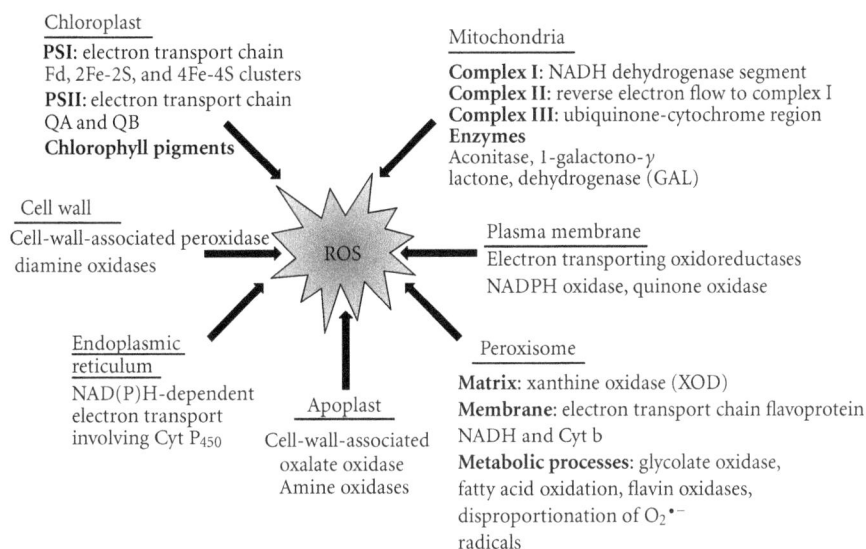

Figure 2: Sites of production of reactive oxygen species (ROS) in plants. ROS are produced at several locations in the cell-like chloroplast, mitochondria, plasma membrane, peroxisomes, apoplast, endoplasmic reticulum, and cell wall.

2.2. Sites of Production of ROS. ROS are produced in both unstressed and stressed cells at several locations in chloroplasts, mitochondria, plasma membranes, peroxisomes, apoplast, endoplasmic reticulum, and cell walls (Figure 2). ROS are always formed by the inevitable leakage of electrons onto O_2 from the electron transport activities of chloroplasts, mitochondria, and plasma membranes or as a byproduct of various metabolic pathways localized in different cellular compartments.

2.2.1. Chloroplasts. In chloroplasts, various forms of ROS are generated from several locations. ETCs in PSI and PSII are the main sources of ROS in chloroplasts. Production of ROS by these sources is enhanced in plants by conditions limiting CO_2 fixation, such as drought, salt, and temperature stresses, as well as by the combination of these conditions with high-light stress. Under normal conditions, the electron flow from the excited PS centers to NADP which is reduced to NADPH which, then, enters the Calvin cycle and reduces the final electron acceptor, CO_2. In case of overloading of the ETC, due to decreased NADP supply resulting from stress conditions, there is leakage of electron from ferredoxin to O_2, reducing it to $O_2^{\bullet-}$ [54]. This process is called Mehler reaction:

$$2O_2 + 2Fd_{red} \longrightarrow 2O_2^{\bullet-} + 2Fd_{ox} \qquad (8)$$

Leakage of electrons to O_2 may also occur from 2Fe-2S and 4Fe-4S clusters in the ETC of PSI. In PSII, acceptor side of ETC contains QA and QB. Leakage of electron from this site to O_2 contributes to the production of $O_2^{\bullet-}$ [55].

The formation of $O_2^{\bullet-}$ by O_2 reduction is a rate-limiting step. Once formed $O_2^{\bullet-}$ generates more aggressive ROS. It may be protonated to $HO_2^{\bullet}$ on the internal, "lumen" membrane surface or dismutated enzymatically (by SOD) or

spontaneously to H_2O_2 on the external "stromal" membrane surface. At Fe-S centers where Fe^{2+} is available, H_2O_2 may be transformed through the Fenton reaction into the much more dangerous $OH^{\bullet}$.

2.2.2. Mitochondria. Mitochondria can produce ROS in several sites of ETC. In mitochondria direct reduction of oxygen to $O_2^{\bullet-}$ occurs in the flavoprotein region of NADH dehydrogenase segment (complex I) of the respiratory chain [56]. When NAD^+-linked substrates for complex I are limited, electron transport can occur from complex II to complex I (reverse electron flow). This process has been shown to increase ROS production at complex I and is regulated by ATP hydrolysis [57]. Ubiquinone-cytochrome region (complex III) of the ETC also produces $O_2^{\bullet-}$ from oxygen. It is believed that fully reduced ubiquinone donates an electron to cytochrome C_1 and leaves an unstable highly reducing ubisemiquinone radical which is favorable for the electron leakage to O_2 and, hence, to $O_2^{\bullet-}$ formation [58]. In plants, under normal aerobic conditions, ETC and ATP syntheses are tightly coupled; however, various stress factors lead to inhibition and modification of its component, leading to over reduction of electron carriers and, hence, formation of ROS [4, 59].

Several enzymes present in mitochondrial matrix can produce ROS. Some of them produce ROS directly, for example aconitase, whereas some others like 1-galactono-γ lactone dehydrogenase (GAL), are able to feed electrons to ETC [60, 61]. $O_2^{\bullet-}$ is the primary ROS formed by monovalent reduction in the ETC. It is converted quickly either by the MnSOD (mitochondrial form of SOD) or APX into the relatively stable and membrane-permeable H_2O_2. H_2O_2 can be further converted to extremely active hydroxyl radical ($OH^{\bullet}$) in the Fenton reaction.

2.2.3. Endoplasmic Reticulum. In endoplasmic reticulum, NAD(P)H-dependent electron transport involving Cyt P_{450} produces $O_2^{\bullet-}$ [7]. Organic substrate, RH, reacts first with Cyt P_{450} and then is reduced by a flavoprotein to form a radical intermediate (Cyt $P_{450}R^-$). Triplet oxygen can readily react with this radical intermediate as each has one unpaired electron. This oxygenated complex (Cyt P_{450}-ROO^-) may be reduced by cytochrome b or occasionally the complexes may decompose releasing $O_2^{\bullet-}$.

2.2.4. Peroxisomes. Peroxisomes are probably the major sites of intracellular H_2O_2 production, as a result of their essentially oxidative type of metabolism [3]. The main metabolic processes responsible for the generation of H_2O_2 in different types of peroxisomes are the glycolate oxidase reaction, the fatty acid β-oxidation, the enzymatic reaction of flavin oxidases, and the disproportionation of $O_2^{\bullet-}$ radicals [62]. During photorespiration, the oxidation of glycolate by glycolate oxidase in peroxisomes accounts for the majority of H_2O_2 production [63]. Like mitochondria and chloroplasts, peroxisomes also produce $O_2^{\bullet-}$ as a consequence of their normal metabolism. In peroxisomes from pea leaves and watermelon cotyledons, at least, two sites of $O_2^{\bullet-}$ generation have been identified using biochemical and electron spin resonance spectroscopy (ESR) methods: one in the organelle matrix, the generating system being XOD, which catalyses the oxidation of xanthine or hypoxanthine to uric acid, and produces $O_2^{\bullet-}$ in the process and another site in the peroxisomal membranes where a small ETC composed of a flavoprotein NADH and Cyt b is involved. Three integral peroxisomal membrane polypeptides (PMPs) with molecular masses of 18, 29, and 32 kDa were found to be involved in $O_2^{\bullet-}$ production. While the 18- and 32-kDa PMPs use NADH as electron donor for $O_2^{\bullet-}$ production, the 29-kDa PMP was clearly dependent on NADPH and was able to reduce cytochrome c with NADPH as electron donor [64]. Among the three integral polypeptides, the main producer of $O_2^{\bullet-}$ was the 18-kDa PMP which was proposed to be a cytochrome possibly belonging to the b-type group. The PMP32 very probably corresponds to the MDHAR, and the third $O_2^{\bullet-}$-generating polypeptide, PMP29, could be related to the peroxisomal NADPH:cytochrome P450 reductase [64]. The $O_2^{\bullet-}$ produced is subsequently converted into H_2O_2 by SOD.

2.2.5. Plasma Membranes. Electron transporting oxidoreductases are ubiquitous at plasma membranes and lead to generation of ROS at plasma membrane. Production of ROS was studied using EPR spin-trapping techniques and specific dyes in isolated plasma membranes from the growing and the nongrowing zones of hypocotyls and roots of etiolated soybean seedlings as well as coleoptiles and roots of etiolated maize seedlings [5]. NAD(P)H mediated the production of $O_2^{\bullet-}$ in all plasma membrane samples. It was suggested that in soybean plasma membranes, $O_2^{\bullet-}$ production could be attributed to the action of at least two enzymes, an NADPH oxidase, and, in the presence of menadione, a quinone reductase [5]. NADPH oxidase catalyses transfer of electrons from cytoplasmic NADPH to O_2 to form $O_2^{\bullet-}$. $O_2^{\bullet-}$ is dismutated to H_2O_2 either spontaneously or by SOD activity. NADPH oxidase has been proposed to play a key role in the production and accumulation of ROS in plants under stress conditions [28, 42, 65].

2.2.6. Cell Walls. Cell walls are also regarded as active sites for ROS production. Role of cell-wall-associated peroxidase in H_2O_2 generation has been shown. In horseradish, peroxidase associated with isolated cell walls catalyzes the formation of H_2O_2 in the presence of NADH. The reaction is stimulated by various monophenols, especially of coniferyl alcohol. Malate dehydrogenase was found to be the sole candidate for providing NADH [66]. The generation of ROS by cell-wall-located peroxidases has been shown during hypersensitive response (HR) triggered in cotton by the bacterium *Xanthomonas campestris* pv. malvacearum [67] and potassium (K) deficiency stress in Arabidopsis [68]. Diamine oxidases are also involved in production of activated oxygen in the cell wall using diamine or polyamines (putrescine, spermidine, cadaverine, etc.) to reduce a quinone that autooxidizes to form peroxides [54].

2.2.7. Apoplast. Cell-wall-located enzymes have been proved to be responsible for apoplastic ROS production [5, 28]. The cell-wall-associated oxalate oxidase, also known as germin, releases H_2O_2 and CO_2 from oxalic acid [69]. This enzyme was reported to be involved in apoplastic hydrogen peroxide accumulation during interactions between different cereals species and fungi [70]. Amine oxidase-like enzymes may contribute to defense responses occurring in the apoplast following biotic stress, mainly through H_2O_2 production [71]. Amine oxidases catalyze the oxidative deamination of polyamines (i.e., putrescine, spermine, and spermidine) using FAD as a cofactor [71]. Heyno and coworkers [5], based on their study, concluded that apoplastic $OH^{\bullet}$ generation depends fully, or for the most part, on peroxidase localized in the cell wall.

2.3. Role of ROS as Messengers. At low/moderate concentration, ROS have been implicated as second messengers in intracellular signaling cascades that mediate several plant responses in plant cells, including stomatal closure [19, 20, 65], programmed cell death [7, 72], gravitropism [73], and acquisition of tolerance to both biotic and abiotic stresses [42, 74]. Figure 3 shows the role of ROS as second messenger in hormone mediated cellular responses in plants. Plants can sense, transduce and translate ROS signal into appropriate cellular responses with the help of some redox-sensitive proteins, calcium mobilization, protein phosphorylation, and gene expression. ROS can be sensed directly also by key signaling proteins such as a tyrosine phosphatase through oxidation of conserved cysteine residues (reviewed in [75]). ROS can also modulate the activities of many components in signaling, such as protein phosphatases, protein kinases and transcription factors [76] and communicate with other signal molecules and the pathway forming part of the signaling network that controls response downstream of ROS

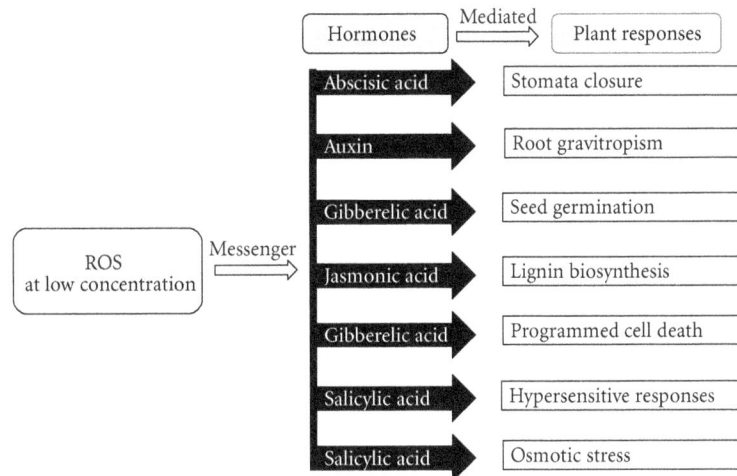

FIGURE 3: Reactive oxygen species (ROS) as second messengers in several plant hormone responses, including stomatal closure, root gravitropism, seed germination, lignin biosynthesis, programmed cell death, hypersensitive responses, and osmotic stress.

[19]. The strength, lifetime and size of the ROS signaling pool depends on balance between oxidant production and removal by the antioxidant. Using mutants deficient in key ROS-scavenging enzymes, Miller and coworkers [74] identified a signaling pathway that is activated in cells in response to ROS accumulation. Interestingly, many of the key players in this pathway, including different zinc finger proteins and WRKY transcription factors, are also central regulators of abiotic stress responses involved in temperature, salinity and osmotic stresses.

ROS are considered second messengers in the abscisic acid (ABA) transduction pathway in guard cells [19, 20]. ABA induced H_2O_2 is an essential signal in mediating stomatal closure to reduce water loss through the activation of calcium-permeable channels in the plasma membrane [77]. Jannat and coworkers [78] observed that ABA-inducible cytosolic H_2O_2 elevation functions in ABA-induced stomatal closure, while constitutive increase of H_2O_2 does not cause stomatal closure. Role of ROS as second messenger in root gravitropism has been demonstrated. Based on their work, Joo and coworkers [73] proposed that gravity induces asymmetric movement of auxin within 60 min, and, then, the auxin stimulates ROS generation to mediate gravitropism. Further, scavenging of ROS by antioxidants (N-acetylcysteine, ascorbic acid, and Trolox) inhibited root gravitropism [73]. ROS appear to be involved in dormancy alleviation. In dormant barley grains under control condition, gibberellic acid (GA) signaling and ROS content are low, while ABA signaling is high, resulting in dormancy. Exogenous H_2O_2 does not appear to alter ABA biosynthesis and signaling, but has a more pronounced effect on GA signaling, inducing a change in hormonal balance that results in germination [79]. ROS have been shown to play a key role in PCD in barley aleurone cells, initiated by GA. Bethke and Jones [72] observed that GA-treated aleurone protoplasts are less tolerant to internally generated or exogenously applied H_2O than ABA-treated protoplasts

and suggested that ROS are components of the hormonally regulated cell death pathway in barley aleurone cells.

Plants have evolved a complex regulatory network to mediate biotic and abiotic stress responses based on ROS synthesis, scavenging, and signaling. Transient production of ROS is detected in the early events of plant-pathogen interactions and plays an important signaling role in pathogenesis signal transduction regulators. This production-called oxidative burst could be considered as a specific signal during the interaction process [80]. In HR, SA is thought to potentiate ROS signaling [81]. ROS are shown to act as a second messenger for the induction of defense genes in tomato plants in response to wounding [82]. ROS were generated near cell walls of vascular bundle cells of tomato leaves in response to wounding and resulted H_2O_2 from wound-inducible polygalacturonase acted as a second messenger for the activation of defense genes in mesophyll cells, but not for signaling pathway genes in vascular bundle cells [82].

Lignin is important for the plant's response to environmental stress. Denness and coworkers [83] characterized a genetic network enabling plants to regulate lignin biosynthesis in response to cell wall damage through dynamic interactions between Jasmonic acid and ROS. ROS have been shown to play important roles in osmotic stress, low temperature, and heavy metal signal transduction pathway [75, 84, 85]. Genes involved in osmotic stress signaling have been shown to be upregulated by ROS, including the transcription factor DREB2A and a histidine kinase [18]. In Arabidopsis culture cells, it was reported that the MAPK AtMPK6 that can be activated by low temperature and osmotic stress could also be activated by oxidative stress [84]. Borsani and coworkers [86] suggested that the increased osmotic stress tolerance of transgenic Arabidopsis expressing a salicylate hydroxylase (NahG) gene, might result from decreased SA-mediated ROS generation. Zhao and coworkers [87] reported that ROS play important roles in drought-induced abscisic acid synthesis in plant and suggested that they may be the signals through

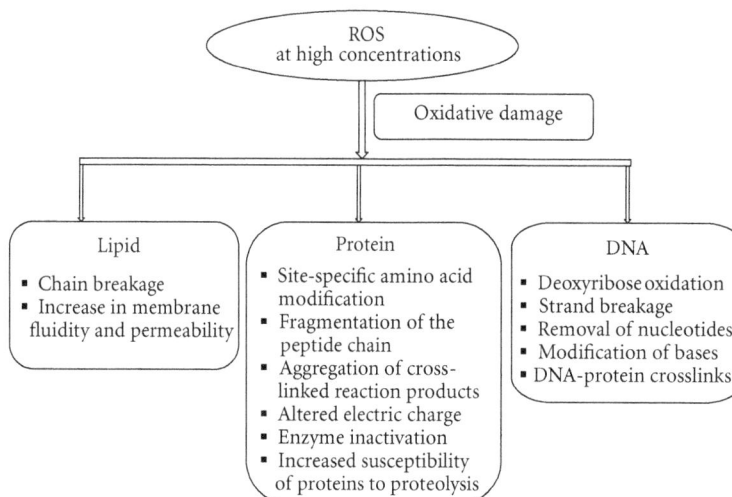

FIGURE 4: Reactive oxygen species (ROS) induced oxidative damage to lipids, proteins, and DNA.

which the plant can "sense" the drought condition. Using pharmacological inhibitors, it is demonstrated that metals Cd^{2+} and Cu^{2+} induce MAP kinase activation *via* distinct ROS-generating systems [85].

2.4. ROS and Oxidative Damage to Biomolecules. Production and removal of ROS must be strictly controlled in order to avoid oxidative stress. When the level of ROS exceeds the defense mechanisms, a cell is said to be in a state of "oxidative stress". However, the equilibrium between production and scavenging of ROS is perturbed under a number of stressful conditions such as salinity, drought, high light, toxicity due to metals, pathogens, and so forth. Enhanced level of ROS can cause damage to biomolecules such as lipids, proteins and DNA (Figure 4). These reactions can alter intrinsic membrane properties like fluidity, ion transport, loss of enzyme activity, protein cross-linking, inhibition of protein synthesis, DNA damage, and so forth ultimately resulting in cell death.

2.4.1. Lipids. When ROS level reaches above threshold, enhanced lipid peroxidation takes place in both cellular and organellar membranes, which, in turn, affect normal cellular functioning. Lipid peroxidation aggravates the oxidative stress through production of lipid-derived radicals that themselves can react with and damage proteins and DNA. The level of lipid peroxidation has been widely used as an indicator of ROS mediated damage to cell membranes under stressful conditions. Increased peroxidation (degradation) of lipids has been reported in plants growing under environmental stresses [8, 10, 12, 13]. Increase in lipid peroxidation under these stresses parallels with increased production of ROS. Malondialdehyde (MDA) is one of the final products of peroxidation of unsaturated fatty acids in phospholipids and is responsible for cell membrane damage [43]. Two common sites of ROS attack on the phospholipid molecules are the unsaturated (double) bond between two carbon atoms and the ester linkage between glycerol and the fatty

acid. The polyunsaturated fatty acids (PUFAs) present in membrane phospholipids are particularly sensitive to attack by ROS. A single $^\bullet OH$ can result in peroxidation of many polyunsaturated fatty acids because the reactions involved in this process are part of a cyclic chain reaction. The overall process of lipid peroxidation involves three distinct stages: initiation, progression, and termination steps. The initial phase of lipid peroxidation includes activation of O_2 which is rate limiting. $O_2^{\bullet-}$ and $^\bullet OH$ can react with methylene groups of PUFA forming conjugated dienes, lipid peroxy radicals and hydroperoxides [88]:

$$PUFA - H + X^\bullet \longrightarrow PUFA + X - H. \qquad (9)$$

$$PUFA + O_2 \longrightarrow PUFA - OO^\bullet. \qquad (10)$$

The peroxy radical formed is highly reactive and able to propagate the chain reaction:

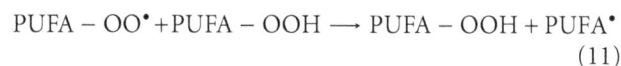

$$PUFA - OO^\bullet + PUFA - OOH \longrightarrow PUFA - OOH + PUFA^\bullet \qquad (11)$$

The formation of conjugated diene occurs when free radicals attack the hydrogens of methylene groups separating double bonds and, thereby, rearrangement of the bonds occurs [89]. The lipid hydroperoxides produced (PUFA-OOH) can undergo reductive cleavage by reduced metals, such as Fe^{2+}, according to the following reaction:

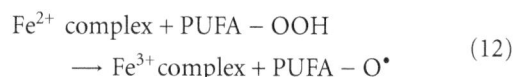

$$Fe^{2+} \text{ complex} + PUFA - OOH \\ \longrightarrow Fe^{3+} \text{complex} + PUFA - O^\bullet \qquad (12)$$

Several reactive species including: lipid alkoxyl radicals, aldehydes (malonyldialdehyde, acrolein and crotonaldehyde), alkanes, lipid epoxides, and alcohols can be easily formed by the decomposition of lipid hydroperoxide [90]. The lipid alkoxy radical produced, (PUFA-$O^\bullet$), can initiate additional chain reactions [91]:

$$PUFA - O^\bullet + PUFA - H \longrightarrow PUFA - OH + PUFA^\bullet \qquad (13)$$

Peroxidation of polyunsaturated fatty acid by ROS attack can lead to chain breakage and, thereby, increase in membrane fluidity and permeability.

2.4.2. Proteins.

The attack of ROS on proteins may cause modification of proteins in a variety of ways, some are direct and others indirect. Direct modification involves modulation of a protein's activity through nitrosylation, carbonylation, disulphide bond formation, and glutathionylation. Proteins can be modified indirectly by conjugation with breakdown products of fatty acid peroxidation [92]. As a consequence of excessive ROS production, site-specific amino acid modification, fragmentation of the peptide chain, aggregation of cross-linked reaction products, altered electric charge and increased susceptibility of proteins to proteolysis occur. Tissues injured by oxidative stress generally contain increased concentrations of carbonylated proteins which is widely used marker of protein oxidation [93]. Enhanced modification of proteins has been reported in plants under various stresses [8, 11, 12, 94]. The amino acids in a peptide differ in their susceptibility to attack by ROS. Thiol groups and sulphur containing amino acids are very susceptible sites for attack by ROS. Activated oxygen can abstract an H atom from cysteine residues to form a thiyl radical that will cross-link to second thiyl radical to form disulphide bridge. Several metals, including Cd, Pb, and Hg have been shown to cause the depletion of protein bound thiol groups [95]. Oxygen also can be added to a methionine to form methionine sulphoxide derivative [96]. Tyrosine is readily cross-linked to form bityrosine products in the presence of ROS [97].

Oxidation of iron-sulphur centers by $O_2^{\bullet-}$ is irreversible and leads to enzyme inactivation [98]. In these cases, the metal (Fe) binds to a divalent cation-binding site on the protein. The metal (Fe), then, reacts in a Fenton reaction to form a $^{\bullet}OH$ that rapidly oxidizes an amino acid residue at or near the cation-binding site of the protein [99]. Oxidized proteins serve as better substrates for proteolytic digestion. It has been suggested that protein oxidation could predispose it to ubiquitination, which, in turn, would be a target for proteasomal degradation [100]. The incubation of pea leaf crude extracts with increasing H_2O_2 concentrations, Cd-treated plants and peroxisomes purified from pea leaves showed increase in carbonyl content. Oxidized proteins were more efficiently degraded, and the proteolytic activity increased 20% due to the metal treatment [94]. Several studies have revealed that after a certain degree further damage leads to extensively cross-linked and aggregated products, which are not only poor substrates for degradation, but also can inhibit proteases to degrade other oxidized proteins [101].

2.4.3. DNA.

ROS are a major source of DNA damage [102]. ROS can cause oxidative damages to nuclear, mitochondrial, and chloroplastic DNA. DNA is cell's genetic material and any damage to the DNA can result in changes in the encoded proteins, which may lead to malfunctions or complete inactivation of the encoded proteins. Oxidative attack on DNA results in deoxyribose oxidation, strand breakage, removal of nucleotides, variety of modifications in the organic bases of the nucleotides, and DNA-protein crosslinks. Further, changes in the nucleotides of one strand can result in the mismatches with the nucleotides in the other strand, yielding subsequent mutations. Enhanced DNA degradation has been observed in plants exposed to various environmental stresses such as salinity [103] and metal toxicity [17]. Both the sugar and base moieties of DNA are susceptible to oxidation by ROS. Oxidative attack to DNA bases generally involves $^{\bullet}OH$ addition to double bonds, while sugar damage mainly results from hydrogen abstraction from deoxyribose [104]. The hydroxyl radical is known to react with all purine and pyrimidine bases and, also, the deoxyribose backbone [105]. $^{\bullet}OH$ generates various products from the DNA bases which mainly include C-8 hydroxylation of guanine to form 8-oxo-7,8 dehydro-2′- deoxyguanosine, hydroxymethyl urea, urea, thymine glycol, thymine and adenine ring-opened, and saturated products [106]. 8-Hydroxyguanine is the most commonly observed product. $^{1}O_2$ only reacts with guanine, whereas H_2O_2 and $O_2^{\bullet-}$ do not react with bases at all [104, 107]. ROS-induced DNA damages include various mutagenic alterations as well. For example, mutation arising from selective modification of G:C sites, especially, indicates oxidative attack on DNA by ROS. ROS attack DNA bases indirectly through reactive products generated by ROS attack to other macromolecules such as lipid [108].

ROS attack to DNA sugars leads to single-strand breaks. ROS abstract hydrogen atom from the C4′ position of deoxyribose, leading to generation of a deoxyribose radical that further reacts to produce DNA strand breakage [109]. Under physiological conditions, neither H_2O_2 alone nor $O_2^{\bullet-}$ can cause in vitro strand breakage. Therefore, it was concluded that the toxicity associated with these ROS in vivo is most likely the result of Fenton reaction. When $^{\bullet}OH$ attacks on either DNA or proteins associated with it, DNA protein crosslinks are formed [110]. DNA protein crosslinks cannot be readily repaired and may be lethal if replication or transcription precedes repair. Mitochondrial and chloroplast DNA are more susceptible to oxidative damage than nuclear DNA due to the lack of protective protein, histones, and close locations to the ROS producing systems in the former [111]. Even though repair system exists for damaged DNA, excessive changes caused by ROS lead to permanent damage to the DNA with potentially detrimental effects for the cell.

3. Antioxidative Defense System in Plants

Plants possess complex antioxidative defense system comprising of nonenzymatic and enzymatic components to scavenge ROS. In plant cells, specific ROS producing and scavenging systems are found in different organelles such as chloroplasts, mitochondria, and peroxisomes. ROS scavenging pathways from different cellular compartments are coordinated [112]. Under normal conditions, potentially toxic oxygen metabolites are generated at a low level and there is an appropriate balance between production and quenching of ROS. The balance between production and quenching of ROS may be perturbed by a number of adverse environmental factors, giving rise to rapid increases in intracellular ROS levels [63, 113], which can induce oxidative damage to lipids, proteins, and nucleic acids. In order to avoid the oxidative damage, higher plants raise the level of endogenous antioxidant defense [113]. Various components of antioxidative defense system involved in ROS scavenging have been manipulated, overexpressed or downregulated to add to the present knowledge and understanding the role of the antioxidant systems.

3.1. Nonenzymatic Components of Antioxidative Defense System.
Nonenzymic components of the antioxidative defense system include the major cellular redox buffers ascorbate (AsA) and glutathione (γ-glutamyl-cysteinyl-glycine, GSH) as well as tocopherol, carotenoids, and phenolic compounds. They interact with numerous cellular components and in addition to crucial roles in defense and as enzyme cofactors, these antioxidants influence plant growth and development by modulating processes from mitosis and cell elongation to senescence and cell death [114]. Mutants with decreased nonenzymic antioxidant contents have been shown to be hypersensitive to stress [115, 116].

3.1.1. Ascorbate.
Ascorbate (AsA) is the most abundant, low molecular weight antioxidant that has a key role in defense against oxidative stress caused by enhanced level of ROS. AsA is considered powerful antioxidant because of its ability to donate electrons in a number of enzymatic and nonenzymatic reactions. AsA has been shown to play important role in several physiological processes in plants, including growth, differentiation, and metabolism. The majority of the AsA pool in plants is contributed by D-mannose/L-galactose commonly called Smirnoff-Wheeler pathway which proceeds *via* GDP-D-mannose, GDP-L- galactose, L-galactose, and L-galactono-1,4-lactone [117]. AsA is also synthesized *via* uronic acid intermediates, such as D-galacturonic acid [118]. In this pathway D-galacturonic acid is reduced to L-galactonic acid by galacturonic acid reductase, which is subsequently converted to L-galactono-1,4-lactone. The L-galactono-1,4-lactone is further oxidized to AsA by L-galactono-1,4-lactone dehydrogenase (GALDH) enzyme. It is synthesized in the mitochondria by L-galactono-γ-lactone dehydrogenase and is transported to the other cell components through a proton-electrochemical gradient or through facilitated diffusion. It is detected in the majority of plant cell types, organelles and apoplast in plants [119] and is found to be particularly abundant in photosynthetic tissues [120]. Most of AsA, almost more than 90%, is localized in cytoplasm, but unlike other soluble antioxidants a substantial portion is exported to the apoplast, where it is present in millimolar concentration. Apoplastic AsA is believed to represent the first line of defense against potentially damaging external oxidants [121]. AsA protects critical macromolecules from oxidative damage. Under normal physiological condition, AsA mostly exists in reduced state in chloroplast where it also acts as a cofactor of violaxanthin de-epoxidase, thus, sustaining dissipation of excess excitation energy [122]. It provides membrane protection by directly reacting with $O_2^{\bullet-}$, H_2O_2 and regenerating α-tocopherol from tocopheroxyl radical and preserves the activities of the enzymes that contain prosthetic transition metal ions [21]. AsA has a key role in removal of H_2O_2 *via* AsA-GSH cycle [49]. Oxidation of AsA occurs in two sequential steps, first producing monodehydroascorbate (MDHA) and subsequently dehydroascorbate (DHA). In the AsA-GSH cycle, two molecules of AsA are utilized by APX to reduce H_2O_2 to water with concomitant generation of MDHA. MDHA is a radical with a short life time and can spontaneously dismutate into DHA and AsA or is reduced to AsA by NADP(H) dependent enzyme MDHAR [123]. DHA is also highly unstable at pH values greater than 6.0 and is decomposed to tartarate and oxalate [21]. To prevent this, DHA is rapidly reduced to AsA by the enzyme DHAR using reducing equivalents from GSH [124].

AsA level has been reported to alter in response to various stresses [8, 11, 13, 14, 125, 126]. The level of AsA under environmental stresses depends on the balance between the rates and capacity of AsA biosynthesis and turnover related to antioxidant demand [127]. Overexpression of enzymes involved in AsA biosynthesis confers abiotic stress tolerance in plants. GDP-Mannose 3',5'-epimerase (GME) catalyses the conversion of GDP-D-mannose to GDP-L-galactose, an important step in the Smirnoff-Wheeler pathway of AsA biosynthesis in higher plants. Overexpression of two members of the *GME* gene family resulted in increased accumulation of ascorbate and improved tolerance to abiotic stresses in tomato plants [128]. Overexpression of strawberry D-galacturonic acid reductase which participates in AsA biosynthetic pathway involving D-galacturonic acid as intermediate and reduces D-galacturonic acid to L-galactonic acid, leads to accumulation of AsA and enhanced abiotic stress tolerance in potato plants [129]. Increased AsA content has been shown to confer oxidative stress tolerance in Arabidopsis [130]. The *vtc-1* mutant, deficient in the activity of GDP-mannose pyrophosphorylase, an enzyme found in the initial part of the ascorbate biosynthetic pathway before it becomes committed to ascorbate synthesis [117] was found to be more sensitive to supplementary UV-B treatment than wild type plants [115].

3.1.2. Glutathione.
Tripeptide glutathione (γ-glutamyl-cysteinyl-glycine, GSH) is one of the crucial low molecular weight nonprotein thiol that plays an important role in intracellular defense against ROS-induced oxidative damage. It has been detected virtually in all cell compartments such as cytosol, chloroplasts, endoplasmic reticulum, vacuoles, and mitochondria [131]. GSH is synthesized in the cytosol and chloroplasts of plant cells by compartment specific isoforms of γ-glutamyl-cysteinyl synthetase (γ-ECS) and glutathione synthetase (GS). The balance between the GSH and glutathione disulfide (GSSG) is a central component in maintaining cellular redox state. Due to its reducing power, GSH plays an important role in diverse biological processes, including cell growth/division, regulation of sulfate transport, signal transduction, conjugation of metabolites, enzymatic regulation, synthesis of proteins and nucleic acids,

synthesis of phytochelatins for metal chelation, detoxification of xenobiotics, and the expression of the stress-responsive genes [48]. GSH functions as an antioxidant in many ways. It can react chemically with $O_2^{\bullet-}$, $^{\bullet}OH$, H_2O_2 and, therefore, can function directly as a free radical scavenger. GSH can protect macromolecules (i.e., proteins, lipids, DNA) either by the formation of adducts directly with reactive electrophiles (glutathiolation) or by acting as proton donor in the presence of ROS or organic free radicals, yielding GSSG [132]. It can participate in regeneration of another potential antioxidant AsA, *via* the AsA-GSH cycle. GSH recycles AsA from it's oxidized to reduced form by the enzyme DHAR [133]. GSH can also reduce DHA by a nonenzymic mechanism at pH > 7 and at GSH concentrations greater than 1 mM. This may be a significant pathway in chloroplasts, where in the presence of light, pH remains around 8 and GSH concentration may be as high as 5 mM [134]. Generation and maintenance of reduced GSH pool, either by *de novo* synthesis or *via* recycling by GR, using NADPH as a cofactor and electron donor, is of vital importance for the cell. The role of GSH in the antioxidative defense system provides a rationale for its use as a stress marker. When apple trees were subjected to progressive drought, the initial response was a little oxidation of the GSH pool, followed by increased GSH concentrations. When the stress increased, GSH concentrations dropped and redox state became more oxidized, which marked the degradation of the system [135]. Similar to drought stress, altered ratio of GSH/GSSG has been observed in plants under various stresses like salinity [136], chilling [126], and metal toxicity [11, 13–15]. Overexpression of enzyme glutathione synthetase involved in GSH biosynthesis showed no effect on GSH level and was not sufficient to improve ozone tolerance [137] and resistance to photoinhibition [138] in hybrid poplar (*Populus tremula* × *P. alba*). However, overexpression of γ-ECS showed reduced sensitivity towards cadmium stress in Indian mustard [139] and enhanced tolerance towards chloroacetanilide herbicides in poplar plants [140]. Eltayeb and coworkers [141] observed greater protection against oxidative damages imposed by various environmental stresses in transgenic potato with higher level of reduced glutathione.

3.1.3. Tocopherols. Tocopherols (α, β, γ, and δ) represent a group of lipophilic antioxidants involved in scavenging of oxygen free radicals, lipid peroxy radicals, and $^1O_2$ [142]. Relative antioxidant activity of the tocopherol isomers *in vivo* is α > β > γ > δ which is due to the methylation pattern and the amount of methyl groups attached to the phenolic ring of the polar head structure [143]. Hence, α-tocopherol with its three methyl substituents has the highest antioxidant activity of tocopherols [144]. Tocopherols are synthesized only by photosynthetic organisms and are present in only green parts of plants. The tocopherol biosynthetic pathway utilizes two compounds homogentisic acid (HGA) and phytyl diphosphate (PDP) as precursors. At least 5 enzymes 4-hydroxyphenylpyruvate dioxygenase (HPPD), homogentisate phytyl transferases (VTE2),

2-methyl-6-phytylbenzoquinol methyltransferase (VTE3), tocopherol cyclase (VTE1), γ-tocopherol methyltransferase (VTE4) are involved in the biosynthesis of tocopherols, excluding the bypass pathway of phytyl-tail synthesis and utilization [145]. Tocopherols are known to protect lipids and other membrane components by physically quenching and chemically reacting with O_2 in chloroplasts, thus protecting the structure and function of PSII [146]. Tocopherols prevent the chain propagation step in lipid autooxidation which makes it an effective free radical trap. Fully substituted benzoquinone ring and fully reduced phytyl chain of tocopherol act as antioxidants in redox interactions with $^1O_2$ [105, 147]. $^1O_2$ oxygen quenching by tocopherols is highly efficient, and it is estimated that a single α-tocopherol molecule can neutralize up to 220 $^1O_2$ molecules *in vitro* before being degraded [143]. Regeneration of the oxidized tocopherol back to its reduced form can be achieved by AsA, GSH [147] or coenzyme Q [148]. Accumulation of α-tocopherol has been shown to induce tolerance to chilling, water deficit, and salinity in different plant species [149–152]. It was found that metabolic engineering of tocopherol biosynthetic pathway affected endogenous ascorbate and glutathione pools in leaves. Further study suggested that expression levels of genes encoding enzymes of Halliwell-Asada cycle were up-regulated, such as APX, DHAR and MDHAR [145]. Mutants of *Arabidopsis thaliana* with T-DNA insertions in tocopherol biosynthesis genes, tocopherol cyclase (*vte1*) and γ-tocopherol methyltransferase (*vte4*) showed higher concentration of protein carbonyl groups and GSSG compared to the wild type, indicating the development of oxidative stress [116]. Transgenic rice plants with Os-VTE1 RNA interference (OsVTE1-RNAi) were more sensitive to salt stress whereas, in contrast, transgenic plants overexpressing OsVTE1 (OsVTE1-OX) showed higher tolerance to salt stress [153]. OsVTE1-OX plants also accumulated less H_2O_2 than control plants.

3.1.4. Carotenoids. Carotenoids also belong to the group of lipophilic antioxidants and are able to detoxify various forms of ROS [154]. Carotenoids are found in plants as well as microorganisms. In plants, carotenoids absorb light in the region between 400 and 550 nm of the visible spectrum and pass the captured energy to the Chl [155]. As an antioxidant, they scavenge $^1O_2$ to inhibit oxidative damage and quench triplet sensitizer (3Chl*) and excited chlorophyll (Chl*) molecule to prevent the formation of $^1O_2$ to protect the photosynthetic apparatus. Carotenoids also serve as precursors to signaling molecules that influence plant development and biotic/abiotic stress responses [156]. The ability of carotenoids to scavenge, prevent or minimize the production of triplet chlorophyll may be accounted for by their chemical specificity. Carotenoids contain a chain of isoprene residues bearing numerous conjugated double bonds which allows easy energy uptake from excited molecules and dissipation of excess energy as heat [7]. Gomathi and Rakkiyapan [157] observed that high carotenoids content favors better adaptation of sugarcane plants under saline condition.

3.1.5. Phenolic Compounds. Phenolics are diverse secondary metabolites (flavonoids, tannins, hydroxycinnamate esters, and lignin) which possess antioxidant properties. They are abundantly found in plant tissues [158]. Polyphenols contain an aromatic ring with –OH or OCH_3 substituents which together contribute to their biological activity, including antioxidant action. They have been shown to outperform well-known antioxidants, AsA and α-tocopherol, in *in vitro* antioxidant assays because of their strong capacity to donate electrons or hydrogen atoms. Polyphenols can chelate transition metal ions, can directly scavenge molecular species of active oxygen, and can inhibit lipid peroxidation by trapping the lipid alkoxyl radical. They also modify lipid packing order and decrease fluidity of the membranes [159]. These changes could strictly hinder diffusion of free radicals and restrict peroxidative reactions. Moreover, it has been shown that, especially, flavonoids and phenylpropanoids are oxidized by peroxidase, and act in H_2O_2-scavenging, phenolic/AsA/POD system. There is some evidence of induction of phenolic metabolism in plants as a response to multiple stresses [160]. Janas and coworkers [161] observed that ROS could serve as a common signal for acclimation to Cu^{2+} stress and could cause accumulation of total phenolic compounds in dark-grown lentil roots. A mutant *Arabidopsis thaliana* L., having a single gene defect which led to a block in the synthesis of a group of flavonoids, displayed a dramatic increase in sensitivity to UV-B radiation compared with wild-type plants [162]. Transgenic potato plant with increased concentration of flavonoid showed improved antioxidant capacity [163].

3.2. Enzymatic Components. The enzymatic components of the antioxidative defense system comprise of several antioxidant enzymes such as superoxide dismutase (SOD), catalase (CAT), guaiacol peroxidase (GPX), enzymes of ascorbate-glutathione (AsA-GSH) cycle ascorbate peroxidase (APX), monodehydroascorbate reductase (MDHAR), dehydroascorbate reductase (DHAR), and glutathione reductase (GR) [21]. These enzymes operate in different subcellular compartments and respond in concert when cells are exposed to oxidative stress. Table 1 shows various antioxidant enzymes that play important role in scavenging stress-induced ROS generated in plants.

3.2.1. Superoxide Dismutase. Superoxide dismutase (SOD, 1.15.1.1) plays central role in defense against oxidative stress in all aerobic organisms [175]. The enzyme SOD belongs to the group of metalloenzymes and catalyzes the dismutation of $O_2^{\bullet-}$ to O_2 and H_2O_2. It is present in most of the subcellular compartments that generate activated oxygen. Three isozymes of SOD copper/zinc SOD (Cu/Zn-SOD), manganese SOD (Mn-SOD), and iron SOD (Fe-SOD) are reported in plants [176, 177]. All forms of SOD are nuclear encoded and targeted to their respective subcellular compartments by an amino terminal targeting sequence [178]. MnSOD is localized in mitochondria, whereas Fe-SOD is localized in chloroplasts [179]. Cu/Zn-SOD is present in three isoforms, which are found in the cytosol, chloroplast,

and peroxisome and mitochondria [178, 180–182]. Eukaryotic Cu/Zn-SOD is cyanide sensitive and presents as dimer, whereas the other two (Mn-SOD and Fe-SOD) are cyanide insensitive and may be dimer or tetramers [175, 182].

SOD activity has been reported to increase in plants exposed to various environmental stresses, including drought and metal toxicity [8, 13]. Increased activity of SOD is often correlated with increased tolerance of the plant against environmental stresses. It was suggested that SOD can be used as an indirect selection criterion for screening drought-resistant plant materials [22]. Overproduction of SOD has been reported to result in enhanced oxidative stress tolerance in plants [183].

3.2.2. Catalase. Among antioxidant enzymes, catalase (CAT, 1.11.1.6) was the first enzyme to be discovered and characterized. It is a ubiquitous tetrameric heme-containing enzyme that catalyzes the dismutation of two molecules of H_2O_2 into water and oxygen. It has high specificity for H_2O_2, but weak activity against organic peroxides. Plants contain several types of H_2O_2-degrading enzymes, however, CATs are unique as they do not require cellular reducing equivalent. CATs have a very fast turnover rate, but a much lower affinity for H_2O_2 than APX. The peroxisomes are major sites of H_2O_2 production. CAT scavenges H_2O_2 generated in this organelle during photorespiratory oxidation, β-oxidation of fatty acids, and other enzyme systems such as XOD coupled to SOD [3, 184, 185]. Though there are frequent reports of CAT being present in cytosol, chloroplast, and mitochondria, the presence of significant CAT activity in these is less well established [186]. To date, all angiosperm species studied, contain three CAT genes. Willekens et al. [187] proposed a classification of CAT based on the expression profile of the tobacco genes. Class I CATs are expressed in photosynthetic tissues and are regulated by light. Class II CATs are expressed at high levels in vascular tissues, whereas Class III CATs are highly abundant in seeds and young seedlings.

H_2O_2 has been implicated in many stress conditions. When cells are stressed for energy and are rapidly generating H_2O_2 through catabolic processes, H_2O_2 is degraded by CAT in an energy efficient manner [188]. Environmental stresses cause either enhancement or depletion of CAT activity, depending on the intensity, duration, and type of the stress [8, 10, 189]. In general, stresses that reduce the rate of protein turnover also reduce CAT activity. Stress analysis revealed increased susceptibility of CAT-deficient plants to paraquat, salt and ozone, but not to chilling [190]. In transgenic tobacco plants, having 10% wild-type, CAT activity showed accumulation of GSSG and a 4-fold decrease in AsA, indicating that CAT is critical for maintaining the redox balance during the oxidative stress [190]. Overexpression of a CAT gene from *Brassica juncea* introduced into tobacco, enhanced its tolerance to Cd induced oxidative stress [191].

3.2.3. Guaiacol Peroxidase. Guaiacol peroxidase (GPX, EC 1.11.1.7), a heme containing protein, preferably oxidizes aromatic electron donor such as guaiacol and pyragallol at the expense of H_2O_2. It is widely found in animals, plants,

TABLE 1: Activation of antioxidant enzymes in response to oxidative stress induced by various environmental stresses.

Stresses	Antioxidant enzymes	Plant species	References
Drought	SOD, GPX, APX, MDHAR, DHAR and GR	*Oryza sativa*	[8]
	SOD, CAT and GPX	*Beta vulgaris*	[164]
	SOD, APX and GR	*Triticum sativum*	[165]
Salinity	SOD, CAT, GPX, APX, GR	*Oryza sativa*	[166]
	CAT, SOD and GR	*Olea europaea*	[167]
	GPX	*Oryza sativa*	[168]
Chilling	APX, MDHAR, DHAR, GR and SOD	*Zea mays*	[169]
		Fragaria×ananassa	[170]
MetalsAl	SOD, GPX and APX	*Oryza sativaGlycine max*	[15, 171]
Ni	SOD, GPX and APX	*Oryza sativa*	[11]
As	SOD, GPX and APX	*Oryza sativa*	[13]
Mn	SOD, GPX, APX and GR	*Oryza sativa*	[14]
UV-B	SOD, APX, CAT and GPX	*Picea asperata*	[10]
	GPX andAPX	*Arabidopsis thaliana*	[172]
Pathogen *Odium lini* fungus	GPX and CAT	*Linum usitatissimum*	[173]
Bean yellow mosaic virus	POD, CAT, APX and SOD	*Vicia faba*	[174]

and microbes. These enzymes have four conserved disulfide bridges and contain two structural Ca^{2+} ions [192]. Many isoenzymes of GPX exist in plant tissues localized in vacuoles, the cell wall, and the cytosol [193]. GPX is associated with many important biosynthetic processes, including lignification of cell wall, degradation of IAA, biosynthesis of ethylene, wound healing, and defense against abiotic and biotic stresses [194]. GPXs are widely accepted as stress "enzyme." GPX can function as effective quencher of reactive intermediary forms of O_2 and peroxy radicals under stressed conditions [195]. Various stressful conditions of the environment have been shown to induce the activity of GPX [6, 8, 10, 16, 166, 189, 196]. Radotic and coworkers [196] correlated increased activity of GPX to oxidative reactions under metal toxicity conditions and suggested its potential as biomarker for sublethal metal toxicity in plants. Recently, Tayefi-Nasrabadi and coworkers [197] also concluded that greater protection of salt-tolerant safflower plants from salt-induced oxidative damage results, at least in part, through the increase of the GPX activity, catalytic efficiency and induction of specific isoenzymes compared to salt-sensitive cultivar.

3.2.4. Enzymes of Ascorbate-Glutathione Cycle. The change in the ratio of AsA to DHA and GSH to GSSG is crucial for the cell to sense oxidative stress and respond accordingly. The AsA-GSH cycle also referred to as Halliwell-Asada pathway is the recycling pathway of AsA and GSH regeneration which also detoxifies H_2O_2. The AsA-GSH cycle involves successive oxidation and reduction of AsA, GSH, and NADPH catalyzed by the enzymes APX, MDHAR, DHAR, and GR. The AsA-GSH cycle is present in at least four different subcellular locations, including the cytosol, chloroplast, mitochondria,

and peroxisomes [198]. AsA-GSH cycle plays an important role in combating oxidative stress induced by environmental stresses [8, 199].

(1) Ascorbate Peroxidase. Ascorbate peroxidase (APX, EC 1.1.11.1) is a central component of AsA-GSH cycle, and plays an essential role in the control of intracellular ROS levels. APX uses two molecules of AsA to reduce H_2O_2 to water with a concomitant generation of two molecules of MDHA. APX is a member of Class I super family of heme peroxidases [200] and is regulated by redox signals and H_2O_2 [201]. Based on amino acid sequences, five chemically and enzymatically distinct isoenzymes of APX have been found at different subcellular localization in higher plants. These are cytosolic, stromal, thylakoidal, mitochondrial and peroxisomal isoforms [198, 202–204]. APX found in organelles scavenges H_2O_2 produced within the organelles, whereas cytosolic APX eliminates H_2O_2 produced in the cytosol, apoplast or that diffused from organelles [205]. The chloroplastic and cytosolic APX isoforms are specific for AsA as electron donor and the cytosolic isoenzymes are less sensitive to depletion of AsA than the chloroplastic isoenzymes, including stromal and thylakoid bound enzymes [203, 206].

APX is regarded as one of the most widely distributed antioxidant enzymes in plant cells and isoforms of APX have much higher affinity for H_2O_2 than CAT, making APXs efficient scavengers of H_2O_2 under stressful conditions [207]. Many workers have reported enhanced activity of APX in response to abiotic stresses such as drought, salinity, chilling, metal toxicity, and UV irradiation [8, 10, 11, 15, 136, 208]. Overexpression of a cytosolic APX-gene derived from pea

(*Pisum sativum* L.) in transgenic tomato plants (*Lycopersicon esculentum* L.) ameliorated oxidative injury induced by chilling and salt stress [209]. Similarly, overexpression of the tApx gene in either tobacco or in Arabidopsis increased tolerance to oxidative stress [210].

(2) Monodehydroascorbate Reductase. MDHA radical produced in APX catalyzed reaction has a short lifetime, and if not rapidly reduced, it disproportionates to AsA and DHA [211]. Monodehydroascorbate reductase (MDHAR, 1.6.5.4) is a FAD enzyme that catalyzes the regeneration of AsA from the MDHA radical using NAD(P)H as the electron donor [212]. It is the only known enzyme to use an organic radical (MDA) as a substrate and is also capable of reducing phenoxyl radicals which are generated by horseradish peroxidase with H_2O_2 [213]. MDHAR activity is widespread in plants. The isoenzymes of MDHAR have been reported to be present in several cellular compartments such as chloroplasts [214], cytosol and mitochondria and peroxisomes [198, 215]. In chloroplasts, MDHAR could have two physiological functions: the regeneration of AsA from MDHA and the mediation of the photoreduction of dioxygen to $O_2^{\bullet -}$ when the substrate MDHA is absent [216]. Characterization of membrane polypeptides from pea leaf peroxisomes also revealed MDHAR to be involved in $O_2^{\bullet -}$ generation [64].

Several studies have shown increased activity of MDHAR in plants subjected to environmental stresses [8, 11, 15, 208]. Overexpression of Arabidopsis MDHAR gene in tobacco confers enhanced tolerance to salt and polyethylene glycol stresses [217]. Tomato chloroplastic MDHAR overexpressed in transgenic Arabidopsis enhanced its tolerance to temperature and methyl viologen-mediated oxidative stresses [218].

(3) Dehydroascorbate Reductase. Dehydroascorbate reductase (DHAR, EC 1.8.5.1) catalyzes the reduction of DHA to AsA using GSH as the reducing substrate [211] and, thus, plays an important role in maintaining AsA in its reduced form. Despite the possibility of enzymic and nonenzymic regeneration of AsA directly from MDHA, some DHA is always produced when AsA is oxidized in leaves and other tissues. DHA, a very short-lived chemical, can either be hydrolyzed irreversibly to 2,3-diketogulonic acid or recycled to AsA by DHAR. Overexpression of DHAR in tobacco leaves, maize, and potato is reported to increase AsA content suggesting that DHAR plays important roles in determining the pool size of AsA [219, 220]. DHAR is a monomeric thiol enzyme abundantly found in dry seeds, roots and etiolated as well as green shoots. DHAR has been purified from chloroplast as well as nonchloroplast sources in several plant species, including spinach leaves [221] and potato tuber [222].

Environmental stresses such as drought, metal toxicity, and chilling increase the activity of the DHAR in plants [8, 11, 15, 125, 208, 223]. Consistent upregulation of the gene encoding cytosolic DHAR was found in *L. japonicas*, which was found to be more tolerant to salt stress than other legumes. This upregulation of DHAR was correlated

to its role in AsA recycling in the apoplast [224]. Transgenic potato overexpressing Arabidopsis cytosolic AtDHAR1 showed higher tolerance to herbicide, drought, and salt stresses [225].

(4) Glutathione Reductase (GR). When acting as an antioxidant by participating in enzymic as well as nonenzymic oxidation-reduction cycles, GSH is oxidized to GSSG. In AsA-GSH cycle, GSH is oxidized in a reaction catalyzed by DHAR. Glutathione reductase (GR, EC 1.6.4.2), a NAD(P)H-dependent enzyme catalyzes the reduction of GSSG to GSH and, thus, maintains high cellular GSH/GSSG ratio. GR belongs to a group of flavoenzymes and contains an essential disulfide group [226]. The catalytic mechanism involves two steps: first the flavin moiety is reduced by NADPH, the flavin is oxidized and a redox active disulfide bridge is reduced to produce a thiolate anion and a cysteine. The second step involves the reduction of GSSG *via* thiol-disulfide interchange reactions [226]. If the reduced enzyme is not reoxidized by GSSG, it can suffer a reversible inactivation. Although it is located in the chloroplasts, cytosol, mitochondria, and peroxisomes, around 80% of GR activity in photosynthetic tissues is accounted for by chloroplastic isoforms [227]. In chloroplast, GSH and GR are involved in detoxification of H_2O_2 generated by Mehler reaction.

Several authors have reported increased activity of GR under environmental stresses [8, 11, 15, 125, 223]. Pastori and Trippi [228] observed correlation between the oxidative stress resistance and activity of GR and suggested that oxidative stress caused by paraquat or H_2O_2 could stimulate GR *de novo* synthesis, probably at the level of translation by preexisting mRNA. Antisense-mediated depletion of tomato chloroplast GR has been shown to enhance susceptibility to chilling stress [229]. Overexpression of GR in *N. tabacum* and Populus plants leads to higher foliar AsA contents and improved tolerance to oxidative stress [138, 230].

Due to the complexity of ROS detoxification system, overexpressing one component of antioxidative defense system may or may not change the capacity of the pathway as a whole [231, 232]. Several studies have shown that overexpression of combinations of antioxidant enzymes in transgenic plants has synergistic effect on stress tolerance [233, 234]. Kwon et al. [234] demonstrated that simultaneous expression of Cu/Zn-SOD and APX genes in tobacco chloroplasts enhanced tolerance to methyl viologen (MV) stress compared to expression of either of these genes alone. Similarly, enhanced tolerance to multiple environmental stresses has been developed by simultaneous overexpression of the genes of SOD and APX in the chloroplasts [235, 236], SOD and CAT in cytosol [231] and SOD and GR in cytosol [233]. Further, simultaneous expression of multiple antioxidant enzymes, such as Cu/Zn-SOD, APX, and DHAR, in chloroplasts has shown to be more effective than single or double expression for developing transgenic plants with enhanced tolerance to multiple environmental stresses [26]. Therefore, in order to achieve tolerance to multiple environmental stresses, increased emphasis is now given to produce transgenic plants overexpressing multiple antioxidants.

4. Overproduction of ROS under Stressful Conditions

The production of ROS in plants under normal growth conditions is low. However, in response to various environmental stresses, ROS are drastically increased in plants disturbing the normal balance of $O_2^{\bullet-}$, $^{\bullet}OH$, and H_2O_2 in the intracellular environment [113]. The effects of various environmental stresses such as drought, salinity, chilling, metal toxicity, UV-B radiation, and pathogen attack on ROS production are discussed below.

4.1. Drought. Under drought stress, ROS production is enhanced in several ways. Inhibition of carbon dioxide (CO_2) assimilation, coupled with the changes in photosystem activities and photosynthetic transport capacity under drought stress results in accelerated production of ROS *via* the chloroplast Mehler reaction [237]. During drought stress, CO_2 fixation is limited due to stomatal closure which, in turn, leads to reduced $NADP^+$ regeneration through the Calvin cycle. Due to lack of electron acceptor, over reduction of the photosynthetic ETC occurs which leads to a higher leakage of electrons to O_2 by the Mehler reaction. Biehler and Fock [238] reported 50% more leakage of photosynthetic electrons to the Mehler reaction in drought stressed wheat plants, compared to unstressed plants. Photosynthetic activity is inhibited in plant tissues due to an imbalance between light capture and its utilization under drought stress [239]. Dissipation of excess light energy in the PSII core and antenna leads to generation of ROS which are potentially dangerous under drought stress conditions [1]. Under drought stress, the photorespiratory pathway is also enhanced, especially, when RUBP oxygenation is maximal due to limitation in CO_2 fixation [63]). Noctor and collaborators [63] have estimated that photorespiration is likely to account for over 70% of total H_2O_2 production under drought stress conditions.

$O_2^{\bullet-}$ initiates a chain reaction leading to the production of more toxic radical species, which may cause damage far in excess of the initial reaction products. Under drought stress one of the real threats towards the chloroplast is the production of the $^{\bullet}OH$ in the thylakoids through "iron-catalysed" reduction of H_2O_2 by both SOD and AsA. Increased production of ROS leads to oxidative stress in growing plants. Rice seedlings subjected to drought showed increased concentration of $O_2^{\bullet-}$, increased level of lipid peroxidation, chlorophyll bleaching, loss of some antioxidants (AsA, GSH, α-tocopherol, and carotenoids), total soluble protein, and thiols [8, 208]. To combat danger posed by ROS, plants possess different scavenging enzymes and metabolites. Enhanced activity of enzymes of antioxidative defense system has been reported under drought stress in several plant species [8, 164, 165, 208]. Comparative study of the antioxidant responses in drought tolerant and drought sensitive genotypes revealed higher antioxidant capacity in tolerant genotypes. In contrast to drought susceptible wheat genotype HD 2329, drought tolerant wheat genotype C 306 had higher APX and CAT activity, higher AsA content and lower H_2O_2 and MDA content [240]. In another study, the drought tolerant maize genotype Giza 2 was suggested to be comparatively tolerant to water stress compared to drought sensitive Trihybrid 321 owing to the lower increase in H_2O_2 and MDA content along with higher increase in SOD, CAT, and POX activities [189]. Similarly, among two apple rootstocks *Malus prunifolia* (drought-tolerant) and *M. hupehensis* (drought-sensitive), *M. hupehensis* was more vulnerable to drought than *M. prunifolia*, resulting in larger increases in the levels of H_2O_2, $O_2^{\bullet-}$, and MDA. The activities of SOD, POD, APX, GR, and DHAR and levels of AsA and GSH increased to a greater extent in *M. prunifolia* than in *M. hupehensis* in response to drought [241]. APX serves as an important component of antioxidative defense system under drought [8]. In rice plants, increase in the capacity of AsA regeneration system by *de novo* synthesis of MDHAR, DHAR, and GR has been shown to be one of the primary responses to water deficit so as to mitigate oxidative stress [8, 208].

4.2. Salinity. Salinity stress results in an excessive generation of ROS [12, 242]. High salt concentrations lead to overproduction of the ROS- $O_2^{\bullet-}$, $^{\bullet}OH$, H_2O_2, and $^1O_2$ by impairment of the cellular electron transport within different subcellular compartments such as chloroplasts and mitochondria, as well as from induction of metabolic pathways such as photorespiration. Salt stress can lead to stomatal closure, which reduces CO_2 availability in the leaves and inhibits carbon fixation which, in turn, causes exposure of chloroplasts to excessive excitation energy and overreduction of photosynthetic electron transport system leading to enhanced generation of ROS and induced oxidative stress. Low chloroplastic CO_2/O_2 ratio also favors photorespiration leading to increased production of ROS such as H_2O_2 [242]. Elevated CO_2 mitigates the oxidative stress caused by salinity, involving lower ROS generation and a better maintenance of redox homeostasis as a consequence of higher assimilation rates and lower photorespiration [243]. Salinity-induced ROS disrupt normal metabolism through lipid peroxidation, denaturing proteins, and nucleic acids in several plant species [12, 242, 244]. Differential genomic and proteomic screenings carried out in *Physcomitrella patens* plants showed that they responded to salinity stress by upregulating a large number of genes involved in antioxidant defense mechanism [245] suggesting that the antioxidative system may play a crucial role in protecting cells from oxidative damage following exposure to salinity stress in *P. patens*. Salinity-induced oxidative stress and possible relationship between the status of the components of antioxidative defense system and the salt tolerance in Indica rice (*Oryza sativa* L.) genotypes were studied by Mishra et al. [166]. Seedlings of salt-sensitive cultivar showed a substantial increase in the rate of $O_2^{\bullet-}$ production, elevated levels of H_2O_2, MDA, declined levels of thiol, AsA and GSH and lower activity of antioxidant enzymes compared to salt-tolerant seedlings. It was suggested that a higher status of antioxidants AsA and GSH and a coordinated higher activity of the enzymes SOD, CAT, GPX, APX, and GR can serve as the major determinants in the model for depicting salt tolerance in Indica rice seedlings [166]. Similarly, study of immediate responses (enzymatic and nonenzymatic) to salinity-induced oxidative stress in two major rice (*Oryza sativa* L.) cultivars, salt sensitive Pusa Basmati 1 (PB) and salt-tolerant Pokkali (PK), revealed a lesser extent of membrane damage (lipid peroxidation), lower levels of H_2O_2, higher activity of the ROS scavenging enzyme, CAT and enhanced levels of antioxidants like ASA and GSH in PK compared to PB [246]. Comparative study using cultivated tomato *Lycopersicon esculentum* Mill. *cv.* M82 (Lem) and its wild salt-tolerant relative *L. pennellii* (Corr.) D'Arcy accession Atico (Lpa) showed better protection of Lpa roots from salt-induced oxidative damage, at least partially, from the increased activities of the SOD, CAT, APX, MDHAR, and increased contents of AsA and GSH [247]. In salt-stressed root of Lem, a gradual increase in the membrane lipid peroxidation was observed, whereas no change in lipid peroxidation was observed in Lpa. Salt-tolerant *Plantago maritima* showed a lower level of MDA and a better protection mechanism against oxidative damage caused by salt stress by increasing activities of SOD, CAT, GR, and APX than the salt-sensitive *P. media* [248]. NADP-dehydrogenases and peroxidase have been suggested as key antioxidative enzymes in olive plants under salt stress conditions [167]. Mittal and Dubey [168] observed a correlation between peroxidase activity and salt tolerance in rice seedling.

4.3. Chilling. Chilling stress is a key environmental factor limiting growth and productivity of crop plants. Chilling leads to the overproduction of ROS by exacerbating imbalance between light absorption and light use by inhibiting Calvin-Benson cycle activity [249], enhancing photosynthetic electron flux to O_2 and causing overreduction of respiratory ETC [9]. Chilling stress also causes significant reductions in *rbc*L and *rbc*S transcripts, RUBISCO content and initial RUBISCO activity, leading to higher electron flux to O_2 [250]. H_2O_2 accumulation in chloroplast was negatively correlated with the initial RUBISCO activity and photosynthetic rate [250]. Chilling-induced oxidative stress evident by increased accumulation of ROS, including H_2O_2 and $O_2^{\bullet-}$, lipid peroxidation, and protein oxidation is a significant factor in relation to chilling injury in plants [169, 251, 252]. Protein carbonyl content, an indication of oxidative damage, was increased 2-fold in maize seedlings when exposed to chilling temperatures [251]. Lipoxygenase activity as well as lipid peroxidation was increased in maize leaves during low temperatures, suggesting that lipoxygenase-mediated peroxidation of membrane lipids contributes to the oxidative damage occurring in chill-stressed maize leaves [169]. Responses to chilling-induced oxidative stress include alteration in activities of enzymes of antioxidant defense system. The activities of antioxidative enzymes APX, MDHAR, DHAR, GR, and SOD increased during chilling periods in maize [169] and strawberry leaves [170]. However, if the duration of chilling stress is too long, the defense system may not remove overproduced ROS effectively, which may result in severe damage or

even death [252]. Nonenzymic antioxidants (AsA, GSH, carotenoids, and α-tocopherol) also play important role in cold response. Under cold stress conditions, low-molecular weight antioxidants, especially, that of reduced AsA, have been suggested to be an important component in plant cell defense [126]. Many comparative studies using chilling-tolerant and sensitive genotypes have shown greater antioxidant capacity in chilling-tolerant species compared to sensitive ones [253–255]. In rice, higher activities of defense enzymes and higher content of antioxidant under stress were associated with tolerance to chilling [255]. The responses of antioxidative system of rice to chilling were investigated in a tolerant cultivar, Xiangnuo-1, and a susceptible cultivar, IR-50. The electrolyte leakage and malondialdehyde content of Xiangnuo-1 were little affected by chilling treatment, but those of IR-50 increased. Activities of SOD, CAT, APX, and GR and AsA content of Xiangnuo-1 remained high, while those of IR-50 decreased under chilling stress. GR activity was also found to increase within 24 h in chilling-tolerant *Zea diploperennis*, but it decreased slightly in chilling-susceptible *Z. mays cv.* LG11 [253].

4.4. Metal Toxicity. The increasing levels of metals into the environment drastically affect plant growth and metabolism, ultimately, leading to severe losses in crop yields [256, 257]. One of the consequences of the presence of the toxic metals within the plant tissues is the formation of ROS, which can be initiated directly or indirectly by the metals and, consequently, leading to oxidative damage to different cell constituents [6, 11, 14, 15, 258]. Under metal stress condition, net photosynthesis (Ph_n) decreases due to damage to photosynthetic metabolism, including photosynthetic electron transport (Ph_{et}) [259]. For example, copper has been shown to negatively affect components of both the light reactions (e.g., PSII, thylakoid membrane structure, and chlorophyll content) [259] and CO_2-fixation reactions [260]. These alterations in photosynthetic metabolism lead to overproduction of ROS such as $O_2^{\bullet-}$, $^\bullet OH$, and H_2O_2. The induction of ROS production due to metals (cadmium and zinc) in *Nicotiana tabacum* L. cv. Bright Yellow 2 (TBY-2) cells in suspension cultures showed properties comparable to the elicitor-induced oxidative burst in other plant cells [261]. Redox-active metals, such as iron, copper, and chromium, undergo redox cycling producing ROS, whereas redox-inactive metals, such as lead, cadmium, mercury, and others, deplete cell's major antioxidants, particularly thiol-containing antioxidants and enzymes [6, 11, 14–16, 262–264]. If metal-induced production of ROS is not adequately counterbalanced by cellular antioxidants, oxidative damage of lipids, proteins, and nucleic acids ensues [13–15, 43, 46, 265, 266]. Significant enhancement in lipid peroxidation and decline in protein thiol contents were observed when rice seedlings were subjected to Al, Ni, and Mn toxicity [11, 14, 15].

The increased activity of antioxidative enzymes in metal stressed plants appears to serve as an important component of antioxidant defense mechanism of plants to combat metal-induced oxidative injury [6]. Responses of metal exposure to plants vary depending on plant species, tissues, stages of development, type of metal and its concentration. One of the key responses includes triggering of a series of defense mechanisms which involve enzymatic and nonenzymatic components [6, 11, 13–16, 262]. Various groups of workers have reported increased activities of antioxidant enzymes like GPX, SOD, APX, MDHAR, DHAR, and GR as well as nonenzymic antioxidants in metal-treated plants and suggested involvement of antioxidant defense system in the adaptive response to metal ions [6, 11, 13–16, 171]. However, results suggest that activation of antioxidant enzymes in response to oxidative stress induced by metals is not enough to confer tolerance to metal accumulation. Comparative study of antioxidative response of two maize lines differing in Al tolerance suggested that better protection of the Al tolerant maize roots from Al-induced oxidative damage results, at least partially, from the increased activity of their antioxidative system. After 24 h of Al exposure, a gradual increase in the membrane lipid peroxidation in Al-stressed root of the susceptible maize line was accompanied by decreased activities of the antioxidant enzymes SOD and POD. In contrast, increased activities of the SOD and POD were found in Al-treated roots of the tolerant maize line, in which the level of membrane lipid peroxidation remained almost unchanged [267]. Comparative antioxidant profiling of tolerant (TPM-1) and sensitive (TM-4) variety of *Brassica juncea* L. performed after exposure to arsenate [As(V)] and arsenite [As(III)] showed in general, better response of antioxidant enzymes and the level of glutathione in TPM-1 than in TM-4 [268]. These responses presumably allowed TPM-1 to tolerate higher As concentrations as compared with that of TM-4 [268].

4.5. UV-B Radiations.

UV-B radiation on plants is now of major concern to plant biologists due to the threat to productivity in global agriculture [269]. Enhanced UV-B significantly inhibits net photosynthetic rate. It has been shown that UV-B treatment results in decrease in the light-saturated rate of CO_2 assimilation, accompanied by decreases in carboxylation velocity and RUBISCO content and activity [270]. He and coworkers [271] observed marked decrease in the ratios of variable to maximum chlorophyll fluorescence yield and in the quantum yield of photosynthetic oxygen evolution in pea and rice leaves. Limited CO_2 assimilation due to UV-B leads to excessive production of ROS which, in turn, cause oxidative damage in plants [10, 272]. Rao and coworkers [172] suggested that UV-B exposure generates activated oxygen species by increasing NADPH-oxidase activity. Plants must adapt to the deleterious effects of UV-B radiation because they are dependent on sunlight for photosynthesis and, therefore, cannot avoid exposure to UV-B radiation. Plants possess antioxidative enzymatic scavengers SOD, POD, CAT, and APX and nonenzymatic antioxidants like AsA, GSH, and carotenoids to keep the balance between the production and removal of ROS. In *Picea asperata* seedlings although enhanced UV-B (30%) increased the efficiency of antioxidant defense system consisting of UV-B absorbing compounds, carotenoids, and antioxidant

enzymes SOD, APX, CAT, and GPX [10], it induced overproduction of ROS and oxidative stress eventually. Peroxidase-related enzymes were found to be preferentially induced by UV-B exposure in Arabidopsis [172]. Gao and Zhang [115] observed that AsA-deficient mutant *vtc1* was more sensitive to supplementary UV-B treatment than wild-type plants and, therefore, suggested that AsA could be considered as an important antioxidant for UV-B radiation.

4.6. Pathogens.

One of the earliest cellular responses following successful pathogen recognition is oxidative burst involving production of ROS. Recognition of a variety of pathogens leads to generation of $O_2^{\bullet-}$, or its dismutation product H_2O_2 in apoplast [273, 274]. Radwan and coworkers [174] observed higher H_2O_2 and MDA concentrations in *Vicia faba* leaves infected with bean yellow mosaic virus than those of the corresponding controls. Several enzymes have been implicated in apoplastic ROS production following successful pathogen recognition. The use of inhibitors pointed to plasma membrane NADPH oxidases and cell wall peroxidases as the two most likely biochemical sources [274]. The expression of these enzymes is induced following recognition of bacterial and fungal pathogens [275, 276]. Although the primary oxidative burst following pathogen recognition occurs in the apoplast, ROS can be produced in other cellular compartments like mitochondria and chloroplast. Abdollahi and Ghahremani [277] studied the role of chloroplasts in the interaction between *Erwinia amylovora* and host plants by using uracil as chloroplast ETC inhibitor. Uracil presence significantly reduced ROS generation during pathogen-host interaction, and ROS generation corresponded with the appearance of necrosis in all cultivars [277]. Liu and coworkers [278] showed that activation of the SIPK/Ntf4/WIPK cascade by pathogens actively promotes the generation of ROS in chloroplasts, which plays an important role in the signaling for and/or execution of HR cell death in plants. They concluded that chloroplast burst occur earlier than NADPH oxidase burst and mitochondria-generated ROS might be essential in accelerating the cell death process.

Differential regulation of antioxidant enzymes, in part mediated by SA, may contribute to increases in ROS and activation of defenses following infection [81, 279]. In tobacco, the reduction of CAT and APX activities resulted in plants hyperresponsive to pathogens [279]. Significant increase in the activities of POD and CAT was observed in leaves of flax lines infected with powdery mildew [173]. Increase in POD activity was much pronounced in tolerant lines than susceptible lines. Enhanced activities of POD, CAT, APX, and SOD were observed in *Vicia faba* leaves infected with bean yellow mosaic virus indicating that the ROS-scavenging systems can have an important role in managing ROS generated in response to pathogens [174].

5. Concluding Remarks

ROS are unavoidable by products of normal cell metabolism. ROS are generated by electron transport activities of chloroplast, mitochondria, and plasma membrane or as a byproduct of various metabolic pathways localized in different cellular compartments. Under normal growth condition, ROS production in various cell compartments is low. However, various environmental stresses such as drought, salinity, chilling, metal toxicity, and UV-B, if prolonged over to a certain extent, disrupt the cellular homeostasis and enhance the production of ROS. ROS play two divergent roles in plants; in low concentrations they act as signaling molecules that mediate several plant responses in plant cells, including responses under stresses, whereas in high concentrations they cause exacerbating damage to cellular components. Enhanced level of ROS causes oxidative damage to lipid, protein, and DNA leading to altered intrinsic membrane properties like fluidity, ion transport, loss of enzyme activity, protein crosslinking, inhibition of protein synthesis, DNA damage, ultimately resulting in cell death. In order to avoid the oxidative damage, higher plants possess a complex antioxidative defense system comprising of nonenzymatic and enzymatic components. Although rapid progress has been made in recent years, there are many uncertainties and gaps in our knowledge of ROS formation and their effect on plants mainly due to short half-life and high reactivity of ROS. Study of formation and fate of ROS using advanced analytical techniques will help in developing broader view of the role of ROS in plants. Future progress in genomics, metabolomics, and proteomics will help in clear understanding of biochemical networks involved in cellular responses to oxidative stress. Improved understanding of these will be helpful in producing plants with in-built capacity of enhanced levels of tolerance to ROS using biotechnological approach.

References

[1] C. H. Foyer and J. Harbinson, "Oxygen metabolism and the regulation of photosynthetic electron transport," in *Causes of Photooxidative Stresses and Amelioration of Defense Systems in Plants*, C. H. Foyer and P. Mullineaux, Eds., pp. 1–42, CRC Press, Boca Raton, Fla, USA, 1994.

[2] C. H. Foyer, "Oxygen metabolism and electron transport in photosynthesis," in *Molecular Biology of Free Radical Scavenging Systems*, J. Scandalios, Ed., pp. 587–621, Cold Spring Harbor Laboratory Press, New York, NY, USA, 1997.

[3] L. A. Del Río, L. M. Sandalio, F. J. Corpas, J. M. Palma, and J. B. Barroso, "Reactive oxygen species and reactive nitrogen species in peroxisomes. Production, scavenging, and role in cell signaling," *Plant Physiology*, vol. 141, no. 2, pp. 330–335, 2006.

[4] O. Blokhina and K. V. Fagerstedt, "Reactive oxygen species and nitric oxide in plant mitochondria: origin and redundant regulatory systems," *Physiologia Plantarum*, vol. 138, no. 4, pp. 447–462, 2010.

[5] E. Heyno, V. Mary, P. Schopfer, and A. Krieger-Liszkay, "Oxygen activation at the plasma membrane: relation between superoxide and hydroxyl radical production by isolated membranes," *Planta*, vol. 234, no. 1, pp. 35–45, 2011.

[6] K. Shah, R. G. Kumar, S. Verma, and R. S. Dubey, "Effect of cadmium on lipid peroxidation, superoxide anion generation and activities of antioxidant enzymes in growing rice seedlings," *Plant Science*, vol. 161, no. 6, pp. 1135–1144, 2001.

[7] R. Mittler, "Oxidative stress, antioxidants and stress tolerance," *Trends in Plant Science*, vol. 7, no. 9, pp. 405–410, 2002.

[8] P. Sharma and R. S. Dubey, "Drought induces oxidative stress and enhances the activities of antioxidant enzymes in growing rice seedlings," *Plant Growth Regulation*, vol. 46, no. 3, pp. 209–221, 2005.

[9] W. H. Hu, X. S. Song, K. Shi, X. J. Xia, Y. H. Zhou, and J. Q. Yu, "Changes in electron transport, superoxide dismutase and ascorbate peroxidase isoenzymes in chloroplasts and mitochondria of cucumber leaves as influenced by chilling," *Photosynthetica*, vol. 46, no. 4, pp. 581–588, 2008.

[10] C. Han, Q. Liu, and Y. Yang, "Short-term effects of experimental warming and enhanced ultraviolet-B radiation on photosynthesis and antioxidant defense of *Picea asperata* seedlings," *Plant Growth Regulation*, vol. 58, no. 2, pp. 153–162, 2009.

[11] R. Maheshwari and R. S. Dubey, "Nickel-induced oxidative stress and the role of antioxidant defence in rice seedlings," *Plant Growth Regulation*, vol. 59, no. 1, pp. 37–49, 2009.

[12] G. Tanou, A. Molassiotis, and G. Diamantidis, "Induction of reactive oxygen species and necrotic death-like destruction in strawberry leaves by salinity," *Environmental and Experimental Botany*, vol. 65, no. 2-3, pp. 270–281, 2009.

[13] S. Mishra, A. B. Jha, and R. S. Dubey, "Arsenite treatment induces oxidative stress, upregulates antioxidant system, and causes phytochelatin synthesis in rice seedlings," *Protoplasma*, vol. 248, no. 3, pp. 565–577, 2011.

[14] S. Srivastava and R. S. Dubey, "Manganese-excess induces oxidative stress, lowers the pool of antioxidants and elevates activities of key antioxidative enzymes in rice seedlings," *Plant Growth Regulation*, pp. 1–16, 2011.

[15] P. Sharma and R. S. Dubey, "Involvement of oxidative stress and role of antioxidative defense system in growing rice seedlings exposed to toxic concentrations of aluminum," *Plant Cell Reports*, vol. 26, no. 11, pp. 2027–2038, 2007.

[16] S. Verma and R. S. Dubey, "Lead toxicity induces lipid peroxidation and alters the activities of antioxidant enzymes in growing rice plants," *Plant Science*, vol. 164, no. 4, pp. 645–655, 2003.

[17] B. Meriga, B. K. Reddy, K. R. Rao, L. A. Reddy, and P. B. K. Kishor, "Aluminium-induced production of oxygen radicals, lipid peroxidation and DNA damage in seedlings of rice (*Oryza sativa*)," *Journal of Plant Physiology*, vol. 161, no. 1, pp. 63–68, 2004.

[18] R. Desikan, S. A.-H.-Mackerness S., J. T. Hancock, and S. J. Neill, "Regulation of the *Arabidopsis* transcriptome by oxidative stress," *Plant Physiology*, vol. 127, no. 1, pp. 159–172, 2001.

[19] S. Neill, R. Desikan, and J. Hancock, "Hydrogen peroxide signalling," *Current Opinion in Plant Biology*, vol. 5, no. 5, pp. 388–395, 2002.

[20] J. Yan, N. Tsuichihara, T. Etoh, and S. Iwai, "Reactive oxygen species and nitric oxide are involved in ABA inhibition of stomatal opening," *Plant, Cell and Environment*, vol. 30, no. 10, pp. 1320–1325, 2007.

[21] G. Noctor and C. H. Foyer, "Ascorbate and glutathione: keeping active oxygen under control," *Annual Review of Plant Biology*, vol. 49, pp. 249–279, 1998.

[22] M. Zaefyzadeh, R. A. Quliyev, S. M. Babayeva, and M. A. Abbasov, "The effect of the interaction between genotypes and drought stress on the superoxide dismutase and chlorophyll content in durum wheat landraces," *Turkish Journal of Biology*, vol. 33, no. 1, pp. 1–7, 2009.

[23] Q. Chen, M. Zhang, and S. Shen, "Effect of salt on malondialdehyde and antioxidant enzymes in seedling roots of Jerusalem artichoke (*Helianthus tuberosus* L.)," *Acta Physiologiae Plantarum*, vol. 33, no. 2, pp. 273–278, 2010.

[24] R. D. Allen, R. P. Webb, and S. A. Schake, "Use of transgenic plants to study antioxidant defenses," *Free Radical Biology and Medicine*, vol. 23, no. 3, pp. 473–479, 1997.

[25] M. Faize, L. Burgos, L. Faize et al., "Involvement of cytosolic ascorbate peroxidase and Cu/Zn-superoxide dismutase for improved tolerance against drought stress," *Journal of Experimental Botany*, vol. 62, no. 8, pp. 2599–2613, 2011.

[26] Y. P. Lee, S. H. Kim, J. W. Bang, H. S. Lee, S. S. Kwak, and S. Y. Kwon, "Enhanced tolerance to oxidative stress in transgenic tobacco plants expressing three antioxidant enzymes in chloroplasts," *Plant Cell Reports*, vol. 26, no. 5, pp. 591–598, 2007.

[27] K. Asada and M. Takahashi, "Production and scavenging of active oxygen in photosynthesis," in *Photoinhibition: Topics of Photosynthesis*, D. J. Kyle, C. B. Osmond, and C. J. Arntzen, Eds., pp. 227–287, Elsevier, Amsterdam, The Netherlands, 9th edition, 1987.

[28] K. Apel and H. Hirt, "Reactive oxygen species: metabolism, oxidative stress, and signal transduction," *Annual Review of Plant Biology*, vol. 55, pp. 373–399, 2004.

[29] A. Krieger-Liszkay, "Singlet oxygen production in photosynthesis," *Journal of Experimental Botany*, vol. 56, no. 411, pp. 337–346, 2005.

[30] S. Hatz, J. D. C. Lambert, and P. R. Ogilby, "Measuring the lifetime of singlet oxygen in a single cell: addressing the issue of cell viability," *Photochemical and Photobiological Sciences*, vol. 6, no. 10, pp. 1106–1116, 2007.

[31] S. Hackbarth, J. Schlothauer, A. Preuß, and B. Röder, "New insights to primary photodynamic effects—singlet oxygen kinetics in living cells," *Journal of Photochemistry and Photobiology B*, vol. 98, no. 3, pp. 173–179, 2010.

[32] D. Wagner, D. Przybyla, R. Op Den Camp et al., "The genetic basis of singlet oxygen-induced stress response of *Arabidopsis thaliana*," *Science*, vol. 306, no. 5699, pp. 1183–1185, 2004.

[33] H. Kasai, "Analysis of a form of oxidative DNA damage, 8-hydroxy-2'-deoxyguanosine, as a marker of cellular oxidative stress during carcinogenesis," *Mutation Research*, vol. 387, no. 3, pp. 147–163, 1997.

[34] A. Krieger-Liszkay, C. Fufezan, and A. Trebst, "Singlet oxygen production in photosystem II and related protection mechanism," *Photosynthesis Research*, vol. 98, no. 1-3, pp. 551–564, 2008.

[35] B. Halliwell and J. M. C. Gutteridge, "Oxygen toxicity, oxygen radicals, transition metals and disease," *Biochemical Journal*, vol. 219, no. 1, pp. 1–14, 1984.

[36] M. Valko, H. Morris, and M. T. D. Cronin, "Metals, toxicity and oxidative stress," *Current Medicinal Chemistry*, vol. 12, no. 10, pp. 1161–1208, 2005.

[37] B. Halliwell, "Generation of hydrogen peroxide, superoxide and hydroxyl radicals during the oxidation of dihydroxyfumaric acid by peroxidase," *Biochemical Journal*, vol. 163, no. 3, pp. 441–448, 1977.

[38] J. A. Imlay, "Pathways of oxidative damage," *Annual Review of Microbiology*, vol. 57, pp. 395–418, 2003.

[39] J. M. McCord, J. D. Crapo, and I. Fridovich, "Superoxide dismutase assay. a review of methodology," in *Superoxide and Superoxide Dismutase*, A. M. Michelson, J. M. McCord, and I. Fridovich, Eds., pp. 11–17, Academic press, London, UK, 1977.

[40] R. Mittler and B. A. Zilinskas, "Purification and characterization of pea cytosolic ascorbate peroxidase," *Plant Physiology*, vol. 97, no. 3, pp. 962–968, 1991.

[41] G. P. Bienert, A. L. B. Møller, K. A. Kristiansen et al., "Specific aquaporins facilitate the diffusion of hydrogen peroxide across membranes," *Journal of Biological Chemistry*, vol. 282, no. 2, pp. 1183–1192, 2007.

[42] M. A. Torres, J. L. Dangl, and J. D. G. Jones, "*Arabidopsis* gp91[phox] homologues Atrbohd and Atrbohf are required for accumulation of reactive oxygen intermediates in the plant defense response," *Proceedings of the National Academy of Sciences of the United States of America*, vol. 99, no. 1, pp. 517–522, 2002.

[43] B. Halliwell and J. M. C. Gutteridge, *Free Radicals in Biology and Medicine*, Oxford University Press, Oxford, UK, 2rd edition, 1989.

[44] W. M. Kaiser, "Reversible inhibition of the calvin cycle and activation of oxidative pentose phosphate cycle in isolated intact chloroplasts by hydrogen peroxide," *Planta*, vol. 145, no. 4, pp. 377–382, 1979.

[45] R. C. Leegood and D. A. Walker, "Regulation of fructose-1,6-bisphosphatase activity in leaves," *Planta*, vol. 156, no. 5, pp. 449–456, 1982.

[46] J. Dat, S. Vandenabeele, E. Vranová, M. Van Montagu, D. Inzé, and F. Van Breusegem, "Dual action of the active oxygen species during plant stress responses," *Cellular and Molecular Life Sciences*, vol. 57, no. 5, pp. 779–795, 2000.

[47] A. Rigo, R. Stevanato, A. Finazzi Agro', and G. Rotilio, "An attempt to evaluate the rate of the Haber Weiss reaction by using •OH radical scavengers," *FEBS Letters*, vol. 80, no. 1, pp. 130–132, 1977.

[48] C. H. Foyer, H. Lopez-Delgado, J. F. Dat, and I. M. Scott, "Hydrogen peroxide- and glutathione-associated mechanisms of acclimatory stress tolerance and signalling," *Physiologia Plantarum*, vol. 100, no. 2, pp. 241–254, 1997.

[49] E. Pinto, T. C. S. Sigaud-Kutner, M. A. S. Leitão, O. K. Okamoto, D. Morse, and P. Colepicolo, "Heavy metal-induced oxidative stress in algae," *Journal of Phycology*, vol. 39, no. 6, pp. 1008–1018, 2003.

[50] H. Ishida, Y. Nishimori, M. Sugisawa, A. Makino, and T. Mae, "The large subunit of ribulose-1,5-bisphosphate carboxylase/oxygenase is fragmented into 37-kDa and 16-kDa polypeptides by active oxygen in the lysates of chloroplasts from primary leaves of wheat," *Plant and Cell Physiology*, vol. 38, no. 4, pp. 471–479, 1997.

[51] S. Luo, H. Ishida, A. Makino, and T. Mae, "Fe^{2+}-catalyzed site-specific cleavage of the large subunit of ribulose 1,5-bisphosphate carboxylase close to the active site," *Journal of Biological Chemistry*, vol. 277, no. 14, pp. 12382–12387, 2002.

[52] E. F. Elstner, "Oxygen activation and oxygen toxicity," *Annual Review of Plant Biology*, vol. 33, pp. 73–96, 1982.

[53] E. F. Elstner, "Metabolism of activated oxygen species," in *Biochemistry of Plants*, D. D. Davies, Ed., pp. 253–315, Academic Press, London, UK, 1987.

[54] E. F. Elstner, "Mechanisms of oxygen activation in different compartments of plant cells," in *Active Oxygen/Oxidative Stress and Plant Metabolism*, E. J. Pell and K. L. Steffen, Eds., pp. 13–25, American Society of Plant Physiologists, Rockville, Md, USA, 1991.

[55] R. E. Cleland and S. C. Grace, "Voltammetric detection of superoxide production by photosystem II," *FEBS Letters*, vol. 457, no. 3, pp. 348–352, 1999.

[56] A. Arora, R. K. Sairam, and G. C. Srivastava, "Oxidative stress and antioxidative system in plants," *Current Science*, vol. 82, no. 10, pp. 1227–1238, 2002.

[57] J. F. Turrens, "Mitochondrial formation of reactive oxygen species," *Journal of Physiology*, vol. 552, no. 2, pp. 335–344, 2003.

[58] M. P. Murphy, "How mitochondria produce reactive oxygen species," *Biochemical Journal*, vol. 417, no. 1, pp. 1–13, 2009.

[59] G. Noctor, R. De Paepe, and C. H. Foyer, "Mitochondrial redox biology and homeostasis in plants," *Trends in Plant Science*, vol. 12, no. 3, pp. 125–134, 2007.

[60] A. Y. Andreyev, Y. E. Kushnareva, and A. A. Starkov, "Mitochondrial metabolism of reactive oxygen species," *Biochemistry (Moscow)*, vol. 70, no. 2, pp. 200–214, 2005.

[61] A. G. Rasmusson, D. A. Geisler, and I. M. Møller, "The multiplicity of dehydrogenases in the electron transport chain of plant mitochondria," *Mitochondrion*, vol. 8, no. 1, pp. 47–60, 2008.

[62] A. Baker and A. I. Graham, *Plant Peroxisomes: Biochemistry, Cell Biology and Biotechnological Applications*, Kluwer Academic Publishers, Dordrecht, The Netherlands, 2002.

[63] G. Noctor, S. Veljovic-Jovanovic, S. Driscoll, L. Novitskaya, and C. H. Foyer, "Drought and oxidative load in the leaves of C_3 plants: a predominant role for photorespiration?" *Annals of Botany*, vol. 89, pp. 841–850, 2002.

[64] E. López-Huertas, F. J. Corpas, L. M. Sandalio, and L. A. Del Río, "Characterization of membrane polypeptides from pea leaf peroxisomes involved in superoxide radical generation," *Biochemical Journal*, vol. 337, no. 3, pp. 531–536, 1999.

[65] J. M. Kwak, I. C. Mori, Z. M. Pei et al., "NADPH oxidase AtrbohD and AtrbohF genes function in ROS-dependent ABA signaling in *Arabidopsis*," *EMBO Journal*, vol. 22, no. 11, pp. 2623–2633, 2003.

[66] G. G. Gross, "Cell wall-bound malate dehydrogenase from horseradish," *Phytochemistry*, vol. 16, no. 3, pp. 319–321, 1977.

[67] C. Martinez, J. L. Montillet, E. Bresson et al., "Apoplastic peroxidase generates superoxide anions in cells of cotton cotyledons undergoing the hypersensitive reaction to *Xanthomonas campestris* pv. *malvacearum* 18," *Molecular Plant-Microbe Interactions*, vol. 11, no. 11, pp. 1038–1047, 1998.

[68] M. J. Kim, S. Ciani, and D. P. Schachtman, "A peroxidase contributes to ros production during *Arabidopsis* root response to potassium deficiency," *Molecular Plant*, vol. 3, no. 2, pp. 420–427, 2010.

[69] P. Wojtaszek, "Oxidative burst: an early plant response to pathogen infection," *Biochemical Journal*, vol. 322, no. 3, pp. 681–692, 1997.

[70] B. G. Lane, "Oxalate, germins, and higher-plant pathogens," *IUBMB Life*, vol. 53, no. 2, pp. 67–75, 2002.

[71] A. Cona, G. Rea, R. Angelini, R. Federico, and P. Tavladoraki, "Functions of amine oxidases in plant development and defence," *Trends in Plant Science*, vol. 11, no. 2, pp. 80–88, 2006.

[72] P. C. Bethke and R. L. Jones, "Cell death of barley aleurone protoplasts is mediated by reactive oxygen species," *Plant Journal*, vol. 25, no. 1, pp. 19–29, 2001.

[73] Jung Hee Joo, Yun Soo Bae, and June Seung Lee, "Role of auxin-induced reactive oxygen species in root gravitropism," *Plant Physiology*, vol. 126, no. 3, pp. 1055–1060, 2001.

[74] G. Miller, V. Shulaev, and R. Mittler, "Reactive oxygen signaling and abiotic stress," *Physiologia Plantarum*, vol. 133, no. 3, pp. 481–489, 2008.

[75] L. Xiong, K. S. Schumaker, and J. K. Zhu, "Cell signaling during cold, drought, and salt stress," *Plant Cell*, vol. 14, pp. S165–S183, 2002.

[76] Y. Cheng and C. Song, "Hydrogen peroxide homeostasis and signaling in plant cells," *Science in China. Series C, Life sciences*, vol. 49, no. 1, pp. 1–11, 2006.

[77] Z. M. Pel, Y. Murata, G. Benning et al., "Calcium channels activated by hydrogen peroxide mediate abscisic acid signalling in guard cells," *Nature*, vol. 406, no. 6797, pp. 731–734, 2000.

[78] R. Jannat, M. Uraji, M. Morofuji et al., "Roles of intracellular hydrogen peroxide accumulation in abscisic acid signaling in *Arabidopsis* guard cells," *Journal of Plant Physiology*, vol. 168, no. 16, pp. 1919–1926, 2011.

[79] E. Bahin, C. Bailly, B. Sotta, I. Kranner, F. Corbineau, and J. Leymarie, "Crosstalk between reactive oxygen species and hormonal signalling pathways regulates grain dormancy in barley," *Plant, Cell and Environment*, vol. 34, no. 6, pp. 980–993, 2011.

[80] A. K. Nanda, E. Andrio, D. Marino, N. Pauly, and C. Dunand, "Reactive oxygen species during plant-microorganism early interactions," *Journal of Integrative Plant Biology*, vol. 52, no. 2, pp. 195–204, 2010.

[81] D. F. Klessig, J. Durner, R. Noad et al., "Nitric oxide and salicylic acid signaling in plant defense," *Proceedings of the National Academy of Sciences of the United States of America*, vol. 97, no. 16, pp. 8849–8855, 2000.

[82] M. L. Orozco-Cárdenas, J. Narváez-Vásquez, and C. A. Ryan, "Hydrogen peroxide acts as a second messenger for the induction of defense genes in tomato plants in response to wounding, systemin, and methyl jasmonate," *Plant Cell*, vol. 13, no. 1, pp. 179–191, 2001.

[83] L. Denness, J. F. McKenna, C. Segonzac et al., "Cell wall damage-induced lignin biosynthesis is regulated by a reactive oxygen species- and jasmonic acid-dependent process in *Arabidopsis*," *Plant Physiology*, vol. 156, no. 3, pp. 1364–1374, 2011.

[84] T. Yuasa, K. Ichimura, T. Mizoguchi, and K. Shinozaki, "Oxidative stress activates ATMPK6, an *Arabidopsis* homologue of map kinase," *Plant and Cell Physiology*, vol. 42, no. 9, pp. 1012–1016, 2001.

[85] C. M. Yeh, P. S. Chien, and H. J. Huang, "Distinct signalling pathways for induction of MAP kinase activities by cadmium and copper in rice roots," *Journal of Experimental Botany*, vol. 58, no. 3, pp. 659–671, 2007.

[86] O. Borsani, P. Díaz, M. F. Agius, V. Valpuesta, and J. Monza, "Water stress generates an oxidative stress through the induction of a specific Cu/Zn superoxide dismutase in *Lotus corniculatus* leaves," *Plant Science*, vol. 161, no. 4, pp. 757–763, 2001.

[87] Z. Zhao, G. Chen, and C. Zhang, "Interaction between reactive oxygen species and nitric oxide in drought-induced abscisic acid synthesis in root tips of wheat seedlings," *Australian Journal of Plant Physiology*, vol. 28, no. 10, pp. 1055–1061, 2001.

[88] N. Smirnoff, "Antioxidant systems and plant response to the environment," in *Environment and Plant Metabolism: Flexibility and Acclimation*, N. Smirnoff, Ed., pp. 217–243, Bios Scientific Publishers, Oxford, UK, 1995.

[89] R. O. Recknagal and E. A. Glende, "Oxygen radicals in biological systems," in *Methods in Enzymology*, L. Packer, Ed., vol. 105, pp. 331–337, Academic Press, New York, NY, USA, 1984.

[90] K. J. A. Davies, "Oxidative stress, antioxidant defenses, and damage removal, repair, and replacement systems," *IUBMB Life*, vol. 50, no. 4-5, pp. 279–289, 2000.

[91] G. R. Buettner, "The pecking order of free radicals and antioxidants: lipid peroxidation, α-tocopherol, and ascorbate," *Archives of Biochemistry and Biophysics*, vol. 300, no. 2, pp. 535–543, 1993.

[92] Y. Yamauchi, A. Furutera, K. Seki, Y. Toyoda, K. Tanaka, and Y. Sugimoto, "Malondialdehyde generated from peroxidized linolenic acid causes protein modification in heat-stressed plants," *Plant Physiology and Biochemistry*, vol. 46, no. 8-9, pp. 786–793, 2008.

[93] I. M. Møller and B. K. Kristensen, "Protein oxidation in plant mitochondria as a stress indicator," *Photochemical and Photobiological Sciences*, vol. 3, no. 8, pp. 730–735, 2004.

[94] M. C. Romero-Puertas, J. M. Palma, M. Gómez, L. A. Del Río, and L. M. Sandalio, "Cadmium causes the oxidative modification of proteins in pea plants," *Plant, Cell and Environment*, vol. 25, no. 5, pp. 677–686, 2002.

[95] S. J. Stohs and D. Bagchi, "Oxidative mechanisms in the toxicity of metal ions," *Free Radical Biology and Medicine*, vol. 18, no. 2, pp. 321–336, 1995.

[96] N. Brot and H. Weissbach, "The biochemistry of methionine sulfoxide residues in proteins," *Trends in Biochemical Sciences*, vol. 7, no. 4, pp. 137–139, 1982.

[97] K. J. Davies, "Protein damage and degradation by oxygen radicals. I. general aspects," *Journal of Biological Chemistry*, vol. 262, no. 20, pp. 9895–9901, 1987.

[98] P. R. Gardner and I. Fridovich, "Superoxide sensitivity of the *Escherichia coli* 6-phosphogluconate dehydratase," *Journal of Biological Chemistry*, vol. 266, no. 3, pp. 1478–1483, 1991.

[99] E. R. Stadtman, "Oxidation of proteins by mixed-function oxidation systems: implication in protein turnover, ageing and neutrophil function," *Trends in Biochemical Sciences*, vol. 11, no. 1, pp. 11–12, 1986.

[100] E. Cabiscol, E. Piulats, P. Echave, E. Herrero, and J. Ros, "Oxidative stress promotes specific protein damage in *Saccharomyces cerevisiae*," *Journal of Biological Chemistry*, vol. 275, no. 35, pp. 27393–27398, 2000.

[101] T. Grune, T. Reinheckel, and K. J. A. Davies, "Degradation of oxidized proteins in mammalian cells," *FASEB Journal*, vol. 11, no. 7, pp. 526–534, 1997.

[102] J. A. Imlay and S. Linn, "DNA damage and oxygen radical toxicity," *Science*, vol. 240, no. 4857, pp. 1302–1309, 1988.

[103] T. Liu, J. Van Staden, and W. A. Cress, "Salinity induced nuclear and DNA degradation in meristematic cells of soybean (*Glycine max* (L.)) roots," *Plant Growth Regulation*, vol. 30, no. 1, pp. 49–54, 2000.

[104] M. Dizdaroglu, "Chemistry of free radical damage to DNA and nucleoproteins," in *DNA and Free Radicals*, B. Halliwell and O. I. Aruoma, Eds., pp. 19–39, Ellis Horwood, London, UK, 1993.

[105] B. Halliwell and J. M. C. Gutteridge, *Free Radicals in Biology and Medicine*, Oxford University Press, Oxford, UK, 3rd edition, 1999.

[106] H. Tsuboi, K. Kouda, H. Takeuchi et al., "8-Hydroxy-deoxyguanosine in urine as an index of oxidative damage to DNA in the evaluation of atopic dermatitis," *British Journal of Dermatology*, vol. 138, no. 6, pp. 1033–1035, 1998.

[107] B. Halliwell and O. I. Aruoma, "DNA damage by oxygen-derived species. Its mechanism and measurement in mammalian systems," *FEBS Letters*, vol. 281, no. 1-2, pp. 9–19, 1991.

[108] S. P. Fink, G. R. Reddy, and L. J. Marnett, "Mutagenicity in *Escherichia coli* of the major DNA adduct derived from the endogenous mutagen malondialdehyde," *Proceedings of the National Academy of Sciences of the United States of America*, vol. 94, no. 16, pp. 8652–8657, 1997.

[109] M. D. Evans, M. Dizdaroglu, and M. S. Cooke, "Oxidative DNA damage and disease: induction, repair and significance," *Mutation Research*, vol. 567, no. 1, pp. 1–61, 2004.

[110] N. L. Oleinick, Song-mao Chiu, N. Ramakrishnan, and Liang-yan Xue, "The formation, identification, and significance of DNA-protein cross-links in mammalian cells," *British Journal of Cancer, supplement*, vol. 8, pp. 135–140, 1987.

[111] C. Richter, "Reactive oxygen and DNA damage in mitochondria," *Mutation Research, DNAging Genetic Instability and Aging*, vol. 275, no. 3-6, pp. 249–255, 1992.

[112] C. H. Pang and B. S. Wang, "Oxidative stress and salt tolerance in plants," in *Progress in Botany*, U. Lüttge, W. Beyschlag, and J. Murata, Eds., pp. 231–245, Springer, Berlin, Germany, 2008.

[113] P. Sharma, A. B. Jha, and R. S. Dubey, "Oxidative stress and antioxidative defense system in plants growing under abiotic Stresses," in *Handbook of Plant and Crop Stress*, M. Pessarakli, Ed., pp. 89–138, CRC Press, Taylor and Francis Publishing Company, Fla, USA, 3rd edition, 2010.

[114] M. C. De Pinto and L. De Gara, "Changes in the ascorbate metabolism of apoplastic and symplastic spaces are associated with cell differentiation," *Journal of Experimental Botany*, vol. 55, no. 408, pp. 2559–2569, 2004.

[115] Q. Gao and L. Zhang, "Ultraviolet-B-induced oxidative stress and antioxidant defense system responses in ascorbate-deficient *vtc1* mutants of *Arabidopsis thaliana*," *Journal of Plant Physiology*, vol. 165, no. 2, pp. 138–148, 2008.

[116] N. M. Semchuk, O. V. Lushchak, J. Falk, K. Krupinska, and V. I. Lushchak, "Inactivation of genes, encoding tocopherol biosynthetic pathway enzymes, results in oxidative stress in outdoor grown *Arabidopsis thaliana*," *Plant Physiology and Biochemistry*, vol. 47, no. 5, pp. 384–390, 2009.

[117] G. L. Wheeler, M. A. Jones, and N. Smirnoff, "The biosynthetic pathway of vitamin C in higher plants," *Nature*, vol. 393, no. 6683, pp. 365–369, 1998.

[118] F. A. Isherwood, Y. T. Chen, and L. W. Mapson, "Synthesis of L-ascorbic acid in plants and animals," *The Biochemical Journal*, vol. 56, no. 1, pp. 1–15, 1954.

[119] H. B. Shao, L. Y. Chu, Z. H. Lu, and C. M. Kang, "Primary antioxidant free radical scavenging and redox signaling pathways in higher plant cells," *International Journal of Biological Sciences*, vol. 4, no. 1, pp. 8–14, 2008.

[120] N. Smirnoff, J. A. Running, and S. Gatzek, "Ascorbate biosynthesis: a diversity of pathways," in *Vitamin C: Its Functions and Biochemistry in Animals and Plants*, H. Asard, J. M. May, and N. Smirnoff, Eds., pp. 7–29, BIOS Scientific, New York, NY, USA, 2004.

[121] J. D. Barnes, Y. Zheng, and T. M. Lyons, "Plant resistance to ozone: the role of ascorbate," in *Air Pollution and Plant Biotechnology*, K. Omasa, H. Saji, S. Youssefian, and N. Kondo, Eds., pp. 235–254, Springer, Tokyo, Japan, 2002.

[122] N. Smirnoff, "Ascorbic acid: metabolism and functions of a multi-facetted molecule," *Current Opinion in Plant Biology*, vol. 3, no. 3, pp. 229–235, 2000.

[123] C. Miyake and K. Asada, "Ferredoxin-dependent photoreduction of the monodehydroascorbate radical in spinach thylakoids," *Plant and Cell Physiology*, vol. 35, no. 4, pp. 539–549, 1994.

[124] K. Asada, "Radical production and scavenging in the chloroplasts," in *Photosynthesis and the Environment*, N. R. Baker, Ed., pp. 123–150, Kluwer, Dordrecht, The Netherlands, 1996.

[125] J. A. Hernández, M. A. Ferrer, A. Jiménez, A. R. Barceló, and F. Sevilla, "Antioxidant systems and O_2^-/H_2O_2 production in the apoplast of pea leaves. Its relation with salt-induced necrotic lesions in minor veins," *Plant Physiology*, vol. 127, no. 3, pp. 817–831, 2001.

[126] M. S. Radyuk, I. N. Domanskaya, R. A. Shcherbakov, and N. V. Shalygo, "Effect of low above-zero temperature on the content of low-molecular antioxidants and activities of antioxidant enzymes in green barley leaves," *Russian Journal of Plant Physiology*, vol. 56, no. 2, pp. 175–180, 2009.

[127] M. M. Chaves, J. S. Pereira, J. Maroco et al., "How plants cope with water stress in the field. Photosynthesis and growth," *Annals of Botany*, vol. 89, pp. 907–916, 2002.

[128] C. Zhang, J. Liu, Y. Zhang et al., "Overexpression of SlGMEs leads to ascorbate accumulation with enhanced oxidative stress, cold, and salt tolerance in tomato," *Plant Cell Reports*, vol. 30, no. 3, pp. 389–398, 2011.

[129] Hemavathi, C. P. Upadhyaya, K. E. Young et al., "Overexpression of strawberry d-galacturonic acid reductase in potato leads to accumulation of vitamin C with enhanced abiotic stress tolerance," *Plant Science*, vol. 177, no. 6, pp. 659–667, 2009.

[130] Z. Wang, Y. Xiao, W. Chen, K. Tang, and L. Zhang, "Increased vitamin C content accompanied by an enhanced recycling pathway confers oxidative stress tolerance in *Arabidopsis*," *Journal of Integrative Plant Biology*, vol. 52, no. 4, pp. 400–409, 2010.

[131] C. H. Foyer and G. Noctor, "Redox sensing and signalling associated with reactive oxygen in chloroplasts, peroxisomes and mitochondria," *Physiologia Plantarum*, vol. 119, no. 3, pp. 355–364, 2003.

[132] K. Asada, "Production and action of active oxygen species in photosynthetic tissues," in *Causes of Photooxidative Stress and Amelioration of Defense Systems in Plants*, C. H. Foyer and P. M. Mullineaux, Eds., pp. 77–104, CRC Press, Boca Raton, Fla, USA, 1994.

[133] F. A. Loewus, "Ascorbic acid and its metabolic products," in *The Biochemistry of Plants*, J. Preiss, Ed., pp. 85–107, ,Academic Press, New York, NY, USA, 1988.

[134] C. H. Foyer and B. Halliwell, "The presence of glutathione and glutathione reductase in chloroplasts: a proposed role in ascorbic acid metabolism," *Planta*, vol. 133, no. 1, pp. 21–25, 1976.

[135] M. Tausz, H. Šircelj, and D. Grill, "The glutathione system as a stress marker in plant ecophysiology: is a stress-response concept valid?" *Journal of Experimental Botany*, vol. 55, no. 404, pp. 1955–1962, 2004.

[136] M. Hefny and D. Z. Abdel-Kader, "Antioxidant-enzyme system as selection criteria for salt tolerance in forage sorghum genotypes (*Sorghum bicolor* L. Moench)," in *Salinity and Water Stress*, M. Ashraf, M. Ozturk, and H. R. Athar, Eds., pp. 25–36, Springer, The Netherlands, 2009.

[137] M. Strohm, M. Eiblmeier, C. Langebartels et al., "Responses of transgenic poplar (*Populus tremula* × *P. alba*) overexpressing glutathione synthetase or glutathione reductase to acute ozone stress: visible injury and leaf gas exchange," *Journal of Experimental Botany*, vol. 50, no. 332, pp. 365–374, 1999.

[138] C. H. Foyer, N. Souriau, S. Perret et al., "Overexpression of glutathione reductase but not glutathione synthetase leads to increases in antioxidant capacity and resistance to

[139] photoinhibition in poplar trees," *Plant Physiology*, vol. 109, no. 3, pp. 1047–1057, 1995.

[139] Y. L. Zhu, E. A. H. Pilon-Smits, A. S. Tarun, S. U. Weber, L. Jouanin, and N. Terry, "Cadmium tolerance and accumulation in Indian mustard is enhanced by overexpressing γ-glutamylcysteine synthetase," *Plant Physiology*, vol. 121, no. 4, pp. 1169–1177, 1999.

[140] G. Gullner, T. Kömives, and H. Rennenberg, "Enhanced tolerance of transgenic poplar plants overexpressing γ-glutamylcysteine synthetase towards chloroacetanilide herbicides," *Journal of Experimental Botany*, vol. 52, no. 358, pp. 971–979, 2001.

[141] A. E. Eltayeb, S. Yamamoto, M. E. E. Habora et al., "Greater protection against oxidative damages imposed by various environmental stresses in transgenic potato with higher level of reduced glutathione," *Breeding Science*, vol. 60, no. 2, pp. 101–109, 2010.

[142] T. Diplock, L. J. Machlin, L. Packer, and W. A. Pryor, "Vitamin E: biochemistry and health implications," *Annals of the New York Academy of Sciences*, vol. 570, pp. 372–378, 1989.

[143] K. Fukuzawa, A. Tokumura, S. Ouchi, and H. Tsukatani, "Antioxidant activities of tocopherols on Fe^{2+}-ascorbate-induced lipid peroxidation in lecithin liposomes," *Lipids*, vol. 17, no. 7, pp. 511–514, 1982.

[144] A. Kamal-Eldin and L. Å. Appelqvist, "The chemistry and antioxidant properties of tocopherols and tocotrienols," *Lipids*, vol. 31, no. 7, pp. 671–701, 1996.

[145] Y. Li, Y. Zhou, Z. Wang, X. Sun, and K. Tang, "Engineering tocopherol biosynthetic pathway in *Arabidopsis* leaves and its effect on antioxidant metabolism," *Plant Science*, vol. 178, no. 3, pp. 312–320, 2010.

[146] B. N. Ivanov and S. Khorobrykh, "Participation of photosynthetic electron transport in production and scavenging of reactive oxygen species," *Antioxidants and Redox Signaling*, vol. 5, no. 1, pp. 43–53, 2003.

[147] M. J. Fryer, "The antioxidant effect of thylakoid vitamin-E (α-tocopherol)," *Plant, Cell and Environment*, vol. 15, no. 4, pp. 381–392, 1992.

[148] V. E. Kagan, J. P. Fabisiak, and P. J. Quinn, "Coenzyme Q and vitamin E need each other as antioxidants," *Protoplasma*, vol. 214, no. 1-2, pp. 11–18, 2000.

[149] K. Yamaguchi-Shinozaki and K. Shinozaki, "A novel *cis* element in an *Arabidopsis* gene is involved in responsiveness to drought, low-temperature, or high-salt stress," *Plant Cell*, vol. 6, no. 2, pp. 251–264, 1994.

[150] S. Munné-Bosch, K. Schwarz, and L. Alegre, "Enhanced formation of α-tocopherol and highly oxidized abietane diterpenes in water-stressed rosemary plants," *Plant Physiology*, vol. 121, no. 3, pp. 1047–1052, 1999.

[151] J. Guo, X. Liu, X. Li, S. Chen, Z. Jin, and G. Liu, "Overexpression of VTE1 from *Arabidopsis* resulting in high vitamin E accumulation and salt stress tolerance increase in tobacco plant," *Chinese Journal of Applied and Environmental Biology*, vol. 12, no. 4, pp. 468–471, 2006.

[152] S. O. Bafeel and M. M. Ibrahim, "Antioxidants and accumulation of α-tocopherol induce chilling tolerance in *Medicago sativa*," *International Journal of Agriculture and Biology*, vol. 10, no. 6, pp. 593–598, 2008.

[153] S. Q. Ouyang, S. J. He, P. Liu, W. K. Zhang, J. S. Zhang, and S. Y. Chen, "The role of tocopherol cyclase in salt stress tolerance of rice (*Oryza sativa*)," *Science China Life Sciences*, vol. 54, no. 2, pp. 181–188, 2011.

[154] J. Young, "The photoprotective role of carotenoids in higher plants," *Physiologia Plantarum*, vol. 83, no. 4, pp. 702–708, 1991.

[155] D. Sieferman-Harms, "The light harvesting function of carotenoids in photosynthetic membrane," *Plant Physiology*, vol. 69, no. 3, pp. 561–568, 1987.

[156] F. Li, R. Vallabhaneni, J. Yu, T. Rocheford, and E. T. Wurtzel, "The maize phytoene synthase gene family: overlapping roles for carotenogenesis in endosperm, photomorphogenesis, and thermal stress tolerance," *Plant Physiology*, vol. 147, no. 3, pp. 1334–1346, 2008.

[157] R. Gomathi and P. Rakkiyapan, "Comparative lipid peroxidation, leaf membrane thermostability, and antioxidant system in four sugarcane genotypes differing in salt tolerance," *International Journal of Plant Physiology and Biochemistry*, vol. 3, no. 4, pp. 67–74, 2011.

[158] S. G. Grace and B. A. Logan, "Energy dissipation and radical scavenging by the plant phenylpropanoid pathway," *Philosophical Transactions of the Royal Society B*, vol. 355, no. 1402, pp. 1499–1510, 2000.

[159] A. Arora, T. M. Byrem, M. G. Nair, and G. M. Strasburg, "Modulation of liposomal membrane fluidity by flavonoids and isoflavonoids," *Archives of Biochemistry and Biophysics*, vol. 373, no. 1, pp. 102–109, 2000.

[160] A. Michalak, "Phenolic compounds and their antioxidant activity in plants growing under heavy metal stress," *Polish Journal of Environmental Studies*, vol. 15, no. 4, pp. 523–530, 2006.

[161] K. M. Janas, R. Amarowicz, J. Zielińska-Tomaszewska, A. Kosińska, and M. M. Posmyk, "Induction of phenolic compounds in two dark-grown lentil cultivars with different tolerance to copper ions," *Acta Physiologiae Plantarum*, vol. 31, no. 3, pp. 587–595, 2009.

[162] R. Lois and B. B. Buchanan, "Severe sensitivity to ultraviolet radiation in an *Arabidopsis* mutant deficient in flavonoid accumulation. II.Mechanisms of UV-resistance in *Arabidopsis*," *Planta*, vol. 194, no. 4, pp. 504–509, 1994.

[163] M. Lukaszewicz, I. Matysiak-Kata, J. Skala, I. Fecka, W. Cisowski, and J. Szopa, "Antioxidant capacity manipulation in transgenic potato tuber by changes in phenolic compounds content," *Journal of Agricultural and Food Chemistry*, vol. 52, no. 6, pp. 1526–1533, 2004.

[164] S. Sayfzadeh and M. Rashidi, "Response of antioxidant enzymes activities of sugar beet to drought stress," *ARPN Journal of Agricultural and Biological Science*, vol. 6, no. 4, pp. 27–33, 2011.

[165] C. Sgherri, B. Stevanovic, and F. Navari-Izzo, "Role of phenolic acids during dehydration and rehydration of *Ramonda serbica*," *Physiologia Plantarum*, vol. 122, no. 4, pp. 478–485, 2000.

[166] P. Mishra, B. Kumari, and R. S. Dubey, "Differential responses of antioxidative defense system to prolonged salinity stress in salt-tolerant and salt-sensitive Indica rice (*Oryza sativa* L.) seedlings," *Protoplasma*. In press.

[167] R. Valderrama, F. J. Corpas, A. Carreras et al., "The dehydrogenase-mediated recycling of NADPH is a key antioxidant system against salt-induced oxidative stress in olive plants," *Plant, Cell and Environment*, vol. 29, no. 7, pp. 1449–1459, 2006.

[168] R. Mittal and R. S. Dubey, "Behaviour of peroxidases in rice: changes in enzymatic activity and isoforms in relation to salt tolerance," *Plant Physiology and Biochemistry*, vol. 29, no. 1, pp. 31–40, 1991.

[169] M. J. Fryer, J. R. Andrews, K. Oxborough, D. A. Blowers, and N. R. Baker, "Relationship between CO_2 assimilation, photosynthetic electron transport, and active O_2 metabolism in leaves of maize in the field during periods of low temperature," *Plant Physiology*, vol. 116, no. 2, pp. 571–580, 1998.

[170] Y. Zhang, Y. Luo, Y. X. Hou, H. Jiang, Q. Chen, and R. H. Tang, "Chilling acclimation induced changes in the distribution of H_2O_2 and antioxidant system of strawberry leaves," *Agricultural Journal*, vol. 3, no. 4, pp. 286–291, 2008.

[171] I. Cakmak and W. J. Horst, "Effect of aluminium on lipid peroxidation, superoxide dismutase, catalase, and peroxidase activities in root tips of soybean (*Glycine max*)," *Physiologia Plantarum*, vol. 83, no. 3, pp. 463–468, 1991.

[172] M. V. Rao, G. Paliyath, and D. P. Ormrod, "Ultraviolet-B- and ozone-induced biochemical changes in antioxidant enzymes of *Arabidopsis thaliana*," *Plant Physiology*, vol. 110, no. 1, pp. 125–136, 1996.

[173] N. A. Ashry and H. I. Mohamed, "Impact of secondary metabolites and related enzymes in flax resistance and/or susceptibility to powdery mildew," *African Journal of Biotechnology*, vol. 11, no. 5, pp. 1073–1077, 2012.

[174] D. E. M. Radwan, K. A. Fayez, S. Y. Mahmoud, and G. Lu, "Modifications of antioxidant activity and protein composition of bean leaf due to *Bean yellow mosaic virus* infection and salicylic acid treatments," *Acta Physiologiae Plantarum*, vol. 32, no. 5, pp. 891–904, 2010.

[175] J. G. Scandalios, "Oxygen stress and superoxide dismutases," *Plant Physiology*, vol. 101, no. 1, pp. 7–12, 1993.

[176] I. Fridovich, "Superoxide dismutases. An adaptation to a paramagnetic gas," *Journal of Biological Chemistry*, vol. 264, no. 14, pp. 7761–7764, 1989.

[177] M. L. Racchi, F. Bagnoli, I. Balla, and S. Danti, "Differential activity of catalase and superoxide dismutase in seedlings and *in vitro* micropropagated oak (Quercus robur L.)," *Plant Cell Reports*, vol. 20, no. 2, pp. 169–174, 2001.

[178] C. Bowler, M. Van Montagu, and D. Inzé, "Superoxide dismutase and stress tolerance," *Annual Review of Plant Physiology and Plant Molecular Biology*, vol. 43, no. 1, pp. 83–116, 1992.

[179] C. Jackson, J. Dench, A. L. Moore, B. Halliwell, C. H. Foyer, and D. O. Hall, "Subcellular localisation and identification of superoxide dismutase in the leaves of higher plants," *European Journal of Biochemistry*, vol. 91, no. 2, pp. 339–344, 1978.

[180] S. Kanematsu and K. Asada, "Cuzn-superoxide dismutases in rice: occurrence of an active, monomeric enzyme and two types of isozyme in leaf and non-photosynthetic tissues," *Plant and Cell Physiology*, vol. 30, no. 3, pp. 381–391, 1989.

[181] P. Bueno, J. Varela, G. Gimenez-Gallego, and L. A. Del Rio, "Peroxisomal copper, zinc superoxide dismutase. Characterization of the isoenzyme from watermelon cotyledons," *Plant Physiology*, vol. 108, no. 3, pp. 1151–1160, 1995.

[182] L. A. Del Río, G. M. Pastori, J. M. Palma et al., "The activated oxygen role of peroxisomes in senescence," *Plant Physiology*, vol. 116, no. 4, pp. 1195–1200, 1998.

[183] A. S. Gupta, J. L. Heinen, A. S. Holaday, J. J. Burke, and R. D. Allen, "Increased resistance to oxidative stress in transgenic plants that overexpress chloroplastic Cu/Zn superoxide dismutase," *Proceedings of the National Academy of Sciences of the United States of America*, vol. 90, no. 4, pp. 1629–1633, 1993.

[184] G. Scandalios, L. Guan, and A. N. Polidoros, "Catalases in plants: gene structure, properties, regulation and expression,"

in *Oxidative Stress and the Molecular Biology of Antioxidants Defenses*, J. G. Scandalios, Ed., pp. 343–406, Cold Spring Harbor Laboratory Press, New York, NY, USA, 1997.

[185] F. J. Corpas, J. M. Palma, L. M. Sandalio, R. Valderrama, J. B. Barroso, and L. A. del Río, "Peroxisomal xanthine oxidoreductase: characterization of the enzyme from pea (*Pisum sativum* L.) leaves," *Journal of Plant Physiology*, vol. 165, no. 13, pp. 1319–1330, 2008.

[186] A. Mhamdi, G. Queval, S. Chaouch, S. Vanderauwera, F. Van Breusegem, and G. Noctor, "Catalase function in plants: a focus on *Arabidopsis* mutants as stress-mimic models," *Journal of Experimental Botany*, vol. 61, no. 15, pp. 4197–4220, 2010.

[187] H. Willekens, D. Inze, M. Van Montagu, and W. Van Camp, "Catalases in plants," *Molecular Breeding*, vol. 1, no. 3, pp. 207–228, 1995.

[188] N. Mallick and F. H. Mohn, "Reactive oxygen species: response of algal cells," *Journal of Plant Physiology*, vol. 157, no. 2, pp. 183–193, 2000.

[189] R. Moussa and S. M Abdel-Aziz, "Comparative response of drought tolerant and drought sensitive maize genotypes to water stress," *Australian Journal of Crop Sciences*, vol. 1, no. 1, pp. 31–36, 2008.

[190] H. Willekens, S. Chamnongpol, M. Davey et al., "Catalase is a sink for H_2O_2 and is indispensable for stress defence in C-3 plants," *EMBO Journal*, vol. 16, no. 16, pp. 4806–4816, 1997.

[191] Z. Guan, T. Chai, Y. Zhang, J. Xu, and W. Wei, "Enhancement of Cd tolerance in transgenic tobacco plants overexpressing a Cd-induced catalase cDNA," *Chemosphere*, vol. 76, no. 5, pp. 623–630, 2009.

[192] D. J. Schuller, N. Ban, R. B. Van Huystee, A. McPherson, and T. L. Poulos, "The crystal structure of peanut peroxidase," *Structure*, vol. 4, no. 3, pp. 311–321, 1996.

[193] K. Asada, "Ascorbate peroxidase: a hydrogen peroxide scavenging enzyme in plants," *Physiologia Plantarum*, vol. 85, no. 2, pp. 235–241, 1992.

[194] K. Kobayashi, Y. Kumazawa, K. Miwa, and S. Yamanaka, "ε-(γ-Glutamyl)lysine cross-links of spore coat proteins and transglutaminase activity in *Bacillus subtilis*," *FEMS Microbiology Letters*, vol. 144, no. 2-3, pp. 157–160, 1996.

[195] J. Vangronsveld and H. Clijsters, "Toxic effects of metals," in *Plants and the Chemical Elements. Biochemistry, Uptake, Tolerance and Toxicity*, M. E. Farago, Ed., pp. 150–177, VCH Publishers, Weinheim, Germany, 1994.

[196] K. Radotić, T. Dučić, and D. Mutavdžić, "Changes in peroxidase activity and isoenzymes in spruce needles after exposure to different concentrations of cadmium," *Environmental and Experimental Botany*, vol. 44, no. 2, pp. 105–113, 2000.

[197] H. Tayefi-Nasrabadi, G. Dehghan, B. Daeihassani, A. Movafegi, and A. Samadi, "Some biochemical properties of guaiacol peroxidases as modified by salt stress in leaves of salt-tolerant and salt-sensitive safflower (*Carthamus tinctorius* L.cv.) cultivars," *African Journal of Biotechnology*, vol. 10, no. 5, pp. 751–763, 2011.

[198] A. Jiménez, J. A. Hernández, L. A. Del Río, and F. Sevilla, "Evidence for the presence of the ascorbate-glutathione cycle in mitochondria and peroxisomes of pea leaves," *Plant Physiology*, vol. 114, no. 1, pp. 275–284, 1997.

[199] J. E. Pallanca and N. Smirnoff, "The control of ascorbic acid synthesis and turnover in pea seedlings," *Journal of Experimental Botany*, vol. 51, no. 345, pp. 669–674, 2000.

[200] K. G. Welinder, "Superfamily of plant, fungal and bacterial peroxidases," *Current Opinion in Structural Biology*, vol. 2, no. 3, pp. 388–393, 1992.

[201] W. R. Patterson and T. L. Poulos, "Crystal structure of recombinant pea cytosolic ascorbate peroxidase," *Biochemistry*, vol. 34, no. 13, pp. 4331–4341, 1995.

[202] R. Madhusudhan, T. Ishikawa, Y. Sawa, S. Shigeoka, and H. Shibata, "Characterization of an ascorbate peroxidase in plastids of tobacco BY-2 cells," *Physiologia Plantarum*, vol. 117, no. 4, pp. 550–557, 2003.

[203] P. Sharma and R. S. Dubey, "Ascorbate peroxidase from rice seedlings: properties of enzyme isoforms, effects of stresses and protective roles of osmolytes," *Plant Science*, vol. 167, no. 3, pp. 541–550, 2004.

[204] Y. Nakano and K. Asada, "Purification of ascorbate peroxidase in spinach chloroplasts; its inactivation in ascorbate-depleted medium and reactivation by monodehydroascorbate radical," *Plant and Cell Physiology*, vol. 28, no. 1, pp. 131–140, 1987.

[205] R. Mittler and B. A. Zilinskas, "Molecular cloning and characterization of a gene encoding pea cytosolic ascorbate peroxidase," *Journal of Biological Chemistry*, vol. 267, no. 30, pp. 21802–21807, 1992.

[206] T. Ishikawa, K. Yoshimura, K. Sakai, M. Tamoi, T. Takeda, and S. Shigeoka, "Molecular characterization and physiological role of a glyoxysome-bound ascorbate peroxidase from spinach," *Plant and Cell Physiology*, vol. 39, no. 1, pp. 23–34, 1998.

[207] J. Wang, H. Zhang, and R. D. Allen, "Overexpression of an *Arabidopsis* peroxisomal ascorbate peroxidase gene in tobacco increases protection against oxidative stress," *Plant and Cell Physiology*, vol. 40, no. 7, pp. 725–732, 1999.

[208] Y. C. Boo and J. Jung, "Water deficit - Induced oxidative stress and antioxidative defenses in rice plants," *Journal of Plant Physiology*, vol. 155, no. 2, pp. 255–261, 1999.

[209] Y. Wang, M. Wisniewski, R. Meilan, M. Cui, R. Webb, and L. Fuchigami, "Overexpression of cytosolic ascorbate peroxidase in tomato confers tolerance to chilling and salt stress," *Journal of the American Society for Horticultural Science*, vol. 130, no. 2, pp. 167–173, 2005.

[210] Y. Yabuta, T. Motoki, K. Yoshimura, T. Takeda, T. Ishikawa, and S. Shigeoka, "Thylakoid membrane-bound ascorbate peroxidase is a limiting factor of antioxidative systems under photo-oxidative stress," *Plant Journal*, vol. 32, no. 6, pp. 915–925, 2002.

[211] T. Ushimaru, Y. Maki, S. Sano, K. Koshiba, K. Asada, and H. Tsuji, "Induction of Enzymes Involved in the Ascorbate-Dependent Antioxidative System, Namely, Ascorbate Peroxidase, Monodehydroascorbate Reductase and Dehydroascorbate Reductase, after Exposure to Air of Rice (*Oryza sativa*) Seedlings Germinated under Water," *Plant and Cell Physiology*, vol. 38, no. 5, pp. 541–549, 1997.

[212] M. A. Hossain and K. Asada, "Monodehydroascorbate reductase from cucumber is a flavin adenine dinucleotide enzyme," *Journal of Biological Chemistry*, vol. 260, no. 24, pp. 12920–12926, 1985.

[213] Y. Sakihama, J. Mano, S. Sano, K. Asada, and H. Yamasaki, "Reduction of phenoxyl radicals mediated by monodehydroascorbate reductase," *Biochemical and Biophysical Research Communications*, vol. 279, no. 3, pp. 949–954, 2000.

[214] M. A. Hossain, Y. Nakano, and K. Asada, "Monodehydroascorbate reductase in spinach chloroplasts and its participation in regeneration of ascorbate for scavenging hydrogen peroxide," *Plant and Cell Physiology*, vol. 25, no. 3, pp. 385–395, 1984.

[215] D. A. Dalton, L. M. Baird, L. Langeberg et al., "Subcellular localization of oxygen defense enzymes in soybean (*Glycine*

max [L.] Merr.) root nodules," *Plant Physiology*, vol. 102, no. 2, pp. 481–489, 1993.

[216] C. Miyake, U. Schreiber, H. Hormann, S. Sano, and K. Asada, "The FAD-enzyme monodehydroascorbate radical reductase mediates photoproduction of superoxide radicals in spinach thylakoid membranes," *Plant and Cell Physiology*, vol. 39, no. 8, pp. 821–829, 1998.

[217] A. E. Eltayeb, N. Kawano, G. H. Badawi et al., "Overexpression of monodehydroascorbate reductase in transgenic tobacco confers enhanced tolerance to ozone, salt and polyethylene glycol stresses," *Planta*, vol. 225, no. 5, pp. 1255–1264, 2007.

[218] F. Li, Q. Y. Wu, Y. L. Sun, L. Y. Wang, X. H. Yang, and Q. W. Meng, "Overexpression of chloroplastic monodehydroascorbate reductase enhanced tolerance to temperature and methyl viologen-mediated oxidative stresses," *Physiologia Plantarum*, vol. 139, no. 4, pp. 421–434, 2010.

[219] Z. Chen, T. E. Young, J. Ling, S. C. Chang, and D. R. Gallie, "Increasing vitamin C content of plants through enhanced ascorbate recycling," *Proceedings of the National Academy of Sciences of the United States of America*, vol. 100, no. 6, pp. 3525–3530, 2003.

[220] A. Qin, Q. Shi, and X. Yu, "Ascorbic acid contents in transgenic potato plants overexpressing two dehydroascorbate reductase genes," *Molecular Biology Reports*, vol. 38, no. 3, pp. 1557–1566, 2011.

[221] M. A. Hossain and K. Asada, "Purification of dehydroascorbate reductase from spinach and its characterization as a thiol enzyme," *Plant and Cell Physiology*, vol. 25, no. 1, pp. 85–92, 1984.

[222] S. Dipierro and G. Borraccino, "Dehydroascorbate reductase from potato tubers," *Phytochemistry*, vol. 30, no. 2, pp. 427–429, 1991.

[223] S. Yoshida, M. Tamaoki, T. Shikano et al., "Cytosolic dehydroascorbate reductase is important for ozone tolerance in *Arabidopsis thaliana*," *Plant and Cell Physiology*, vol. 47, no. 2, pp. 304–308, 2006.

[224] M. C. Rubio, P. Bustos-Sanmamed, M. R. Clemente, and M. Becana, "Effects of salt stress on the expression of antioxidant genes and proteins in the model legume Lotus japonicus," *New Phytologist*, vol. 181, no. 4, pp. 851–859, 2009.

[225] A. E. Eltayeb, S. Yamamoto, M. E.E. Habora, L. Yin, H. Tsujimoto, and K. Tanaka, "Transgenic potato overexpressing *Arabidopsis* cytosolic *AtDHAR1* showed higher tolerance to herbicide, drought and salt stresses," *Breeding Science*, vol. 61, no. 1, pp. 3–10, 2011.

[226] S. Ghisla and V. Massey, "Mechanisms of flavoprotein-catalyzed reactions," *European Journal of Biochemistry*, vol. 181, no. 1, pp. 1–17, 1989.

[227] E. A. Edwards, S. Rawsthorne, and P. M. Mullineaux, "Subcellular distribution of multiple forms of glutathione reductase in leaves of pea (*Pisum sativum* L.)," *Planta*, vol. 180, no. 2, pp. 278–284, 1990.

[228] G. M. Pastori and V. S. Trippi, "Oxidative stress induces high rate of glutathione reductase synthesis in a drought-resistant maize strain," *Plant and Cell Physiology*, vol. 33, no. 7, pp. 957–961, 1992.

[229] D. -F. Shu, L. -Y. Wang, M. Duan, Y. -S. Deng, and Q. -W. Meng, "Antisense-mediated depletion of tomato chloroplast glutathione reductase enhances susceptibility to chilling stress," *Plant Physiology and Biochemistry*, vol. 49, no. 10, pp. 1228–1237, 2011.

[230] M. Aono, A. Kubo, H. Saji, K. Tanaka, and N. Kondo, "Enhanced tolerance to photooxidative stress of transgenic *Nicotiana tabacum* with high chloroplastic glutathione reductase activity," *Plant and Cell Physiology*, vol. 34, no. 1, pp. 129–135, 1993.

[231] M. J. Tseng, C. W. Liu, and J. C. Yiu, "Tolerance to sulfur dioxide in transgenic Chinese cabbage transformed with both the superoxide dismutase containing manganese and catalase genes of *Escherichia coli*," *Scientia Horticulturae*, vol. 115, no. 2, pp. 101–110, 2008.

[232] S. C. Lee, S. Y. Kwon, and S. R. Kim, "Ectopic expression of a cold-responsive CuZn superoxide dismutase gene, SodCc1, in transgenic rice (*Oryza sativa* L.)," *Journal of Plant Biology*, vol. 52, no. 2, pp. 154–160, 2009.

[233] M. Aono, H. Saji, A. Sakamoto, K. Tanaka, N. Kondo, and Tanaka, "Paraquat tolerance of transgenic *Nicotiana tabacum* with enhanced activities of glutathione reductase and superoxide dismutase," *Plant and Cell Physiology*, vol. 36, no. 8, pp. 1687–1691, 1995.

[234] S. Y. Kwon, Y. J. Jeong, H. S. Lee et al., "Enhanced tolerances of transgenic tobacco plants expressing both superoxide dismutase and ascorbate peroxidase in chloroplasts against methyl viologen-mediated oxidative stress," *Plant, Cell and Environment*, vol. 25, no. 7, pp. 873–882, 2002.

[235] S. Lim, Y. H. Kim, S. H. Kim et al., "Enhanced tolerance of transgenic sweetpotato plants that express both CuZnSOD and APX in chloroplasts to methyl viologen-mediated oxidative stress and chilling," *Molecular Breeding*, vol. 19, no. 3, pp. 227–239, 2007.

[236] S. S. Kwak, S. Lim, L. Tang, S. Y. Kwon, and H. S. Lee, "Enhanced tolerance of transgenic crops expressing both SOD and APX in chloroplasts to multiple environmental stress," in *Salinity and Water Stress*, M. Ashraf, M. Ozturk, and H. R. Athar, Eds., pp. 197–203, Springer, Netherland, 2009.

[237] K. Asada, "The water-water cycle in chloroplasts: scavenging of active oxygens and dissipation of excess photons," *Annual Review of Plant Biology*, vol. 50, pp. 601–639, 1999.

[238] K. Biehler and H. Fock, "Evidence for the contribution of the Mehler-peroxidase reaction in dissipating excess electrons in drought-stressed wheat," *Plant Physiology*, vol. 112, no. 1, pp. 265–272, 1996.

[239] C. H. Foyer and G. Noctor, "Oxygen processing in photosynthesis: regulation and signalling," *New Phytologist*, vol. 146, no. 3, pp. 359–388, 2000.

[240] R. K. Sairam, P. S. Deshmukh, and D. C. Saxena, "Role of antioxidant systems in wheat genotypes tolerance to water stress," *Biologia Plantarum*, vol. 41, no. 3, pp. 387–394, 1998.

[241] S. Wang, D. Liang, C. Li, Y. Hao, F. Ma, and H. Shu, "Influence of drought stress on the cellular ultrastructure and antioxidant system in leaves of drought-tolerant and drought-sensitive apple rootstocks," *Plant Physiology and Biochemistry*, vol. 51, pp. 81–89, 2012.

[242] J. A. Hernández, A. Jiménez, P. Mullineaux, and F. Sevilla, "Tolerance of pea (*Pisum sativum* L.) to long-term salt stress is associated with induction of antioxidant defences," *Plant, Cell and Environment*, vol. 23, no. 8, pp. 853–862, 2000.

[243] U. Perez-Lopez, A. Robredo, M. Lacuesta et al., "The oxidative stress caused by salinity in two barley cultivars is mitigated by elevated CO_2," *Physiologia Plantarum*, vol. 135, no. 1, pp. 29–42, 2009.

[244] N. Karray-Bouraoui, F. Harbaoui, M. Rabhi et al., "Different antioxidant responses to salt stress in two different

provenances of *Carthamus tinctorius* L.," *Acta Physiologiae Plantarum*, vol. 33, no. 4, pp. 1435–1444, 2011.

[245] X. Wang, P. Yang, Q. Gao et al., "Proteomic analysis of the response to high-salinity stress in Physcomitrella patens," *Planta*, vol. 228, no. 1, pp. 167–177, 2008.

[246] H. Vaidyanathan, P. Sivakumar, R. Chakrabarty, and G. Thomas, "Scavenging of reactive oxygen species in NaCl-stressed rice (*Oryza sativa* L.) - Differential response in salt-tolerant and sensitive varieties," *Plant Science*, vol. 165, no. 6, pp. 1411–1418, 2003.

[247] A. Shalata, V. Mittova, M. Volokita, M. Guy, and M. Tal, "Response of the cultivated tomato and its wild salt-tolerant relative *Lycopersicon pennellii* to salt-dependent oxidative stress: the root antioxidative system," *Physiologia Plantarum*, vol. 112, no. 4, pp. 487–494, 2001.

[248] A. Hediye Sekmen, I. Türkan, and S. Takio, "Differential responses of antioxidative enzymes and lipid peroxidation to salt stress in salt-tolerant Plantago maritima and salt-sensitive Plantago media," *Physiologia Plantarum*, vol. 131, no. 3, pp. 399–411, 2007.

[249] B. A. Logan, D. Kornyeyev, J. Hardison, and A. S. Holaday, "The role of antioxidant enzymes in photoprotection," *Photosynthesis Research*, vol. 88, no. 2, pp. 119–132, 2006.

[250] Y. H. Zhou, J. Q. Yu, W. H. Mao, L. F. Huang, X. S. Song, and S. Nogués, "Genotypic variation of Rubisco expression, photosynthetic electron flow and antioxidant metabolism in the chloroplasts of chill-exposed cucumber plants," *Plant and Cell Physiology*, vol. 47, no. 2, pp. 192–199, 2006.

[251] T. K. Prasad, "Role of catalase in inducing chilling tolerance in pre-emergent maize seedlings," *Plant Physiology*, vol. 114, no. 4, pp. 1369–1376, 1997.

[252] Y. Zhang, H. R. Tang, and Y. Luo, "Variation in antioxidant enzyme activities of two strawberry cultivars with short-term low temperature stress," *World Journal of Agricultural Sciences*, vol. 4, no. 4, pp. 458–462, 2008.

[253] L. S. Jahnke, M. R. Hull, and S. P. Long, "Chilling stress and oxygen metabolizing enzymes in *Zea mays and Zea diploperennis*," *Plant, Cell and Environment*, vol. 14, no. 1, pp. 97–104, 1991.

[254] D. M. Hodges, C. J. Andrews, D. A. Johnson, and R. I. Hamilton, "Antioxidant compound responses to chilling stress in differentially sensitive inbred maize lines," *Physiologia Plantarum*, vol. 98, no. 4, pp. 685–692, 1996.

[255] M. Huang and Z. Guo, "Responses of antioxidative system to chilling stress in two rice cultivars differing in sensitivity," *Biologia Plantarum*, vol. 49, no. 1, pp. 81–84, 2005.

[256] D. E. Salt, M. Blaylock, N. P. B. A. Kumar et al., "Phytoremediation: a novel strategy for the removal of toxic metals from the environment using plants," *Biotechnology*, vol. 13, no. 5, pp. 468–474, 1995.

[257] S. Mishra and R. S. Dubey, "Heavy metal toxicity induced alterations in photosynthetic metabolism in plants," in *Handbook of Photosynthesis*, M. Pessarakli, Ed., pp. 845–863, CRC Press, Taylor and Francis Publishing Company, Fla, USA, 2nd edition, 2005.

[258] S. Gallego, M. Benavides, and M. Tomaro, "Involvement of an antioxidant defence system in the adaptive response to heavy metal ions in *Helianthus annuus* L. cells," *Plant Growth Regulation*, vol. 36, no. 3, pp. 267–273, 2002.

[259] F. Vinit-Dunand, D. Epron, B. Alaoui-Sossé, and P. M. Badot, "Effects of copper on growth and on photosynthesis of mature and expanding leaves in cucumber plants," *Plant Science*, vol. 163, no. 1, pp. 53–58, 2002.

[260] M. Moustakas, T. Lanaras, L. Symeonidis, and S. Karataglis, "Growth and some photosynthetic characteristics of field grown *Avena sativa* under copper and lead stress," *Photosynthetica*, vol. 30, no. 3, pp. 389–396, 1994.

[261] A. Źróbek-Sokolnik, H. Asard, K. Górska-Koplińska, and R. J. Górecki, "Cadmium and zinc-mediated oxidative burst in tobacco BY-2 cell suspension cultures," *Acta Physiologiae Plantarum*, vol. 31, no. 1, pp. 43–49, 2009.

[262] S. M. Gallego, M. P. Benavides, and M. L. Tomaro, "Effect of heavy metal ion excess on sunflower leaves: evidence for involvement of oxidative stress," *Plant Science*, vol. 121, no. 2, pp. 151–159, 1996.

[263] J. E. J. Weckx and H. M. M. Clijsters, "Oxidative damage and defense mechanisms in primary leaves of *Phaseolus vulgaris* as a result of root assimilation of toxic amounts of copper," *Physiologia Plantarum*, vol. 96, no. 3, pp. 506–512, 1996.

[264] Y. Yamamoto, A. Hachiya, and H. Matsumoto, "Oxidative damage to membranes by a combination of aluminum and iron in suspension-cultured tobacco cells," *Plant and Cell Physiology*, vol. 38, no. 12, pp. 1333–1339, 1997.

[265] S. S. Sharma and K. J. Dietz, "The relationship between metal toxicity and cellular redox imbalance," *Trends in Plant Science*, vol. 14, no. 1, pp. 43–50, 2009.

[266] L. M. Sandalio, M. Rodríguez-Serrano, L. A. del Río, and M. C. Romero-Puertas, "Reactive oxygen species and signaling in cadmium toxicity," in *Reactive Oxygen Species in Plant Signaling*, L. A. Rio and A. Puppo, Eds., pp. 175–189, Springer, Berlin, Germany, 2009.

[267] A. Giannakoula, M. Moustakas, T. Syros, and T. Yupsanis, "Aluminum stress induces up-regulation of an efficient antioxidant system in the Al-tolerant maize line but not in the Al-sensitive line," *Environmental and Experimental Botany*, vol. 67, no. 3, pp. 487–494, 2010.

[268] S. Srivastava, A. K. Srivastava, P. Suprasanna, and S. F. D'souza, "Comparative antioxidant profiling of tolerant and sensitive varieties of *Brassica juncea* L. to arsenate and arsenite exposure," *Bulletin of Environmental Contamination and Toxicology*, vol. 84, no. 3, pp. 342–346, 2010.

[269] M. Blumthaler and W. Ambach, "Indication of increasing solar ultraviolet-B radiation flux in alpine regions," *Science*, vol. 248, no. 4952, pp. 206–208, 1990.

[270] D. J. Allen, I. F. Mckee, P. K. Farage, and N. R. Baker, "Analysis of limitations to CO_2 assimilation on exposure of leaves of two *Brassica napus* cultivars to UV-B," *Plant, Cell and Environment*, vol. 20, no. 5, pp. 633–640, 1997.

[271] J. He, L. K. Huang, W. S. Chow, M. L. Whitecross, and J. M Anderson, "Effects of supplementary ultraviolet-B radiation on rice and pea plants," *Australian Journal of Plant Physiology*, vol. 20, no. 2, pp. 129–142, 1993.

[272] A. Strid, W. S. Chow, and J. M. Anderson, "UV-B damage and protection at the molecular level in plants," *Photosynthesis Research*, vol. 39, no. 3, pp. 475–489, 1994.

[273] N. Doke, "Generation of superoxide anion by potato tuber protoplasts during the hypersensitive response to hyphal wall components of Phytophthora infestans and specific inhibition of the reaction by suppressors of hypersensitivity," *Physiological Plant Pathology*, vol. 23, no. 3, pp. 359–367, 1983.

[274] J. J. Grant, B. W. Yun, and G. J. Loake, "Oxidative burst and cognate redox signalling reported by luciferase imaging: identification of a signal network that functions independently of ethylene, SA and Me-JA but is dependent on MAPKK activity," *Plant Journal*, vol. 24, no. 5, pp. 569–582, 2000.

[275] J. M. Chittoor, J. E. Leach, and F. F. White, "Differential induction of a peroxidase gene family during infection of rice by *Xanthomonas oryzae pv. oryzae*," *Molecular Plant-Microbe Interactions*, vol. 10, no. 7, pp. 861–871, 1997.

[276] K. Sasaki, T. Iwai, S. Hiraga et al., "Ten rice peroxidases redundantly respond to multiple stresses including infection with rice blast fungus," *Plant and Cell Physiology*, vol. 45, no. 10, pp. 1442–1452, 2004.

[277] H. Abdollahi and Z. Ghahremani, "The role of chloroplasts in the interaction between *Erwinia amylovora* and host plants," *Acta Horticulturae*, vol. 896, pp. 215–221, 2011.

[278] Y. Liu, D. Ren, S. Pike, S. Pallardy, W. Gassmann, and S. Zhang, "Chloroplast-generated reactive oxygen species are involved in hypersensitive response-like cell death mediated by a mitogen-activated protein kinase cascade," *Plant Journal*, vol. 51, no. 6, pp. 941–954, 2007.

[279] R. Mittler, E. H. Herr, B. L. Orvar et al., "Transgenic tobacco plants with reduced capability to detoxify reactive oxygen intermediates are hyperresponsive to pathogen infection," *Proceedings of the National Academy of Sciences of the United States of America*, vol. 96, no. 24, pp. 14165–14170, 1999.

5

Aluminium Toxicity Targets in Plants

Sónia Silva

CESAM and Department of Biology, University of Aveiro, 3810-193 Aveiro, Portugal

Correspondence should be addressed to Sónia Silva, soniasilva@ua.pt

Academic Editor: Helena Oliveira

Aluminium (Al) is the third most abundant metallic element in soil but becomes available to plants only when the soil pH drops below 5.5. At those conditions, plants present several signals of Al toxicity. As reported by literature, major consequences of Al exposure are the decrease of plant production and the inhibition of root growth. The root growth inhibition may be directly/indirectly responsible for the loss of plant production. In this paper the most remarkable symptoms of Al toxicity in plants and the latest findings in this area are addressed. Root growth inhibition, ROS production, alterations on root cell wall and plasma membrane, nutrient unbalances, callose accumulation, and disturbance of cytoplasmic Ca^{2+} homeostasis, among other signals of Al toxicity are discussed, and, when possible, the behavior of Al-tolerant versus Al-sensitive genotypes under Al is compared.

1. Introduction

Aluminium (Al) ranks third in abundance among the Earth's crust elements, after oxygen and silicon, and is the most abundant metallic element. A large amount of Al is incorporated into aluminosilicate soil minerals, and very small quantities appear in the soluble form, capable of influencing biological systems [1].

Al bioavailability, and in consequence, toxicity, is mainly restricted to acid environments. Acid soils (with a pH of 5.5 or lower) are among the most important limitations to agricultural production. The production of staple food crops, in particular grain crops, is negatively influenced by acid soils [2]. Some agricultural practices, as removal of products from the farm, leaching of nitrogen below the plant root zone, inappropriate use of nitrogenous fertilizers, and build-up in organic matter, are causing further acidification of agricultural soils.

When pH drops below 5.5, aluminosilicate clays and aluminium hydroxide minerals begin to dissolve, releasing aluminium-hydroxy cations and $Al(H2O)_6^{3+}$ (Al^{3+}), that then exchange with other cations. On that conditions, Al^{3+} also forms the mononuclear species $AlOH^{2+}$, $Al(OH)_2^{+}$, $Al(OH)_3$, and $Al(OH)_4$ [3]. The mononuclear Al^{3+} species and Al_{13} are considered as the most toxic forms [4, 5].

Although some crops (e.g., pineapple, tea) are considered tolerant to high levels of exchangeable Al, for most crops it is a serious constraint. Species and genotypes within species greatly differ in their tolerance to Al. For most crops, fertilization and attempts of soil correction (e.g., liming) may not be enough per se to reduce Al toxicity (e.g., as the soil reaction remains strongly acid), and in most target countries these strategies may also be jeopardized by economical constrains [6]. Therefore, it is imperative to fully understand the mechanisms that are used by the Al-tolerant species to cope Al toxicity, as well which genotypes, within the most resistant/tolerant cereal species, are more suitable to grow in acidic soils in order to increase world cereal production. Furthermore, the development of new cultivars (or the reinvestment in ancient genotypes from Al rich regions) with increased Al-tolerance is fundamental and economic solution to increase world food production.

2. Aluminium Toxicity

2.1. Root Growth. A major consequence of Al toxicity is the inhibition of root growth, and this outcome has been reported during the last century (e.g., [7]) for innumerous species [8–15]. Consequently, root growth inhibition has been widely used to assess Al toxicity.

Root growth is the combination of cell division and elongation. Only during the last decade, researchers started to look at the cell cycle (de)regulation induced by Al, with some works focusing unbalances on mitosis phase and very few on other interphase phases (e.g., [15]). Decrease of mitotic activity was reported as a consequence of Al exposure in root tips of several species as wheat [16, 17], maize [18, 19], barley [20], and bean. [19]. Some authors defended that inhibition of cell elongation was the primary mechanism leading to root growth inhibition [21, 22]. The reason for that is that root growth inhibition could occur within a short time period—30 min in Al-sensitive maize [23]—and that cell division is a slow process (cell cycle takes usually several hours to be completed). However, Doncheva et al. [18] reported inhibition of cell division (decrease of S-phase cells) in the proximal meristem after 5 min Al exposure and inhibition of root cell division in the apical meristem within 10 or 30 minutes. Furthermore, Al can accumulate in the nuclei of cells in the meristematic region of the root tip within 30 minutes [15]. Therefore, whereas inhibition of cell elongation or cell division is the primary mechanism leading to root growth inhibition is still unclear. More recently, Yi et al. [24] reported that Al exposure led to abnormal progress through mitosis and induced micronuclei formation in *Vicia faba* roots, which is in agreement with Al-induced chromosome aberrations found in wheat roots [25] and Al-induced chromosome stickiness and breaks in *Oryza sativa* [26]. From the literature review, it is evident that Al leads to cell cycle unbalances, but many questions still remain to clarify. For example it still remains unclear how and where Al exerts its influence throughout the cell cycle, if these changes are species and region dependent (most studies are performed in root apices), how the putative changes are exerted through time, and/or if they may be reversible after Al removal.

The root growth inhibition and increase in root diameter observed in roots exposed to Al [27] suggested that plant cytoskeleton could be a cellular target of Al phytotoxicity [28]. Blancaflor et al. [28] and Horst et al. [29] studied Al-induced effects on microtubules and actin microfilaments and showed that microtubules and microfilaments are altered, in their stability, organization, and polymerization, when exposed to Al. Also, in *Triticum turgidum* Al treatment led to disorganization of actin filaments and formation of actin deposits [30]. Zhang et al. [31] showed that Al inhibited actin and profilin genes. Profilin, as an actin-binding protein, provides cells with the ability to remodel the cytoskeleton [32]. In *Arabidopsis thaliana* a decrease in profilin expression resulted in an elongation defect [33]. Furthermore, Sivaguru et al. [34] and Čiamporová [21] showed that organization of cytoskeleton is most sensitive in the distal transition zone of the root apex, providing evidence that this zone represents a potential target with respect to Al toxicity.

The most sensitive root zone to Al toxicity is under great attention. Earlier, it was hypothesized that root cap played a major role in the mechanism of Al toxicity/protection [35]. However, Ryan et al. [9] demonstrated that the removal of the root cap had no effect on the Al-induced inhibition of root growth in maize. Furthermore, the same authors

also suggested that the meristem is the primary site of Al toxicity. Later, Sivaguru and Horst [36], applying Al to 1 mm root segments, reported that Al accumulation in the distal transition zone (DTZ: 1-2 mm) led to a rapid inhibition of the root elongation and suggested that this root zone is the primary target of Al in an Al-sensitive maize cultivar.

2.2. Oxidative Stress. Al-induced oxidative stress and changes in cell wall properties have been suggested as the two major factors leading to Al toxicity [22, 37]. Oxidative stress occurs when any condition disrupts the cellular redox homeostasis. The reactive oxygen species (ROS) have the capacity to oxidize cellular components such as lipids, proteins, enzymes, and nucleic acids, leading to cell death. Metals are known to act as catalysts in ROS production and to induce oxidative damage in plants. Al itself is not a transition metal and cannot catalyze redox reactions; however, Al exposure leads to oxidative stress [37–43]. Because aluminium ions form electrostatic bonds preferentially with oxygen donor ligands (e.g., carboxylate or phosphate groups), cell wall pectin and the outer surface of the plasma membrane seem to be major targets of aluminium [37]. Al binding to biomembranes leads to rigidification [44], which seems to facilitate the radical chain reactions by iron (Fe) ions and enhance the peroxidation of lipids [38].

Al induction lipid peroxidation has been reported for some species, including barley [45], sorghum [46], triticale [42], rice [40], greengram [47], and wheat [48]. Yamamoto et al. [37] found that, for *Pisum sativum* seedlings treated with Al in a simple Ca solution, Al accumulation, lipid peroxidation, and callose production had a similar distribution on the root apex surface and were accompanied by root growth inhibition. However, the loss of membrane integrity was only detected at the periphery of the cracks on the surface of the root apex. Furthermore, Yamamoto et al. [38] concluded that the Al enhancement of lipid peroxidation is an early symptom of Al accumulation and appears to cause partly callose production, but not root growth inhibition. Later, however, in maize, Al treatment did not induce lipid peroxidation, indicating that lipids are not the primary cellular target of oxidative stress in maize [39]. So, it seems that cellular target of oxidative stress depends on plant species.

Plant cells are equipped with a defensive system composed by enzymatic antioxidants such as catalase (CAT), ascorbate peroxidase (APX), guaiacol peroxidase (G-POX), superoxide dismutase (SOD), monodehydroascorbate reductase (MDHAR), dehydroascorbate reductase (DHAR), glutathione-S-transferase (GST), and gluthatione reductase (GR) and nonenzymatic antioxidants such as ascorbate (AsA), glutathione (GSH), α-tocopherol, and carotenoids that help to detoxify the ROS. Some works reported ROS production and alterations in the antioxidant system as a consequence of Al exposure. In pea seedlings, ROS production is detected in root apex after two hours of Al exposure and increased with time exposure [38]. In maize roots, Al treatment also led to increase in ROS production rate in all epidermal cells, only within 10 min of Al exposure and continued to increase during Al exposure [41]. APX and

SOD activity increased in roots of both Al-resistant and Al-sensitive triticale cultivars (with higher magnitude in the sensitive one), but changes were detected first in the sensitive cultivar (6 h) and then in the resistant (12 h) [42]. Boscolo et al. [39] reported for maize root tips an increase of SOD and APX activities. Furthermore, these authors found that SOD and APX activity is inversely proportional to root growth rate and, therefore, suggested that the increase of O_2^- and H_2O_2 production is related to Al toxicity. An increase in SOD, APX, and GR activities was reported for greengram seedlings, whereas a decrease in CAT activity and glutathione and ascorbate contents was also found at higher Al concentrations [47]. These authors justified the decrease in CAT activity due to the fact that this enzyme is photosensitive and, therefore, needs constant synthesis and suggested that glutathione and ascorbate may be able to detoxify the ROS directly [47]. Devi et al. [49] found an increase in manganese superoxide dismutase (MnSOD) activity in both sensitive and tolerant cell lines of tobacco and in AsA and GSH contents, mostly in the tolerant line. These data indicated that AsA and GSH seem to be in part responsible for the tolerance mechanisms of the tolerant line to Al. Activities of SOD, CAT, and APX also increased in roots of plants and in cultured tea cells exposed to Al [50]. owever, However, plants of this species provide a complex scenario compared with other models, as aluminium may show a stimulatory effect on plant growth. That increase seemed to result in increased membrane integrity, since lipid peroxidation reduced with Al exposure [50].

These findings reporting increase of antioxidants (enzymatic and nonenzymatic) are accompanied with others that prove gene regulation associated with oxidative stress. For example, Ezaki et al. [51] expressed nine genes derived from *Arabidopsis*, tobacco, wheat, and yeast in *Arabidopsis* ecotype Landsberg. An *Arabidopsis* blue-copper-binding protein gene (*AtBCB*), a tobacco glutathione-S-transferase gene (*parB*), a tobacco peroxidase gene (*NtPox*), and a tobacco GDP-dissociation inhibitor gene (*NtGDI1*) conferred a degree of resistance to Al: significative differences in relative root growth and decrease in Al content and oxidative damages. They also showed that overexpression of three Al-induced genes in plants conferred oxidative stress resistance. Furthermore, overexpression of the *parB* gene simultaneously conferred resistance to both Al and oxidative stresses. Therefore, Ezaki and coworkers concluded that some of the genes induced during Al exposure and oxidative stresses play protective roles against both stresses. Cançado et al. [52] identified a maize Al-inducible cDNA encoding a glutathione-S-transferase (GST). Expression of that gene (GST27.2) was upregulated in response to various Al concentrations in both Al-tolerant and Al-sensitive maize lines. Recently, using Al-sensitive *Medicago truncatula* cultivar Jemalong genotype A17, 324 genes were upregulated and 267 genes were downregulated after Al exposure [53]. Upregulated genes were enriched in transcripts involved in cell-wall modification and abiotic and biotic stress responses, while downregulated genes were enriched in transcripts involved in primary metabolism, secondary metabolism, protein synthesis and processing, and the cell cycle. Known markers of Al-induced

gene expression including genes associated with oxidative stress and cell wall stiffening were differentially regulated in that study [53]. For maize plants, Al exposure led to alteration in gene expression, mostly in the Al-sensitive genotype. Although Al-sensitive genotype showed changes in the expression of more genes, several Al-regulated genes exhibited higher expression in the tolerant genotype [54]. So, it is clear that expression of some genes confers Al resistance and contributes to reduce oxidative stress.

2.3. Cell Wall, Plasma Membrane, and Nutrient Unbalances. Al accumulation is primarily and predominantly in the root apoplast (30–90% of the total absorbed Al) (e.g., [42, 55]) of peripheral cells and is only very slowly translocated to more central tissues [19, 56, 57]. The primary binding of Al^{3+} in the apoplast is probably the pectin matrix, with its negatively charged carboxylic groups [57, 58].

Several works reported increases of pectin levels in Al-sensitive genotypes [29, 43, 57–60], and some also detected increase in Al contents in the same sensitive genotypes [29, 57, 60]. These findings indicated that pectin plays a major role in the binding of Al and suggested that some of the additional Al accumulation in sensitive genotypes bound in the newly formed cell wall pectin [43, 57, 58]. Binding of Al to the pectin matrix and other cell wall constituents could alter cell wall characteristics and functions such as extensibility [61], porosity, and enzyme activities thus leading to inhibition of root growth [57]. Another mechanism for Al toxicity targeted to the apoplast invokes a rapid and irreversible displacement of Ca^{2+} from cell wall components by Al ions [22, 61]. Accumulation of Al occurs predominantly in the root apoplast. Nevertheless, Al accumulates also in the symplast and with a fast rate [19]. Recently, Xia et al. [62] reported a transporter, *Nrat1* (Nramp aluminium transporter 1), specific for Al^{3+} localized at the plasma membrane of all rice root tips cells, except epidermal cells. Those authors referred that the elimination of the *Nrat1* enhanced Al sensitivity, decreased Al uptake, increased Al binding to cell wall and concluded that this transporter is required for prior step of final Al detoxification through sequestration of Al into vacuoles. Furthermore, given its physicochemical properties, Al can interact strongly with the negatively charged plasma membrane. For instance, Al can displace other cations (e.g., Ca^{2+}) that may form bridges between the phospholipid head groups of the membrane bilayer [63]. Furthermore, Al interaction with plasma membrane could lead to depolarization of the transmembranar potential (e.g., [64]) and/or reduction of H^+-ATPase (e.g., [65]) which, in turn, can alter the activities of ions near the plasma membrane surface and impede the formation and maintenance of the transmembrane H^+ gradient [2]. Moreover, Al changes in plasma membrane can modify the uptake of several cations (e.g., Ca^{2+}, Mg^{2+}, K^+, NH_4^+) [8, 66–68]. These changes are related to direct Al^{3+} interactions with plasma membrane ion channels [69] and changes in membrane potential.

Nutritional unbalances induced by Al exposure were reported for several plant species. Eleven families of pteridophytes presented different nutritional unbalances (mostly in Ca, Mg, P, K) depending on Al accumulation [70], and in maize, Al had negative effects on the uptake of macro- and micronutrients, with Ca and Mg being the macro- and Mn and Zn the micronutrients more affected [68]. Also, the maize Al-tolerant genotypes accumulated higher concentration of Ca, Mg [68], and K [71] than the sensitive genotypes. In wheat, both sensitive and tolerant genotypes presented a decrease in K and Mg contents in roots, whereas Ca, Al, Si contents increased [72]. However, the sensitive wheat genotype showed more nutritional unbalances and Al accumulation than the tolerant one in both roots and shoots [72]. Al exposure led to an increase of Ca accumulation in rye-sensitive genotype, contrarily to the tolerant rye genotype [73]. However, other studies reported different results in Al-induced nutritional imbalances in maize: Lidon et al. [74] referred that all elements in roots, except K, Mn, and Zn, increased in Al-treated roots and that in shoots Ca and Mg had little variation. Reference [67] reported that only the specific absorption rate of B was correlated to the Al-induced root growth inhibition. Al exposure led to decrease in K, Mg, Ca, and P contents and uptake in rice plants, and, as observed in maize, the tolerant cultivar presented less negative effects in nutrient content than the sensitive one [75]. In tomato cultivars, Al exposure decreased the content of Ca, K, Mg, Mn, Fe, and Zn in roots, stems, and leaves [76]. Zobel et al. [27] related changes in fine root diameter with changes in concentration of some nutrients, as N, P, and Al. It seems that the differential tolerance to Al may be due to their differences in uptake, ability to keep adequate concentrations and to use the nutrients efficiently. Differences in nutrient uptake, accumulation, and translocation are evident between plant species and within each species. Furthermore, since each author utilized different Al concentrations, diverse nutritive solutions and time exposures, it is difficult to make a general and accurate model of Al-induced nutritional unbalances.

2.4. Cytoplasmic Ca^{2+}. Disturbance of cytoplasmic Ca^{2+} homeostasis is believed to be the primary target of Al toxicity [77] and may be involved in the inhibition of the cell division or root elongation by causing potential disruptions of Ca^{2+}-dependent biochemical and physiological processes [34, 77, 78].

In wheat root apices, [44] found that Al inhibits Ca^{2+}-dependent phospholipase C, which acts on the lipid substrate phosphatidylinositol-4,5-biphosphate. The authors hypothesized that phosphoinositide signaling pathway might be the initial target of Al. In accordance, Zhang et al. [31] found Al-induced inhibition of genes related to phosphoinositide signaling pathway and hypothesized that the gene inhibition could result in disruption of this pathway. Also, it was reported that components of the actin-based cytoskeleton interact directly with phospholipase C in oat [79].

Most works reported an increase in cytoplasmic Ca^{2+} when plants were exposed to Al [13, 80, 81]. However, Jones et al. [82] reported a decrease in cytoplasmic Ca^{2+} in tobacco cell cultures in the presence of Al. Furthermore, Zhang and Rengel [80] reported an increase in cytoplasmic Ca^{2+} in two lines with different tolerance to Al and correlated it with the inhibition of root growth in both lines. Moreover, Ma et al. [13] correlated cytoplasmic Ca^{2+} to root growth response. Moreover, alteration in cytoplasmic Ca^{2+} homeostasis can occur within few minutes (20–30 minutes) in root hair tips of *Arabidopsis thaliana* [82].

It is certain that Al exposure influences cytoplasmic Ca^{2+} homeostasis, but it is still unclear if it is a primary cause of Al-induced inhibition of root growth or a secondary effect. The source of Ca^{2+} for the increase of cytosolic Ca^{2+} activity could be extracellular and/or intracellular but is still insufficiently documented, as well the effects on increased cytosolic Ca^{2+} (for review see [77]).

2.5. Callose. The induction of callose $(1,3-\beta$-D-glucan) formation in Al-exposed roots has been reported in many plant species (e.g., [20, 41, 67, 83–86]). Al-induced callose formation in root tips is recognized as an excellent indicator of Al sensibility [81, 86–90], and some works negatively correlated root elongation with callose formation during Al exposure (e.g., [86, 91]). Recently, it was reported that Al induced callose accumulation not only in the root meristematic regions but also in mature zones, in both wheat and rye genotypes [72, 73]. In maize roots, Jones et al. [41] found a close spatial and temporal coordination between Al accumulation and callose production in roots. Also, in wheat, callose accumulation in root tissues was progressive with Al-exposure, and, contrarily to the tolerant genotype, the sensitive one presented callose deposition at inner cell layers [72, 73]. Still, Tahara et al. [86] reported that, in some Myrtaceae species, induction of callose formation was not accompanied by root growth inhibition and suggested that callose formation is a more sensitive indicator to Al than root elongation.

Since Al induces a transient rise of cytosolic Ca^{2+}, an increase of callose accumulation under Al stress is not unexpected. Cytosolic Ca^{2+} is one of the prerequisites for the induction of callose synthesis, but not the only factor modulating increases in callose synthesis and deposition [81]. Callose formation, as response to Al, is described in sensitive and, to a lesser extent, in tolerant roots [85, 87]. In a less extent, callose deposition has been considered as a mechanism to prevent Al from penetrating into the apoplast. Also, this accumulation is reported to inhibit the symplastic transport and cell communication by blocking plasmodesmata, avoiding Al-induced lesions in the symplast [92]. However, callose deposition in sensitive roots has also been shown to lead to uncontrolled rigidity of cell walls [41] leading ultimately to protoplast degradation.

2.6. Others. Al-induced effects/damages are first detected in the root system [18, 93]. Changes in the root system may affect nutrient uptake, which can lead to nutritional deficiencies in shoots and leaves [94]. Except for Al-accumulator plants, Al accumulates more in roots than in leaves [95]. In some species, Al-induced alterations in leaves were considered indirect, since Al accumulation was not detected in leaves [94]. Nevertheless, alterations in leaves induced by Al exposure were reported for many species. Several works reported leaves biomass reduction [96], thickness [95], lipid peroxidation [97], nutritional imbalances [98], changes in the photosynthetic performance [99], and changes in chlorophyll contents [96, 97, 99, 100], among others. Reductions in carbon dioxide (CO_2) assimilation rate due to Al toxicity are reported for several species [94, 99–101], and some works indicated that Al exposure induced damage of the photosystem II [97, 102]. Very few works focused on the consequence of Al treatment in the carbohydrate metabolism. The effects of Al exposure on Ribulose-1,5-bisphosphate carboxylase/oxygenase (RuBisCo) content and activity are still unclear, and the few reports available were performed in citrus [99, 100] and in wild rice [103].

3. Conclusions

Most studies on Al toxicity are performed with different media composition, Al concentration, and period of exposure. Also, there is a large variation between genotypes. This battery of nonharmonized experimental data needs caution during interpretation, mostly concerning generalizations of functional models. So, it would be important to uniform the experimental procedures in order to better comprehend the plant response to Al exposure and the mechanisms of Al tolerance.

Acknowledgments

FCT/MCT supported this work (POCI/AGR/58174/2004) and S. Silva was supported by (SFRH/BPD/74299/2010) grants.

References

[1] H. M. May and D. K. Nordstrom, "Assessing the solubilities and reactions kinetics of aluminuous mineral in soils," in *Soil Acidity*, B. Ulrich and M. E. Summer, Eds., pp. 125–148, Springer, Berlin, Germany, 1991.

[2] L. V. Kochian, M. A. Piñeros, and O. A. Hoekenga, "The physiology, genetics and molecular biology of plant aluminum resistance and toxicity," *Plant and Soil*, vol. 274, no. 1-2, pp. 175–195, 2005.

[3] S. K. Panda and H. Matsumoto, "Molecular physiology of aluminum toxicity and tolerance in plants," *Botanical Review*, vol. 73, no. 4, pp. 326–347, 2007.

[4] T. B. Kinraide, "Identity of the rhizotoxic aluminium species," *Plant and Soil*, vol. 134, no. 1, pp. 167–178, 1991.

[5] L. V. Kochian, "Cellular mechanisms of aluminum toxicity and resistance in plants," *Annual Review of Plant Physiology and Plant Molecular Biology*, vol. 46, pp. 237–260, 1995.

[6] B. Marschner, U. Henke, and G. Wessolek, "Effects of meliorative additives on the adsorption and binding forms of heavy-metals in contaminated topsoil from a former sewage farm," *Zeitschrift fur Pflanzenernahrung und Bodenkunde*, vol. 158, pp. 9–14, 1995.

[7] W. S. Eisenmenger, "Toxicity of aluminum on seedlings and action of certain ions in the elimination of the toxic effects," *Plant Physiology*, vol. 10, no. 1, pp. 1–25, 1935.

[8] F. Jan, "Aluminium effects on growth, nutrient net uptake and transport in 3 rice (*Oryza sativa*) cultivars with different sensitivity to aluminium," *Physiologia Plantarum*, vol. 83, no. 3, pp. 441–448, 1991.

[9] P. R. Ryan, J. M. Ditomaso, and L. V. Kochian, "Aluminium toxicity in roots: an investigation of spatial sensitivity and the role of the root cap," *Journal of Experimental Botany*, vol. 44, no. 2, pp. 437–446, 1993.

[10] Z. G. Shen, J. L. Wang, and H. Y. Guan, "Effect of aluminium and calcium on growth of wheat seedlings and germination of seeds," *Journal of Plant Nutrition*, vol. 16, no. 11, pp. 2135–2148, 1993.

[11] S. A. Crawford and S. Wilkens, "Effect of aluminium on root elongation in two Australian perennial grasses," *Australian Journal of Plant Physiology*, vol. 25, no. 2, pp. 165–171, 1998.

[12] S. J. Ahn, M. Sivaguru, H. Osawa, G. C. Chung, and H. Matsumoto, "Aluminum inhibits the H^+-ATpase activity by permanently altering the plasma membrane surface potentials in squash roots," *Plant Physiology*, vol. 126, no. 4, pp. 1381–1390, 2001.

[13] Q. Ma, Z. Rengel, and J. Kuo, "Aluminium toxicity in rye (*Secale cereale*): root growth and dynamics of cytoplasmic Ca^{2+} in intact root tips," *Annals of Botany*, vol. 89, no. 2, pp. 241–244, 2002.

[14] S. Kikui, T. Sasaki, M. Maekawa et al., "Physiological and genetic analyses of aluminium tolerance in rice, focusing on root growth during germination," *Journal of Inorganic Biochemistry*, vol. 99, no. 9, pp. 1837–1844, 2005.

[15] I. R. Silva, T. J. Smyth, D. F. Moxley, T. E. Carter, N. S. Allen, and T. W. Rufty, "Aluminum accumulation at nuclei of cells in the root tip. Fluorescence detection using lumogallion and confocal laser scanning microscopy," *Plant Physiology*, vol. 123, no. 2, pp. 543–552, 2000.

[16] G. Frantzios, B. Galatis, and P. Apostolakos, "Aluminium effects on microtubule organization in dividing root-tip cells of *Triticum turgidum* II. Cytokinetic cells," *Journal of Plant Research*, vol. 114, no. 1114, pp. 157–170, 2001.

[17] Y. Li, G. X. Yang, L. T. Luo et al., "Aluminium sensitivity and tolerance in model and elite wheat varieties," *Cereal Research Communications*, vol. 36, no. 2, pp. 257–267, 2008.

[18] S. Doncheva, M. Amenós, C. Poschenrieder, and J. Barceló, "Root cell patterning: a primary target for aluminium toxicity in maize," *Journal of Experimental Botany*, vol. 56, no. 414, pp. 1213–1220, 2005.

[19] S. Marienfeld, N. Schmohl, M. Klein, W. H. Schröder, A. J. Kuhn, and W. J. Horst, "Localisation of aluminium in root tips of *Zea mays* and *Vicia faba*," *Journal of Plant Physiology*, vol. 156, no. 5-6, pp. 666–671, 2000.

[20] S. Budikova and and K. Durcekova, "Aluminium accumulation in roots of Al-sensitive barley cultivar changes root cell structure and induces callose synthesis," *Biologia*, vol. 59, pp. 215–220, 2004.

[21] M. Čiamporová, "Morphological and structural responses of plant roots to aluminium at organ, tissue, and cellular levels," *Biologia Plantarum*, vol. 45, no. 2, pp. 161–171, 2002.

[22] S. J. Zheng and J. L. Yang, "Target sites of aluminum phytotoxicity," *Biologia Plantarum*, vol. 49, no. 3, pp. 321–331, 2005.

[23] M. Llugany, C. Poschenrieder, and J. Barcelo, "Monitoring of aluminium-induced inhibition of root elongation in four maize cultivars differing in tolerance to aluminium and proton toxicity," *Physiologia Plantarum*, vol. 93, no. 2, pp. 265–271, 1995.

[24] M. Yi, H. Yi, H. Li, and L. Wu, "Aluminum induces chromosome aberrations, micronuclei, and cell cycle dysfunction in root cells of *Vicia faba*," *Environmental Toxicology*, vol. 25, no. 2, pp. 124–129, 2010.

[25] N. V. Bulanova, B. I. Synzynys, and G. V. Koz'min, "Aluminum induces chromosome aberrations in cells of wheat root meristem," *Russian Journal of Genetics*, vol. 37, no. 12, pp. 1455–1458, 2001.

[26] S. Mohanty, A. B. Das, P. Das, and P. Mohanty, "Effect of a low dose of aluminum on mitotic and meiotic activity, 4C DNA content, and pollen sterility in rice, *Oryza sativa* L. cv. Lalat," *Ecotoxicology and Environmental Safety*, vol. 59, no. 1, pp. 70–75, 2004.

[27] R. W. Zobel, T. B. Kinraide, and V. C. Baligar, "Fine root diameters can change in response to changes in nutrient concentrations," *Plant and Soil*, vol. 297, no. 1-2, pp. 243–254, 2007.

[28] E. B. Blancaflor, D. L. Jones, and S. Gilroy, "Alterations in the cytoskeleton accompany aluminum-induced growth inhibition and morphological changes in primary roots of maize," *Plant Physiology*, vol. 118, no. 1, pp. 159–172, 1998.

[29] W. J. Horst, N. Schmohl, M. Kollmeier, F. Baluška, and M. Sivaguru, "Does aluminium affect root growth of maize through interaction with the cell wall—plasma membrane—cytoskeleton continuum?" *Plant and Soil*, vol. 215, no. 2, pp. 163–174, 1999.

[30] G. Frantzios, B. Galatis, and P. Apostolakos, "Aluminium causes variable responses in actin filament cytoskeleton of the root tip cells of *Triticum turgidum*," *Protoplasma*, vol. 225, no. 3-4, pp. 129–140, 2005.

[31] J. Zhang, Z. He, H. Tian, G. Zhu, and X. Peng, "Identification of aluminium-responsive genes in rice cultivars with different aluminium sensitivities," *Journal of Experimental Botany*, vol. 58, no. 8, pp. 2269–2278, 2007.

[32] K. Krishnan and P. D. J. Moens, "Structure and functions of profilins," *Biophysical Reviews*, vol. 1, pp. 71–81, 2009.

[33] S. Ramachandran, H. E. M. Christensen, Y. Ishimaru et al., "Profilin plays a role in cell elongation, cell shape maintenance, and flowering in Arabidopsis," *Plant Physiology*, vol. 124, no. 4, pp. 1637–1647, 2000.

[34] M. Sivaguru, F. Baluška, D. Volkmann, H. H. Felle, and W. J. Horst, "Impacts of aluminum on the cytoskeleton of the maize root apex. Short-term effects on the distal part of the transition zone," *Plant Physiology*, vol. 119, no. 3, pp. 1073–1082, 1999.

[35] R. J. Bennet and C. M. Breen, "The aluminium signal: new dimensions to mechanisms of aluminium tolerance," *Plant and Soil*, vol. 134, no. 1, pp. 153–166, 1991.

[36] M. Sivaguru and W. J. Horst, "The distal part of the transition zone is the most aluminum-sensitive apical root zone of maize," *Plant Physiology*, vol. 116, no. 1, pp. 155–163, 1998.

[37] Y. Yamamoto, Y. Kobayashi, and H. Matsumoto, "Lipid peroxidation is an early symptom triggered by aluminum, but not the primary cause of elongation inhibition in Pea roots," *Plant Physiology*, vol. 125, no. 1, pp. 199–208, 2001.

[38] Y. Yamamoto, Y. Kobayashi, S. R. Devi, S. Rikiishi, and H. Matsumoto, "Oxidative stress triggered by aluminum in plant roots," *Plant and Soil*, vol. 255, no. 1, pp. 239–243, 2003.

[39] P. R. S. Boscolo, M. Menossi, and R. A. Jorge, "Aluminum-induced oxidative stress in maize," *Phytochemistry*, vol. 62, no. 2, pp. 181–189, 2003.

[40] M. C. Kuo and C. H. Kao, "Aluminum effects on lipid peroxidation and antioxidative enzyme activities in rice leaves," *Biologia Plantarum*, vol. 46, no. 1, pp. 149–152, 2003.

[41] D. L. Jones, E. B. Blancaflor, L. V. Kochian, and S. Gilroy, "Spatial coordination of aluminium uptake, production of reactive oxygen species, callose production and wall rigidification in maize roots," *Plant, Cell and Environment*, vol. 29, no. 7, pp. 1309–1318, 2006.

[42] Q. Liu, J. L. Yang, L. S. He, Y. Y. Li, and S. J. Zheng, "Effect of aluminum on cell wall, plasma membrane, antioxidants and root elongation in triticale," *Biologia Plantarum*, vol. 52, no. 1, pp. 87–92, 2008.

[43] Q. Liu, L. Zhu, L. Yin, C. Hu, and L. Chen, "Cell wall pectin and its binding capacity contribute to aluminium resistance in buckwheat," in *Proceedings of the 2nd International Conference on Bioinformatics and Biomedical Engineering (ICBBE '08)*, pp. 4508–4511, Shanghai, China, May 2006.

[44] D. L. Jones and L. V. Kochian, "Aluminum interaction with plasma membrane lipids and enzyme metal binding sites and its potential role in Al cytotoxicity," *FEBS Letters*, vol. 400, no. 1, pp. 51–57, 1997.

[45] T. Guo, G. Zhang, M. Zhou, F. Wu, and J. Chen, "Effects of aluminum and cadmium toxicity on growth and antioxidant enzyme activities of two barley genotypes with different Al resistance," *Plant and Soil*, vol. 258, no. 1-2, pp. 241–248, 2004.

[46] P. H. P. Peixoto, J. Cambraia, R. Sant'Anna, P. R. Mosquim, and M. A. Moreira, "Aluminum effects on lipid peroxiadation and on the activities of enzymes of oxidative metabolism in sorghum," *Revista Brasileira de Fisiologia Vegetal*, vol. 11, pp. 137–143, 1999.

[47] S. K. Panda, L. B. Singha, and M. H. Khan, "Does aluminium phytotoxicity induce oxidative stress in greengram (*Vigna radiata*)?" *Bulgarian Journal of Plant Physiology*, vol. 29, pp. 77–86, 2003.

[48] M. A. Hossain, A. K. M. Z. Hossain, T. Kihara, H. Koyama, and T. Hara, "Aluminum-induced lipid peroxidation and lignin deposition are associated with an increase in H_2O_2 generation in wheat seedlings," *Soil Science and Plant Nutrition*, vol. 51, no. 2, pp. 223–230, 2005.

[49] S. R. Devi, Y. Yamamoto, and H. Matsumoto, "An intracellular mechanism of aluminum tolerance associated with high antioxidant status in cultured tobacco cells," *Journal of Inorganic Biochemistry*, vol. 97, no. 1, pp. 59–68, 2003.

[50] F. Ghanati, A. Morita, and H. Yokota, "Effects of aluminum on the growth of tea plant and activation of antioxidant system," *Plant and Soil*, vol. 276, no. 1-2, pp. 133–141, 2005.

[51] B. Ezaki, R. C. Gardner, Y. Ezaki, and H. Matsumoto, "Expression of aluminum-induced genes in transgenic Arabidopsis plants can ameliorate aluminum stress and/or oxidative stress," *Plant Physiology*, vol. 122, no. 3, pp. 657–665, 2000.

[52] G. M. A. Cançado, V. E. De Rosa, J. H. Fernandez, L. G. Maron, R. A. Jorge, and M. Menossi, "Glutathione S-transferase and aluminum toxicity in maize," *Functional Plant Biology*, vol. 32, no. 11, pp. 1045–1055, 2005.

[53] D. Chandran, N. Sharopova, S. Ivashuta, J. S. Gantt, K. A. VandenBosch, and D. A. Samac, "Transcriptome profiling identified novel genes associated with aluminum toxicity, resistance and tolerance in *Medicago truncatula*," *Planta*, vol. 228, no. 1, pp. 151–166, 2008.

[54] L. G. Maron, M. Kirst, C. Mao, M. J. Milner, M. Menossi, and L. V. Kochian, "Transcriptional profiling of aluminum toxicity and tolerance responses in maize roots," *New Phytologist*, vol. 179, no. 1, pp. 116–128, 2008.

[55] Z. Rengel and R. J. Reid, "Uptake of Al across the plasma membrane of plant cells," *Plant and Soil*, vol. 192, no. 1, pp. 31–35, 1997.

[56] S. Marienfeld, H. Lehmann, and R. Stelzer, "Ultrastructural investigations and EDX-analyses of Al-treated oat (*Avena sativa*) roots," *Plant and Soil*, vol. 171, no. 1, pp. 167–173, 1995.

[57] N. Schmohl and W. J. Horst, "Cell wall pectin content modulates aluminium sensitivity of *Zea mays* (L.) cells grown in suspension culture," *Plant, Cell and Environment*, vol. 23, no. 7, pp. 735–742, 2000.

[58] Y. C. Chang, Y. Yamamoto, and H. Matsumoto, "Accumulation of aluminium in the cell wall pectin in cultured tobacco (*Nicotiana tabacum* L.) cells treated with a combination of aluminium and iron," *Plant, Cell and Environment*, vol. 22, no. 8, pp. 1009–1017, 1999.

[59] Le Van Hoa, S. Kuraishi, and N. Sakurai, "Aluminum-induced rapid root inhibition and changes in cell-wall components of squash seedlings," *Plant Physiology*, vol. 106, no. 3, pp. 971–976, 1994.

[60] A. K. M. Zakir Hossain, H. Koyama, and T. Hara, "Growth and cell wall properties of two wheat cultivars differing in their sensitivity to aluminum stress," *Journal of Plant Physiology*, vol. 163, no. 1, pp. 39–47, 2006.

[61] A. Tabuchi and H. Matsumoto, "Changes in cell-wall properties of wheat (*Triticum aestivum*) roots during aluminum-induced growth inhibition," *Physiologia Plantarum*, vol. 112, no. 3, pp. 353–358, 2001.

[62] J. Xia, N. Yamaji, T. Kasai, and J. F. Ma, "Plasma membrane-localized transporter for aluminum in rice," *Proceedings of the National Academy of Sciences of the United States of America*, vol. 107, no. 43, pp. 18381–18385, 2010.

[63] M. A. Akeson, D. N. Munns, and R. G. Burau, "Adsorption of Al^{3+} to phosphatidylcholine vesicles," *Biochimica et Biophysica Acta*, vol. 986, no. 1, pp. 33–40, 1989.

[64] T. B. Kinraide, P. R. Ryan, and L. V. Kochian, "Interactive effects of Al^{3+}, H^+, and other cations on root elongation considered in terms of cell-surface electrical potential," *Plant Physiology*, vol. 99, no. 4, pp. 1461–1468, 1992.

[65] S. J. Ahn, M. Sivaguru, G. C. Chung, Z. Rengel, and H. Matsumoto, "Aluminium-induced growth inhibition is associated with impaired efflux and influx of H^+ across the plasma membrane in root apices of squash (*Cucurbita pepo*)," *Journal of Experimental Botany*, vol. 53, no. 376, pp. 1959–1966, 2002.

[66] B. E. Nichol, L. A. Oliveira, A. D. M. Glass, and M. Y. Siddiqi, "The effects of aluminum on the influx of calcium, potassium, ammonium, nitrate, and phosphate in an aluminum-sensitive cultivar of barley (*Hordeum vulgare* L.)," *Plant Physiology*, vol. 101, no. 4, pp. 1263–1266, 1993.

[67] C. Poschenrieder, M. Llugany, and J. Barcelo, "Short-term effects of pH and aluminium on mineral nutrition in maize varieties differing in proton and aluminium tolerance," *Journal of Plant Nutrition*, vol. 18, no. 7, pp. 1495–1507, 1995.

[68] E. D. Mariano and W. G. Keltjens, "Long-term effects of aluminum exposure on nutrient uptake by maize genotypes differing in aluminum resistance," *Journal of Plant Nutrition*, vol. 28, no. 2, pp. 323–333, 2005.

[69] M. A. Piñeros and L. V. Kochian, "A patch-clamp study on the physiology of aluminum toxicity and aluminum tolerance in maize. Identification and characterization of Al^{3+} anion channels," *Plant Physiology*, vol. 125, no. 1, pp. 292–305, 2001.

[70] E. Olivares, E. Peña, E. Marcano et al., "Aluminum accumulation and its relationship with mineral plant nutrients in 12 pteridophytes from Venezuela," *Environmental and Experimental Botany*, vol. 65, no. 1, pp. 132–141, 2009.

[71] A. Giannakoula, M. Moustakas, P. Mylona, I. Papadakis, and T. Yupsanis, "Aluminum tolerance in maize is correlated with increased levels of mineral nutrients, carbohydrates and proline, and decreased levels of lipid peroxidation and Al accumulation," *Journal of Plant Physiology*, vol. 165, no. 4, pp. 385–396, 2008.

[72] S. Silva, O. Pinto-Carnide, P. Martins-Lopes, M. Matos, H. Guedes-Pinto, and C. Santos, "Differential aluminium changes on nutrient accumulation and root differentiation in an Al sensitive vs. tolerant wheat," *Environmental and Experimental Botany*, vol. 68, no. 1, pp. 91–98, 2010.

[73] S. Silva, C. Santos, M. Matos, and O. Pinto-Carnide, "Al toxicity mechanisms in tlerant and sensitive rye genotypes," *Environmental and Experimental Botany*, vol. 75, pp. 89–97, 2011.

[74] F. C. Lidon, H. G. Azinheira, and M. G. Barreiro, "Aluminum toxicity in maize: biomass production and nutrient uptake and translocation," *Journal of Plant Nutrition*, vol. 23, no. 2, pp. 151–160, 2000.

[75] R. J. de Mendonça, J. Cambraia, J. A. de Oliveira, and M. A. Oliva, "Aluminum effects on the uptake and utilization of macronutrients in two rice cultivars," *Pesquisa Agropecuaria Brasileira*, vol. 38, no. 7, pp. 843–848, 2003.

[76] L. Simon, T. J. Smalley, J. Benton Jones Jnr, and F. T. Lasseigne, "Aluminum toxicity in tomato .1. Growth and mineral-nutrition," *Journal of Plant Nutrition*, vol. 17, no. 2-3, pp. 293–306, 1994.

[77] Z. Rengel and W. H. Zhang, "Role of dynamics of intracellular calcium in aluminium-toxicity syndrome," *New Phytologist*, vol. 159, no. 2, pp. 295–314, 2003.

[78] S. Silva, *Aluminium Toxicology in Wheat and Rye [Ph.D. thesis]*, Biologia Universidade de Aveiro, Aveiro, Portugal, 2011.

[79] C. H. Huang and R. C. Crain, "Phosphoinositide-specific phospholipase C in oat roots: association with the actin cytoskeleton," *Planta*, vol. 230, no. 5, pp. 925–933, 2009.

[80] W. H. Zhang and Z. Rengel, "Aluminium induces an increase in cytoplasmic calcium in intact wheat root apical cells," *Australian Journal of Plant Physiology*, vol. 26, no. 5, pp. 401–409, 1999.

[81] P. Bhuja, K. McLachlan, J. Stephens, and G. Taylor, "Accumulation of 1,3-β-D-glucans, in response to aluminum and cytosolic calcium in *Triticum aestivum*," *Plant and Cell Physiology*, vol. 45, no. 5, pp. 543–549, 2004.

[82] D. L. Jones, S. Gilroy, P. B. Larsen, S. H. Howell, and L. V. Kochian, "Effect of aluminum on cytoplasmic Ca^{2+} homeostasis in root hairs of *Arabidopsis thaliana* (L.)," *Planta*, vol. 206, no. 3, pp. 378–387, 1998.

[83] A. C. Jorns, C. Hechtbuchholz, and A. H. Wissemeier, "Aluminum-induced callose formation in root-tips of Norway spruce (*Picea-Abies* (L) Karst)," *Zeitschrift Fur Pflanzernernahrung Und Bodenkunde*, vol. 154, pp. 349–353, 1991.

[84] K. A. Schreiner, J. Hoddinott, and G. J. Taylor, "Aluminum-induced deposition of (1,3)-beta-glucans (Callose) in *Triticum aestivum* L.," *Plant and Soil*, vol. 162, no. 2, pp. 273–280, 1994.

[85] D. Eticha, C. Thé, C. Welcker, L. Narro, A. Staß, and W. J. Horst, "Aluminium-induced callose formation in root apices: inheritance and selection trait for adaptation of tropical maize to acid soils," *Field Crops Research*, vol. 93, no. 2-3, pp. 252–263, 2005.

[86] K. Tahara, M. Norisada, T. Hogetsu, and K. Kojima, "Aluminum tolerance and aluminum-induced deposition of callose and lignin in the root tips of *Melaleuca* and *Eucalyptus*

species," *Journal of Forest Research*, vol. 10, no. 4, pp. 325–333, 2005.

[87] W. J. Horst, A. K. Püschel, and N. Schmohl, "Induction of callose formation is a sensitive marker for genotypic aluminium sensitivity in maize," *Plant and Soil*, vol. 192, no. 1, pp. 23–30, 1997.

[88] N. Massot, M. Llugany, C. Poschenrieder, and J. Barceló, "Callose production as indicator of aluminum toxicity in bean cultivars," *Journal of Plant Nutrition*, vol. 22, no. 1, pp. 1–10, 1999.

[89] B. Meriga, B. K. Reddy, G. Jogeswar, L. A. Reddy, and P. B. K. Kishor, "Alleviating effect of citrate on aluminium toxicity of rice (*Oryza sativa* L.) seedlings," *Current Science*, vol. 85, no. 3, pp. 383–386, 2003.

[90] Y. Hirano, E. G. Pannatier, S. Zimmermann, and I. Brunner, "Induction of callose in roots of Norway spruce seedlings after short-term exposure to aluminum," *Tree Physiology*, vol. 24, no. 11, pp. 1279–1283, 2004.

[91] N. E. Nagy, L. S. Dalen, D. L. Jones, B. Swensen, C. G. Fossdal, and T. D. Eldhuset, "Cytological and enzymatic responses to aluminium stress in root tips of Norway spruce seedlings," *New Phytologist*, vol. 163, no. 3, pp. 595–607, 2004.

[92] M. Sivaguru, T. Fujiwara, J. Samaj et al., "Aluminum-induced 133-b-D-glucan inhibits cell-to-cell trafficking of molecules through plasmodesmata. A new mechanism of aluminum toxicity in plants," *Plant Physiology*, vol. 124, no. 3, pp. 991–1005, 2000.

[93] J. Barceló and C. Poschenrieder, "Fast root growth responses, root exudates, and internal detoxification as clues to the mechanisms of aluminium toxicity and resistance: a review," *Environmental and Experimental Botany*, vol. 48, no. 1, pp. 75–92, 2002.

[94] M. Moustakas, G. Ouzounidou, E. P. Eleftheriou, and R. Lannoye, "Indirect effects of aluminium stress on the function of the photosynthetic apparatus," *Plant Physiology and Biochemistry*, vol. 34, no. 4, pp. 553–560, 1996.

[95] A. Konarska, "Effects of aluminum on growth and structure of red pepper (*Capsicum annuum* L.) leaves," *Acta Physiologiae Plantarum*, vol. 32, no. 1, pp. 145–151, 2010.

[96] R. Azmat and S. Hasan, "Photochemistry of light harvesting pigments and some biochemical changes under aluminium stress," *Pakistan Journal of Botany*, vol. 40, no. 2, pp. 779–784, 2008.

[97] X. B. Zhang, P. Liu, Y. S. Yang, and G. D. Xu, "Effect of Al in soil on photosynthesis and related morphological and physiological characteristics of two soybean genotypes," *Botanical Studies*, vol. 48, no. 4, pp. 435–444, 2007.

[98] T. R. Guo, G. P. Zhang, and Y. H. Zhang, "Physiological changes in barley plants under combined toxicity of aluminum, copper and cadmium," *Colloids and Surfaces B*, vol. 57, no. 2, pp. 182–188, 2007.

[99] H. X. Jiang, L. S. Chen, J. G. Zheng, S. Han, N. Tang, and B. R. Smith, "Aluminum-induced effects on Photosystem II photochemistry in Citrus leaves assessed by the chlorophyll a fluorescence transient," *Tree Physiology*, vol. 28, no. 12, pp. 1863–1871, 2008.

[100] L. S. Chen, Y. P. Qi, B. R. Smith, and X. H. Liu, "Aluminum-induced decrease in CO_2 assimilation in citrus seedlings is unaccompanied by decreased activities of key enzymes involved in CO_2 assimilation," *Tree Physiology*, vol. 25, no. 3, pp. 317–324, 2005.

[101] F. C. Lidon, M. G. Barreiro, J. C. Ramalho, and J. A. Lauriano, "Effects of aluminum toxicity on nutrient accumulation in maize shoots: implications on photosynthesis," *Journal of Plant Nutrition*, vol. 22, no. 2, pp. 397–416, 1999.

[102] M. Reyes-Díaz, C. Inostroza-Blancheteau, R. Millaleo et al., "Long-term aluminum exposure effects on physiological and biochemical features of highbush blueberry cultivars," *Journal of the American Society for Horticultural Science*, vol. 135, no. 3, pp. 212–222, 2010.

[103] Y. Cao, Y. Lou, Y. Han et al., "Al toxicity leads to enhanced cell division and changed photosynthesis in *Oryza rufipogon* L.," *Molecular Biology Reports*, vol. 38, no. 8, pp. 4839–4846, 2011.

6

Exploring Diversification and Genome Size Evolution in Extant Gymnosperms through Phylogenetic Synthesis

J. Gordon Burleigh,[1] W. Brad Barbazuk,[1] John M. Davis,[2] Alison M. Morse,[2] and Pamela S. Soltis[3]

[1] Department of Biology, University of Florida, Gainesville, FL 32611, USA
[2] School of Forest Resources and Conservation, University of Florida, Gainesville, FL 32611, USA
[3] Florida Museum of Natural History, University of Florida, Gainesville, FL 32611, USA

Correspondence should be addressed to J. Gordon Burleigh, gburleigh@ufl.edu

Academic Editor: Hiroyoshi Takano

Gymnosperms, comprising cycads, *Ginkgo*, Gnetales, and conifers, represent one of the major groups of extant seed plants. Yet compared to angiosperms, little is known about the patterns of diversification and genome evolution in gymnosperms. We assembled a phylogenetic supermatrix containing over 4.5 million nucleotides from 739 gymnosperm taxa. Although 93.6% of the cells in the supermatrix are empty, the data reveal many strongly supported nodes that are generally consistent with previous phylogenetic analyses, including weak support for Gnetales sister to Pinaceae. A lineage through time plot suggests elevated rates of diversification within the last 100 million years, and there is evidence of shifts in diversification rates in several clades within cycads and conifers. A likelihood-based analysis of the evolution of genome size in 165 gymnosperms finds evidence for heterogeneous rates of genome size evolution due to an elevated rate in *Pinus*.

1. Introduction

Recent advances in sequencing technology offer the possibility of identifying the genetic mechanisms that influence evolutionarily important characters and ultimately drive diversification. Within angiosperms, large-scale phylogenetic analyses have identified complex patterns of diversification (e.g., [1–3]), and numerous genomes are at least partially sequenced. Yet the other major clade of seed plants, the gymnosperms, have received far less attention, with few comprehensive studies of diversification and no sequenced genomes. Note that throughout this paper "gymnosperms" specifies only the approximately 1000 extant species within cycads, *Ginkgo*, Gnetales, and conifers. These comprise the *Acrogymnospermae* clade described by Cantino et al. [4].

Many gymnosperms have exceptionally large genomes (e.g., [5–7]), and this has hindered whole-genome sequencing projects, especially among economically important *Pinus* species. This large genome size is interesting because one suggested mechanism for rapid increases in genome size, polyploidy, is rare among gymnosperms [8]. Recent sequencing efforts have elucidated some of genomic characteristics associated with the large genome size in *Pinus*. Morse et al. [9] identified a large retrotransposon family in *Pinus*, that, with other retrotransposon families, accounts for much of the genomic complexity. Similarly, recent sequencing of 10 BAC (bacterial artificial chromosome) clones from *Pinus taeda* identified many conifer-specific LTR (long terminal repeat) retroelements [10]. These studies suggest that the large genome size may be caused by rapid expansion of retrotransposons and may be limited to conifers, Pinaceae, or *Pinus*. Other studies have quantified patterns of genome size among gymnosperms, especially within *Pinus* and the other Pinaceae [6, 7, 11–14]. These studies have largely focused on finding morphological, biogeographic, or life history correlates of genome size, but the rates and patterns of genome size evolution in gymnosperms are largely unknown.

This study first synthesizes the available phylogenetically informative sequences to build a phylogenetic hypothesis of gymnosperms that reflects the recent advances in sequencing and computational phylogenetics. The resulting tree provides a starting point for large-scale evolutionary and ecological analyses of gymnosperms and will hopefully be a resource to promote and guide future phylogenetic and comparative studies. We use the tree to examine large-scale patterns of diversification of the extant gymnosperm lineages and also to examine rates of genome size evolution.

2. Methods and Materials

2.1. Supermatrix Phylogenetic Inference. We constructed a phylogenetic hypothesis of gymnosperms from available, phylogenetically informative sequence data in GenBank that was available on June 30, 2009. We first downloaded from GenBank all core nucleotide sequence data from gymnosperms (Coniferophyta, Cycadophyta, Ginkgophyta, and Gnetophyta). Additionally, we downloaded sequences from the "basal angiosperm" lineages (e.g., *Amborella*, Nymphaeales, Chloranthaceae, and Austrobaileyales) to represent the angiosperms and a diverse sampling of Moniliformopses taxa (including species from *Equisetum*, *Psilotum*, *Ophioglossum*, *Botrychium*, *Angiopteris*, and *Adiantum*) to use as outgroups.

To identify sets of homologous sequences from the GenBank data, we clustered sequences less than 10,000 bp in length based on results from an all-by-all pairwise BLAST analysis. The all-by-all blastn search was done with blastall using the default parameters [15]. Significant BLAST hits had a maximum e-value of $1.0e^{-10}$ and at least 50% overlap of both the target and query sequences. A Perl script identified the largest clusters of sequences in which each sequence has a significant BLAST hit against at least one other sequence in the cluster. We only considered clusters containing loci that had been used previously for phylogenetic analyses. This included plastid and mitochondrial loci as well as nuclear 18S rDNA, 26S rDNA, and internal transcribed spacer (ITS) sequences. Among these clusters, those containing sequences from at least 15 taxa were aligned using Muscle [16], and the resulting alignments were manually checked and adjusted. The resulting alignments were edited for inclusion in the supermatrix by removing hybrid taxa and those that lacked a specific epithet and also keeping only a single sequence per species. The final cluster alignments were then concatenated to make a single phylogenetic supermatrix (e.g., [17]).

2.2. Phylogenetic and Dating Analysis. To estimate the optimal topology and molecular branch lengths for the gymnosperms, we performed maximum likelihood (ML) phylogenetic analysis on the full supermatrix alignment using RAxML-VI-HPC version 7.0.4 [18]. All ML analyses used the general time reversible (GTR) nucleotide substitution model with the default settings for the optimization of individual per-site substitution rates and classification of these rates into rate categories. To assess uncertainty in the topology and branch length estimates, we ran 100 nonparametric bootstrap replicates on the original data set [19].

We transformed the optimal and bootstrap trees to chronograms, ultrametric trees in which the branch lengths

represent time, using penalized likelihood [20] implemented in r8s version 1.71 [21]. We used a smoothing parameter of 10000, which was chosen based on cross-validation of the fossil constraints. For the r8s analysis, we used the same time constraints on seed plant clades used by Won and Renner [22]. The most recent common ancestor of seed plants was constrained to a maximum age of 385 million years ago (mya). The most recent common ancestor of the extant gymnosperms was fixed at 315 mya and *Gnetum* at 110 mya. The following clades were given minimum age constraints: angisperms: 125 mya, cycads: 270 mya, Cupressaceae: 90 mya, Araucariaceae: 160 mya, Gnetales + Pinaceae: 225 mya, Pinaceae: 90 mya, and Gnetales: 125 mya.

2.3. Diversification Analysis. To examine the general patterns of diversification through time among the extant gymnosperm lineages, we first made lineage through time plots using the R package APE [23]. To account for uncertainty in the dating estimates, we plotted each bootstrap tree after it had been transformed into a chronogram and all nongymnosperm taxa were removed.

Since there appears to be much variance in the divergence time estimates among trees, and branch length estimates are often unreliable, especially when estimated from such a sparse, heterogeneous sequence matrix, we used a test for changes in diversification rate that relies on tree shape, not branch lengths. Specifically, we used the whole-tree, topology-based test described by Moore et al. [24] to detect nodes associated with significant shifts in diversification rate based on the Δ_1 statistic. The analyses were performed using the apTreeshape R package [25]. We used only the optimal tree estimate and again, pruned all non-gymnosperm taxa from the tree prior to analysis.

2.4. Rates of Genome Size Evolution. We first assembled a set of mean genome size data for all gymnosperms present in the phylogenetic tree (in pg DNA) from the Kew C-value database [26]. This includes data from the studies of Murray [6] and Grotkopp et al. [14]. When there were multiple estimates available from a single species, we used the mean of the estimates. We tested for shifts in rates of genome size evolution using Brownie v. 2.1.2 [27]. We used the censored rate test, which tests for differences in rates of evolution of a continuous character (genome size) in one clade versus another clade or paraphyletic group based on a Brownian motion model. We made the following comparisons: conifers + Gnetales versus cycads + *Ginkgo*, Pinaceae versus other conifers + Gnetales, non-Pinaceae conifers + Gnetales versus cycads + *Ginkgo*, *Pinus* versus other Pinaceae, the non-*Pinus* Pinceae versus the other conifers + Gnetales, and *Pinus* subgenus *Strobus* subgenus versus *Pinus* subgenus *Pinus*. To account for topological and branch length uncertainty, we performed all hypothesis tests in Brownie on each bootstrap tree and weighted the results across replicates. The penalized likelihood analysis in r8s collapsed some branch lengths to 0, and Brownie does not work with 0 branch lengths in the tree. Thus, prior to the Brownie analysis, all 0 branch lengths were changed to 0.1.

3. Results

3.1. Phylogenetic Data. The alignment from the complete supermatrix contains sequences from 950 taxa (739 gymnosperms, 108 angiosperms, and 103 nonseed plant outgroups) and is 74,105 characters in length. The 739 gymnosperm taxa include at least one representative from every family as well as from 88 genera. In total, the matrix contains 4,511,144 nucleotides and 93.6% missing data. The number of nucleotides per taxon ranges from 252 to 33,138 (average = 4,749; median = 3,355).

3.2. Phylogenetic Inference. In the 950-taxon trees, 63.3% (601) of the nodes have at least 50% bootstrap support, 41.7% (396) have at least 70% support, 25.8% (245) have at least 90% support, and 9.7% (92) have 100% support. The seed plants have 100% support, and the angiosperms are sister to a clade of all gymnosperms (Figure 1). Within gymnosperms, a clade of *Ginkgo* + cycads (bootstrap support (BS) = 66%) is sister to a clade consisting of conifers + Gnetales (BS = 96%). Gnetales are sister to Pinaceae within the conifers, although the "Gne-Pine" clade has only 57% support. Within the major groups of gymnosperms (conifers, Gnetales, and cycads), family-level and generic relationships generally are congruent with those inferred in other analyses (Figure 1). Of the 54 gymnosperm genera represented by more than one species in the tree, 47 have at least 50% bootstrap support, 36 have at least 90% bootstrap support, and 26 have 100% bootstrap support. A full version of the bootstrap consensus tree is available as Supplemental Material.

3.3. Diversification. Although the lineage through time plots display much variation among bootstrap replicates (Figure 2), the general trend among the bootstrap trees is similar, with what appears to be high and possibly increasing diversification over the last 100 million years. Still, lineage through time plots are imprecise and difficult to interpret. If this trend of high recent diversification were true, we would expect to find evidence of increased rates of diversification in some relatively young clades.

The Δ_1 statistic indicated a significant shift in the rates of diversification at 10 nodes in the tree. Several are within the cycads. This includes the node dividing *Cycas* and *Epicycas* species from the other cycads ($P = 0.0474$) and its daughter node separating *Cycas*, *Epicycas*, and *Dioon* from the other cycads ($P = 0.157$). Also, two basal-most nodes of *Zamia* show significant shifts in diversification rates ($P = 0.014$ and 0.316). Within conifers, there is a significant shift in diversification at the most recent common ancestor of *Podocarpus* ($P = 0.017$). Also, there are significant shifts in diversification at the two basal nodes of Cupressaceae ($P = 0.0326$ and 0.0366) and within Cupressaceae, at the most recent common ancestor of *Callitris, Neocallitropsis, Actinostrobus, Widdringtonia, Fitzroya, Diselma,* and *Austrocedrus* ($P = 0.0387$). Finally, there is a significant shift in two of the basal nodes of *Picea* ($P = 0.0166$, $P = 0.0029$).

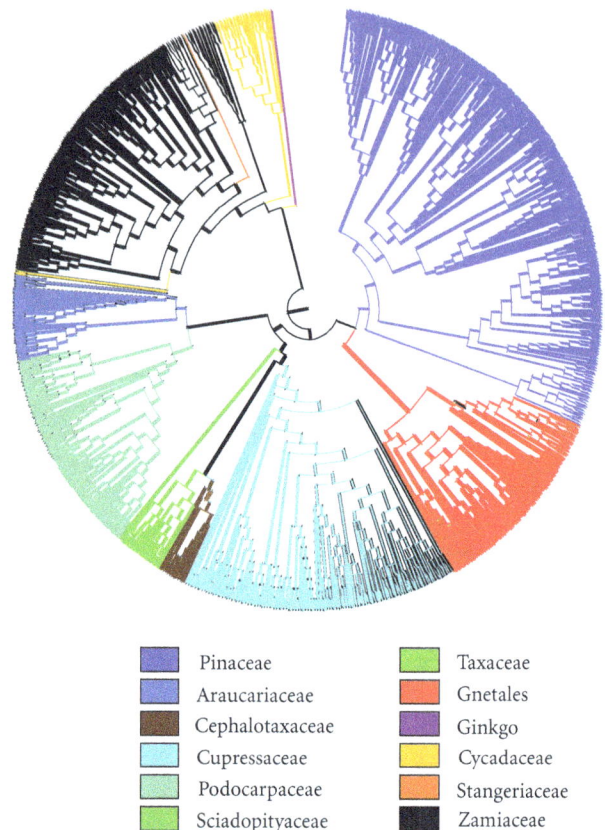

Figure 1: Overview of the ML tree of 739 gymnosperm taxa; angiosperms and outgroups have been removed. Colors represent the different families of conifers (Pinaceae, Araucariaceae, Cephalotaxaceae, Cupressaceae, Podocarpaceae, Sciadopityaceae, and Taxaceae), Gnetales, *Ginkgo*, and the families of cycads (Cycadaceae, Stangeriaceae, and Zamiaceae). A full bootstrap consensus tree is available as Supplementary Material available online at doi:10.1155/2011/292857.

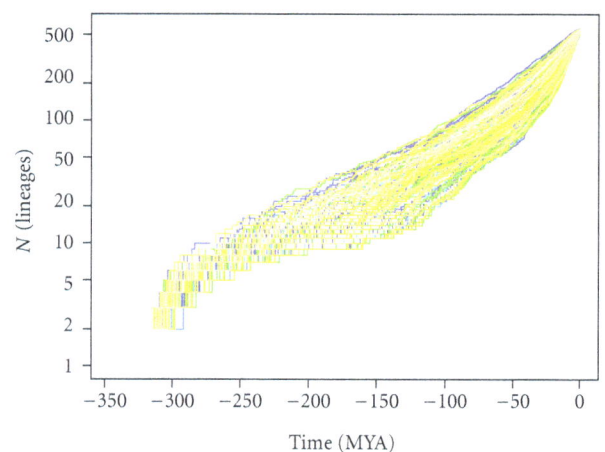

Figure 2: Lineage through time plot for the gymnosperm species. All bootstrap trees, with ultrametric branch lengths from r8s, were pruned to include only the gymnosperm taxa. Each line represents a single ML bootstrap tree. The graph shows the pattern of diversification of the gymnosperm taxa in the tree through time, as the tree grew from a single lineage at the root to the current sampling of 739 species.

3.4. Genome Size Evolution. Based on the large size of genomes of *Pinus* species, we hypothesized that there may be an increase in the rate of genome size evolution (Figure 3). We performed a series of likelihood ratio tests to examine the patterns of rate variation throughout gymnosperms, with a focus on testing for rate variation associated with conifers, Pinaceae, and *Pinus* (Table 1). In all comparisons in which *Pinus* (or a group containing *Pinus*) was compared to another group, the group with *Pinus* showed significantly elevated rates of genome size evolution (Table 1). We detected no significant shifts in rates of evolution between any groups that did not contain *Pinus*, and there was no significant difference in rates of evolution between the two *Pinus* subgenera (*Pinus* and *Strobus*; Table 1).

4. Discussion

The analyses of gymnosperm diversification and genome size evolution demonstrate the dynamic evolutionary processes of the extant gymnosperms, which sharply contrasts with their reputation as ancient, relictual species. The lineage through time plots are consistent with high, and possible growing, rates of diversification within the last 100 million years, concurrent with major radiations of angiosperms (e.g., [1, 2, 28]) and extant ferns [29]. There is evidence of numerous significant shifts in diversification within both cycads and conifers, and there is strong evidence for a recent, large increase in the rate of genome size evolution in *Pinus*. Although *Pinus* is a species-rich genus, we find no links between increased rates of diversification and shifts in rates of genome size evolution.

Advances in sequencing technology and computational biology over the past decade enable phylogenetic estimates comprising large sections the plant diversity. This study demonstrates that it is possible to construct credible phylogenetic hypotheses including nearly three quarters of the extant gymnosperm species. Unlike supertree approaches (e.g., [14]), the supermatrix methods easily incorporate branch length estimates and estimates of topological and branch length uncertainty. Still, until there is far more data per taxon, estimates of the gymnosperm phylogeny will continue to improve, and thus, it is important to consider error and uncertainty in phylogenetic estimates when using these trees to infer evolutionary processes. There are other reasons to interpret this gymnosperm tree with caution. For example, both heterogeneity in the patterns of molecular evolution and missing data can lead to erroneous estimates of trees and branch lengths in ML phylogenetic analyses (e.g., [30, 31]). Furthermore, our analysis does not attempt to incorporate evolutionary processes, such as incomplete lineage sorting or gene duplication and loss or reticulation, that may cause incongruence between the gene trees and the species phylogeny (e.g., [32]). Although this study used thousands of sequences, it does not incorporate the evolutionary perspective of low-copy nuclear genes.

Still, in many cases, evolutionary or ecological analyses that use phylogenetic trees may be robust to topological and branch length error (e.g., [33]), and the large tree of

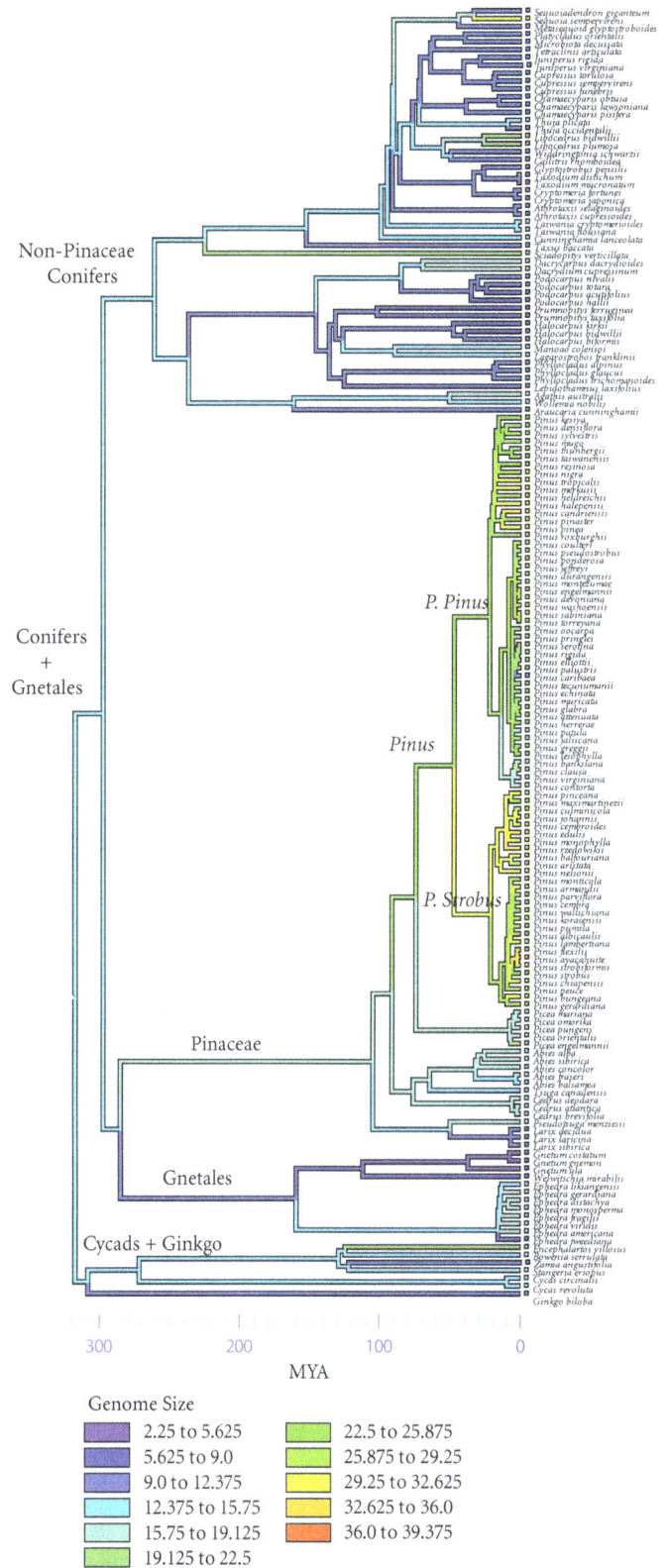

FIGURE 3: Ancestral state reconstruction of genome size (in pg DNA) on a chronogram 165 gymnosperm taxa. Different genome sizes are represented by different colors, with the ancestral genome sizes estimated with squared change parsimony.

TABLE 1: Rate estimates from the two rate parameter models from Brownie. *Indicate that the single rate model was rejected based on the Chi-squared P value (**$P < 0.005$; ***$P < 0.0005$). Significance was also assessed using AIC.

Comparison	Rates of Genome Evolution
Conifers + Gnetales	1.878***
Cycads + *Ginkgo*	0.095
Pinaceae	2.715***
Other conifers + Gnetales	0.178
Other conifers + Gnetales	0.178
Cycads + *Ginkgo*	0.095
Pinus	3.234**
Other Pinaceae	0.431
Other Pinaceae	0.431
Other conifers + Gnetales	0.178
Pinus Strobus subgenus	2.66
Pinus Pinus subgenus	3.56

gymnosperms enables sophisticated and comprehensive tests of evolutionary and ecological hypotheses. We demonstrate this with our diversification analysis, the results of which emphasize numerous, independent shifts in diversification rate throughout gymnosperms and apparently recent, high rates of diversification (Figure 2). Estimates of diversification may be affected by taxonomic sampling and inaccurate branch length estimates. However, we might expect that adding the remaining species, which would likely fit near the tips of the tree, would result in increased estimates of recent diversification. Thus, our analyses suggest the intriguing perspective that the extant gymnosperms are a vibrant, growing clade, and not simply the sole survivors of ancient diversity. Greater sampling and a more robust tree will provide a more complete view of gymnosperm diversification. With better branch length estimates, it will be possible to use more powerful likelihood-based approaches to identify clades with increasing and decreasing diversification rates [34]. With more complete taxon sampling, it may be possible to identify characters associated with changing speciation and extinction rates ([35], but see [36]).

One of the great challenges of evolutionary genomics is to identify the mechanisms of genome evolution that drive diversification. Some of the mechanisms that cause changes in genome size, such as whole-genome duplications or activity of retrotransposons, can have implications on diversification rates. Our analysis of the rates of genome size evolution demonstrate that *Pinus* is unique among gymnosperms. That is, the highly elevated rates of change in genome size appear to be limited to *Pinus*. However, in gymnosperms, we find no evidence of increases in diversification associated with *Pinus*, which displays a significantly elevated rate of genome size evolution. Furthermore, we find no obvious evidence for increase in rates of genome size evolution in clades associated with shifts in diversification. While our analysis failed to link genome size and diversification, this comparative approach for identifying shifts in genome size can inform our search for the specific drivers of the increased genomic complexity in *Pinus*, and this ultimately can help inform strategies for sequencing and assembling the first *Pinus* genomes.

Supplementary Materials

The nucleotide and C-value data matrices along with all trees are available on the Dryad data repository (http://datadryad.org/).

Acknowledgment

This work was funded by a University of Florida Research Opportunity Fund Seed Grant.

References

[1] S. Magallón and A. Castillo, "Angiosperm diversification through time," *American Journal of Botany*, vol. 96, no. 1, pp. 349–365, 2009.

[2] C. D. Bell, D. E. Soltis, and P. S. Soltis, "The age and diversification of the angiosperms re-revisited," *American Journal of Botany*, vol. 97, no. 8, pp. 1296–1303, 2010.

[3] S. A. Smith, J. M. Beaulieu, A. Stamatakis, and M. J. Donoghue, "Understanding angiosperm diversification using small and large phylogenetic trees," *American Journal of Botany*, vol. 98, no. 3, pp. 404–414, 2011.

[4] P. D. Cantino, J. A. Doyle, S. W. Graham et al., "Towards a phylogenetic nomenclature of Tracheophyta," *Taxon*, vol. 56, no. 3, pp. 822–846, 2007.

[5] D. Ohri and T. N. Khoshoo, "Genome size in gymnosperms," *Plant Systematics and Evolution*, vol. 153, no. 1-2, pp. 119–132, 1986.

[6] B. G. Murray, "Nuclear DNA amounts in gymnosperms," *Annals of Botany*, vol. 82, pp. 3–15, 1998.

[7] M. R. Ahuja and D. B. Neale, "Evolution of genome size in conifers," *Silvae Genetica*, vol. 54, no. 3, pp. 126–137, 2005.

[8] T. N. Khoshoo, "Polyploidy in gymnosperms," *Evolution*, vol. 13, no. 1, pp. 24–39, 1958.

[9] A. M. Morse, D. G. Peterson, M. N. Islam-Faridi et al., "Evolution of genome size and complexity in *Pinus*," *PLoS ONE*, vol. 4, no. 2, Article ID e4332, 2009.

[10] A. Kovach, J. L. Wegrzyn, G. Parra et al., "The *Pinus taeda* genome is characterized by diverse and highly diverged repetitive sequences," *BMC Genomics*, vol. 11, no. 1, article 420, 2010.

[11] K. L. Joyner, X.-R. Wang, J. S. Johnston, H. J. Price, and C. G. Williams, "DNA content for Asian pines parallels new world relatives," *Canadian Journal of Botany*, vol. 79, no. 2, pp. 192–196, 2001.

[12] S. E. Hall, W. S. Dvorak, J. S. Johnston, H. J. Price, and C. G. Williams, "Flow cytometric analysis of DNA content for tropical and temperate new world pines," *Annals of Botany*, vol. 86, no. 6, pp. 1081–1086, 2000.

[13] I. Wakamiya, R. J. Newton, J. S. Johnston, and H. J. Price, "Genome size and environmental factors in the genus *Pinus*," *American Journal of Botany*, vol. 80, no. 11, pp. 1235–1241, 1993.

[14] E. Grotkopp, M. Rejmánek, M. J. Sanderson, and T. L. Rost, "Evolution of genome size in pines (*Pinus*) and its life-history correlates: supertree analyses," *Evolution*, vol. 58, no. 8, pp. 1705–1729, 2004.

[15] S. F. Altschul, W. Gish, W. Miller, E. W. Myers, and D. J. Lipman, "Basic local alignment search tool," *Journal of Molecular Biology*, vol. 215, no. 3, pp. 403–410, 1990.

[16] J. D. Thompson, D. G. Higgins, and T. J. Gibson, "CLUSTAL W: improving the sensitivity of progressive multiple sequence alignment through sequence weighting, position-specific gap penalties and weight matrix choice," *Nucleic Acids Research*, vol. 22, no. 22, pp. 4673–4680, 1994.

[17] A. de Queiroz and J. Gatesy, "The supermatrix approach to systematics," *Trends in Ecology and Evolution*, vol. 22, no. 1, pp. 34–41, 2007.

[18] A. Stamatakis, "RAxML-VI-HPC: maximum likelihood-based phylogenetic analyses with thousands of taxa and mixed models," *Bioinformatics*, vol. 22, no. 21, pp. 2688–2690, 2006.

[19] J. Felsenstein, "Confidence limits on phylogenies: an approach using the bootstrap," *Evolution*, vol. 39, no. 4, pp. 783–791, 1985.

[20] M. J. Sanderson, "Estimating absolute rates of molecular evolution and divergence times: a penalized likelihood approach," *Molecular Biology and Evolution*, vol. 19, no. 1, pp. 101–109, 2002.

[21] M. J. Sanderson, "R8s: inferring absolute rates of molecular evolution and divergence times in the absence of a molecular clock," *Bioinformatics*, vol. 19, no. 2, pp. 301–302, 2003.

[22] H. Won and S. S. Renner, "Dating dispersal and radiation in the gymnosperm *Gnetum* (Gnetales)—clock calibration when outgroup relationships are uncertain," *Systematic Biology*, vol. 55, no. 4, pp. 610–622, 2006.

[23] E. Paradis, J. Claude, and K. Strimmer, "APE: analyses of phylogenetics and evolution in R language," *Bioinformatics*, vol. 20, no. 2, pp. 289–290, 2004.

[24] B. R. Moore, K. M. A. Chan, and M. J. Donoghue, "Detecting diversification rate variation in supertrees," in *Phylogenetic Supertrees: Combining Information to Reveal the Tree of Life*, O. R. P. Bininda-Emonds, Ed., pp. 487–533, Kluwer Academic, Dodrecht, The Netherlands, 2004.

[25] N. Bortolussi, E. Durand, M. G. B. Blum, and O. François, "Aptreeshape: statistical analysis of phylogenetic treeshape," *Bioinformatics*, vol. 22, no. 3, pp. 363–364, 2006.

[26] M. D. Bennett and I. J. Leitch, "Plant DNA C-values database," 2005, http://data.kew.org/cvalues/.

[27] B. C. O'Meara, C. Ané, M. J. Sanderson, and P. C. Wainwright, "Testing for different rates of continuous trait evolution using likelihood," *Evolution*, vol. 60, no. 5, pp. 922–933, 2006.

[28] H. Wang, M. J. Moore, P. S. Soltis et al., "Rosid radiation and the rapid rise of angiosperm-dominated forests," *Proceedings of the National Academy of Sciences of the United States of America*, vol. 106, no. 10, pp. 3853–3858, 2009.

[29] H. Schneider, E. Schuettpelz, K. M. Pryer, R. Cranfill, S. Magallón, and R. Lupia, "Ferns diversified in the shadow of angiosperms," *Nature*, vol. 428, no. 6982, pp. 553–557, 2004.

[30] B. Kolaczkowski and J. W. Thornton, "Performance of maximum parsimony and likelihood phylogenetics when evolution is heterogenous," *Nature*, vol. 431, no. 7011, pp. 980–984, 2004.

[31] A. R. Lemmon, J. M. Brown, K. Stanger-Hall, and E. M. Lemmon, "The effect of ambiguous data on phylogenetic estimates obtained by maximum likelihood and bayesian inference," *Systematic Biology*, vol. 58, no. 1, pp. 130–145, 2009.

[32] W. P. Maddison, "Gene trees in species trees," *Systematic Biology*, vol. 46, no. 3, pp. 523–536, 1997.

[33] E. A. Stone, "Why the phylogenetic regression appears robust to tree misspecification," *Systematic Biology*, vol. 60, no. 3, pp. 245–260, 2011.

[34] D. L. Rabosky, "LASER: a maximum likelihood toolkit for detecting temporal shifts in diversification rates," *Evolutionary Bioinformatics*, vol. 2, pp. 247–250, 2006.

[35] W. P. Maddison, P. E. Midford, and S. P. Otto, "Estimating a binary character's effect on speciation and extinction," *Systematic Biology*, vol. 56, no. 5, pp. 701–710, 2007.

[36] D. L. Rabosky, "Extinction rates should not be estimated from molecular phylogenies," *Evolution*, vol. 64, no. 6, pp. 1816–1824, 2010.

7

Saline Agriculture in the 21st Century: Using Salt Contaminated Resources to Cope Food Requirements

Bruno Ladeiro

MAP-BioPlant, Porto, Portugal

Correspondence should be addressed to Bruno Ladeiro, brunoc.f.ladeiro@hotmail.com

Academic Editor: Conceição Santos

With the continue increase of the world population the requirements for food, freshwater, and fuel are bigger every day. This way an urgent necessity to develop, create, and practice a new type of agriculture, which has to be environmentally sustainable and adequate to the soils, is arising. Among the stresses in plant agriculture worldwide, the increase of soil salinity is considered the major stress. This is particularly emerging in developing countries that present the highest population growth rates, and often the high rates of soil degradation. Therefore, salt-tolerant plants provide a sensible alternative for many developing countries. These plants have the capacity to grow using land and water unsuitable for conventional crops producing food, fuel, fodder, fibber, resin, essential oils, and pharmaceutical products. In addition to their production capabilities they can be used simultaneously for landscape reintegration and soil rehabilitation. This review will cover important subjects concerning saline agriculture and the crop potential of halophytes to use salt-contaminated resources to manage food requirements.

1. Human Population Growth and Agriculture Challenges in the 21st Century

It is estimated that in November 2011, the mankind has reached the 7 thousand million people, and the United Nations (2008) predicted a population increase up to 8.01 thousand million people in 2025. This represents a duplication of human population in approximately 50 years. So, agriculture strategies for feeding all people represent one of the most important challenges in the 21st century. Therefore, there is enormous demographic and economic pressure to rise, within the next 40 years, leading to an increase of the crop production by about 50% years in a sustainable manner to fulfil the world food necessities [1, 2]. Some other facts are aggravating this demand, namely, (a) the increase of land occupation for biofuel supply that deviates arable soils from food crops [3], (b) the challenges posed by increasing occasional episodes of extreme environmental conditions and even natural disasters often associated with the generally called "climate change events", (c) some excesses in soil pressure posed by postgreen revolution practices; (d) the raise of soil degradation and/or the increase of saline soils is growing dramatically, reducing the area of

arable land. The duality faced by mankind concerning food supply was summarized by Rudel [4] stating *"our ability to supply the growing global demand for food, fiber, and fuel, while maintaining a landscape able to provide a full suite of environmental services, hinges on our ability to produce more on less land. This pressure is particularly high in developing countries"*.

Agriculture is the first human activity and represents the major use of land across the world [5]. Agriculture can be defined as an "artificial management to enhance the food value of cultivated land" and represents "the major land use across the globe" [6]. Some emerging strategies to enhance food production involve, for example, increasing yield and cropping intensities, genetic modification [7] which may involve converging strategies of seed and germplasm preservation and improvement [2], and/or developing strategies for use of contaminated or dried soils, meanwhile abandoned as nonproductive.

This increase is particularly emerging in developing countries that present the highest population growth rates, and often the high rates of soil degradation (the Mediterranean Basin [8], Australia, Central Asia, and the Middle East and North of Africa). This pressure is so important that it is considered in the United Nations Millennium Development Goals [9]. In 2004 the Food and Agriculture Organization of the United Nations (FAO) estimated that there was 852 million food-deprived people worldwide in 2000–2002, with being the most critical places the sub-Saharan Africa and Southern Asia. However, in 2004, FAO also highlighted that despite in most places these hunger issues translate a real lack of food and deficient agricultural techniques. This problem is however the convergence of more complex problems [5]. This is evident by the statistical data pointing to approximately 9 million hungry people in industrialized countries where in theory there is no lack of food [10]. In 2008, FAO expected an increase of food requirements by 20% in developed countries and 60% in developing countries.

It is therefore recognised that food requirements are increasing quicker than crop production, and researchers and politicians increasingly highlight the urgent need to improve alternative agricultural strategies [5, 6, 11, 12]. In other words, to produce more and better it is necessary to invest in better technology within the scopes of the Millennium Development Goals strengthening practices leading to sustainable agriculture without destroying lands and natural resources [5].

Despite all these emerging food problems, in 2011 FAO reported a study highlighting the losses occurring along the entire food chain suggesting that roughly one-third of food produced for human consumption is lost or wasted globally, which amounts to about 1.3 billion tons per year. In this study, FAO revealed that much more food is wasted in the industrialized world than in developing countries, estimating a per capita food waste by consumers in Europe and North-America of 95–115 kg/year, while this figure in Sub-Saharan Africa and South/Southeast Asia is only 6–11 kg/year.

The causes of food losses and waste in low-income countries are mainly connected to financial, managerial, and technical limitations in harvesting techniques, storage, and cooling facilities in difficult climatic conditions, infrastructure, packaging, and marketing systems. While in medium- and high-income countries food is to a significant extent wasted at the consumption stage, meaning that it is discarded even if it is still suitable for human consumption.

This way, FAO, [13], states that one of the first mean to cope with world food requirements is to also promote food loss reduction which alone has a considerable potential to increase the efficiency of the whole food chain. It is important to emphasise that the reducing food losses should not be a forgotten priority.

Agriculture improvement and adjustment to the 21st challenges also needs to take into consideration the type of land and water available [11, 14].

The area of irrigated land increased enormously, between 1960' and 1980's decades. Irrigation schemes cover only 15% of the cultivated land worldwide, though it contributes with two-third of the world's food production [15]. It is also estimated that irrigation limits production in approximately 600 million hectares of "potentially suitable arable land" [11]. Taken together these data, one must be careful on predicting agriculture improvement in quantity and quality. However,

some interesting work has been done using halophytes to complement agriculture needs [16, 17].

It is estimated that the agricultural production increase will need around 202×10^6 ha of arable soil in developing countries, but only approximately 93×10^6 ha seem available [14, 18]. These data support the need to develop strategies of sustainable agriculture contrarily to what it should be expected if real sustainable agriculture practices dominated. Millions of ha of agricultural land are lost every year, mostly during the last 50–60 years of agricultural development (due to, e.g., unsustainable irrigation practices, excessive use of fertilizers, soil contamination, urban pressure, climate changes). This issue was well addressed by the USA Department of Agriculture [19] which estimated that approximately 10 million ha/year of arable soils are lost in the globe.

Fortunately, it is increasing the number of countries that subscribe the principles inherent of sustainable development and address the Millennium Development Goals, so pressure to use sustainable strategies is rising (e.g., soil and water conservation), despite the increasing pressure for food supply and urbanization.

2. Insufficient Freshwater, Salt Contamination, and Soil Degradation

Despite it may be considered as having an ubiquitous distribution in all the continents of the world, most of the arid and semiarid regions are located in developing countries [20]. These problems also are present in regions of the United States of America, Australia, Israel, or the Mediterranean Basin [21], and some of these countries have long tradition in circumventing soil degradation using that land for agriculture. For example, in the USA a large amount of waste land is being used for fields biodiesel plant production. In Australia, on the other hand, a major problem arises with dry land salinity due to rising water tables resulting from clearing the original native vegetation due to changing the type of plant population [15]. On counterpart, other countries, like Israel, possess crop production practices with unconventional water resources irrigation, and the use of brackish water deserves particular attention. Also, in Tunisia strong investigations were addressed to this thematic [22, 23] and one can find agriculture practices based in alternative plant species, most of them are halophytes, which are able to tolerate high temperatures and/or low water availability.

Salinity is one of the most widespread soil degradation processes on the Earth. Soil salinisation affects an estimated 1 to 3 million hectares in Europe, mainly in the Mediterranean countries. It is regarded as a major cause of desertification and therefore is a serious form of soil degradation being salinisation and sodification among the major degradation processes endangering the potential use of European soils. For instance, in Spain 3% of the 3.5 million hectares of irrigated land is severely affected, reducing markedly its agricultural potential while another 15% is under serious risk.

Other examples of salt-affected soil in Europe are the Caspian Basin, the Ukraine, and the Carpathian Basin (Hungary) [24].

The availability of freshwater is a major limiting factor in sustainable agriculture. The decrease of water availability is found in these developing regions of burgeoning population pressure, and limits the area of arable land and crop production for these people. As stated by Galvani [5], when it comes to extreme environments, such as arid and semiarid areas, pressure must be put in major adjustments in alternative agriculture [25].

3. Soil and Water Availability and Saline Agriculture

An innovative strategy for enhancing land and water availability is the use of salted soils and salted water, in a strategy designated as saline agriculture. This strategy is not new, as, for example, the use of seawater for crop production in coastal deserts has already been suggested in the last three decades [26–29].

This way, a possible definition for saline agriculture can be as follows.

Profitable and improved agricultural practices using saline land and saline irrigation water with the purpose to achieve better production through a sustainable and integrated use of genetic resources (plants, animals, fish, insects, and microorganisms) avoiding expensive soil recovery measures [30].

The saline water that may be used in halophyte crop irrigation can be, for example, seawater, salt-contaminated phreatic sheets, brackish water (from e.g., Estuaries), drainage water from other plantations irrigation, drainage water from humanized areas, for example, sewage [31, 32], or even water derived from aquaculture waste [33]. It was suggested that around half of the irrigation systems are susceptible to salt contamination or waterlogging, probably due to low quality of used water, leaching, and rising water tables [34]. It is therefore clear that, facing the human population pressure, the technological advances, and the increase of salinized soils and reduction of arable land usable by conventional agriculture, the use of these salinized soils in alternative agriculture may be regarded as a strategy to cope with food demand [17].

4. Saline Agriculture: An Opportunity for Saline Soils Use

Soil salinization has been worldwide recognized as being among the most important problems for crop production in arid and semi-arid regions [35]. As reported above, some of the emerging regions in risk of increasing levels of salinization of their soils are, for example, the Mediterranean Basin [36], Australia, Central Asia, the Middle East, and Northern Africa [32, 35].

Soil salinization has numerous origins, namely, natural causes provoked by, for example, the microscopic salt particles carried by the wind to inland from the oceans, or, as discussed above, some anthropic causes (secondary salinization) [32], among which irrigation water quality is one of the most important.

The use of salinized land through drainage/irrigation without using high-quality water, but instead also some salinized water, may be, therefore, the solution but demands exploration of the potential of halophytes as new emerging crops and changing mankind habits to incorporate this new crops in daily diet [25].

There is a need for more studies on the potential use of halophytes in saline agriculture and their use and incorporation in the consumers' diet. Also, the adequate conditions for increasing to industrial levels the halophyte species production (e.g., physiological studies, organoleptic and nutritional properties, etc.) deserve more attention. As stated by Koyro et al. [37] there are some requisites for the selection of tolerant plants with promising yields and characteristics that make them interesting as crops in saline agriculture: (a) screen of literature for their natural habitats, and so forth; (b) after selecting the species, determining the salinity threshold [15].

For halophytes succeed as irrigated crops, four basic conditions must be gathered:

(1) high yield potential; (2) the irrigation needs must not exceed the conventional crops and be harmless to the soil; (3) the products from halophyte crops must be able to replace the conventional crop products; (4) high-salinity agriculture must be applicable to the existing agricultural infrastructure [38].

5. Which Halophyte Crops Can We Use?

The potentiality of using halophytes in saline agriculture has been explored in the last decades. The use of halophytes in commercial cultures/exploitation, though still limited, is already being applied for some species. Also, the project "Greening Eritrea" from the Seawater Foundation [39] represents an example of how to convert a desertified region into a useful soil.

Halophytes can be improved into new, salt-resistant crops, or used as a source of genes to be introduced into conventional crop species that in general have their economical production decreased as soil salt levels increase. We'll discuss here some of the potential of halophytic species to use as emergent or already used crops in arid and semi-arid regions, in a perspective of sustainable development. The sustainable use of halophytes has multiple purposes as stimulating productive ecosystems and regreening degraded areas. About 2,600 halophytic species are known and only few are extensively studied for their potential in agriculture and as biological resources with economical potential as sources of oils, flavours, gums, resins, oils, pharmaceuticals, and fibbers [5], or with environmental potential for protection and conservation of ecosystems (e.g., improvement of soil structure and fertility, habitat for wildlife, source of biomass for the production of biodiesel) [40, 41].

We'll discuss below some of these species.

5.1. Food Yielding Halophytes. *Aster tripolium* (also known as Sea Aster or Sea Spinach) belongs to the family of the Asteraceae (Compositae). This Northern European plant, present in salt marches and estuaries, is very productive and can be cut several times with a regrowth of young shoots every 3-4 weeks. This is a familiar plant in The Netherlands since it was known as a famine food during harsh times and has become a delicacy nowadays. According to Brock et al. [42], *Aster tripolium* grows in temperate regions in its natural form, close to the coast mainly in the salt meadows. Also, some interesting research is being conducted concerning the response of *Aster tripolium* and *Puccinellia maritima* to atmospheric carbon dioxide enrichment and their interactions with flooding and salinity [43].

Salicornia bigelovii (Chenopodiaceae) is a very well-studied species. As a typical halophyte it is a succulent (having CAM metabolism) plant being cultivated for its oilseed (both for human and animal use) and straw. The residual seed meal is very rich in protein (approximately 33–34% crude protein). This oilseed halophyte has a yield and seed quality similar to the soybean, reaching yields of 2 t/ha of seed containing 28% oil and 31% protein [38]. In fact, field trials with *Salicornia bigelovii* conducted in Puerto Penasco, Mexico (a coastal desert environment), demonstrated a production of 18 t/ha of biomass and 2 t/ha of seed over a 200-day growing cycle [18] in which the seed contained 31% protein and 28% oil rich in polyunsaturated fatty acids (being the linoleic acid 74% of the total). The growth rates of animals fed with *Salicornia bigelovii* are equivalent to those fed with conventional forages on equal amounts such as alfalfa and wheat straw [18].

5.2. Oilseeds. Seeds of various halophytes, such as *Suaeda fruticosa*, *Arthrocnemum macrostachyum*, *Salicornia bigelovii*, *S. brachiata*, *Halogeton glomeratus*, *Kochia scoparia*, and *Haloxylon stocksii* possess a sufficient quantity of high quality edible oil with unsaturation ranging from 70–80%. Seeds of *Salvadora oleoides* and *S. persica* contain 40–50% fat and are a good source of lauric acid—a potential substitute for coconut oil.

Also *Diplotaxis tenuifolia* is a promising species for saline agriculture, as it lives naturally in saline/dry ecosystems, or with strong influence of sea. This plant has a potential for food (salads) and forages. *Diplotaxis tenuifolia* exposed to salt showed halophyte-like behaviour similar to those of known halophyte plants as *Cakile maritima* [44] for example or fodder beet [45]. In a recent study by Guerra [46], *Diplotaxis* plants survived, grew, and reproduced in all salinity up to 300 mM NaCl, despite a small growth reduction was found in the higher concentration, with good values of growth at 100 mM and with no loss of nutritional value. This suggests that this species has high potential in large scale production of salt water/soils.

Diplotaxis L. (DC.) genus is original from the Mediterranean Basin in dry, nutrient-rich, and sandy soils [47, 48], despite nowadays is relatively distributed around the world [49, 50]. It can be grown with other halophytes as *Cakile maritima* in dunes and in road verges usually at a small distance

from the sea [46]. Some studies have been developed with this species, for example in The Netherlands and in Portugal. This last has a coast with unique characteristics, potential use of large-scale production of this and other halophytes, and opening perspectives for an emerging market opportunity.

5.3. Fuel Wood and Timber. More than a billion people in developing countries rely on wood for cooking and heating. Quite often fuel wood is obtained from salt-tolerant trees and shrubs, which may include species of *Prosopis*, *Tamarix*, *Salsola*, *Acacia*, *Suaeda*, *Kochia*, *Capparis*, *Casuarina*, *Pithecellobium*, *Parkinsonia*, and *Salvadora*. In addition species like *Dalbergia sissoo*, *Pongamia pinnata*, *Populus euphratica*, and *Tamarix spp.* could provide good-quality wood. In coastal areas the mangroves species of *Rhizophora*, *Ceriops*, *Avicennia* and *Aegiceras* are good fuel woods and also contribute to charcoal production.

On the other hand, the agricultural applications on biofuels to cope with the global energy requirements have been increasing since high oil prices are creating new markets for agricultural commodities that can be used as feedstock for the production of bio-fuels. This way, bio-fuels are being promoted as contributing to a wide range of policy objectives, as providing greater energy security with regard to liquid fuels, increasing rural incomes, lowering greenhouse gas emissions, and providing economic opportunities for developing countries [51].

Fuels such as biodiesel can be produced from biomass ranging from cow manure to wood chips. The advantage of developing biofuel from halophytes as opposed to other types of biomass is that saltwater plants are not dependent on fresh water, which is in increasingly short supply, and can instead be irrigated using plentiful seawater supplies.

Suitable areas around the world for cultivating halophytes include the Sahara desert, Western Australia, Southwest USA, parts of the Middle East, and parts of Peru. Scientists claim that an area smaller than the Sahara desert could yield enough biomass to replace the world's fossil fuel requirements. Furthermore, in the case of halophyte production, as these plants are grown in the desert, they will produce a cooler, wetter, land surface, which could lead to rainfall in areas of the world where rainwater is in short supply [52–54].

This way, the innovation and development of biofuels and the controversies around the sustainability of conventional crops as a feedstock for biofuel will likely help halophytes to get timely focus and advantage since the most currently used biofuel sources are conventional crops such as corn, sugar cane, rape oilseed, and palm oil and do not meet for most of the sustainability related issues. Specific examples include bioethanol, made from sugar and starch crops, and biodiesel, made from vegetable oils, animal fats, and other recycled greases.

Aspects such as competition with food, land use, energy efficiency, pressure on other important resources including freshwater, rain forest, and in some cases political instability are issues of ongoing controversy around conventional biofuel which in turn can lead to an increase on food prices [53, 55, 56].

5.4. Source of Chemicals. A kind of soda is obtained in large quantities from *Suaeda, Salicornia, Salsola,* and *Haloxylon* species, used in soap making and in glass industry. Seeds of *Annona glabra* are a source of insecticide [57].

5.5. Ornamental. Many halophytes are useful ornamentals; these include among others *Aster tripolium, Limoniastrum monopetalum, Batis maritima, Tamarix nilotica, Tamarix amnicola, Cistanche fistulosum, Atriplex halimus, Sesuvium portulacastrum,* and *Noronhia emarginata* [57].

5.6. Environmental Protection. Some species of halophytes can give an important contribute to the coastline protection and restoration of coastal ecosystems such as *Spartina alterniflora, Spartina maritime,* and *Avicennia marina* [58, 59]. Research is being undertaken to explore the potential of *Avicennia germinans* (black mangrove) to restore coastal and back-barrier salt marshes by taking advantage of the plants woody structure and extensive root system, providing sustainability and habitat [60].

6. Environmental and Economic Impact of Saline Agriculture

The economic analysis of saline water irrigation has three main aspects: (1) concerning the reclamation of saline and sodic soils preceding cultivation, (2) the constant use of saline water for irrigation and, (3) the reuse of drainage water for irrigation and the drainage installations [61].

According to a governmental report on "Saline agriculture farmer participatory development project in Pakistan" performed between 2002 and 2008 [30], the impact on the economy and environment was demonstrated.

In regard to the *economic benefits*, tree plantations are sustainable sources of raw materials required for a variety of industries, for example, pulp or paper, match manufacturing, sports goods, panel products, furniture timber, plywood, saw wood, fiberboard, and fuel wood. Vast quantities of dung can be saved by using fuel wood from plantings, enriching agricultural fields.

Concerning the *environmental management* impact, trees can be an important help to recover salt-affected land. Vegetation over saline soils tends to decrease salt concentration in the top soil due to increased infiltration and reduced capillary rise of water. This approach can allow farmers to get instant economic returns by growing field crops and also immediate economic benefits from saline wasteland with the help of trees. Such planting is preferred to various expensive engineering methods since it is cheaper and lasting. The vital impact of trees on the microclimate, soil erosion, and floods is well known. Moreover, large plantations create more favorable conditions to obtain rainfall.

An additional and interesting advantage demonstrated on this report is the ability of the trees to act as scavenging pollutants by removing water condensation nuclear particles and reducing fog containing gases harmful for life since the branches, leaves, and stems filter and precipitate dust (carried by the wind) [30].

Saline agriculture can also be a potential strategy for reducing CO_2 in the atmosphere in degraded salt-affected areas. Other studies refer the use of halophyte crops to reclaim saline soils [62] since these plants can reduce the salt content of soil over time [63].

The environmental impact assessment (EIA) needs equally to concentrate on means in which positive impacts can be enhanced and negative impacts mitigated [11].

Yamaguchi and Blumwald [12] consider the "identification of key genetic determinants of stress tolerance" a precondition to the knowledge expansion on salt tolerant crops. These same authors consider two different genetic approaches, first the exploitation of natural genetic variations through marker-assisted breeding and second the generation of transgenic plants, a very popular subject that is being currently addressed by researchers. In fact, the use and improvement of conventional and molecular breading (as well as molecular genetic modification—GM) are subjects of research to adapt our existing food crops to increasing temperatures, decreased water availability in some places and flooding in others, rising salinity [64], and changing pathogen and insect threats [65]. An environmental and important good of such research is to increase the efficiency of crops nitrogen uptake and use, due to nitrogenous compounds in fertilizers being the main contributors to waterway eutrophication and greenhouse gas emissions [64].

7. Conclusion and Future Perspectives

The increase in nutrient demand by the prosperous society and the decreasing availability of arable land and freshwater lead to the problem of agriculture sustainable development. With these perspectives, saline agriculture is coming up as an emerging role.

In the future this agricultural area may be of extreme importance in Mediterranean countries due to the increasing soil degradation in some regions and to their geographic and climatic conditions [46].

Without a comprehensive and long-term strategy adaptable to the prevailing economic, climatic, social, as well as edaphic and hydrogeological conditions, it is not considered possible to meet the future challenges of irrigated agriculture using poor-quality water.

References

[1] N. Borlaug, "Feeding a Hungry world," *Science*, vol. 318, no. 5849, pp. 318–359, 2007.

[2] F. Guillon, C. Larré, F. Petipas et al., "A comprehensive overview of grain development in *Brachypodium distachyon* variety Bd21," *Journal of Experimental Botany*, vol. 63, no. 2, pp. 739–755, 2012.

[3] A. G. Prins, B. Eickhout, M. Banse, H. van Meijl, W. Rienks, and G. Woltjer, "Global impacts of European agricultural and biofuel policies," *Ecology and Society*, vol. 16, no. 1, p. 49, 2011.

[4] T. K. Rudel, L. Schneider, M. Uriarte et al., "Agricultural intensification and changes in cultivated areas, 1970–2005," *Proceedings of the National Academy of Sciences of the United States of America*, vol. 106, no. 49, pp. 20675–20680, 2009.

[5] A. Galvani, "The challenge of the food sufficiency through salt tolerant crops," *Reviews in Environmental Science and Biotechnology*, vol. 6, no. 1–3, pp. 3–16, 2007.

[6] S. M. Howden, J. F. Soussana, F. N. Tubiello, N. Chhetri, M. Dunlop, and H. Meinke, "Adapting agriculture to climate change," *Proceedings of the National Academy of Sciences of the United States of America*, vol. 104, no. 50, pp. 19691–19696, 2007.

[7] T. J. A. Bruce, "GM as a route for delivery of sustainable crop protection," *Journal of Experimental Botany*, vol. 63, no. 2, pp. 537–541, 2012.

[8] R. Choukr-Allah, *Introduction Advanced Course on Halophyte Utilization in Agriculture*, 1993.

[9] http://www.un.org/millenniumgoals/.

[10] Food and Agriculture Organization, *Monitoring progress towards the World Food Summit and Millennium Development Goals*, Food and Agriculture Organization, Roma, Italia, 2004.

[11] Food and Agriculture Organization, "The use of saline waters for crop production," Tech. Rep., Food and Agriculture Organization, Rome, Italy, 1992.

[12] T. Yamaguchi and E. Blumwald, "Developing salt-tolerant crop plants: challenges and opportunities," *Trends in Plant Science*, vol. 10, no. 12, pp. 615–620, 2005.

[13] Study conducted for the International Congress SAVE FOOD! at Interpack 2011 Düsseldorf, Germany—Extent, Causes and Prevention—by Jenny Gustavsson, Christel Cederberg Ulf Sonesson, Swedish Institute for Food and Biotechnology (SIK) Gothenburg, Sweden and Robert van Otterdijk, Alexandre Meybeck, FAO, Rome, Italy, 2011.

[14] M. A. Khan and N. C. Duke, "Halophytes—a resource for the future," *Wetlands Ecology and Management*, vol. 9, no. 6, pp. 455–456, 2001.

[15] R. Munns, "Comparative physiology of salt and water stress," *Plant, Cell and Environment*, vol. 25, no. 2, pp. 239–250, 2002.

[16] L. Bernstein, "Effects of salinity and sodicity on plant growth," *Annual Review of Phytopathology*, vol. 13, pp. 295–312, 1975.

[17] V. P. S. Shekhawat, A. Kumar, and K.-H. Neumann, "Bioreclamation of secondary salinized soils using halophytes," in *Biosaline Agriculture and Salinity Tolerance in Plants*, pp. 147–154, 2006.

[18] E. P. Glenn, L. F. Pitelka, and M. W. Olsen, "The use of halophytes to sequester carbon," *Water, Air, and Soil Pollution*, vol. 64, no. 1-2, pp. 251–263, 1992.

[19] http://www.usda.gov/.

[20] F. Rasouli, A. K. Pouya, and S. A. Cheraghi, "Hydrogeochemistry and water quality assessment of the Kor-Sivand Basin, Fars province, Iran," *Environmental Monitoring and Assessment*, vol. 184, no. 8, pp. 4861–4877, 2012.

[21] R. Setia, P. Smith, P. Marschner, J. Baldock, D. Chittleborough, and J. Smith, "Introducing a decomposition rate modifier in the rothamsted carbon model to predict soil organic carbon stocks in saline soils," *Environmental Science and Technology*, vol. 45, no. 15, pp. 6396–6403, 2011.

[22] A. Lakhdar, M. Rabhi, T. Ghnaya, F. Montemurro, N. Jedidi, and C. Abdelly, "Effectiveness of compost use in salt-affected soil," *Journal of Hazardous Materials*, vol. 171, no. 1–3, pp. 29–37, 2009.

[23] H. Bchini, M. B. Naceur, R. Sayar, H. Khemira, and L. B. Kaab-Bettaeïb, "Genotypic differences in root and shoot growth of barley (*Hordeum vulgare* L.) grown under different salinity levels," *Hereditas*, vol. 147, no. 3, pp. 114–122, 2010.

[24] Stand Management and Natural Hazards Unit, http://eusoils.jrc.ec.europa.eu/, Joint Research Centre—The European Commission's in-house science service, http://ec.europa.eu/dgs/jrc/index.cfm, and Institute for Environment and Sustainability (IES), http://ies.jrc.ec.europa.eu/, European Commission, March 2012.

[25] M. Qadir and J. D. Oster, "Crop and irrigation management strategies for saline-sodic soils and waters aimed at environmentally sustainable agriculture," *Science of the Total Environment*, vol. 323, no. 1–3, pp. 1–19, 2004.

[26] H. Boyko, *Salinity and Aridity: New Approaches to Old Problems*, The Hague, 1966.

[27] E. Epstein, J. D. Norlyn, and D. W. Rush, "Saline culture of crops: a genetic approach," *Science*, vol. 210, no. 4468, pp. 399–404, 1980.

[28] E. Glenn, N. Hicks, J. Riley, and S. Swingle, "Seawater irrigation of halphytes for animal feed," in *Halophytes and Biosaline Agriculture*, Marcel Dekker, New York, NY, USA, 1995.

[29] E. Glenn, S. Miyamoto, D. Moore, J. J. Brown, T. L. Thompson, and P. Brown, "Water requirements for cultivating *Salicornia bigelovii* Torr. with seawater on sand in a coastal desert environment," *Journal of Arid Environments*, vol. 36, no. 4, pp. 711–730, 1997.

[30] Z. Aslam, A. R. Awan, M. Rizwan, A. Gulnaz, and K. A. Malik, *Governmental Report on Saline Agriculture Farmer Participatory Development Project in Pakistan*, Nuclear Institute for Agriculture and Biology (NIAB) Pakistan Atomic Energy Commission, Faisalabad, Pakistan, 2009.

[31] C. M. Grieve and D. L. Suarez, "Purslane (*Portulaca oleracea* L.): a halophytic crop for drainage water reuse systems," *Plant and Soil*, vol. 192, no. 2, pp. 277–283, 1997.

[32] N. P. Yensen, "Halophyte uses for the twenty-first century," in *Ecophysiology of High Salinity Tolerant Plants*, pp. 367–396, Springer, Berlin, Germany, 2006.

[33] E. R. Porto, M. C. C. Amorim, M. T. Dutra, R. V. Paulino, T. L. Brito, and A. N. B. Matos, "Rendimento da Atriplex nummularia irrigada com efluentes da criação de tilápia em rejeito da dessalinização de água," *Revista Brasileira de Engenharia Agrícola e Ambiental*, vol. 10, no. 1, pp. 97–103, 2006.

[34] I. Szabolcs, "Salt affected soils as the ecosystem for halophytes," in *International Workshop on Halophytes for Reclamation of Saline Wastelands and as Resource for Livestock*, Nairobi, Kenya, 1992.

[35] FAO, "Land and plant nutrition management service," http://www.fao.org/ag/AGL/public.stm/, 2008.

[36] B. Nedjimi, Y. Daoud, and M. Touati, "Growth, water relations, proline and ion content of in vitro cultured Atriplex halimus subsp. schweinfurthii as affected by CaCl2," *Communications in Biometry Crop Science*, vol. 1, no. 2, pp. 79–89, 2006.

[37] H.-W. Koyro, N. Geissler, S. Hussin, and B. Huchzermeyer, "Mechanisms of cash crop halophytes to maintain yields and reclaim saline soils in arid areas," in *Ecophysiology of High Salinity Tolerant Plants*, M. A. Khan and D. J. Weber, Eds., pp. 345–366, Springer, Amsterdam, The Netherlands, 2006.

[38] E. P. Glenn, J. J. Brown, and E. Blumwald, "Salt tolerance and crop potential of halophytes," *Critical Reviews in Plant Sciences*, vol. 18, no. 2, pp. 227–255, 1999.

[39] http://www.seawaterfoundation.org/.

[40] R. F. Barnes and J. E. Baylor, "Forages in a changing world," in *Forages: An Introduction to Grassland Agriculture*, pp. 3–13, State University Press, New York, NY, USA, 1995.

[41] J. Y. Zhang and Z. Y. WangM. A. Jenks, P. M. Hasegawa, and S. M. Jain, "Recent advances in molecular breeding of forage crops for improved drought and salt stress tolerance," in *Advances in Molecular Breeding towards Salinity and Drought Tolerance*, pp. 797–817, Springer, 2007.

[42] J. Brock, S. Aboling, R. Stelzer, E. Esch, and J. Papenbrock, "Genetic variation among different populations of *Aster tripolium* grown on naturally and anthropogenic salt-contaminated habitats: implications for conservation strategies," *Journal of Plant Research*, vol. 120, no. 1, pp. 99–112, 2007.

[43] G. M. Lenssen, W. E. van Duin, P. Jak, and J. Rozema, "The response of *Aster tripolium* and *Puccinellia maritima* to atmospheric carbon dioxide enrichment and their interactions with flooding and salinity," *Aquatic Botany*, vol. 50, no. 2, pp. 181–192, 1995.

[44] A. Debez, W. Taamalli, D. Saadaoui et al., "Salt effect on growth, photosynthesis, seed yield and oil composition of the potential crop halophyte *Cakile maritima*," in *Biosaline Agriculture and Salinity Tolerance in Plants*, pp. 55–63, 2006.

[45] B. H. Niazi, *The Response of Fodderbeet to Salinity: Introduction of a Non Conventional Fodder Crop (Fodderbeet) to Salt Affected Lands of Pakistan [Ph.D. thesis]*, Department of Systems Ecology, Institute of Ecological Science, Faculty of Earth and Life Sciences, Vrije Universiteit, Amsterdam, The Netherland, 2007.

[46] C. Guerra, *Salt Effects on Growth, Nutrient and Secondary Compound Contents of Diplotaxis tenuifolia [M.S. thesis]*, University of Aveiro, Aveiro, Portugal, 2008.

[47] L. F. D'Antuono, S. Elementi, and R. Neri, "Glucosinolates in *Diplotaxis* and *Eruca* leaves: diversity, taxonomic relations and applied aspects," *Phytochemistry*, vol. 69, no. 1, pp. 187–199, 2008.

[48] W. Szwed and K. V. Sýkora, "The vegetation of road verges in the coastal dunes of the Netherlands," *Folia Geobotanica*, vol. 31, no. 4, pp. 433–451, 1996.

[49] V. V. Bianco, "Rocket, an ancient underutilized vegetable crop and its potential," in *Rocket Genetic Resources Network*, S. Padulosi, Ed., pp. 38–60, International Plant Genetic Resources Institute, Rome, Italy, 1996.

[50] M. W. Hanson, "Hackney Marshes," Wildlife Survey and Management Plan 7, 2004.

[51] *Current World Fertilizer Trends and Outlook to 2011/12*, Food and Agriculture Organization of the United Nations, Rome, Italy.

[52] Biopact Towards a green energy pact between Europe and Africa, Belgium, http://news.mongabay.com/bioenergy, 2007.

[53] M. Khanna, G. Hochman, D. Rajagopal, S. Sexton, and D. Zilberman, "Sustainability of food, energy and environment with biofuels," *CAB Reviews*, vol. 4, no. 28, pp. 1–10, 2009.

[54] J. W. 'leary and E. P. Glenn, "Global distribution and potential for halophytes," in *Halophytes as a Resource for Livestock and for Rehabilitation of Degraded Lands, Vol. 32*, Tasks for vegetation science, 34, pp. 7–17, 1994.

[55] R. Murphy, J. Woods, M. Black, and M. McManus, "Global developments in the competition for land from biofuels," *Food Policy*, vol. 36, supplement 1, pp. S52–S61, 2011.

[56] M. Harvey and S. Pilgrim, "The new competition for land: food, energy, and climate change," *Food Policy*, vol. 36, no. 1, pp. 40–51, 2011.

[57] M. A. Khan, R. Ansari, B. Gul, and M. Qadir, "Crop diversification through halophyte production on salt-prone land resources," *CAB Reviews*, vol. 1, article 048, pp. 1–9, 2006.

[58] J. C. Callaway and J. B. Zedler, "Restoration of urban salt marshes: lessons from southern California," *Urban Ecosystems*, vol. 7, pp. 107–124, 2004.

[59] H. M. E. Shaer, *Halophytes as Cash Crops for Animal Feeds in Arid and Semi-Arid Regions*, Birkhäuser, Basle, Switzerland, 2006.

[60] L. K. Alleman and M. W. Hester, "Refinement of the fundamental niche of black mangrove (*Avicennia germinans*) seedlings in Louisiana: applications for restoration," *Wetlands Ecology and Management*, vol. 19, no. 1, pp. 47–60, 2011.

[61] B. S. Tanwar, "Saline water management for irrigation," in *Proceedings of the (3rd Revised Draft) International Commission on Irrigation and Drainage (ICID '03)*, pp. 1–140, New Delhi, India, 2003.

[62] C. H. Keiffer and I. A. Ungar, "The effect of extended exposure to hypersaline conditions on the germination of five inland halophyte species," *American Journal of Botany*, vol. 84, no. 1, pp. 104–111, 1997.

[63] E. P. Glenn, J. J. Brown, and E. Blumwald, "Salt tolerance and crop potential of halophytes," *Critical Reviews in Plant Sciences*, vol. 18, no. 2, pp. 227–255, 1999.

[64] N. V. Fedoroff, D. S. Battisti, R. N. Beachy et al., "Radically rethinking agriculture for the 21st century," *Science*, vol. 327, no. 5967, pp. 833–834, 2010.

[65] P. J. Gregory, S. N. Johnson, A. C. Newton, and J. S. I. Ingram, "Integrating pests and pathogens into the climate change/food security debate," *Journal of Experimental Botany*, vol. 60, no. 10, pp. 2827–2838, 2009.

8

Phytotoxicity by Lead as Heavy Metal Focus on Oxidative Stress

Sónia Pinho[1] and Bruno Ladeiro[2]

[1] *Biology Department, University of Aveiro, 3810-193 Aveiro, Portugal*
[2] *MAP-BioPlant, Biology Department, Faculty of Science, University of Porto, 4169-007 Porto, Portugal*

Correspondence should be addressed to Sónia Pinho, sonia.andreia@ua.pt

Academic Editor: Conceição Santos

In the recent years, search for better quality of life in urban areas has been provoking an increase in urban agriculture. However, this new way of agriculture can bring risks to human health since this land is highly contaminated, due to anthropogenic activities. This way, lead (Pb) phytotoxicity approach must be taken into consideration since it can be prejudicial to human health through food chain. Pb is a common environmental contaminant, which originate numerous disturbances in plant physiological processes due to the bioacummulation of this metal pollutant in plant tissues. This review, focus on the uptake and interaction of lead by plants and how it can be introduced in food chain. Special attention was taken to address the oxidative stress by lead regarding the effects produced in plant physiological and biochemical processes. Furthermore, the antioxidant defence system was taken into consideration. Phytoremediation is applied on site or chronic polluted soils. This emerging technique is useful to bioaccumulate, degrade or decrease risks associated with contaminants in soils, water or air through the use of hyperaccumulaters. In addition, the impact of nanoparticles in plant science was also focused in this article since some improving properties in plants have been increasingly investigated.

1. General Introduction

Metals occur naturally in the environment as constituents of the Earth's crust [1]. They tend to accumulate and persist in the ecosystems due to their stability and mainly because they cannot be degraded or destroyed.

Plants absorb numerous elements from soil. Some of the absorbed elements are referred to as essentials because they are required for plants to complete their life cycle. Certain essential transition elements such as iron, manganese, molybdenum, copper, zinc, and nickel are known as micronutrients because they are required by plants in minute quantity [2]. Other transition metals such as silver, gold and cobalt [3, 4], and nontransition elements like aluminum [5] have proven to have a stimulatory effect on plant growth, but are not considered essential. Moreover, it has been documented elsewhere that plants also absorb elements which have no known biological function and are even known to be toxic at low concentrations. Among these are the heavy metals arsenic, cadmium, chromium, mercury, and Pb. However, even micronutrients become toxic for plants when absorbed above certain threshold values [6].

2. Lead (Pb)

Lead (Pb) is a silvery-white highly malleable metal. Among his physical properties, at normal environmental conditions this metal is presented in the solid state; it is dense, ductiles, and very soft with poor electrical conductivity when compared to most other metals. The chemical symbol for lead, Pb, is an abbreviation of the Latin word *plumbum*, meaning soft metal.

Pb is rarely found in native form in nature but it combines with other elements to form a variety of interesting and beautiful minerals. Galena, which is the dominant Pb ore mineral, is blue-white in color when first uncovered but tarnishes to dull gray when exposed to air [7].

Archeological research indicates that Pb has been used by humans for a variety of purposes for more than 5,000 years. In fact, archeological discoveries found glazes on prehistoric ceramics. The Egyptians used grounded Pb ore as eyeliner with therapeutic proprieties and cosmetic kohl, Pb-based pigments were used as part of yellow, red, and white paint. In ancient Rome-Pb was used to build pipes for water transportation [7–9].

Not so long ago, Pb had a widespread use in all anthropogenic activities, for instance, leaded paints, automobile batteries (as lead oxide), ammunitions, molten Pb as coolant, leaded glass, crystal, and fossil fuels. Until recently, tetraethyllead (TEL) was commonly used in petrol fuels as an inexpensive additive used since 1920. TEL was banned in most industrialized countries in the late 1990s to early 2000s due to environmental and health concerns over air and soil pollution (e.g., the areas around roads) and the accumulative neurotoxicity of Pb. This additive compound, however, is still used today in aviation fuel for piston-engine-powered aircraft. Even today Pb is still used in protective coatings with applications for radiation shielding in medical analysis [10, 11].

According to the U.S. Agency for Toxic Substances and Disease Registry, environmental levels of Pb have increased more than 1,000-fold over the past three centuries as a result of human activity. The greatest increase took place between 1950 and 2000 and reflected the increased use of leaded gasoline worldwide.

Pb commonly occurs in mineral deposits along with other base metals, such as copper and zinc which have been mined on all continents except Antarctica.

Currently, approximately 240 mines in more than 40 countries produce Pb. World mine production was estimated to be 4.1 million metric tons in 2010, and the leading producers were China, Australia, the United States, and Peru, in descending order of output. In recent years, Pb was mined domestically in Alaska, Idaho, Missouri, Montana, and Washington. In addition, secondary (recycled) Pb is a significant portion of the global Pb supply.

World consumption of refined Pb was 9.35 million metric tons in 2010. The leading refined Pb consuming countries were China, the United States, and Germany. Demand for Pb worldwide is expected to grow largely because of increased consumption in China, which is being driven by growth in the automobile and electric bicycle markets [12, 13].

According to Geological Society of America (http://geology.com/usgs/lead/) the worldwide supply and reserves of Pb are present on Table 1.

3. Pb in Agriculture and Main Causes of Soil Contamination: The Status in the 21st Century

Accordingly to an increased number of studies, food crops accumulate trace metals in their tissues when grown on contaminated soil with Cd, Pb, and Zn from metal smelting activity, irrigation with wastewater, disposal of solid wastes including sewage sludge, vehicular exhaust, and adjacent industrial activity. Long-term use of these wastewaters on agricultural lands often results in the buildup of elevated levels of heavy metals in soils [15, 16].

In addition, in countries with a high demand for food, contaminated arable land is used for crops like rice, cereal grains, and potatoes [6].

Increasing concern on the lack of suitable land for agriculture is prompting urban farmers to use contaminated

TABLE 1: Pb production and reserves. Data from [14].

Country	Production (1000 m³ ton)	Reserves
USA	400	7000
Australia	620	27 000
Bolivia	90	1600
Canada	65	650
China	1600	13 000
India	95	2600
Ireland	45	600
Mexico	185	5600
Peru	280	6000
Poland	35	1500
Russia	90	9200
South africa	50	300
Sweden	65	1100
Other	330	4000
Total	4100	80 000

land, such as waste disposal sites, to produce food crops. This situation is exacerbated by rapid population growth, urbanization and industrialization [17]. Thus, urban agriculture, practiced widely in developing countries, can be at great risk due to the proximity of these contaminant sources [18, 19].

In urban agriculture, wastewater and solid organic wastes are often the main sources of water and fertilizer used to enhance the yields of stable crops and vegetables. This way, municipal or industrial effluents and solid wastes, often rich in trace metals, contribute significantly to metal loadings in irrigated and waste-amended urban soils. However, studies conducted in soils where the atmospheric deposition was the dominant pathway for Pb contamination, revealed that the Pb concentration in those soils decreased with the increasing distance from the road.

Facing the rising population in urban areas, urban agriculture faces problems regarding the balance of the food needs with the potential hazards arising from the use of contaminated urban sites for food production and effluents for irrigation. Previous studies of metal uptake have focused mainly on crop species grown in the developed world and comparatively little information is available concerning vegetables typically grown in periurban environments in developing countries [20].

4. Edible Vegetables Affected by Pb Contamination

Many researchers have shown that some common vegetables are capable of accumulating high levels of metals from the soils [21, 22].

Studies conducted with edible vegetables species revealed the correlations between the Pb content in the soils and environment and its effects in vegetables.

Othman [23] reported a direct positive correlation of Zn and Pb levels between soils and vegetables. In this study, edible portions of five varieties of green vegetables (collected from several areas in Dar Es Salaam, Africa) were analyzed for Pb, Cd, Cr, Zn, Ni, and Cu [16].

Tangahu and colleagues [24] demonstrated Pb accumulation in plant tissues (mg/g dry weight) of the roots, shoots, and leaves from different species. They suggested that several plants could accumulate Pb in their tissues to more than 50 mg/g dry weight of plant. Among those species are *Brassica campestris* L, *Brassica carinata A. Br.*, *Brassica juncea (L.) Czern*, and *Brassica nigra (L.) Koch* that could accumulate more than 100 mg Pb/g dry weight [24]. Also Uwah [22] suggested that certain species of *Brassica* (cabbage) are hyperaccumulators of heavy metals into their edible tissues.

More studies (De la Rosa et al. [25]) suggested that some wild plants (*Prosopis sp.* and *Salsola kali*) edible by humans and/or animals were recently identified as potential hyperaccumulators of Pb and Cd, respectively.

5. Pb Uptake by Plants

As Pb is not an essential element, plants do not have channels for Pb uptake. Instead, this element is bound to carboxylic groups of mucilage uronic acids on root surfaces [26, 27], but it is still unknown how this element goes into the root tissue. Although some plants species tolerate Pb through complexation and inactivation (*Allium cepa, Hordeum vulgare* and *Zea mays*), other species experience toxicity (*Brassica napus* and *Phaseolus vulgaris*) because Pb hampers some metabolic pathways [28]. In a few plant species, the excess of Pb inhibits seed germination, plant growth, and chlorophyll synthesis, among other effects [6].

Pb is considered to have low solubility and availability for plant uptake because it precipitates as phosphates and sulfates, chemicals commonly found in the rhizosphere of plants [29]. Also, Pb is immobilized in soil when it forms complexes with the organic matter [6].

Several studies have shown that most of the absorbed Pb remains accumulated in the roots, making the root the first barrier for the Pb translocation to the above ground plant parts, [29] acting like a natural barrier. Moreover, the increase in accumulation level is directly proportional to the amount of exogenous Pb.

Uptake behavior is known to depend on total soil concentration, soil physico-chemical conditions, and the species and genotypes of the plants involved [30]. Authors have reported the effect of pH variation in Pb uptake, in different plant species: in low pH soils (3.9) an increased mobility of Pb was observed, resulting in higher uptake. Also, in addition to soil factors and plant species, previous studies have shown that trace metal concentrations may differ between cultivars of individual crop species when grown on the same soil, making the risks associated with contaminated soils, with trace metals, difficult to assess [20].

Once inside the root cortex, Pb moves in the apoplastic space, using the transpiration conductive system [28, 31]. It can also bypass the endodermis and gain symplastic access in the young root zone and in sites of lateral root initiation [32]. Pb has been shown to enter and move within the cytoplasm and proteins mediating cross-membrane movement of Pb have been identified [33, 34]. Most of the Pb absorbed by roots is in the form of extracellular precipitate (as phosphate and carbonate) or is bound to ion exchangeable sites in the cell walls [35]. The unbound Pb is moved through Ca channels accumulating near the endodermis [36].

Previous experimental results suggest that at low concentration, the Casparian strip of the endodermis is a partial barrier for Pb movement into the central cylinder tissue [37]. Depending on the plants exposed, different cellular types can be used to store Pb [36]. Varga et al. [38] found that in roots of wheat, Pb is fixed to the cell wall but it can be removed as a complex using citric acid, However, Marmiroli [39] reported that in European walnut (*Juglans regia*), Pb is retained in the lignocellulosic structure of roots. On the other hand, a small portion can also be translocated upwards to stems, leaves, and probably seeds [6].

Results from the Gardea-Torresdey research group have shown (unpublished data) that in hydroponically grown honey mesquite (*Prosopis sp.*) associated with *Glomus deserticola* and treated with high-Pb concentrations (more than 50 mg Pb L^{-1}), Pb concentrates in the phloem tissues, which suggests the Pb movement through the xylem to leaves, returning through the phloem to the plant body. As described by Cobbett [40], Pb, like other toxic elements, is complexed by the cysteine-rich low molecular weight polypeptides widely known as phytochelatins. However, in *Sesbania drummondii* Pb is transported to stems and leaves in structures similar to Pb-acetate, Pb-nitrate, and Pb-sulfide [41]. In addition, López et al. [42, 43] have reported the formation of different Pb complexes in stems and leaves of alfalfa.

6. Pb in Food Chain

According to Ma [44] and Rossato et al. [45] one way of exposure of humans and mammals to Pb is via the food chain. It has long been recognized that the heavy metal accumulation in soil may result in potential health risk to plants, animals, and humans [46].

Published studies illustrating the transport of Pb in the food chain are scarce, and further research is needed to establish the role of the plant Pb compounds in the transference and metabolism of Pb in the food chain. Other researchers have reported that in humans, two binding polypeptides (thymosin and acyl-coA binding protein) are responsible for the Pb binding in kidneys [47]. Pb in blood serum is bound to proteins or complexed with low-molecular-weight compounds such as sulfhydryl groups (e.g., cysteine, homocysteine) and others as citrate, cysteamine, ergothioneine, glutathione, histidine, and oxylate [48].

Lead (II) acetate (also known as sugar of Pb) was used by the Roman Empire as a sweetener for wine, and some consider this to be the cause of dementia that affected many of the Roman Emperors [6].

Zhuang and colleagues [46] performed a study where they evaluated heavy metal transfer along a plant-insect-chicken food chain on metal contaminated soil. They concluded that chicken fed with insect-larva accumulated significantly high Pb in the liver, suggesting that the accumulation of heavy metals in specific animal organ should not be ignored. In their study they also demonstrated decreases of heavy metals along the soil-plant-insect-chicken food chain. Interestingly, cadmium (Cd) steadily declined with increasing trophic level, but concentrations of zinc (Zn) and copper (Cu) slightly increased from plant to insect larva. An important route to avoid bioaccumulation was the elimination of the four elements in feces of insect and chicken. Metal concentrations in liver, muscle, and blood of chickens were highly variable; however, the highest concentration was in liver and the lowest in blood [46].

Many people could be at risk of adverse health effects from consuming common vegetables cultivated in contaminated soil. The condition of the soil is often unknown or undocumented and therefore, exposure to toxic levels can unconsciously occur [49]. Xu and Thornton [50] suggested the existence of health risks from consuming vegetables with elevated heavy metal concentrations. The populations most affected by heavy metal toxicity are pregnant women or very young children [51]. Low birthweight and severe mental retardation of newborn children have been reported in some cases where the pregnant women ingested toxic amounts of heavy metal through direct or indirect consumption of vegetables [52]. Some of the reported effects of heavy metal poisoning are neurological disorders, central nervous system (CNS) destruction, and cancers of various body organs [16, 48].

Taking the health risks encountered in human diet as a result of high levels of heavy metals in vegetables, agricultural good practices should be implemented. This way, educational and official programs should be implemented and broadcasted to educate farmers on the problems associated with the excessive use of fertilizers and other chemicals, as well as the irrigation of crops with waste and all sorts of polluted water, and the need to grow crops with safe levels of heavy metals [16].

7. Pb Phytotoxicity

Pb is known to negatively affect some of the most classical endpoints of plant toxicity like seed germination rate, seedling growth, dry mass of roots and shoots, photosynthesis, plant water status, mineral nutrition, and enzymatic activities [53]. In general, effects are more pronounced at higher concentrations and continuance. In some cases, lower concentrations can stimulate metabolic processes and the enzymes involved in those processes [36].

These negative effects can be expressed as symptoms in the form of chlorotic spots, necrotic lesions in leaf surface, senescence of the leaf, and stunted growth. Germination of seeds is drastically affected at higher concentrations. Development and growth of root and shoot in seedling stage are also affected roots being more sensitive.

Pb negatively influences growth by reducing the uptake and transport of nutrients in plants, such as Ca, Fe, Mg, Mn, P, and Zn, and by blocking the entry or binding of the ions to ion-carriers making them unavailable for uptake and transport from roots to leaves [54]. Thus, Pb interferes with several physiological and biochemical processes; photosynthesis being one of the most affected [36].

Many European countries have adopted a bioavailability based rationale to improve the reliability of assessments of metal uptake [55]. Current legislation in most countries still uses total soil metal concentration as a simple index of hazard in contaminated soils, even though this approach does not take account of soil characteristics which influence the bioavailability of metallic pollutants in contaminated soil [56].

This has major implications for diet-related risk assessments as these often rely on a generic vegetable approach to predict the transfer of trace elements to the human diet (Section 6).

8. Metals and Oxidative Stress

8.1. General Considerations. Reactive oxygen species (ROS) are formed and degraded by all aerobic organisms, leading to either physiological concentrations required for normal cell function or excessive quantities, a state called oxidative stress [57]. Under normal physiological conditions a balance is maintained between the formation of ROS and the cells protective antioxidant mechanism. However, this balance can be disturbed with many environmental stresses including temperature, salinity, drought, flooding, nutritional imbalances and postanoxia stress, a range of gaseous pollutants (ozone, nitrogen oxides, volatile organic compounds, etc.), heavy metals, pathogens attack, and herbicides which have been indicated to increase oxidative stress, leading to overproduction of ROS overcoming the cellular antioxidant capacity [58, 59].

These stresses lead to a series of changes in the plant resulting in deficient plant growth and development by affecting molecular, biochemical, morphological, and physiological, processes [59]. The changes caused by various stressful conditions are frequently due to a secondary stress (usually osmotic or oxidative) that perturbs the structural and functional stability of membrane proteins and disrupts cellular homeostasis [60, 61].

ROS are molecules with an unpaired electron making them highly reactive, by interacting nonspecifically with a variety of cellular components [62]. All aerobic organisms are totally dependent upon redox reactions, and the transfer of single electrons and many life processes (e.g., oxidative respiration, photorespiration, photosynthesis, lipid metabolism, and cell signaling) involve free radical intermediates, molecular oxygen, and activated oxygen species such as the superoxide radical anion ($O_2^{\bullet-}$), the hydroxyl radical ($HO^{\bullet}$), and peroxyl radicals ($ROO^{\bullet}$), as well as nonradical derivatives of molecular oxygen (O_2), such as hydrogen peroxide (H_2O_2), hypochlorous acid (HOCl), singlet oxygen O_2, and peroxynitrite ($ONOO^-$) [63–65]. Although H_2O_2 per se does not contain any unpaired electrons, it is ascribed to ROS, as it can be easily converted into more aggressive radical species, for example into $HO^{\bullet}$ via Fentoncatalyzed reduction. Moreover, H_2O_2 is membrane permeable and diffusible, proving it suitable for intracellular signaling. Uncontrolled ROS production may ultimately attack macromolecules such as polyunsaturated fatty acids (PUFAs) of the chloroplast membranes, leading to toxic breakdown products and trigger lipid peroxidation [66, 67]. Peroxidation injury of the cell membrane leads to leakage of cellular contents, failure of cell function, rapid desiccation, and, eventually to a breakdown in structural integrity which can lead to necrosis [68].

Scandalios [69] described some damages induced by ROS on biomolecules:

(i) Oxidative damage to lipids occurs via several mechanisms of ROS reacting with fatty acids in the membrane lipid bilayer, leading to membrane leakage and cell death. In foods, lipid peroxidation causes rancidity and development of undesirable odors and flavors.

(ii) In proteins, oxidative damage is due to site-specific amino acids modifications since specific amino acids differ in their susceptibility to ROS attack. Other effects of protein oxidative damage are: fragmentation of the peptide chain, aggregation of cross-linked reaction products, altered electrical charge, increase of susceptibility to proteolysis, oxidation of Fe-S centers by $O_2^{\bullet-}$, destroying enzymatic function, oxidation of specific amino "marks" proteins for degradation by specific proteases and oxidation of specific amino acids (e.g., Try) leading to cross-linking.

(iii) DNA damage by ROS leads to DNA deletions, mutations, translocations, base degradation, single-strand breakage, and cross-linking of DNA to proteins.

In plants, ROS are produced within the cellular compartments like chloroplast, mitochondria, cytosol, plasma membrane, microbodies (peroxisomes and glyoxisomes), and in the cell walls during metabolic pathways as photosynthesis and photorespiration, which is the most obvious oxygenation pathways in the chloroplast [70]. The main types of active O_2 species are superoxide and H_2O_2. In peroxisomes and glyoxisomes, however, just H_2O_2 is produced.

Hydrogen peroxide (H_2O_2) is an interesting form of ROS since it has been considered to be a second messenger for signals generated by ROS due to the capacity to easily diffuse through the membranes and it's relatively long life [70]. Many studies have suggested the existence of a close interaction between intracellular H_2O_2 and cytosolic calcium in response to biotic and abiotic stresses. In fact, environmental stress might trigger a rapid and transient increase in calcium influx, which enhances the generation of H_2O_2. Yang and Poovaiah [71] and other authors have proposed calcium/calmodulin (CAM) a controlling mechanism of H_2O_2 homeostasis in plants. They also verified that increasing cytosolic Ca^{2+} can downregulate H_2O_2 levels by means of Ca^{2+}/CaM-mediated stimulation of catalase activity in tobacco leaves.

The characterization and monitorization of the oxidative stress can be assessed by many parameters: plant membranes integrity evaluation, lipid peroxidation estimation through thiobarbituric acid reactive substances, measurement of redox potential and stress-related metabolites (H_2O_2, ascorbic acid, and glutathione), enzymes like poly (ADP-ribose)-polymerase, screening for heat-shock proteins (HSP), enzymes associated with cell cycle, and evaluation of antioxidant enzymes [72]. According with Wang et al. [73], biological monitoring is a direct test of biological responses to environmental contaminants and has been proposed to complement the information given by chemical analysis. Thus, the use of biochemical or physiological parameters as biomarkers of ecotoxicity is under constant development and has the advantage of delineating effects before observed symptom.

Several techniques can be applied to assess ROS-induced DNA/chromosome injuries such as, flow cytometry (measurement of changes in chromosome number and DNA content), microdensitometry, and fluorescent in situ hybridization (FISH) (look for somatic recombination) or others that detect DNA sequence mutations such as microsatellites, restriction fragment length (RFLP), and amplified fragment length polymorphism (AFLP) [72].

Plants respond in different ways to heavy metal ion stress including exclusion, chelation, compartmentalization, and expression of stress protein genes. This way, being Pb one of the main sources of environmental pollution, previous studies have shown that Pb inhibits metabolic processes such as nitrogen assimilation, photosynthesis, respiration, water uptake, and transcription. In fact, Pb causes two types of unfavorable processes in biological systems. Firstly, Pb inactivates several enzymes by binding with their SH-groups. Secondly, Pb ions can lead to oxidative stress by intensifying the processes of reactive oxygen species (ROS) production. These processes are mutually connected and stimulate each other by destructively affecting cell structure and metabolism, resulting in a possible decreased efficiency of oxidation-reduction enzymes or the electron transport system leading to fast production of ROS in the cell. Pb can exert a negative effect on mitochondria by decreasing the number of mitochondrial cristae, which in turn can lower the capacity of oxidative phosphorylation during photosynthesis and respiration [74].

9. Plant Protection Mechanisms against Oxidative Stress: Antioxidant Defense System

Plants have different defense strategies to cope with the toxicity of heavy metals. The primary defense strategy consists in avoiding the metal entry into the cell by excluding or binding it to a cell wall. The secondary defense system is composed of various antioxidants to combat the increased production of ROS caused by metals [45].

These antioxidants are substances that (either directly or indirectly) protect cells against adverse effects of xenobiotics, drugs, carcinogens, and toxic radical reactions.

In plant cells, the antioxidant defense system is essentially constituted by superoxide dismutase (SOD), catalase (CAT), ascorbate peroxidase (APX), glutathione (GSH), ascorbate (vitamin C), tocopherol (vitamin E), and carotenoids among others. These species are distributed through the cell and are present in vacuoles and chloroplasts in higher amounts.

The following distribution for the main antioxidant components was suggested by Scandalios [69]: 73% in the vacuole (ascorbate, glutathione, and peroxidase); 17% in chloroplasts (carotenoids, α-tocopherol, ascorbate, ascorbate peroxidase, glutathione, glutathione reductase, Cu/Zn-SOD, monodehydroascorbate radical reductase, and dehydroascorbate reductase); 5% in the cytosol (ascorbate peroxidase, CuZn-SOD, catalase, peroxidase, glutathione, ascorbate, glutathione reductase, and monodehydroascorbate radical reductase); 4% in the apoplast (peroxidase and ascorbate); 1% in the mitochondria (catalase, glutathione, glutathione reductase, Mn-SOD, and monodehydroascorbate radical reductase) and peroxisomes (catalase; Cu/Zn-SOD).

Besides the antioxidative system, stress proteins (also called heat-shock proteins, HSPs) are also activated in plant species under adverse conditions [73], and the accumulation of some organic compounds in plants such as polyamines (diamine, putrescine, triamine spermidine, and tetramine spermin) and L-proline play significant roles in plant adaptation to a variety of environmental stresses [75, 76].

Clearly, plant response to Pb contamination is a key research problem, and a special effort is being undertaken in seeking factors affecting the reduction of Pb absorption or toxicity in plants [74].

Selenium (Se) is one of the potential antagonists to Pb. Recent publications indicate that Se addition may also alter the total content of heavy metals in animal tissues by reducing their uptake by plants [77–81]. Magdalena Mroczek-Zdyrska and Wójcik [74] showed that cell viability was enhanced at low concentrations whereas at high concentrations Se was pro-oxidant and increased the lipid peroxidation and cell membrane injury. On the other hand, addition of Se controlled the accumulation of Pb and Cd in lettuce and enhanced absorption of some nutritional elements (Fe, Mn, Cu, Ca, and Mg) [77].

Rossato et al. [45] discussed that Pb stress triggered an efficient defense mechanism against oxidative stress in *Pluchea sagittalis*, but its magnitude was depending on the plant organ and of their physiological status.

10. Phytoremediation

10.1. General Considerations. Heavy metals, with soil residence times of thousands of years, pose numerous health dangers to higher organisms. They are known to affect plant growth, ground cover and to have a negative impact on soil microflora. It is well known that heavy metals cannot be chemically degraded and need to be physically removed or transformed into nontoxic compounds [24].

The generic term "phytoremediation" consists of the Greek prefix phyto (plant), attached to the Latin root remedium (to correct or remove an evil) [82, 83]. Generally, according to Erakhrumen and Agbontalor [82], phytoremediation is defined as an emerging technology using selected plants to clean up the contaminated environment from hazardous contaminants to improve the environment quality [24].

The uptake mechanisms through phytoremediation technology are divided between organic and inorganic contaminants. For organics, it involves phytostabilization, rhizodegradation, rhizofiltration, phytodegradation, and phytovolatilization. For inorganics, mechanisms involved are phytostabilization, rhizofiltration, phytoaccumulation, and phytovolatilization [24].

Plants have developed highly specific and very efficient mechanisms to obtain essential micronutrients from the environment, even when these are present at low ppm levels. Plant roots, are able to solubilize and take up micronutrients from very low levels in the soil, even from nearly insoluble precipitates. Plants have also developed highly specific mechanisms to translocate and store micronutrients. The same mechanisms are also involved in the uptake, translocation, and storage of toxic elements, whose chemical properties simulate those of essential elements. Thus, micronutrient uptake mechanisms are of great interest to phytoremediation [84].

Metal accumulating plant species can concentrate heavy metals like Cd, Zn, Co, Mn, Ni, and Pb up to 100 or 1000 times more than those taken up by nonaccumulator (excluder) plants. In most cases, bacteria and fungi living in the rhizosphere closely associated with plants may contribute to mobilize metal ions, increasing the bioavailable fraction [24].

There are several factors affecting the uptake mechanisms like: plant species characteristic, properties of medium agronomical practices developed to enhance remediation (pH adjustment, addition of chelators, and fertilizers), and addition of chelating agent [24].

11. Phytoremediation Advantages and Limitations

Phytoremediation has several advantages but remains controversial in some aspects. We will describe below some of the main advantages and limitations of this strategy applied to metals (e.g., Pb).

Advantages can be: low cost (is lower than traditional processes); applicablity for a wide range of contaminants; effective in contaminant reduction; environmental friendly method and less disruptive than current techniques of physical and chemical processes (e.g., metal precipitation or otherwise attached to an insoluble form through adsorption or ion exchange [85]; solidification and stabilization are other possibilities [86]); plants can be easily monitored; possibility of recovery and reuse of valuable metals (by companies specializing in "phytomining"); aesthetically pleasing.

Limitations of phytoremediation technology are: surface area and depth occupied by the roots; slow growth and low biomass production require a long-term commitment—it is a time-consuming method; the age of plant; the survival of the plants is affected by the toxicity of the contaminated land and the general soil condition; climatic condition; soil chemistry; the contaminant concentration; bioaccumulation of contaminants, especially metals, into the plants which, then, pass them into the food chain from primary level consumers upwards or requires the safe disposal of the affected plant material; the impacts of contaminated vegetation—with plant-based systems of remediation, it is not possible to completely prevent the leakage of contaminants into the groundwater (without the complete removal of the contaminated ground, which in itself does not resolve the problem of contamination) [24].

Heavy metals uptake, by plants using phytoremediation technology, seems to be a prosperous way to remediate heavy-metals-contaminated environment. In fact it has some advantages compared with other commonly used conventional technologies. However, several factors must be considered in order to accomplish a high performance of remediation result being the most important factor a suitable plant species which can be used to uptake the contaminant. Even if the phytoremediation technique seems to be one of the best alternatives, it also has some limitations. Further research is needed.

12. Nanotechnology Applications in Plant Science

Nanomaterials and nanotechnology have been widely applied all over the world in this last decade [87].

Despite nanotechnology being mainly focused on animal science and medical research (in regard of biological applications), nanotechnology can also be applied to plant science research in order to analyze plant genomics and gene function as well as improvement of crop species [87]. However, in 1996, USEPA (United States Environmental Protection Agency) evidenced several negative effects of nanoparticles (NSPs) on growth and development of plantlets.

More recently, some phytotoxicity studies applied in higher plants, using nanoparticles, have been developed. Some examples are given as follows: improvement of the level of seed germination and root growth; increase of source of iron (or other micronutrients); enhancement of Rubisco carboxilase activity; effect on growth of specific species. The species that have been studied are *Raphanus sativus, Brassica napus, Lolium multiflorum, Lactuca sativa, Cucumis sativus, Lolium perenne* (using ZnO nanoparticles), *Zea mays* (magnetic nanoparticles), *Spinacia olerace* (TiO$_2$ nanoparticles), and *Phaseolus vulgaris* (nano alumin particules), *Triticum aestivum* (Cu nanoparticles) among others [88–91]. In order to understand the possible benefits of applying nanotechnology to agriculture, the first step should be to analyze the level of penetration and transport of nanoparticles in plants. It is established that these particles tagged to agrochemicals or to other substances could reduce the injury to plant tissues and the amount of chemicals released into the environment. Some contact is however inescapable, due to the strong interaction of plants with soil growth substrates [87].

Deposition of atmospheric particulate matter on the leaves leads to remarkable alteration in the transpiration rates, thermal balance, and photosynthesis. Da Silva et al. [92] showed that nanoparticles may enter leaf surface.

Since nanoparticles are introduced into the soil as a result of human activities, among the many fields that nanotechnology takes into consideration, it is also important to recall the interactions between nanoparticles, plants, and soil. There are many gaps in our knowledge on the ecotoxicity of NSPs and there are many unresolved problems and new challenges concerning the biological effects of these NSPs [87].

The elements for acceptable catalytic metal nanoparticles have been restricted to groups VIII and IB of the periodic table, especially palladium, platinum (Pt), silver, and Au

[93]. The majority of studies involving Pb nanoparticles were driven to electrochemical materials such as exploration of electrically conductive adhesives (ECAs) for surface mount technology and flip chip applications as Pb-free alternatives [94]. fabrication of pure Pb nanoparticles with nonoxidized surfaces due to Pb particles being readily oxidized even at ambient temperature and in high vacuum [93]; Synthesis of lead dioxide nanoparticles by pulsed current electrochemical method to use as the cathode of lead-acid batteries [95].

References

[1] M. M. Lasat, "Phytoextraction of metals from contaminated soil: a review of plant/soil/metal interaction and assessment of pertinent agronomic issues," *Journal of Hazardous Substance Research*, vol. 2, no. 5, pp. 1–25, 2000.

[2] D. I. Arnon and P. R. Stout, "The essentially of certain elements in minute quantity for plants with special reference to copper," *Plant Physiology*, vol. 14, no. 2, pp. 371–375, 1939.

[3] A. Gomez, *The nanoparticle formation and uptake of precious metals by living alfalfa plants [M.S. thesis]*, University of Texas at El Paso, 2002.

[4] L. Taiz and E. Zeiger, *Plant Physiology*, Sinauer, Sunderland, Mass, USA, 2nd edition, 1998.

[5] F. Ghanati, A. Morita, and H. Yokota, "Effects of aluminum on the growth of tea plant and activation of antioxidant system," *Plant and Soil*, vol. 276, no. 1-2, pp. 133–141, 2005.

[6] J. R. Peralta-Videa, M. L. Lopez, M. Narayan, G. Saupe, and J. Gardea-Torresdey, "The biochemistry of environmental heavy metal uptake by plants: implications for the food chain," *International Journal of Biochemistry and Cell Biology*, vol. 41, no. 8-9, pp. 1665–1677, 2009.

[7] F. M. Johnson, "The genetic effects of environmental lead," *Mutation Research*, vol. 410, no. 2, pp. 123–140, 1998.

[8] T. Rehren, "A review of factors affecting the composition of early Egyptian glasses and faience: alkali and alkali earth oxides," *Journal of Archaeological Science*, vol. 35, no. 5, pp. 1345–1354, 2008.

[9] F. Retief and L. P. Cilliers, "Lead poisoning in ancient Rome," *Acta Theologica Suplemmentum*, vol. 7, 2005.

[10] R. D. Prengaman, "The impact of the new 36 V lead-acid battery systems on lead consumption," *Journal of Power Sources*, vol. 116, no. 1-2, pp. 14–22, 2003.

[11] N. Vural and Y. Duydu, "Biological monitoring of lead in workers exposed to tetraethyllead," *Science of the Total Environment*, vol. 171, no. 1–3, pp. 183–187, 1995.

[12] H. Roberts, "Changing patterns in global lead supply and demand," *Journal of Power Sources*, vol. 116, no. 1-2, pp. 23–31, 2003.

[13] D. Smith, "Lead market: outlook for the global market and prices," *Journal of Power Sources*, vol. 78, no. 1, pp. 188–192, 1999.

[14] USGS (United States Geological Survey), "Mineral Commodity Summary," January 2011.

[15] R. K. Rattan, S. P. Datta, and A. K. Singh, "Effect of long term application of sewage effluents on available water status in soils under Keshopure Effluent Irrigation Scheme in Delhi," *Journal of Water Management*, no. 9, pp. 21–26, 2001.

[16] E. I. Uwah, N. P. Ndahi, F. I. Abdulrahman, and V. O. Ogugbuaja, "Heavy metal levels in spinach (*Amaranthus caudatus*) and lettuce (*Lactuca sativa*) grown in Maiduguri, Nigeria," *Journal of Environmental Chemistry and Ecotoxicology*, vol. 3, no. 10, pp. 264–271, 2011.

[17] G. Nabulo, *Assessing risks to human health from peri-urban agriculture in Uganda [Ph.D. thesis]*, University of Nottingham, 2009.

[18] Y. B. Ho and K. M. Tai, "Elevated levels of lead and other metals in roadside soil and grass and their use to monitor aerial metal depositions in Hong Kong," *Environmental Pollution*, vol. 49, no. 1, pp. 37–51, 1988.

[19] R. Garcia and E. Millán, "Assessment of Cd, Pb and Zn contamination in roadside soils and grasses from Gipuzkoa (Spain)," *Chemosphere*, vol. 37, no. 8, pp. 1615–1625, 1998.

[20] G. Nabulo, C. R. Black, and S. D. Young, "Trace metal uptake by tropical vegetables grown on soil amended with urban sewage sludge," *Environmental Pollution*, vol. 159, no. 2, pp. 368–376, 2011.

[21] Z.-T. Xiong, "Lead uptake and effects on seed germination and plant growth in a Pb hyperaccumulator Brassica pekinensis Rupr," *Bulletin of Environmental Contamination and Toxicology*, vol. 60, no. 2, pp. 285–291, 1998.

[22] E. I. Uwah, "Concentration levels of some heavy metal pollutants in soil, and carrot (*Daucus carota*) obtained in Maiduguri, Nigeria," *Continental Journal of Applied Sciences*, vol. 4, pp. 76–88, 2009.

[23] O. C. Othman, "Heavy metals in green vegetables and soils from vegetable gardens in Dar Es Salaam, Tanzania," *Tanzania Journal of Science*, no. 27, pp. 37–48, 2001.

[24] B. V. Tangahu, S. R. S. Abdullah, H. Basri, M. Idris, N. Anuar, and M. Mukhlisin, "A review on heavy metals (As, Pb, and Hg) uptake by plants through phytoremediation," *International Journal of Chemical Engineering Pages*, 31 pages, 2011.

[25] G. De La Rosa, J. R. Peralta-Videa, M. Montes, J. G. Parsons, I. Cano-Aguilera, and J. L. Gardea-Torresdey, "Cadmium uptake and translocation in tumbleweed (*Salsola kali*), a potential Cd-hyperaccumulator desert plant species: ICP/OES and XAS studies," *Chemosphere*, vol. 55, no. 9, pp. 1159–1168, 2004.

[26] J. L. Morel, M. Mench, and A. Guckert, "Measurement of Pb^{2+}, Cu^{2+} and Cd^{2+} binding with mucilage exudates from maize (*Zea mays* L.) roots," *Biology and Fertility of Soils*, vol. 2, no. 1, pp. 29–34, 1986.

[27] P. Sharma and R. S. Dubey, "Lead toxicity in plants," *Brazilian Journal of Plant Physiology*, vol. 17, no. 1, pp. 35–52, 2005.

[28] M. Wierzbicka, "Comparison of lead tolerance in *Allium cepa* with other plant species," *Environmental Pollution*, vol. 104, no. 1, pp. 41–52, 1999.

[29] M. J. Blaylock and J. W. Huang, "Phytoextraction of metals," in *PhyToremediation of Toxic Metals: Using Plants To Clean Up the Environment*, I. Raskin and B. D. Ensley, Eds., pp. 53–71, John Wiley & Sons, New York, NY, USA, 2000.

[30] P. D. Alexander, B. J. Alloway, and A. M. Dourado, "Genotypic variations in the accumulation of Cd, Cu, Pb and Zn exhibited by six commonly grown vegetables," *Environmental Pollution*, vol. 144, no. 3, pp. 736–745, 2006.

[31] A. Hanć, D. Barałkiewicz, A. Piechalaka, B. Tomaszewska, B. Wagner, and E. Bulska, "An analysis of long-distance root to leaf transport of lead in pisum sativum plants by laser ablation-ICP-MS," *International Journal of Environmental Analytical Chemistry*, vol. 89, no. 8–12, pp. 651–659, 2009.

[32] S. O. Eun, H. S. Youn, and Y. Lee, "Lead disturbs microtubule organization in the root meristem of *Zea mays*," *Physiologia Plantarum*, vol. 110, no. 3, pp. 357–365, 2000.

[33] L. E. Kerper and P. M. Hinkle, "Cellular uptake of lead is activated by depletion of intracellular calcium stores," *Journal of Biological Chemistry*, vol. 272, no. 13, pp. 8346–8352, 1997.

[34] T. Arazi, R. Sunkar, B. Kaplan, and H. Fromm, "A tobacco plasma membrane calmodulin-binding transporter confers Ni^{2+} tolerance and Pb^{2+} hypersensitivity in transgenic plants," *Plant Journal*, vol. 20, no. 2, pp. 171–182, 1999.

[35] S. V. Sahi, N. L. Bryant, N. C. Sharma, and S. R. Singh, "Characterization of a lead hyperaccumulator shrub, *Sesbania drummondii*," *Environmental Science and Technology*, vol. 36, no. 21, pp. 4676–4680, 2002.

[36] E. Gomes, *Genotoxity and cytotoxicity of Cr (VI) and Pb^{2+} in pisum sativum [Ph.D. thesis]*, University of Aveiro, Aveiro, Portugal, 2011.

[37] I. V. Seregin, L. K. Shpigun, and V. B. Ivanov, "Distribution and toxic effects of cadmium and lead on maize roots," *Russian Journal of Plant Physiology*, vol. 51, no. 4, pp. 525–533, 2004.

[38] A. Varga, G. Záray, F. Fodor, and E. Cseh, "Study of interaction of iron and lead during their uptake process in wheat roots by total-reflection X-ray fluorescence spectrometry," *Spectrochimica Acta B*, vol. 52, no. 7, pp. 1027–1032, 1997.

[39] M. Marmiroli, G. Antonioli, E. Maestri, and N. Marmiroli, "Evidence of the involvement of plant ligno-cellulosic structure in the sequestration of Pb: an X-ray spectroscopy-based analysis," *Environmental Pollution*, vol. 134, no. 2, pp. 217–227, 2005.

[40] C. S. Cobbett, "Phytochelatins and their roles in heavy metal detoxi," *Current Opinion in Plant Biology*, vol. 123, no. 3, pp. 211–216, 2000.

[41] N. C. Sharma, J. L. Gardea-Torresdey, J. Parsons, and S. V. Sahi, "Chemical speciation and cellular deposition of lead in *Sesbania drummondii*," *Environmental Toxicology and Chemistry*, vol. 23, no. 9, pp. 2068–2073, 2004.

[42] M. L. López, J. R. Peralta-Videa, J. G. Parsons, T. Benitez, and J. L. Gardea-Torresdey, "Gibberellic acid, kinetin, and the mixture indole-3-acetic acid-kinetin assisted with EDTA-induced lead hyperaccumulation in alfalfa plants," *Environmental Science and Technology*, vol. 41, no. 23, pp. 8165–8170, 2007.

[43] M. L. López, J. R. Peralta-Videa, J. G. Parsons, J. L. Gardea-Torresdey, and M. Duarte-Gardea, "Effect of indole-3-acetic acid, kinetin, and ethylenediaminetetraacetic acid on plant growth and uptake and translocation of lead, micronutrients, and macronutrients in alfalfa plants," *International Journal of Phytoremediation*, vol. 11, no. 2, pp. 131–149, 2009.

[44] W.-C. Ma, "Lead in mammals," in *Environmental Contaminants in Wildlife*, W. N. Beyer, G. Heinz, and A. W. Redmon-Norwood, Eds., SETAC Special Publication Series, pp. 281–296, CRC Lewis, Boca Raton, Fla, USA, 1996.

[45] L. V. Rossato, F. T. Nicoloso, J. G. Farias et al., "Effects of lead on the growth, lead accumulation and physiological responses of Pluchea sagittalis," *Ecotoxicology*, vol. 21, pp. 111–123, 2012.

[46] P. Zhuang, H. Zou, and W. Shu, "Biotransfer of heavy metals along a soil-plant-insect-chicken food chain: field study," *Journal of Environmental Sciences*, vol. 21, no. 6, pp. 849–853, 2009.

[47] D. R. Smith, M. W. Kahng, B. Quintanilla-Vega, and B. A. Fowler, "High-affinity renal lead-binding proteins in environmentally-exposed humans," *Chemico-Biological Interactions*, vol. 115, no. 1, pp. 39–52, 1998.

[48] ATSDR, "Toxicological profile for lead," 2007, http://www.atsdr.cdc.gov/.

[49] K. J. I. Nirmal, S. Hiren, and N. K. Rita, "Characterization of heavy metals in vegetables using Inductive Coupled Plasma Analyzer (ICPA)," *Journal of Applied Sciences and Environmental Management*, vol. 11, no. 3, pp. 75–79, 2007.

[50] J. Xu and I. Thornton, "Arsenic in garden soils and vegetable crops in Cornwall, England: implications for human health,"

Environmental Geochemistry and Health, vol. 7, no. 4, pp. 131–133, 1985.

[51] D. Y. Boon and P. N. Soltanpour, "Lead, cadmium, and zinc contamination of Aspen garden soils and vegetation," *Journal of Environmental Quality*, vol. 21, no. 1, pp. 82–86, 1992.

[52] K. R. Mahaffey, S. G. Capar, B. C. Gladen, and B. A. Fowler, "Concurrent exposure to lead, cadmium, and arsenic. Effects on toxicity and tissue metal concentrations in the rat," *Journal of Laboratory and Clinical Medicine*, vol. 98, no. 4, pp. 463–481, 1981.

[53] O. Munzuroglu and H. Geckil, "Effects of metals on seed germination, root elongation, and coleoptile and hypocotyl growth in *Triticum aestivum* and *Cucumis sativus*," *Archives of Environmental Contamination and Toxicology*, vol. 43, no. 2, pp. 203–213, 2002.

[54] Z.-T. Xiong, "Bioaccumulation and physiological effects of excess lead in a roadside pioneer species *Sonchus oleraceus* L.," *Environmental Pollution*, vol. 97, no. 3, pp. 275–279, 1997.

[55] A. Prueb, "Action values for mobile (NH$_4$NO$_3$) trace elements in soils based on the German national standard DIN, 19730," in *Contaminated Soils. Proceedings of 3rd International Conference on the Biogeochemistry of Trace Elements*, R. Prost, Ed., pp. 415–423, INRA, Paris, France, 1997.

[56] S. P. Datta and S. D. Young, "Predicting metal uptake and risk to the human food chain from leaf vegetables grown on soils amended by long-term application of sewage sludge," *Water, Air, and Soil Pollution*, vol. 163, no. 1–4, pp. 119–136, 2005.

[57] J. Nordberg and E. S. J. Arnér, "Reactive oxygen species, antioxidants, and the mammalian thioredoxin system," *Free Radical Biology and Medicine*, vol. 31, no. 11, pp. 1287–1312, 2001.

[58] B. Vinocur and A. Altman, "Recent advances in engineering plant tolerance to abiotic stress: achievements and limitations," *Current Opinion in Biotechnology*, vol. 16, no. 2, pp. 123–132, 2005.

[59] W. Wang, B. Vinocur, and A. Altman, "Plant responses to drought, salinity and extreme temperatures: towards genetic engineering for stress tolerance," *Planta*, vol. 218, no. 1, pp. 1–14, 2003.

[60] K. Shinozaki and K. Yamaguchi-Shinozaki, "Molecular responses to dehydration and low temperature: differences and cross-talk between two stress signaling pathways," *Current Opinion in Plant Biology*, vol. 3, no. 3, pp. 217–223, 2000.

[61] J. K. Zhu, "Plant salt tolerance," *Trends in Plant Science*, vol. 6, no. 2, pp. 66–71, 2001.

[62] M. Ashraf, "Biotechnological approach of improving plant salt tolerance using antioxidants as markers," *Biotechnology Advances*, vol. 27, no. 1, pp. 84–93, 2009.

[63] I. Fridovich, "Fundamental aspects of reactive oxygen species, or what's the matter with oxygen?" *Annals of the New York Academy of Sciences*, vol. 893, pp. 13–18, 1999.

[64] D. J. Betteridge, "What is oxidative stress?" *Metabolism*, vol. 49, no. 2, pp. 3–8, 2000.

[65] P. Jezek and L. Hlavata, "Mitochondria in homeostasis of reactive oxygen species in cell, tissues, and organism," *International Journal of Biochemistry & Cell Biology*, vol. 37, pp. 2478–2503, 2005.

[66] E. E. Benson, "Do free radicals have a role in plant tissue culture recalcitrance?" *In Vitro Cellular and Developmental Biology*, vol. 36, no. 3, pp. 163–170, 2000.

[67] H. K. Ledford and K. K. Niyogi, "Singlet oxygen and photo-oxidative stress management in plants and algae," *Plant, Cell and Environment*, vol. 28, no. 8, pp. 1037–1045, 2005.

[68] J. G. Scandalios, "Oxygen stress and superoxide dismutases," *Plant Physiology*, vol. 101, no. 1, pp. 7–12, 1993.

[69] J. G. Scandalios, "Oxidative stress: molecular perception and transduction of signals triggering antioxidant gene defenses," *Brazilian Journal of Medical and Biological Research*, vol. 38, no. 7, pp. 995–1014, 2005.

[70] A. Arora, R. K. Sairam, and G. C. Srivastava, "Oxidative stress and antioxidative system in plants," *Current Science*, vol. 82, no. 10, pp. 1227–1238, 2002.

[71] T. Yang and B. W. Poovaiah, "Hydrogen peroxide homeostasis: activation of plant catalase by calcium/calmodulin," *Proceedings of the National Academy of Sciences of the United States of America*, vol. 99, no. 6, pp. 4097–4102, 2002.

[72] A. C. Cassells and R. F. Curry, "Oxidative stress and physiological, epigenetic and genetic variability in plant tissue culture: implications for micropropagators and genetic engineers," *Plant Cell, Tissue and Organ Culture*, vol. 64, no. 2-3, pp. 145–157, 2001.

[73] C. Wang, X. Gu, X. Wang et al., "Stress response and potential biomarkers in spinach (*Spinacia oleracea* L.) seedlings exposed to soil lead," *Ecotoxicology and Environmental Safety*, vol. 74, no. 1, pp. 41–47, 2011.

[74] M. Mroczek-Zdyrska and M. Wójcik, "The influence of selenium on root growth and oxidativestress induced by lead in vicia faba L. minor plants," *Biological Trace Element Research*, vol. 147, no. 1–3, pp. 320–328, 2012.

[75] G. P. Wei, L. F. Yang, Y. L. Zhu, and G. Chen, "Changes in oxidative damage, antioxidant enzyme activities and polyamine contents in leaves of grafted and non-grafted eggplant seedlings under stress by excess of calcium nitrate," *Scientia Horticulturae*, vol. 120, no. 4, pp. 443–451, 2009.

[76] O. Tatar and M. N. Gevrek, "Influence of water stress on proline accumulation, lipid peroxidation and water content of wheat," *Asian Journal of Plant Sciences*, vol. 7, no. 4, pp. 409–412, 2008.

[77] P. P. He, X. Z. Lv, and G. Y. Wang, "Effects of Se and Zn supplementation on the antagonism against Pb and Cd in vegetables," *Environment International*, vol. 30, no. 2, pp. 167–172, 2004.

[78] A. Fargašová, J. Pastierová, and K. Svetková, "Effect of Se-metal pair combinations (Cd, Zn, Cu, Pb) on photosynthetic pigments production and metal accumulation in *Sinapis alba* L. seedlings," *Plant, Soil and Environment*, vol. 52, no. 1, pp. 8–15, 2006.

[79] Z. Pedrero, Y. Madrid, H. Hartikainen, and C. Cámara, "Protective effect of selenium in broccoli (*Brassica oleracea*) plants subjected to cadmium exposure," *Journal of Agricultural and Food Chemistry*, vol. 56, no. 1, pp. 266–271, 2008.

[80] R. Feng, C. Wei, S. Tu, and X. Sun, "Interactive effects of selenium and arsenic on their uptake by *Pteris vittata* L. under hydroponic conditions," *Environmental and Experimental Botany*, vol. 65, no. 2-3, pp. 363–368, 2009.

[81] M. Filek, R. Keskinen, H. Hartikainen et al., "The protective role of selenium in rape seedlings subjected to cadmium stress," *Journal of Plant Physiology*, vol. 165, no. 8, pp. 833–844, 2008.

[82] A. Erakhrumen and A. Agbontalor, "Review phytoremediation: an environmentally sound technology for pollution prevention, control and remediation in developing countries," *Educational Research and Review*, vol. 2, no. 7, pp. 151–156, 2007.

[83] U.S. Environmental Protection Agency, "Introduction to Phytoremediation," National Risk Management Research

Laboratory, EPA/600/R-99/107, 2000, http://www.clu-in.org/download/remed/introphyto.pdf.

[84] U.S. Department of Energy, "Plume Focus Area, December. Mechanisms of plant uptake, translocation, and storage of toxic elements," Summary Report of a workshop on phytoremediation research needs, 1994, http://www.osti.gov/.

[85] T. Maruyama, S. A. Hannah, and J. M. Cohen, "Metal removal by physical and chemical treatment processes," *Journal of the Water Pollution Control Federation*, vol. 47, no. 5, pp. 962–975, 1975.

[86] M. Korać, Ž. Kamberović, and B. Tomović, "Treatment of heavy metals contaminated solid wastes-Stabilization," in *Proceedings of European Metallurgical Conference (EMC '07)*, 2007.

[87] R. C. Monica and R. Cremonini, "Nanoparticles and higher plants," *Caryologia*, vol. 62, no. 2, pp. 161–165, 2009.

[88] D. Lin and B. Xing, "Phytotoxicity of nanoparticles: inhibition of seed germination and root growth," *Environmental Pollution*, vol. 150, no. 2, pp. 243–250, 2007.

[89] M. Racuciu and D. E. Creanga, "TMA-OH coated magnetic nanoparticles internalized in vegetal tissues," *Romanian Journal of Physics*, vol. 52, pp. 395–402, 2007.

[90] F. Gao, F. Hong, C. Liu et al., "Mechanism of nano-anatase TiO_2 on promoting photosynthetic carbon reaction of spinach: inducing complex of Rubisco-Rubisco activase," *Biological Trace Element Research*, vol. 111, no. 1–3, pp. 239–253, 2006.

[91] R. Doshi, W. Braida, C. Christodoulatos, M. Wazne, and G. O'Connor, "Nano-aluminum: transport through sand columns and environmental effects on plants and soil communities," *Environmental Research*, vol. 106, no. 3, pp. 296–303, 2008.

[92] L. C. Da Silva, M. A. Oliva, A. A. Azevedo, and J. M. D. Araújo, "Responses of restinga plant species to pollution from an iron pelletization factory," *Water, Air, and Soil Pollution*, vol. 175, no. 1–4, pp. 241–256, 2006.

[93] J. Kano, T. Kizuka, F. Shikanai, and S. Kojima, "Pure lead nanoparticles with stable metallic surfaces, on perovskite lead strontium titanate particles," *Nanotechnology*, vol. 20, no. 29, Article ID 295704, 2009.

[94] H. Jiang, K. S. Moon, and C. P. Wong, "Synthesis of Ag-Cu alloy nanoparticles for lead-free interconnect materials," in *Proceedings of the 10th International Symposium on Advanced Packaging Materials: Processes, Properties and Interfaces*, pp. 173–177, March 2005.

[95] H. Karami and M. Alipour, "Synthesis of lead dioxide nanoparticles by the pulsed current electrochemical method," *International Journal of Electrochemical Science*, vol. 4, no. 11, pp. 1511–1527, 2009.

9

Boron-Mediated Plant Somatic Embryogenesis: A Provocative Model

Dhananjay K. Pandey, Arvind K. Singh, and Bhupendra Chaudhary

School of Biotechnology, Gautam Buddha University, Greater Noida 201 308, India

Correspondence should be addressed to Bhupendra Chaudhary, bhupendrach@gmail.com

Academic Editor: Philip J. White

A central question in plant regeneration biology concerns the primary driving forces invoking the acquisition of somatic embryogenesis. Recently, the role of micronutrient boron (B) in the initiation and perpetuation of embryogenesis has drawn considerable attention within the scientific community. This interest may be due in part to the bewildering observation that the system-wide induction of embryogenic potential significantly varied in response to a minimal to optimal supply of B (minimal ≤ 0.1 mM, optimal = 0.1 mM). At the cellular level, certain channel proteins and cell wall-related proteins important for the induction of embryogenesis have been shown to be transcriptionally upregulated in response to minimal B supply suggesting the vital role of B in the induction of embryogenesis. At the molecular level, minimal to no B supply increased the endogenous level of auxin, which subsequently influenced the auxin-inducible somatic embryogenesis receptor kinases, suggesting the role of B in the induction of embryogenesis. Also, minimal B concentration may "turn on" other genetic and/or cellular transfactors reported earlier to be essential for cell-restructuring and induction of embryogenesis. In this paper, both the direct and indirect roles of B in the induction of somatic embryogenesis are highlighted and suggested for future validation.

1. Introduction

1.1. Somatic Embryogenesis in Plants. In plants, somatic embryogenesis is a multistep and complex regeneration process which begins with the formation of proembryonic mass followed by somatic embryo initiation, maturation, and, ultimately, entire plantlet regeneration [1]. Mostly, it refers to the developmental plasticity characteristic of the differentiated cells to regain their totipotency and convert into embryos. In theory, each somatic cell has the potential to convert itself into a somatic embryo, though very few somatic cells are capable of undergoing such complicated morphological transformation under culture conditions. In fact, only certain plant taxa and selected explants types have been shown to be capable of inducing embryogenic potential with *in vitro* cultures. As a result of these complicating factors, knowledge of how to "switch on" all somatic cells with such embryogenic potential is quite limited.

Since the initial descriptions of *in vitro* somatic embryogenesis [2, 3], one important characteristic of somatic embryos is the continuous growth resulting from the absence of developmental arrest [4]. In general, the process has three different stages of embryo development: globular, heart-shaped, and torpedo-shaped stages in dicots; globular, scutellar (transition), and coleoptilar stages in monocots [5, 6]. Again, although each plant cell has the competence to undergo somatic embryogenesis, the acquisition of embryogenic potential is extremely complex and involves intricate genetic mechanisms which are influenced by multiple factors [1]. Here, we strive to answer the question of what factors may be the most important in the induction of embryogenesis in plants. Are there different developmental and genetic conditions providing pertinent microenvironment for the acquisition of embryogenic potential?

The current literature on plant regeneration studies highlighted that the factors considered to be responsible for the induction of somatic embryogenesis are highly incoherent and largely dependent upon plant genotype/cultivar, tissue-type, physiological conditions of the donor plant, and varied cultural regimes [6]. Methodologically, the induction of embryogenic state in the somatic cells may include exposure to plant growth regulators and various physical and chemical treatments, and more precisely, it is often accompanied with cellular stress milieu [7, 8]. Therefore, what could be the most prevalent target procedures for the initiation of such stress conditions? For example, the exogenous auxin (2,4-D) has been used for the upregulation of several stress-induced genes to initiate somatic embryo development across plant taxa [9, 10]. If so, are the hormonal conditions most prevalent in the induction of somatic embryogenesis in plants? Moreover, to induce somatic embryogenesis, the endogenous level of auxin has also been influenced through the manipulation in the cumulative effect of micro- and macronutrient *in vitro*. So, is there any direct role of micromineral nutrition in the process of somatic embryogenesis, or indirectly *via* changing the endogenous hormonal levels those which are largely unexplored *hitherto*? An interesting example is the supply of micronutrient boron (B) *in vitro* in the form of boric acid which appears to be an important factor in the initiation of somatic embryogenesis [11–14].

1.2. Boron: An Essential Nutrient with Diverse Functions. B has been categorized as one of the essential micronutrients for the growth and development of vascular plants [15]. Boron is necessary for very diverse physiological processes such as the synthesis and organization of the cell wall and cell membrane structure [15–18], phenylpropanoid metabolism, and lignin biosynthesis [19, 20]. Across biological systems, B is known to influence pollen tube growth [21], nucleic acid metabolism [22, 23], polysaccharide metabolism [12, 24, 25], auxin metabolism [26, 27], and nitrate reductases activity [28, 29]. Plant response to B-mediated stress conditions is highly diverse and can vary based on the species, tissues, and physiological and environmental conditions [21].

Although within the plant, B is found to be most heavily concentrated in the cell wall [30]; it might also play a pivotal role in a wide variety of other cellular processes such as induction of somatic embryogenesis. This incites our convictions that if large concentration of B is positioned in the cell wall and cell membrane, are these the actual sites associated *directly* or *indirectly* with the induction of somatic embryogenesis? However, this has to be explored in depth experimentally to clearly define the role of B-associated cell walls in the induction of embryogenic potential. Aside from the cell wall, B has been observed to be present in other parts of the plant cell. Using stable isotopes 11B and 10B, the intracellular compartmentation of B in the roots of sunflower has been examined *in vitro* on relatively high and low B concentrations [31]. Though the B supply varied up to 100-fold in two different culture regimes, the accumulation of B in the cytosol and vacuole of low B-supplemented plants was 66% and 37% of the respective accumulation in the high B-supplemented plants. Thus, the presence of B in different cell organelles suggests additional roles of B in plant metabolism aside from its major function in cell wall organization [31]. Therefore, is B also having a role in the initiation and perpetuation of embryogenesis?

A bewildering observation from previous studies is that different plant systems do not respond in the same manner to varied concentrations (ranging from minimal to optimal) of B for the induction of somatic embryogenesis. Here, we suggest a scale for the B concentration supplied *in vitro* for somatic embryogenesis, where minimal ≤0.1 mM, optimal =0.1 mM, maximal =0.1–1.0 mM, and toxic ≥1.0 mM (the B concentrations are followed by these terms, hereafter) [32–34] (Figure 1). For example, in rice using coleoptiles or scutellum explants, the supplementation of optimal B concentration showed utmost somatic embryogenesis, whereas least somatic embryogenesis was observed with the maximal B concentration while using root tissues. Also, the optimal B concentration has direct influence on the development of suspensor of the somatic embryos in *Larix decidua,* whereas minimal B concentration led to the inhibition of suspensor development [35, 36]. Conversely, Mashayekhi et al. [32] reported in cucumber that minimal B concentration induced maximum somatic embryogenesis. Since B concentration required for the induction of somatic embryogenesis varied significantly among different explant types even from the same species, it becomes extremely challenging to suggest a standard B concentration required for the induction of somatic embryogenesis across plant taxa. In such changeable scenario, it would be difficult to develop a molecular network of cellular processes during B-mediated somatic embryogenesis.

Essentially, a minimal (to optimal) amount of B (Figure 1) appears to trigger somatic embryogenesis utmost among plant species having an impact on the cell wall-mediated signalling. Although these results have enriched our understanding to an extent, further experimental work is required to explore the intricacies of B-mediated downstream regulation of somatic embryogenesis in different biological systems. In this review, the role of B in the induction of somatic embryogenesis is thoroughly reviewed and hypothesized at times for future experimental validation.

2. Abundant Channel Protein Expression during Early Somatic Embryogenesis for Optimal B Mobility

In the cell membrane, different channel proteins are responsible for the transport of the various moieties required for cellular growth and metabolism. Aquaglyceroporins (AqGPs) are the pore proteins (coded by NIP) which facilitate efficient and selective flux of small solutes across biological membranes and are also responsible for the transport of B into the cytoplasm [39] (Figure 2). It has been observed that NIPs show a high degree of substrate specificity [40], for example, in *Arabidopsis,* it has been clearly illustrated that there are two subgroups in the NIP family, and only subclass II is responsible for the transportation of boric acid inside the cell [41]. It has been assumed that these pore proteins are important for the initiation of somatic embryogenesis mainly through the supply of micronutrients in the cytoplasm. Ciavatta et al. [42] also implicated abundantly expressed AqGPs in early embryo

Category	Minimal	—— Optimal ——	Maximal ——	Toxic
B concentration	[≤0.1 mM]	[0.1 mM]	[0.1–1 mM]	[≥1 mM]
(mM = milimolar)		(MS medium)		

FIGURE 1: The proposed scale of B concentration used for plant tissue culture and induction of somatic embryogenesis in different genotypes. Varied concentrations of B are broadly categorized into four groups ranging from minimal to toxic. The "optimal" category of B concentration is depicted as the concentration proposed in the MS medium [37] that is widely used for plant regeneration studies.

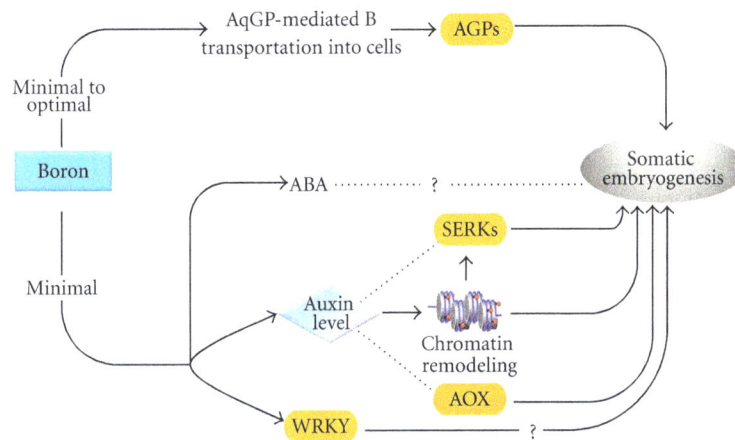

FIGURE 2: A hypothetical model based on the results shown for the involvement of boron on plant somatic embryogenesis. The putative candidate genes and biological processes upregulated in response to boron supplementation are highlighted. The minimal and optimal supplies of boron activate different transcription factors, hormones, and other cellular proteins known to be responsible for the induction of embryogenic potential in the cell (see text for details).

development in *loblolly pine*. The high expression level of NIP1;1 has been observed to be conserved in early somatic and zygotic embryo stages [42]. Further, it has also been shown in *Arabidopsis* that the major intrinsic protein NIP5;1 is essential for efficient B uptake during plant development. Under relatively low B condition, NIP5;1 gene is upregulated in the root elongation zone and the root hair zone, indicating its direct and crucial role in B maintenance during cellular development. However, in NIP5;1 mutant lines, the reduced mRNA level of NIP5;1 gene at a relatively low B concentration suggests the B-dependent transcriptional regulation of the NIP5;1 gene [43]. Conversely, toxic levels of B degrade the mRNA coding for AqGPs, preventing the cell from B toxicity [44]. Thus, it may be concluded that B uptake is maintained by the expression of intrinsic channel proteins, which may be influencing early embryogenic processes. It may be argued here that boric acid can be transported into cell *via* passive diffusion across membrane bilayers, and AqGP is essentially not needed when B is supplied more than an optimal level (>100 μM). For example, a toxic concentration of B (more than 1 mM) significantly decreased the embryogenic potential, which suggests that the excess B is supplied to the cell through diffusion and not through channel proteins [33]. Therefore, AqGPs may only be affecting somatic embryogenesis positively only under relatively low (minimal) B supply.

In such scenario, is there any relationship between specific histological patterns (cell wall) of normal versus embryogenic cells and also the mode of B supply opted by any

such cell? And is the transcript pattern of NIPs localized or varies across species? If so, characterization of embryogenic potential of any particular cell could be directly measured through the expression patterns of NIPs. Currently, it is a question requiring further study whether varied exogenous supply of B will influence the NIP transcript levels, and if it is also responsible for the conversion of normal cells into embryogenic cells. This research effort is being carried out by the authors with cotton (*Gossypium hirsutum* cv. Coker 310) due to its history as an established model system for *in vitro* somatic embryogenesis (Figure 3) [38]. In cotton, the somatic embryogenesis process begins with the initial callusing phase followed by the secondary growth phase that provides a suitable platform for the induction of embryogenic calli. Preliminary results from this study highlight that minimal B supply induces more embryogenesis in cotton cultures than that of optimal B supply (data not shown).

Therefore, it may be hypothesized that NIPs are important for a significant amount of B transport (required for embryogenesis) into the cell, and that as a result, this may also increase levels of AqGPs in the cell membrane. Higher levels of AqGPs may further increase the regeneration potential of somatic cells and ultimately incite somatic embryogenesis (Figure 4). At present, though several interrelated issues important to understanding the precise mechanism of embryogenesis remain unclear and unresolved, yet it may certainly be hypothesized that B-mediated expression of AqGPs plays a vital role in the induction of somatic embryogenesis.

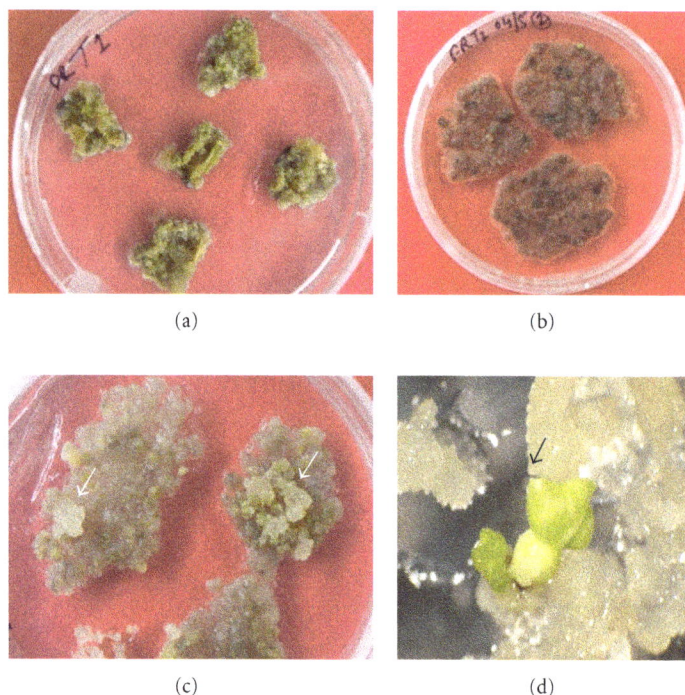

(a)

(b)

(c)

(d)

FIGURE 3: Developmental stages of somatic embryogenesis in the genus *Gossypium*, an established model system for indirect somatic embryogenesis [38]. It involves three major stages: (a) initial callus induction, (b) middle callus growth, and (c) late embryogenic callus induction phase. (d) Fully-grown somatic embryos are developed from the embryogenic callus. The cotton cotyledonary explants were used for initial callus initiation followed by the induction of embryogenic calli (shown by white arrows) and eventually leading to the development of somatic embryos (shown by black arrow).

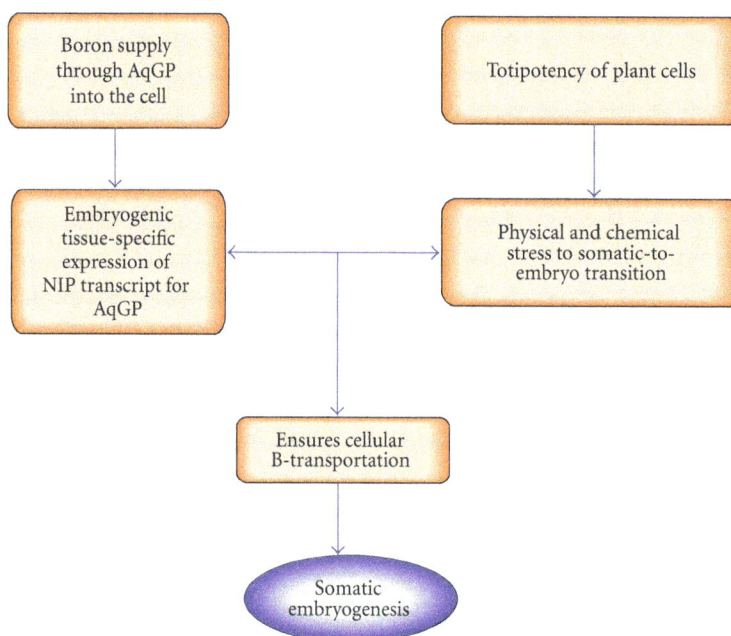

FIGURE 4: Diagrammatic representation of the putative role of AqGPs (NIP) in somatic embryogenesis.

3. Boron Regulates Cell Wall-Related Genes during Somatic Embryogenesis

In general, somatic cells that are competent for the acquisition of embryogenic potential perceive signals from neighbouring cells to trigger somatic embryogenesis and arabinogalactan proteins (AGPs) have been identified to initiate such signals [45]. AGPs are a heterogeneous group of structurally complex macromolecules composed of a polypeptide, a large branched glycan chain, and a lipid [46]. These proteins remain present in the cell wall and plasma membrane, functioning at the cell surface to mediate signal transduction *via* the cell wall—plasma membrane—cytoskeleton continuum [47], which is an essential structural assembly involved in the growth and morphogenesis of higher plants [48]. The structure and function of these proteoglycans are currently of intense interest as there is direct evidence for their involvement in the induction of embryogenesis [49, 50]. The spatiotemporal expression of AGPs has been reported during differentiation of globular stage somatic embryos into torpedo stage embryos in somatic embryogenesis [51]. It may be argued here that during the induction of embryogenesis, will the somatic cells devoid of any AGPs elude its embryogenic competence?

Thompson and Knox [52] and Chapman et al. [51] have shown in *Daucus carota* and *Cichorium* hybrid "474" that scavenging cellular AGPs, through addition of "Yariv" reagent in tissue culture media, inhibit somatic embryogenesis; exogenous addition of AGPs restores such potential, even to the point of significantly increasing somatic embryogenesis. For example, an up to 60-fold increase of somatic embryogenesis was observed in carrot by the addition of activated AGP [53]. These results show that activated AGPs are the extracellular matrix molecules that control and maintain plant cell fate during somatic embryogenesis. Given the important role of AGPs in somatic embryogenesis, what are the essential factors in the cell that maintain the endogenous AGPs level for this complex process to occur?

Interestingly, data from microarray analyses showed that the expression of several genes related to cell wall biosynthesis and cell wall modification, including the AGPs, was highly downregulated in the absence of B, and *vice versa* [54]. The mechanism by which the cell signals B availability to the nucleus remains unknown and will need to be studied further in order to be fully understood. However, it is clear from previous studies that B plays an essential role in the expression of cell wall-related genes responsible for maintaining the structural organization of the cell wall. It may further be speculated that B is the key regulatory factor in determining the availability of AGPs, and consequently commencing the cellular signals necessary for the induction of somatic embryogenesis mainly through cellular communications (Figure 2). At present, no information is available on the B concentration required for the optimal expression levels of AGPs, which may, in turn, induce somatic embryogenesis. Further study whereby embryogenic calli are screened for the AGP expression levels in response to the varied B concentrations is certainly warranted to answer this biological question and may be best carried out using cotton as the model system for this work (Figure 3).

4. Boron Provokes Stress-Mediated Signalling Pathways during Embryogenesis

Boron plays a major role in the regulation of somatic embryogenesis by triggering stress-mediated pathways. Minimal B supply switches on many genetic and/or cellular *trans*factors responsible for the induction of somatic embryogenesis and has been considered to be dependent on the B transportation in the cell. With minimal to optimal B concentrations, *At*NIP channel proteins, as discussed previously, play important role in B transportation and regulate its concentration inside the cell. However, in toxic concentrations, B is transported into the cell mainly through passive diffusion [55–57] and, as a result, may negatively affect embryogenic potential. Tolerance levels to the toxic B concentrations (>1 mM) could be estimated by measuring the expression level of boron excess tolerant1 gene, at least in rice [58], or its homolog in other systems. As B plays a vital role in cellular architecture mainly through cell wall organization, there may be concerns if the minimal B concentration may have an adverse effect on the cellular structure and related metabolic processes. It has been shown that complete B depletion from the culture medium not only induced root development, but also remarkable callusing followed by somatic embryogenesis in carrot and cucumber species, respectively [32, 59]. These results further strengthen our assumption that B stress led to the manipulation of the cellular microenvironment and further assisted in the cell regeneration.

What are the essential molecular components that get up-regulated (=manipulated) in response to the B stress *in vitro* prior to embryogenesis? Under conditions of minimal B, the accumulation of chlorogenic and caffeic acids has shown to result in inhibited IAA oxidase activity, leading to endogenous auxin accumulation [59–64]. It is well understood that the polar transport of auxin is essential for the establishment of bilateral symmetry in dicotyledonous somatic embryos, as well as in zygotic embryos [61, 63, 64]. If auxin is not present during development, somatic embryos may lose their bilateral symmetry during organ formation. Therefore, the threshold level of auxin has to be maintained inside the competent somatic cell for the acquisition of embryogenic potential, and B is important in the maintenance of this level of auxin inside the cell.

As stated above, auxin level is critical in the induction of somatic embryogenesis. It has been shown in the apical and subapical root sections of *Arabidopsis* that a 20-fold increase in IAA oxidase activity was displayed after B was withheld over minimal B-supplied cultures [65]. Thus, it is evident that minimal to no B supply increases the endogenous auxin level, and the altered-auxin level may further influence the downstream signalling for the induction of embryogenic potential (Figure 2). This is mainly considered through the change in the transcript levels of certain signalling genes responsible for the induction of somatic embryos. For example, enhanced endogenous level of auxin upregulates the expression of the somatic embryo receptor kinase (SERK) gene, somatic embryo-related factor (SERF), and Ca^{2+} ion channel-mediated regulatory gene expression, all reported earlier to be essential for somatic embryogenesis [66]. Early auxin-inducible genes in wheat leaf bases, including *Ta*SERKs, are involved in somatic embryogenesis and decreased significantly upon auxin depletion [67]. Thus, it may be concluded that B (*indirectly*) can alter the expression of hormone-regulating genes and various other gene(s) which have an established role in somatic embryogenesis (Figure 2). Although it is uncertain how much B concentration is required initially and even in later stages of development for an absolute induction of embryogenic potential, it may be suggested that for precise induction of somatic embryogenesis *in vitro*, initially minimal B is required, and subsequently may also be culture stage specific (Figures 1 and 3). This could perhaps be best explained in the model system *Gossypium*, where the somatic embryogenesis occurs in three major stages (Figure 3). At the initial callus induction phase, high endogenous auxin is required and may be maintained either by the supply of exogenous auxin or by a minimal concentration of B exerting a stress condition. Later stages of embryogenesis require relatively low levels of auxin as a very high endogenous level may also inhibit somatic embryogenesis. Therefore, the low level of auxin can be maintained by the complete removal of exogenous auxin, and also by the supplementation of an optimal to maximal concentration of B in the culture medium, which would reduce the endogenous auxin level (Figures 3(b) and 3(c)). The latter may be considered to be more promising for the precise manipulation of auxin during the late embryogenic callus induction phase.

As noted elsewhere, at the molecular level, auxin plays an important role in the reprogramming of gene expression in a somatic cell for its induction into a somatic embryo, possibly through DNA methylation and chromatin remodelling [68] (Figure 2). The importance of the latter may be suggested by an increase in the expression of other signalling gene(s) responsible for somatic embryo induction. Boron stress also has a direct impact on levels of abscisic acid (ABA) and ethylene, both of which help to provide favorable conditions for the induction of somatic embryogenesis. The role of ABA in stress-induced somatic embryogenesis has been investigated [69], and it is known that B stress has significant role in increasing ABA level *per se* [59] (Figure 2). Thus, it is in this manner that B indirectly influences the ABA-mediated signalling necessary for somatic embryogenesis. Mashayekhi and Neumann 2006 [59] have also shown the influence of B on embryo development through variation of the phytohormone system in the developing embryo.

Boron deficiency in somatic cells can also lead to accumulation of oxidative free radicals in somatic cells, which can affect membrane function [70] and embryo development within these cells. Moreover, the WRKY transcription factor which has been suggested to play important roles in the regulation of transcriptional reprogramming associated with plant stress responses including embryogenesis and abscisic

acid signalling pathway is highly expressed in response of minimal B concentration [71–74]. It may be reasonably proposed here that expression of this WRKY transcription factor family gene affects and ultimately plays a regulatory role in somatic embryogenesis (Figure 2). However, the precise mechanism of the regulation of this pathway is subject to future research. Boron also plays a significant role in the expression of the mitochondrial alternative oxidase (AOX) gene, which is a gene generally expressed under B-deficient conditions [75]. The AOX gene family has been shown to play a crucial role in somatic embryogenesis, as illustrated by previous work with carrot [76] (Figure 2). It was observed that the carrot AOX genes (DcAOX1a and DcAOX2a) showed upregulation during initiation of somatic embryogenesis [76], thus contributing to the hypothesis that B affects AOX activity and supports the metabolic reorganization that is essential for cell restructuring and *de novo* differentiation.

5. Conclusion

The present study illustrates the importance of the B responsive gene network in the induction of somatic embryogenesis and suggests the basis for the molecular mechanisms involved in embryogenesis. We provide clues here for the direct or indirect role of B in somatic embryogenesis, explaining the molecular changes highlighting cellular mechanisms promoting the conversion of normal cells into embryogenic cell. Notably, the majority of B-mediated processes are diagnosed as having become enhanced during embryogenesis in different tissue types, as opposed to being a result of system-wide phenomena *per se*. This suggestion that expression of embryogenesis-related genes was primarily concomitant with B-mediated signalling pathways is bolstered by the remarkable observation that in independent embryogenesis events, taking place in different systems, this led to the expression of parallel signalling genes. Although this is true at the level of cellular processes, or perhaps metabolism, it is clear that the genesis of these similarities is only partially congruent at the genetic level. That is, different sets of embryogenesis-related genes are up-regulated accompanying only selected plant species/cultivars making the phenomenon highly genotype dependent, and without any precedent. Thus, an exciting prospect for future work will be to dissect this physiological transformation of somatic cells into its responsible constituent genes and to learn the system-wide mutational basis of their altered regulation or function.

Acknowledgments

The authors are thankful to Professor Jonathan F. Wendel and Kara Grupp, Iowa State University, USA for their suggestions in the preparation of this paper. The authors also thank the Department of Science and Technology (DST), Government of India, and Council of Scientific and Industrial Research (CSIR), Government of India for the financial support to carry out this cotton regeneration work.

References

[1] S. Arnold, I. Sabala, P. Bozhkov, J. Dyachok, and L. Filonova, "Developmental pathways of somatic embryogenesis," *Plant Cell, Tissue and Organ Culture*, vol. 69, no. 3, pp. 233–249, 2002.

[2] J. Reinert, "Untersuchungen über die morphogenese an gewebenkulturen," *Berichte der Deutschen Botanischen Gesellschaft-Ges*, vol. 71, p. 15, 1958.

[3] F. Steward, M. Mapes, and K. Hears, "Growth and organized development of cultured cells. II. Growth and division of freely suspended cells," *American Journal of Botany*, vol. 45, pp. 705–708, 1958.

[4] O. Faure, W. Dewitte, A. Nougarède, and H. Van Onckelen, "Precociously germinating somatic embryos of *Vitis vinifera* have lower ABA and IAA levels than their germinating zygotic counterparts," *Physiologia Plantarum*, vol. 102, no. 4, pp. 591–595, 1998.

[5] D. J. Gray, M. E. Compton, R. C. Harrell, and D. J. Cantliffe, "Somatic embryogenesis and the technology of synthetic seed," in *Somatic Embryogenesis and Synthetic Seed I. Biotechnology in Agriculture and Forestry*, Y. P. S. Bajaj, Ed., vol. 30, pp. 126–151, Springer-Verlag, Berlin, Germany, 1995.

[6] M. A. J. Toonen and S. C. De Vries, "Initiation of somatic embryos from single cells," in *Embryogenesis: The Generation of a Plant*, T. L. Wang and A. Cuming, Eds., pp. 173–189, Bios Scientific Publishers, Oxford, UK, 1996.

[7] D. Dudits, L. Bogre, and J. Gyorgyey, "Molecular and cellular approaches to the analysis of plant embryo development from somatic cells in vitro," *Journal of Cell Science*, vol. 99, no. 3, pp. 473–482, 1991.

[8] D. Dudits, J. Györgyey, L. B. Bögre, and L. Bakó, "Molecular biology of somatic embryogenesis," in *In Vitro Embryogenesis In Plants*, T. A. Thorpe, Ed., vol. 267–308, Kluwer Academic Publishers, Dordrecht, The Netherlands, 1995.

[9] S. Davletova, T. Mészáros, P. Miskolczi et al., "Auxin and heat shock activation of a novel member of the calmodulin like domain protein kinase gene family in cultured alfalfa cells," *Journal of Experimental Botany*, vol. 52, no. 355, pp. 215–221, 2001.

[10] J. Györgyey, A. Gartner, K. Németh et al., "Alfalfa heat shock genes are differentially expressed during somatic embryogenesis," *Plant Molecular Biology*, vol. 16, no. 6, pp. 999–1007, 1991.

[11] J. D. Cohen and R. S. Bandurski, "The bound auxins: protection of indole-3-acetic acid from peroxidase-catalyzed oxidation," *Planta*, vol. 139, no. 3, pp. 203–208, 1978.

[12] W. M. Dugger, "Boron in plant metabolism," in *in Encyclopedia of Plant Physiology*, A. Lauchli and L. Bieleski, Eds., vol. 15B, pp. 626–650, New Series, Springer, Berlin, Germany, 1983.

[13] H. Marschner, *Mineral Nutrition of Higher Plants*, Academic Press, London, UK, 2002.

[14] J. G. Paul, R. O. Nable, A. W. H. Lake, M. A. Materne, and A. J. Rathjen, "Response of annual medics (*Medicago ssp.*) and field peas (*Pisum sativum*) to high concentrations of boron: genetic variation and mechanism of tolerance," *Australian Journal of Agricultural Research*, vol. 43, no. 1, pp. 203–213, 1992.

[15] P. H. Brown and H. Hu, "Does boron play only a structural role in the growing tissues of higher plants?" *Plant and Soil*, vol. 196, no. 2, pp. 211–215, 1997.

[16] D. G. Blevins and K. M. Lukaszewski, "Boron in plant structure and function," *Annual Review of Plant Biology*, vol. 49, pp. 481–500, 1998.

[17] W. D. Loomis and R. W. Durst, "Chemistry and biology of boron," *BioFactors*, vol. 3, no. 4, pp. 229–239, 1992.

[18] T. Tanada, "Localization of boron in membranes," *Journal of Plant and Nutrition*, vol. 6, no. 9, pp. 743–749, 1983.

[19] D. H. Lewis, "Are there inter-relations between the metabolic role of boron, synthesis of phenolic phytoalexins and the germination of pollen?" *New Phytologist*, vol. 84, no. 2, pp. 261–270, 1980.

[20] R. Watanabe, W. J. McIlrath, J. Skok, W. Chorney, and S. H. Wender, "Accumulation of scopoletin glucoside in boron-deficient tobacco leaves," *Archives of Biochemistry and Biophysics*, vol. 94, no. 2, pp. 241–243, 1961.

[21] B. Dell and L. Huang, "Physiological response of plants to low boron," *Plant and Soil*, vol. 193, no. 1-2, pp. 103–120, 1997.

[22] A. H. Ali and B. C. Jarvis, "Effect of auxin and boron on nucleic acid metabolism and cell division during adventitious root regeneration," *New Phytologist*, vol. 108, no. 4, pp. 383–391, 1988.

[23] M. S. Cohen and L. S. Albert, "Autoradiographic examination of intact boron deficient squash roots treated with tritiated thymidine," *Plant Physiology*, vol. 54, no. 5, pp. 766–768, 1974.

[24] H. G. Gauch and W. M. Dugger, "The role of boron in the translocation of sucrose," *Plant Physiology*, vol. 28, no. 3, pp. 457–466, 1953.

[25] T. Matoh and M. Kobayashi, "Boron and calcium, essential inorganic constituents of pectic polysaccharides in higher plant cell walls," *Journal of Plant Research*, vol. 111, no. 1101, pp. 179–190, 1998.

[26] L. Coke and W. J. Whittington, "The role of boron in plant growth: IV. Interrelationships between boron and indol-3ylacetic acid in the metabolism of bean radicles," *Journal of Experimental Botany*, vol. 19, no. 2, pp. 295–308, 1968.

[27] B. C. Jarvis, S. Yasmin, and M. T. Coleman, "RNA and protein metabolism during adventitious root formation in stem cuttings of *Phaseolus aureus*," *Physiologia Plantarum*, vol. 64, no. 1, pp. 53–59, 1985.

[28] R. Kastori and N. Petrovic, "Effect of boron on nitrate reductases in tobacco leaves and roots," *Molecular Genomics and Genetics*, vol. 236, pp. 203–208, 1989.

[29] J. M. Ruiz, M. Baghour, G. Bretones, A. Belakbir, and L. Romero, "Nitrogen metabolism in tobacco plants (*Nicotiana tabacum L*): role of boron as a possible regulatory factor," *International Journal of Plant Sciences*, vol. 159, no. 1, pp. 121–126, 1998.

[30] H. Hu and P. H. Brown, "Localization of boron in cell walls of squash and tobacco and its association with pectin. Evidence for a structural role of boron in the cell wall," *Plant Physiology*, vol. 105, no. 2, pp. 681–689, 1994.

[31] H. Pfeffer, F. Dannel, and V. Römheld, "Boron compartmentation in roots of sunflower plants of different boron status: a study using the stable isotopes 10B and 11B adopting two independent approaches," *Physiologia Plantarum*, vol. 113, no. 3, pp. 346–351, 2001.

[32] K. Mashayekhi, M. Sharifani, M. Shahsavand, and H. Kalati, "Induction of somatic embryogenesis in absence of exogenous auxin in cucumber (*Cucumis sativus L.*)," *International Journal of Plant Production*, vol. 2, no. 2, pp. 163–166, 2008.

[33] N. Renukdas, M. L. Mohan, S. S. Khuspe, and S. K. Rawal, "Influence of boron on somatic embryogenesis in papaya," *Biologia Plantarum*, vol. 47, no. 1, pp. 129–132, 2004.

[34] N. A. Sahasrabudhe, M. Nandi, and R. A. Bahulikar, "Influence of boric acid on somatic embryogenesis of a cytosterile line of indica rice," *Plant Cell, Tissue and Organ Culture*, vol. 58, no. 1, pp. 73–75, 1999.

[35] U. Behrendt, "Entwicklungsbiologische untersuchungen zur somatischen embryogenese bei der europischen lrche (*Larix decidua MILL.*)," in *Mathematisch-Natur-wissenschaftliche Fakultt I der*, vol. 125, Humboldt-Universitatzu, Berlin, Germany, 1994.

[36] U. Behrendt and K. Zoglauer, "Boron controls suspensor development in embryogenic cultures of *Larix decidua*," *Physiologia Plantarum*, vol. 97, no. 2, pp. 321–326, 1996.

[37] T. Murashige and F. Skoog, "A revised medium for rapid growth and bioassays with tobacco tissue cultures," *Physiologia Plantarum*, vol. 15, no. 3, pp. 473–497, 1962.

[38] B. Chaudhary, S. Kumar, K. V. S. K. Prasad, G. S. Oinam, P. K. Burma, and D. Pental, "Slow desiccation leads to high-frequency shoot recovery from transformed somatic embryos of cotton (*Gossypium hirsutum* L. cv. Coker 310 FR)," *Plant Cell Reports*, vol. 21, no. 10, pp. 955–960, 2003.

[39] K. L. Fitzpatrick and R. J. Reid, "The involvement of aquaglyceroporins in transport of boron in barley roots," *Plant, Cell and Environment*, vol. 32, no. 10, pp. 1357–1365, 2009.

[40] N. Mitani-Ueno, N. Yamaji, F. J. Zhao, and J. F. Ma, "The aromatic/arginine selectivity filter of NIP aquaporins plays a critical role in substrate selectivity for silicon, boron, and arsenic," *Journal of Experimental Botany*, vol. 62, no. 12, pp. 4391–4398, 2011.

[41] I. S. Wallace and D. M. Roberts, "Distinct transport selectivity of two structural subclasses of the nodulin-like intrinsic protein family of plant aquaglyceroporin channel," *Biochemistry*, vol. 44, no. 51, pp. 16826–16834, 2005.

[42] V. T. Ciavatta, R. Morillon, G. S. Pullman, M. J. Chrispeels, and J. Cairney, "An aquaglyceroporin is abundantly expressed early in the development of the suspensor and the embryo proper of loblolly pine," *Plant Physiology*, vol. 127, no. 4, pp. 1556–1567, 2001.

[43] J. Takano, M. Wada, U. Ludewig, G. Schaaf, N. Von Wirén, and T. Fujiwara, "The Arabidopsis major intrinsic protein NIP5;1 is essential for efficient boron uptake and plant development under boron limitation," *Plant Cell*, vol. 18, no. 6, pp. 1498–1509, 2006.

[44] M. Tanaka, J. Takano, Y. Chiba et al., "Boron-dependent degradation of NIP5,1 mRNA for acclimation to excess boron conditions in Arabidopsis," *The Plant Cell*, vol. 23, no. 9, pp. 3547–3559, 2011.

[45] P. F. McCabe, T. A. Valentine, L. S. Forsberg, and R. I. Pennell, "Soluble signals from cells identified at the cell wall establish a developmental pathway in carrot," *Plant Cell*, vol. 9, no. 12, pp. 2225–2241, 1998.

[46] M. Anna and A. Eugene, "The multiple roles of arabinogalactan proteins in plant development," *Plant Physiology*, vol. 122, no. 1, pp. 3–9, 2000.

[47] H. S. Sardar, J. Yang, and A. M. Showalter, "Molecular interactions of arabinogalactan proteins with cortical microtubules and F-actin in bright yellow-2 tobacco cultured cells," *Plant Physiology*, vol. 142, no. 4, pp. 1469–1479, 2006.

[48] S. E. Wyatts and N. C. Carpita, "The plant cytoskelet on-cell-wall continuum," *Trends in Cell Biology*, vol. 3, no. 122, pp. 413–417, 1994.

[49] M. Kreuger and G. J. van Holst, "Arabinogalactan proteins are essential in somatic embryogenesis of *Daucus carota* L.," *Planta*, vol. 189, no. 2, pp. 243–248, 1993.

[50] N. J. Stacey, K. Roberts, and J. P. Knox, "Patterns of expression of the JIM4 arabinogalactan-protein epitope in cell cultures and during somatic embryogenesis in *Daucus carota* L.," *Planta*, vol. 180, no. 2, pp. 285–292, 1990.

[51] A. Chapman, A. S. Blervacq, J. Vasseur, and J. L. Hilbert, "Arabinogalactan-proteins in *Cichorium* somatic embryogenesis: effect of β-glucosyl Yariv reagent and epitope localisation during embryo development," *Planta*, vol. 211, no. 3, pp. 305–314, 2000.

[52] H. J. M. Thompson and J. P. Knox, "Stage-specific responses of embryogenic carrot cell suspension cultures to arabinogalactan protein-binding β-glucosyl Yariv reagent," *Planta*, vol. 205, no. 1, pp. 32–38, 1998.

[53] A. J. Van Hengel, Z. Tadesse, P. Immerzeel, H. Schols, A. Van Kammen, and S. C. De Vries, "N-acetylglucosamine and glucosamine-containing arabinogalactan proteins control somatic embryogenesis," *Plant Physiology*, vol. 125, no. 4, pp. 1880–1890, 2001.

[54] J. J. Camacho-Cristóbal, M. B. Herrera-Rodríguez, V. M. Beato et al., "The expression of several cell wall-related genes in *Arabidopsis* roots is down-regulated under boron deficiency," *Environmental and Experimental Botany*, vol. 63, no. 1–3, pp. 351–358, 2008.

[55] C. Dordas and P. H. Brown, "Evidence for channel mediated transport of boric acid in squash (*Cucurbita pepo*)," *Plant and Soil*, vol. 235, no. 1, pp. 95–103, 2001.

[56] C. Dordas, M. J. Chrispeels, and P. H. Brown, "Permeability and channel-mediated transport of boric acid across membrane vesicles isolated from squash roots," *Plant Physiology*, vol. 124, no. 3, pp. 1349–1361, 2000.

[57] J. C. R. Stangoulis, R. J. Reid, P. H. Brown, and R. D. Graham, "Kinetic analysis of boron transport in *Chara*," *Planta*, vol. 213, no. 1, pp. 142–146, 2001.

[58] K. Ochiai, A. Shimizu, Y. Okumoto, T. Fujiwara, and T. Matoh, "Suppression of a NAC-like transcription factor gene improves boron-toxicity tolerance in rice," *Plant Physiology*, vol. 156, no. 3, pp. 1457–1463, 2011.

[59] K. Mashayekhi and K. H. Neumann, "Effects of boron on somatic embryogenesis of *Daucus carota*," *Plant Cell, Tissue and Organ Culture*, vol. 84, no. 3, pp. 279–283, 2006.

[60] U. C. Gupta, "Boron," in *Handbook of Plant Nutrition*, A. V. Barker and D. J. Pilbean, Eds., pp. 241–277, Taylor and Francis Publications, Boca Raton, Fla, USA, 2006.

[61] C. M. Liu, Z. H. Xu, and N. H. Chua, "Auxin polar transport is essential for the establishment of bilateral symmetry during early plant embryogenesis," *Plant Cell*, vol. 5, no. 6, pp. 621–630, 1993.

[62] F. LoSchiavo, L. Pitto, G. Giuliano et al., "DNA methylation of embryogenic carrot cell cultures and its variations as caused by mutation, differentiation, hormones and hypomethylating drugs," *Theoretical and Applied Genetics*, vol. 77, no. 3, pp. 325–331, 1989.

[63] F. M. Schiavone and T. J. Cooke, "A geometric analysis of somatic embryo formation in carrot cell cultures," *Canadian Journal of Botany*, vol. 63, no. 9, pp. 1573–1578, 1985.

[64] F. M. Schiavone and T. J. Cooke, "Unusual patterns of somatic embryogenesis in the domesticated carrot: developmental effects of exogenous auxins and auxin transport inhibitors," *Cell Differentiation*, vol. 21, no. 1, pp. 53–62, 1987.

[65] C. W. Bohnsack and L. S. Albert, "Early effectsof boron deficiency on Indoleacetic Acid Oxidase levelsof Squash root tips," *Plant Physiology*, vol. 59, no. 6, pp. 1047–1050, 1977.

[66] T. Takeda, H. Inose, and H. Matsuoka, "Stimulation of somatic embryogenesis in carrot cells by the addition of calcium," *Biochemical Engineering Journal*, vol. 14, no. 2, pp. 143–148, 2003.

[67] B. Singla, J. P. Khurana, and P. Khurana, "Characterization of three somatic embryogenesis receptor kinase genes from

wheat, *Triticum aestivum*," *Plant Cell Reports*, vol. 27, no. 5, pp. 833–843, 2008.

[68] F. Zeng, X. Zhang, S. Jin et al., "Chromatin reorganization and endogenous auxin/cytokinin dynamic activity during somatic embryogenesis of cultured cotton cell," *Plant Cell, Tissue and Organ Culture*, vol. 90, no. 1, pp. 63–70, 2007.

[69] A. Kikuchi, N. Sanuki, K. Higashi, T. Koshiba, and H. Kamada, "Abscisic acid and stress treatment are essential for the acquisition of embryogenic competence by carrot somatic cells," *Planta*, vol. 223, no. 4, pp. 637–645, 2006.

[70] I. Cakmak and V. Römheld, "Boron deficiency-induced impairments of cellular functions in plants," *Plant and Soil*, vol. 193, no. 1-2, pp. 71–83, 1997.

[71] L. Chen, Y. Song, S. Li, L. Zhang, C. Zou, and D. Yu, "The role of WRKY transcription factors in plant abiotic stresses," *Biochemica Biophysica Acta*, vol. 1819, no. 2, pp. 120–128, 2012.

[72] I. Kasajima, Y. Ide, M. Yokota Hirai, and T. Fujiwara, "WRKY6 is involved in the response to boron deficiency in *Arabidopsis thaliana*," *Physiologia plantarum*, vol. 139, no. 1, pp. 80–92, 2010.

[73] M. Lagacé and D. P. Matton, "Characterization of a WRKY transcription factor expressed in late torpedo-stage embryos of *Solanum chacoense*," *Planta*, vol. 219, no. 1, pp. 185–189, 2004.

[74] X. Zou, J. R. Seemann, D. Neuman, and Q. J. Shen, "A WRKY gene from creosote bush encodes an activator of the abscisic acid signaling pathway," *Journal of Biological Chemistry*, vol. 279, no. 53, pp. 55770–55779, 2004.

[75] D. P. Maxwell, Y. Wang, and L. McIntosh, "The alternative oxidase lowers mitochondrial reactive oxygen production in plant cells," *Proceedings of the National Academy of Sciences of the United States of America*, vol. 96, no. 14, pp. 8271–8276, 1999.

[76] A. M. Frederico, M. D. Campos, H. G. Cardoso, J. Imani, and B. Arnholdt-Schmitt, "Alternative oxidase involvement in *Daucus carota* somatic embryogenesis," *Physiologia Plantarum*, vol. 137, no. 4, pp. 498–508, 2009.

10

Chromium as an Environmental Pollutant: Insights on Induced Plant Toxicity

Helena Oliveira

Department of Biology, CESAM, University of Aveiro, 3810-193 Aveiro, Portugal

Correspondence should be addressed to Helena Oliveira, holiveira@ua.pt

Academic Editor: Joanna Deckert

In the past decades the increased use of chromium (Cr) in several anthropogenic activities and consequent contamination of soil and water have become an increasing concern. Cr exists in several oxidation states but the most stable and common forms are Cr(0), Cr(III) and Cr(VI) species. Cr toxicity in plants depends on its valence state. Cr(VI) as being highly mobile is toxic, while Cr(III) as less mobile is less toxic. Cr is taken up by plants through carriers of essential ions such as sulphate. Cr uptake, translocation, and accumulation depend on its speciation, which also conditions its toxicity to plants. Symptoms of Cr toxicity in plants are diverse and include decrease of seed germination, reduction of growth, decrease of yield, inhibition of enzymatic activities, impairment of photosynthesis, nutrient and oxidative imbalances, and mutagenesis.

1. Introduction

Chromium (Cr) is the 17th most abundant element in the Earth's mantle [1]. It occurs naturally as chromite ($FeCr_2O_4$) in ultramafic and serpentine rocks or complexed with other metals like crocoite ($PbCrO_4$), bentorite $Ca_6(Cr,Al)_2(SO_4)_3$ and tarapacaite (K_2CrO_4), vauquelinite ($CuPb_2CrO_4PO_4OH$), among others [2]. Cr is widely used in industry as plating, alloying, tanning of animal hides, inhibition of water corrosion, textile dyes and mordants, pigments, ceramic glazes, refractory bricks, and pressure-treated lumber [1]. Due to this wide anthropogenic use of Cr, the consequent environmental contamination increased and has become an increasing concern in the last years [3].

Chromium exists in several oxidation states, but the most stable and common forms are Cr(0), the trivalent Cr(III), and the hexavalent Cr(VI) species. Cr(0) is the metallic form, produced in industry and is a solid with high fusion point usually used for the manufacturing of steel and other alloys. Cr(VI) in the forms of chromate (CrO_4^{2-}), dichromate (CrO_4^{2-}), and CrO_3 is considered the most toxic forms of chromium, as it presents high oxidizing potential, high solubility, and mobility across the membranes in living organisms and in the environment. Cr(III) in the forms of oxides, hydroxides, and sulphates is less toxic as it is relatively insoluble in water, presents lower mobility, and is mainly bound to organic matter in soil and aquatic environments. Moreover, Cr(III) forms tend to form hydroxide precipitates with Fe at typical ground water pH values. At high concentrations of oxygen or Mn oxides, Cr(III) can be oxidized to Cr(VI) [4, 5].

As Cr(VI) and Cr(III) present different chemical, toxicological, and epidemiological characteristics, they are differently regulated by EPA, which constitutes a unique characteristic of Cr among the toxic metals [6]. Cr(VI) is a powerful epithelial irritant and also considered a human carcinogen [7]. Cr(VI) is also toxic to many plants [8] aquatic animals [9], and microorganisms [10]. Contrarily to Cr(VI), Cr(III) is considered a micronutrient in humans, being necessary for sugar and lipid metabolism [11] and is generally not harmful. In plants, particularly crops, Cr at low concentrations (0.05–1 mg L^{-1}) was found to promote growth and increase yield, but it is not considered essential to plants [5, 12]. In this context, accumulation of chromium in edible plants may represent a potential hazard to animals and humans.

2. Chromium in the Environment

2.1. Chromium in Water. Chromium may enter the natural waters by weathering of Cr-containing rocks, direct discharge from industrial operations, leaching of soils, among others. In the aquatic environment Cr may suffer reduction, oxidation, sorption, desorption, dissolution, and precipitation [6].

The aqueous solubility of Cr(III) is a function of the pH of the water. Under neutral to basic pH, Cr(III) will precipitate and conversely under acidic pH it will tend to solubilize. The forms of Cr(VI) chromate and dichromate are extremely soluble under all pH conditions, but they can precipitate with divalent cations [6]. The recommended limits for Cr concentration in water are $8 \mu g L^{-1}$ for Cr(III) and $1 \mu g L^{-1}$ for Cr(VI). In the effluents in the vicinity of Cr industries the levels of Cr range from 2 to $5 g L^{-1}$ [13].

2.2. Chromium in Soil.

The concentration of Cr in the soils may vary considerably according to the natural composition of rocks and sediments that compose them [6]. The levels of chromium in the soil may increase mainly through anthropogenic deposition, as for example atmospheric deposition [14], also dumping of chromium-bearing liquids and solid wastes as chromium byproducts, ferrochromium slag, or chromium plating baths [6]. Generally, Cr in soil represents a combination of both Cr(III) and (VI). As in aquatic environment, once in the soil or sediment, Cr undergoes a variety of transformations, such as oxidation, reduction, sorption, precipitation, and dissolution [6]. The oxidants present in the soil (e.g., dissolved oxygen and MnO_2) can oxidize Cr(III) to Cr(VI) [15]; however, it seems that oxidation of Cr(III) by dissolved O_2 is residual when compared with MnO_2. The forms of Cr(VI) are on the other hand reduced by iron, vanadium, sulphydes, and organic materials [16]. However, when the reducing capacity of the soil is overcome, Cr(VI) may persist in the soil or sediment for years, especially if the soils are sandy or present low levels of organic matter.

López-Luna et al. [17] compared the toxicity of Cr(VI), Cr(III), and Cr tannery sludge respecting to Cr mobility in the soil and toxicity in wheat, oat, and sorghum plants and found that Cr(VI) was more mobile in soil and caused higher toxicity on those plant seedlings, while tannery sludge was the least toxic [17].

3. Chromium in Plants

3.1. Chromium Uptake.

The pathway of Cr uptake in plants is not yet clearly elucidated. However, being a nonessential element, Cr does not have any specific mechanism for its uptake and is also dependent on Cr speciation. Plant uptake of Cr(III) is a passive process, that is, no energy expenditure is required by the plant [3, 18]. The uptake of Cr(VI) is thought to be an active mechanism performed by carriers for the uptake of essential elements such as sulphate [19, 20]. Cr also competes with Fe, S, and P for carrier binding [8].

Cr(VI) has higher solubility and thus bioavailability is more toxic at lower concentrations than Cr(III), which tends to form stable complexes in the soil [17]. There are conflicting results concerning the uptake and translocation of Cr(VI). While some authors defend that Cr(VI) is reduced to Cr(III) on the root surface [21, 22], others suggest that dissolved Cr(VI) is taken up by plants without reduction [23].

Thus, Cr toxicity is dependent on metal speciation, which is determinant for its uptake, translocation, accumulation. Cr is toxic for agronomic plants at about 0.5 to 5.0 mgm L^{-1} in nutrient solution and 5 to $100 mg g^{-1}$ in soil [24]. Under normal conditions, concentration of Cr in plants is less than $1 \mu g g^{-1}$ [25].

3.2. Chromium Accumulation and Translocation.

Cr accumulates mainly in roots and shoots; however roots accumulate the major part, being usually only a small part translocated to the shoots [12, 26]. In pea plants exposed to Cr there was an increase in concentration of Cr in different parts of the plant with the increase of Cr supply. Accumulation of Cr in the different parts of the plant was in the following order roots > stem > leaves > seed [27]. Corroborating these results are the findings of several works and for instance, Huffman and Allaway [28] found that bean seeds accumulated about 0.1% Cr, while roots accumulated 98%. Furthermore, Liu and coworkers [29] studied hydroponically grown *A. viridis* L. under different concentrations of Cr(VI) and found that Cr was accumulated primarily in roots [29]. Another study performed by Vernay et al. [30] in *Lolium perenne* grown in the presence of $500 \mu M$ of Cr(VI) showed that roots accumulated 10 times more Cr than leaves. Spinach (*Spinacia oleracea* L. cv. "Banarasi") grown in the presence of Cr(VI) showed more accumulation of Cr in the roots than in leaves and stem showed the least accumulation [31]. Also, in celery seedlings grown in the presence of C(III) most Cr was accumulated in roots [32].

López-Luna and coworkers [17] found that roots of wheat, oat, and sorghum accumulated more Cr than shoots; however in spite of that, wheat, oat, and sorghum showed Cr translocation from roots to shoots. Furthermore, Zayed et al. [33] tested Cr(III) and Cr(VI) translocation in several crops and found that translocation of both Cr forms from roots to shoots was very low and accumulation of Cr by roots was 100-fold higher than in shoots, despite of the Cr species. However, Skeffington and coworkers [18] found that more 51Cr was transported from root to shoot when Cr(VI), rather than Cr(III), was supplied to the plant. At high Cr doses (1 mM $CrCl_3$) roots accumulated very high levels of Cr and translocation was mainly to cotyledonary leaves and only small amounts in hypocotyls. Chatterjee and Chatterjee [34] also found low levels of translocation of Cr from roots to the shoots in cauliflower (*Brassica oleracea*) grown on sand with 0.5 mM Cr(III).

These results may conclude that Cr is mainly accumulated in roots, followed by stems and leaves; however only small amounts of Cr are translocated to leaves. This pattern seems independent of the form of Cr tested.

3.3. Plants with Potential of Phytoremediation of Chromium Contamination.

In phytoremediation, hyperaccumulator plants are used to extract and transform toxic metals, as Cr, into nontoxic and immobile compounds [35]. Cr hyperaccumulator plants can accumulate >1,000 mg Cr kg^{-1} (DW), in plant leaves. These plants can tolerate metals through chelation with appropriate high-affinity ligands, biotransformation with reductants, and compartmentalization in the cytoplasm or in the vacuole. Thus, Cr immobilization in vacuoles in plant root cells may represent an important mechanism of Cr detoxification by the plant [8, 36].

The bioconcentration factor (BCF) and translocation factor (TF) are usually used to evaluate plant ability to tolerate and accumulate heavy metals. The BCF is the ratio of metal concentration in the plant tissue to the soil and TF is the ratio of metal concentration in plant shoots to the roots. Plants exhibiting a shoot BCF > 1 are suitable for phytoextraction, and plants with a root BCF > 1 and TF < 1 have the potential for phytostabilization [37].

Rafati and coworkers [37] evaluated the ability to uptake Cr from the soil by different organs of *Populous alba* and *Morus alba*. Leaves accumulated higher levels of Cr than stems or roots. However, neither *P. alba* nor *M. alba* showed potential of Cr phytostabilization, since presented TF > 1 and root BCF < 1; also these plants are not suitable for phytoextraction as they presented a BCF < 1. In another study, Gafoori and coworkers [38] evaluated the potential accumulation of heavy metals, including Cr in *Dyera costulata*. This specie presented high potential to retain high amounts of Cr in leaves, suggesting that this specie has high phytoremediation potential, as presented high translocation factor and low BCF factor. *Pluchea indica* also shown a good potential of phytoremediation, as it presented high levels of Cr accumulation and translocation to the leaves [39]. Mellem and coworkers [40] found that *Amaranthus dubius* tolerate high Cr(VI) concentrations as indicated by the BCF value > 2, showing good potential for phytoremediation. Furthermore, Gardea-Torresdey and coworkers [41] found that *Convolvulus arvensis* L. exposed to 20 mg L^{-1} of Cr(VI) demonstrated capability to accumulate more than 3800 mg of Cr kg^{-1} dw tissue, showing that this specie can be used in phytoremediation of Cr(VI) contaminated soils. Also, the concentration of Cr in leaf tissue (2100 mg kg^{-1} dw) indicates that this plant species could be considered as a potential Cr-hyperaccumulator.

Ipomoea aquatica is a chromium hyperaccumulator that shows no toxicity symptoms when exposed to high levels of Cr(VI). Up to 28 mg L^{-1} Cr(VI), *I. aquatica* exhibits uniform absorption characteristics showing over 75% removal of added Cr(VI). Over 90% Cr(VI) is accumulated in stems and leaves, that is, aerial regions [42]. Furthermore, Mant and coworkers [43] found that *Pennisetum purpureum* and *Brachiaria decumbens* exposed to 20 mg L^{-1} of Cr(III) showed a metal removal efficiency of 78% and 66%. Also, Barbosa and coworkers [44] found that *Genipa americana* has potential for Cr(III) phytoremediation in contaminated watersheds, since its seedlings uptake elevated amounts of Cr(III) from the solution and it presented high capacity of immobilizing and storing the metal on their roots.

3.4. Growth and Development

3.4.1. Germination.
The presence of Cr in the medium may compromise several processes in plants, as for instance plant germination. Thus, the ability to germinate in the presence

of Cr may indicate the degree of tolerance to Cr [45]. Oat seed germination was severely diminished (84%, resp. to the control) in tannery sludge soil with 4000 mg Cr kg^{-1}, while in tannery sludge soil containing 8000 mg Cr kg^{-1} both oat and sorghum seed germination was suppressed [17]. When comparing the sensitivity of sorghum, wheat, and oat germination to Cr, López-Luna and coworkers found that germination of sorghum and wheat were markedly affected at 500–1000 mg Cr(III) kg^{-1} soil respectively, while oat germination was not affected in levels of Cr(III) below 4000 mg kg^{-1} soil [17]. With respect to to Cr(VI) it affected wheat and sorghum germination at the maximum concentration of 500 mg kg^{-1} soil [17]. Germination of *T. aestivum* seeds was also affected by exposure to 100 mg L^{-1} of Cr(VI) [46]. *Echinochloa colona* (L.) seeds showed lower rates of germination when exposed to contaminated medium from chromite minewaste dumps [47]. The effect of Cr contamination on the germination medium was also tested in mungbean (*Vigna radiate* L.) tolerant/sensitive cultivars and results showed that in sensitive plants, germination rate decreased in plants exposed to 96 or 192 μM Cr(VI), while in tolerant plants germination was not affected [48]. Maize seeds exposed to Cr(VI) also presented decreased rates of germination when exposed to concentrations of 100–300 mg L^{-1} of Cr(VI) [49]. Zeid [50] found that germination of beans (*Phaseolus vulgaris*) was reduced in the presence of 5×10^{-2} M Cr(III). Similar results were found by Peralta et al. [45] in alfalfa seeds exposed to Cr(VI). In another study Scoccianti and coworkers [32] found that Cr(III) at concentrations of 0.01 to 10 mM inhibited germination of celery seeds; indeed at 10 mM a total inhibition was detected.

In spite of the findings above, Corradi et al. [51] suggested that Cr(VI) treatment may not affect seed germination, but instead inhibit radicles growth when they emerge and contact Cr solution. Nevertheless, decrease in germination is a common response upon exposure to heavy metals, such as Cd, Pb, and Hg [52–54]. This response of low levels of germination upon Cr exposure can be related with decrease in α and β amylase activities under Cr stress [50]. Amylase hydrolysis of starch is essential for sugar supply to developing embryos. Decrease in amylase activity under Cr treatment decreases sugar availability to developing embryo which may contribute to inhibition of seed germination [46].

3.4.2. Root Growth.
Besides germination, also root growth is frequently affected by heavy metals. Peralta and coworkers [45] showed that 5 mg L^{-1} of Cr(VI) increased root growth comparatively to the control, and at higher doses (20 and 40 mg L^{-1}) there was a dose-inhibition effect. Cr(VI) in concentrations up to 200 mg L^{-1} decreased growth of paddy (*Oriza sativa* L.) [26]. Sensitive mungbean cultivars also showed decreased root growth when exposed to Cr(VI) [55]. Samantary [48] found that there was no root elongation in mungbean exposed to Cr(VI) concentrations between 96 and 1928 μM, but in lower concentrations, sensitive cultivars showed root elongation similar to the control. Also, development of lateral roots and root number was also affected by Cr exposure [48]. Moreover, roots of *Zea mays* L. treated with Cr(VI) were shorter and brownish and presented less

number of roots hairs [56]. López-Luna et al. [17] found that root growth of oat and sorghum was decreased by Cr concentrations in the soil of 100 mg Cr(VI) kg^{-1} soil [17]. Decrease in root growth in presence of Cr(VI) can be explained by inhibition of root cell division and/or elongation, which might have occurred as a result of tissue collapse and consequent incapacity of the roots to absorb water and nutrients from the medium [57] combined with extension of cell cycle [26]. Reduced root surface in Cr(VI) stressed plants may contribute to decreased capacity of plants to search for water in the soil contributing to water stress. Despite of these results, stimulation of growth under low concentrations of chromium was also described (e.g., [58]). For example, Peralta et al. [45] found that roots of alfalfa plants exposed to 5 mg L^{-1} of Cr(VI) grew 166% more than the controls.

3.4.3. Stem Growth.

Stem growth is another parameter usually affected by Cr exposure. Mallick and coworkers [56] found that shoot length of Zea mays L. decreased significantly at 9 μg mL^{-1} Cr(VI) after 7 days. Also, Rout and coworkers [47] found reduction of plant height and shoot growth due to Cr exposure in sensitive mungbean plants. In T. aestivum L. seedlings exposed to 100 mg L^{-1} of Cr(VI) for 7 days, Dey and coworkers [46] found decrease in root length by 63% and in shoot length by 44%, comparatively to the control. Concentrations of Cr(VI) in soil of 500 mg kg^{-1} also affected shoot growth of wheat and oat [17]. This decrease in plant height could be due to the reduced root growth and consequent decreased nutrients and water transport to the higher parts of the plant. Moreover, Cr transport to the aerial part of the plant can directly impact cellular metabolism of shoots contributing to the reduction in plant height.

3.4.4. Leaf Growth.

Cauliflower grown on sand with 0.5 mM Cr(III) showed suppression of growth and leaves were smaller, chlorotic, and wilted comparatively to the control [34]. Leaf area is also usually decreased in response to increase of Cr concentration [53, 59]. Reduction of leaf area can be a consequence of reduction of the number of cells in the leaves stunted by salinization or reduction in cell size [60]. Watermelon plants growing in the presence of Cr(VI) showed reduced number and size of leaves and turned yellow, wilted, and due to loss of turgor hung down from petioles [61]. With continued Cr supply the lamina of affected old leaves became necrotic, permanently wilted, dry, and shed [61].

3.4.5. Yield.

Plant yield is dependent on leaf growth, leaf area, and number. As Cr affects most of the biochemical and physiological process in plants, productivity and yield are also affected. Cr(VI) in irrigation water decreased significantly grain weight and yield (kg ha^{-1}) of paddy (Oriza sativa) up to 80% under 200 mg L^{-1} of Cr [26].

3.5. Physiological Processes

3.5.1. Photosynthesis.

As other heavy metals, Cr may affect plant photosynthesis leading to decrease in productivity and ultimately to death. In a recent work, Rodriguez and coworkers [62] showed that exposure to Cr(VI) induced a reduction of both chloroplast autofluorescence and volume in pea plants. Moreover, both Cr(III) and Cr(VI) can cause ultrastructural changes in the chloroplasts leading to inhibition of photosynthesis [63].

Respecting to pigments, Samantary [48] found chlorophyll degradation in mungbean sensitive cultivars exposed to Cr(VI) and decrease in chlorophyll a and chlorophyll b contents. Furthermore, pea plants grown in sand under different concentrations of Cr(VI) presented reduced chlorophyll contents in leaves [27]. Dey and coworkers [46] also found that total chlorophyll content decreased in shoots of T. aestivum L. with increasing Cr(VI) concentration. Concerning Cr(III), Chatterjee and Chatterjee found a decrease in chlorophyll contents in cauliflower grown on sand with 0.5 mM of Cr(III) [34]. Also, in celery seedlings, Cr(III) reduced chlorophyll contents mostly at concentrations of 1 mM [64].

When comparing the effects of Cr(III) and Cr(VI) on photosynthesis parameters of water hyacinth, Paiva and coworkers [12] found that Cr(III) was much less toxic than Cr(VI), and might eventually increase photosynthesis and chlorophyll content. In another study, Zeid [50] found that low and moderate concentrations of Cr(III) (10^{-6} and 10^{-4} M) in irrigation solution increased pigment content in leaves, but higher Cr(III) concentrations (10^{-2} M) reduced the contents of chlorophyll a, chlorophyll b, and carotenoids.

This general profile of decrease in chlorophyll content at high Cr concentrations suggests that chlorophyll synthesis and/or chlorophyllase activity is being affected. Vajpayee and coworkers [65] showed that Cr affects pigment biosynthesis by, for instance, degrading δ-aminolaevulinic acid dehydrates, an essential enzyme in chlorophyll biosynthesis. Vernay and coworkers [30] also presented evidence that Cr competes with Mg and Fe for assimilation and transport to leaves, affecting therefore pigment biosynthesis. As the levels of reactive oxygen species (ROS) usually increase as a result of Cr exposure (e.g., [63, 66]), Juarez and coworkers [67] showed that ROS damages pigment-protein complexes located in thylakoid membranes followed by pheophitinization of chlorophylls (substitution of Mg^{2+} by H^+ ions) and destruction of thylakoid membranes.

Considering the effects of Cr on plant fluorescence parameters, Liu and coworkers [29] found that A. viridis L. exposure to Cr(VI) resulted in decreased net photosynthetic rate, transpiration rate, stomatal conductance, and intercellular CO_2 concentration. Also, chlorophyll fluorescence parameters F_v/F_m, F'_v/F'_m, ΦPSII, and q_p, decreased in Cr(VI)-treated, but q_N and NPQ showed an increase in Cr(VI)-treated plants [29], indicating that the photochemical apparatus might have been compromised. In another study, Vernay and coworkers [30] found that Cr(VI) affected L. perenne fluorescence parameters associated with PSII. In another study, these authors compared the effects of Cr(VI) and Cr(III) on Datura innoxia and found that Cr(VI) had a more toxic effect on those plants than Cr(III) [68]. In plants stressed with Cr(VI), a decrease in the quantum yield of PSII electron transport (ΦPSII), F'_v/F'_m and q_p was observed [68]. ΦPSII represents the number of electrons transported across a PSII reaction center per mole of quantum absorbed by PSII, F'_v/F'_m represents the excitation capture efficiency of open PSII reaction centers, while (q_p) reflects the number of open reaction centers and it is an indicator of the capacity of photochemical processes [69].

3.5.2. Mineral Nutrition. Cr, being structurally similar to other essential elements, may affect plant mineral nutrition. Mallick et al. [56] found that Cr exposure decreased Cu absorption in *Zea mays* roots, while leaves were not affected. Uptake of both macronutrients (e.g., N, P, K) and micronutrients decreased with increase of Cr(VI) in irrigation of paddy [26]. Also, decreased uptake of the micronutrients Mn, Fe, Cu, and Zn was detected by Liu et al. [29] in *A. viridis* L. exposed to Cr(VI). High content of Cr may displace the nutrients from physiological binding sites and consequently decrease uptake and translocation of essential elements. In watermelon plants grown in the presence of Cr, an increase in concentrations of P and Mn and decrease in Fe, Cu, Zn, and S contents in leaves was observed [61]. In *L. perenne*, Vernay and coworkers [30] found that Cr(VI) exposure affected mineral contents mostly Fe, Ca, and Mg. Cr(VI) also decreased Fe concentration in spinach [31] and sunflower [70]. The decrease in Fe concentration in leaf tissue in response to Cr toxicity is suggestive of Cr(VI) interference in the availability of Fe, leading to impairment of Fe metabolism [71].

3.5.3. Enzymes and Other Compounds. The activity of antioxidant enzymes, namely, peroxidase, catalase (CAT), glucose-6-phosphate dehydrogenase and superoxide dismutase (SOD) increased in case of Cr-sensitive of mungbean exposed to different Cr concentrations. However, the level of antioxidant enzymes decreased in Cr-tolerant cultivars [48]. SOD and CAT activities decreased in *T. aestivum* L. grown in the presence of $K_2Cr_2O_7$ in roots and shoots [46]. CAT activity also decreased in *A. viridis* L. exposed to Cr(VI) but an increase in SOD and guaiacol peroxidase (POX) activity was observed with increase of Cr(VI) concentration [29]. POX decreased in roots and increased in shoots of *T. aestivum* exposed to Cr(VI) [46]. Prado et al. [72] evaluated the metabolic responses to Cr(VI) exposure in floating and submerged leaves of *Salvinia minima* plants and found that Cr affected sucrose contents which were higher in Cr-treated leaves, while glucose contents showed an inverse pattern. Invertase activity also was also affected and suffered a decrease in floating leaves [72]. Zaimoglu and coworkers [73] studied the antioxidant responses of *Brassica juncea* and *Brassica oleracea* to soils enriched with Cr(VI) and found that total enzymatic activity was higher in *B. oleracea* than in *B. juncea*. Cr(VI) and also a decrease in CAT activity in both species [73]. Cellular antioxidants play an important role in protecting *Brassica* sp. to Cr-induced oxidative stress. This high activity of antioxidant enzymes and consequent detoxification of ROS contributes to relative tolerance of these species to Cr(VI). Furthermore, Guédard and coworkers [74] found that leaf fatty acid composition of *Lactuca serriola* was affected by the presence of Cr in metallurgic landfill soil.

3.6. Genotoxicity. Zou and coworkers [58] evaluated the effects of Cr(VI) on root cell growth and division of root tips of *A. viridis* L. and found that the mitotic index decreased with increased concentration of Cr(VI). Furthermore, Cr(VI) also affected chromosome morphology with increase in the frequency of c-mitosis, chromosome bridges, anaphase bridges, and chromosome stickiness [58].

Pea plants grown in the presence of Cr(VI) showed significant variations on cell cycle dynamics and ploidy level in leaves; however roots presented a cell cycle arrest at G2/M phase of the cell cycle; also polyploidization at both 2C and 4C levels was detected [75]. Moreover, in leaves and roots, an increase in DNA damage, assessed both by comet assay, and an increase in full peak coefficient of variation (FPCV) of G0/G1 were also detected [75]. Labra et al. [76] found hypermethylation of DNA and increase in DNA polymorphism in *Brassica napus* in response to Cr(VI) exposure. Cr(VI) also induced genotoxicity detected by AFLP analysis in *Arabidopsis thaliana* (L.) [77]. Furthermore, Knasmüller and coworkers [78] compared Cr(VI) and Cr(III) with respect to their ability to induce micronucleus in *Tradescantia* and found that only in Cr(VI)-exposed plants there was an increases in micronucleus frequencies. Moreover, Wang [79] in a survey to assess the genotoxic effects of Cr in water extracted from contaminated soil found that it was able to induce micronuclei in *Vicia faba* roots. Furthermore, Vannini and coworkers [80] evaluated the molecular changes induced by Cr(III) and Cr(VI) on germination kiwifruit pollen and concluded that neither Cr species induced a genotoxic effects. Both Cr species induced a strong reduction of proteins involved in mitochondrial oxidative phosphorylation and a decline in ATP levels [80].

4. Concluding Remarks

This paper includes an overview of the literature about Cr toxicity in the environment, especially in water and soil and provides new insights about Cr toxicity in plants. Cr exists mainly in three oxidative states Cr(0), Cr(III), and Cr(VI), which are the most stable forms of Cr. As Cr(0) is the metallic form, the forms of Cr(III) and Cr(VI) are the most preponderant in soils and water. Once in water/soil, Cr suffers a variety of transformations such as oxidation, reduction, sorption, desorption, precipitation, and dissolution. While Cr(III) solubility is dependent on pH, Cr(VI) is extremely soluble under all pH conditions. Cr as being a nonessential element for plants does not have any specific mechanism for its uptake. Cr(III) uptake is a passive process, whereas Cr(VI) uptake is performed by carriers of essential elements such as sulphate. Cr accumulates mainly on plant roots, being translocated to shoots in small levels, independently of Cr specie. Despite known toxicity of Cr to plants, there are several plants that hyperaccumulate this metal contributing to its removal from soil/water, showing good potential for application in Cr phytoremediation strategies. Cr affects several processes in plants, namely, seed germination, growth, yield and also physiological processes as photosynthesis impairment and nutrient and oxidative imbalances. Also, it has been shown that Cr is able to induce genotoxicity in several plant species.

Acknowledgment

The work of H. Oliveira was supported by FCT (Grant reference SFRH/BPD/48853/2008).

References

[1] S. Avudainayagam, M. Megharaj, G. Owens, R. S. Kookana, D. Chittleborough, and R. Naidu, "Chemistry of chromium in soils with emphasis on tannery waste sites," *Reviews of Environmental Contamination and Toxicology*, vol. 178, pp. 53–91, 2003.

[2] P. Babula, V. Adam, R. Opatrilova, J. Zehnalek, L. Havel, and R. Kizek, "Uncommon heavy metals, metalloids and their plant toxicity: a review," *Environmental Chemistry Letters*, vol. 6, no. 4, pp. 189–213, 2008.

[3] A. M. Zayed and N. Terry, "Chromium in the environment: factors affecting biological remediation," *Plant and Soil*, vol. 249, no. 1, pp. 139–156, 2003.

[4] T. Becquer, C. Quantin, M. Sicot, and J. P. Boudot, "Chromium availability in ultramafic soils from New Caledonia," *Science of the Total Environment*, vol. 301, no. 1–3, pp. 251–261, 2003.

[5] J. R. Peralta-Videa, M. L. Lopez, M. Narayan, G. Saupe, and J. Gardea-Torresdey, "The biochemistry of environmental heavy metal uptake by plants: implications for the food chain," *International Journal of Biochemistry and Cell Biology*, vol. 41, no. 8-9, pp. 1665–1677, 2009.

[6] D. E. Kimbrough, Y. Cohen, A. M. Winer, L. Creelman, and C. Mabuni, "A critical assessment of chromium in the environment," *Critical Reviews in Environmental Science and Technology*, vol. 29, no. 1, pp. 1–46, 1999.

[7] International Agency for Research on Cancer, "Chromium, nickel and welding," in *IARC Monographs on the Evaluation of Carcinogenic Risks to Humans*, vol. 49, The International Agency for Research on Cancer, Scientific Publications, Lyon , France, 1990.

[8] A. K. Shanker, C. Cervantes, H. Loza-Tavera, and S. Avudainayagam, "Chromium toxicity in plants," *Environment International*, vol. 31, no. 5, pp. 739–753, 2005.

[9] V. Velma, S. S. Vutukuru, and P. B. Tchounwou, "Ecotoxicology of hexavalent chromium in freshwater fish: a critical review," *Reviews on Environmental Health*, vol. 24, no. 2, pp. 129–145, 2009.

[10] F. L. Petrilli and S. De Flora, "Toxicity and mutagenicity of hexavalent chromium on *Salmonella typhimurium*," *Applied and Environmental Microbiology*, vol. 33, no. 4, pp. 805–809, 1977.

[11] Agency for Toxic Substances and Disease Registry, *Toxicological Profile for Chromium*, Health Administration Press, Atlanta, Ga, USA, 2000.

[12] L. B. Paiva, J. G. de Oliveira, R. A. Azevedo, D. R. Ribeiro, M. G. da Silva, and A. P. Vitória, "Ecophysiological responses of water hyacinth exposed to Cr^{3+} and Cr^{6+}," *Environmental and Experimental Botany*, vol. 65, no. 2-3, pp. 403–409, 2009.

[13] Chandra P., Sinha S., and Rai U. N., "Bioremediation of Cr from water and soil by vascular aquatic plants," in *Phytoremediation of Soil and Water Contaminants*, Kruger E. L., Anderson T. A., and Coats J. R., Eds., vol. 664 of *ACS Symposium*, pp. 274–282, DC7 American Chemical Society, Washington, DC, USA, 1997.

[14] I. Rosas, R. Belmomt, A. Baez, and R. Villalobos-Pietrini, "Some aspects of the environmental exposure to chromium residues in Mexico," *Water, Air, and Soil Pollution*, vol. 48, no. 3-4, pp. 463–475, 1989.

[15] S. E. Fendorf and R. J. Zasoski, "Chromium(III) oxidation by δ-MnO_2. 1. Characterization," *Environmental Science and Technology*, vol. 26, no. 1, pp. 79–85, 1992.

[16] E. E. Cary, "Chromium in air, soils, and natural waters," in *Biological and Environmental Aspects of Chromium*, S. Langard, Ed., pp. 49–63, Elsevier Biomedical, New York, NY, USA, 1982.

[17] J. López-Luna, M. C. González-Chávez, F. J. Esparza-García, and R. Rodríguez-Vázquez, "Toxicity assessment of soil amended with tannery sludge, trivalent chromium and hexavalent chromium, using wheat, oat and sorghum plants," *Journal of Hazardous Materials*, vol. 163, no. 2-3, pp. 829–834, 2009.

[18] R. A. Skeffington, P. R. Shewry, and P. J. Peterson, "Chromium uptake and transport in barley seedlings (*Hordeum vulgare* L.)," *Planta*, vol. 132, no. 3, pp. 209–214, 1976.

[19] Y. J. Kim, J. H. Kim, C. E. Lee et al., "Expression of yeast transcriptional activator MSN1 promotes accumulation of chromium and sulfur by enhancing sulfate transporter level in plants," *FEBS Letters*, vol. 580, no. 1, pp. 206–210, 2006.

[20] C. Cervantes, J. C. García, S. Devars et al., "Interactions of chromium with microorganisms and plants," *FEMS Microbiology Reviews*, vol. 25, no. 3, pp. 335–347, 2001.

[21] V. Ramachandran, T. J. D'Souza, and K. B. Mistry, "Uptake and transport of chromium in plants," *Journal of Nuclear Agriculture and Biology*, vol. 9, no. 4, pp. 126–128, 1980.

[22] A. Zayed, C. M. Lytle, J. H. Qian, and N. Terry, "Chromium accumulation, translocation and chemical speciation in vegetable crops," *Planta*, vol. 206, no. 2, pp. 293–299, 1998.

[23] S. Mishra, V. Singh, S. Srivastava et al., "Studies on uptake of trivalent and hexavalent chromium by maize (*Zea mays*)," *Food and Chemical Toxicology*, vol. 33, no. 5, pp. 393–397, 1995.

[24] L. R. Hossner, "Phytoaccumulation of selected heavy metals, uranium, and plutonium in plant systems," *Quarterly Progress Report*, Texas A&M University: College Station, TX, Project UTA96–0043, 1996.

[25] J. J. Mortvedt and P. M. Giordano, "Response of corn to zinc and chromium in municipal wastes applied to soil," *Journal of Environmental Quality*, vol. 4, no. 2, pp. 170–174, 1975.

[26] P. Sundaramoorthy, A. Chidambaram, K. S. Ganesh, P. Unnikannan, and L. Baskaran, "Chromium stress in paddy: (i) nutrient status of paddy under chromium stress; (ii) phytoremediation of chromium by aquatic and terrestrial weeds," *Comptes Rendus Biologies*, vol. 333, no. 8, pp. 597–607, 2010.

[27] K. K. Tiwari, S. Dwivedi, N. K. Singh, U. N. Rai, and R. D. Tripathi, "Chromium (VI) induced phytotoxicity and oxidative stress in pea (*Pisum sativum* L.): biochemical changes and translocation of essential nutrients," *Journal of Environmental Biology*, vol. 30, no. 3, pp. 389–394, 2009.

[28] E. W. Huffman Jr. and H. W. Allaway, "Chromium in plants: distribution in tissues, organelles, and extracts and availability of bean leaf Cr to animals," *Journal of Agricultural and Food Chemistry*, vol. 21, no. 6, pp. 982–986, 1973.

[29] D. Liu, J. Zou, M. Wang, and W. Jiang, "Hexavalent chromium uptake and its effects on mineral uptake, antioxidant defence system and photosynthesis in *Amaranthus viridis* L.," *Bioresource Technology*, vol. 99, no. 7, pp. 2628–2636, 2008.

[30] P. Vernay, C. Gauthier-Moussard, and A. Hitmi, "Interaction of bioaccumulation of heavy metal chromium with water relation, mineral nutrition and photosynthesis in developed leaves of Lolium perenne L.," *Chemosphere*, vol. 68, no. 8, pp. 1563–1575, 2007.

[31] R. Gopal, A. H. Rizvi, and N. Nautiyal, "Chromium alters iron nutrition and water relations of spinach," *Journal of Plant Nutrition*, vol. 32, no. 9, pp. 1551–1559, 2009.

[32] V. Scoccianti, R. Crinelli, B. Tirillini, V. Mancinelli, and A. Speranza, "Uptake and toxicity of Cr(III) in celery seedlings," *Chemosphere*, vol. 64, no. 10, pp. 1695–1703, 2006.

[33] A. Zayed, C. M. Lytle, J. H. Qian, and N. Terry, "Chromium accumulation, translocation and chemical speciation in vegetable crops," *Planta*, vol. 206, no. 2, pp. 293–299, 1998.

[34] J. Chatterjee and C. Chatterjee, "Phytotoxicity of cobalt, chromium and copper in cauliflower," *Environmental Pollution*, vol. 109, no. 1, pp. 69–74, 2000.

[35] S. Bluskov, J. M. Arocena, O. O. Omotoso, and J. P. Young, "Uptake, distribution, and speciation of chromium in *Brassica Juncea*," *International Journal of Phytoremediation*, vol. 7, no. 2, pp. 153–165, 2005.

[36] I. D. Pulford, C. Watson, and S. D. McGregor, "Uptake of chromium by trees: prospects for phytoremediation," *Environmental Geochemistry and Health*, vol. 23, no. 3, pp. 307–311, 2001.

[37] M. Rafati, N. Khorasani, F. Moattar, A. Shirvany, F. Moraghebi, and S. Hosseinzadeh, "Phytoremediation potential of *Populus alba* and *Morus alba* for cadmium, chromuim and nickel absorption from polluted soil," *International Journal of Environmental Research*, vol. 5, no. 4, pp. 961–970, 2011.

[38] M. Gafoori, N. M. Majid, M. M. Islam, and S. Luhat, "Bioaccumulation of heavy metals by *Dyera costulata* cultivated in sewage sludge contaminated soil," *African Journal of Biotechnology*, vol. 10, no. 52, pp. 10674–10682, 2011.

[39] P. Sampanpanish, W. Pongsapich, S. Khaodhiar, and E. Khan, "Chromium removal from soil by phytoremediation with weed plant species in Thailand," *Water, Air, and Soil Pollution*, vol. 6, no. 1-2, pp. 191–206, 2006.

[40] J. J. Mellem, H. Baijnath, and B. Odhav, "Bioaccumulation of Cr, Hg, As, Pb, Cu and Ni with the ability for hyperaccumulation by *Amaranthus dubius*," *African Journal of Agricultural Research*, vol. 7, no. 4, pp. 591–596, 2012.

[41] J. L. Gardea-Torresdey, J. R. Peralta-Videa, M. Montes, G. de la Rosa, and B. Corral-Diaz, "Bioaccumulation of cadmium, chromium and copper by *Convolvulus arvensis* L.: impact on plant growth and uptake of nutritional elements," *Bioresource Technology*, vol. 92, no. 3, pp. 229–235, 2004.

[42] A. Weerasinghe, S. Ariyawnasa, and R. Weerasooriya, "Phytoremediation potential of *Ipomoea aquatica* for Cr(VI) mitigation," *Chemosphere*, vol. 70, no. 3, pp. 521–524, 2008.

[43] C. Mant, S. Costa, J. Williams, and E. Tambourgi, "Phytoremediation of chromium by model constructed wetland," *Bioresource Technology*, vol. 97, no. 15, pp. 1767–1772, 2006.

[44] R. M. T. Barbosa, A. A. F. de Almeida, M. S. Mielke, L. L. Loguercio, P. A. O. Mangabeira, and F. P. Gomes, "A physiological analysis of Genipa americana L.: a potential phytoremediator tree for chromium polluted watersheds," *Environmental and Experimental Botany*, vol. 61, no. 3, pp. 264–271, 2007.

[45] J. R. Peralta, J. L. Gardea-Torresdey, K. J. Tiemann et al., "Uptake and effects of five heavy metals on seed germination and plant growth in alfalfa (*Medicago sativa*) L.," *Bulletin of Environmental Contamination and Toxicology*, vol. 66, no. 6, pp. 727–734, 2001.

[46] S. K. Dey, P. P. Jena, and S. Kundu, "Antioxidative efficiency of *Triticum aestivum* L. exposed to chromium stress," *Journal of Environmental Biology*, vol. 30, no. 4, pp. 539–544, 2009.

[47] G. R. Rout, S. Samantaray, and P. Das, "Effects of chromium and nickel on germination and growth in tolerant and non-tolerant populations of *Echinochloa colona* (L.) link," *Chemosphere*, vol. 40, no. 8, pp. 855–859, 2000.

[48] S. Samantary, "Biochemical responses of Cr-tolerant and Cr-sensitive mung bean cultivars grown on varying levels of chromium," *Chemosphere*, vol. 47, no. 10, pp. 1065–1072, 2002.

[49] M. Labra, E. Gianazza, R. Waitt et al., "*Zea mays* L. protein changes in response to potassium dichromate treatments," *Chemosphere*, vol. 62, no. 8, pp. 1234–1244, 2006.

[50] I. M. Zeid, "Responses of *Phaseolus vulgaris* to chromium and cobalt treatments," *Biologia Plantarum*, vol. 44, no. 1, pp. 111–115, 2001.

[51] M. G. Corradi, A. Bianchi, and A. Albasini, "Chromium toxicity in *Salvia sclarea*. I. Effects of hexavalent chromium on seed germination and seedling development," *Environmental and Experimental Botany*, vol. 33, no. 3, pp. 405–413, 1993.

[52] M. Smiri, A. Chaoui, and E. El Ferjani, "Respiratory metabolism in the embryonic axis of germinating pea seed exposed to cadmium," *Journal of Plant Physiology*, vol. 166, no. 3, pp. 259–269, 2009.

[53] S. Pandey, K. Gupta, and A. K. Mukherjee, "Impact of cadmium and lead on *Catharanthus roseus*—a phytoremediation study," *Journal of Environmental Biology*, vol. 28, no. 3, pp. 655–662, 2007.

[54] O. Munzuroglu and H. Geckil, "Effects of metals on seed germination, root elongation, and coleoptile and hypocotyl growth in *Triticum aestivum* and *Cucumis sativus*," *Archives of Environmental Contamination and Toxicology*, vol. 43, no. 2, pp. 203–213, 2002.

[55] G. R. Rout, S. Samantaray, and P. Das, "Differential chromium tolerance among eight mungbean cultivars grown in nutrient culture," *Journal of Plant Nutrition*, vol. 20, no. 4-5, pp. 473–483, 1997.

[56] S. Mallick, G. Sinam, R. Kumar Mishra, and S. Sinha, "Interactive effects of Cr and Fe treatments on plants growth, nutrition and oxidative status in *Zea mays* L.," *Ecotoxicology and Environmental Safety*, vol. 73, no. 5, pp. 987–995, 2010.

[57] J. Barceló, C. Poschenrieder, and J. Gunsé, "Effect of chromium (VI) on mineral element composition of bush beans," *Journal of Plant Nutrition*, vol. 8, pp. 211–217, 1985.

[58] J. H. Zou, M. Wang, W. S. Jiang, and D. H. Liu, "Effects of hexavalent chromium (VI) on root growth and cell division in root tip cells of *Amaranthus viridis* L.," *Pakistan Journal of Botany*, vol. 38, no. 3, pp. 673–681, 2006.

[59] V. Pandey, V. Dixit, and R. Shyam, "Chromium effect on ROS generation and detoxification in pea (*Pisum sativum*) leaf chloroplasts," *Protoplasma*, vol. 236, no. 1–4, pp. 85–95, 2009.

[60] R. H. Nieman, "Expansion of bean leaves and its suppression by salinity," *Plant Physiology*, vol. 40, pp. 156–161, 1965.

[61] B. K. Dube, K. Tewari, J. Chatterjee, and C. Chatterjee, "Excess chromium alters uptake and translocation of certain nutrients in citrullus," *Chemosphere*, vol. 53, no. 9, pp. 1147–1153, 2003.

[62] E. Rodriguez, C. Santos, R. Azevedo, J. Moutinho-Pereira, C. Correia, and M. C. Dias, "Chromium (VI) induces toxicity at different photosynthetic levels in pea," *Plant Physiology and Biochemistry*, vol. 53, pp. 94–100, 2012.

[63] S. K. Panda and S. Choudhury, "Chromium stress in plants," *Brazilian Journal of Plant Physiology*, vol. 17, no. 1, pp. 95–102, 2005.

[64] V. Scoccianti, R. Crinelli, B. Tirillini, V. Mancinelli, and A. Speranza, "Uptake and toxicity of Cr(III) in celery seedlings," *Chemosphere*, vol. 64, no. 10, pp. 1695–1703, 2006.

[65] P. Vajpayee, S. C. Sharma, R. D. Tripathi, U. N. Rai, and M. Yunus, "Bioaccumulation of chromium and toxicity to photosynthetic pigments, nitrate reductase activity and protein content of *Nelumbo nucifera* Gaertn," *Chemosphere*, vol. 39, no. 12, pp. 2159–2169, 1999.

[66] V. Pandey, V. Dixit, and R. Shyam, "Chromium effect on ROS generation and detoxification in pea (*Pisum sativum*) leaf chloroplasts," *Protoplasma*, vol. 236, no. 1–4, pp. 85–95, 2009.

[67] A. B. Juarez, L. Barsanti, V. Passarelli et al., "In vivo microspectroscopy monitoring of chromium effects on the photosynthetic and photoreceptive apparatus of *Eudorina unicocca* and *Chlorella kessleri*," *Journal of Environmental Monitoring*, vol. 10, no. 11, pp. 1313–1318, 2008.

[68] P. Vernay, C. Gauthier-Moussard, L. Jean et al., "Effect of chromium species on phytochemical and physiological parameters in *Datura innoxia*," *Chemosphere*, vol. 72, no. 5, pp. 763–771, 2008.

[69] K. Maxwell and G. N. Johnson, "Chlorophyll fluorescence—a practical guide," *Journal of Experimental Botany*, vol. 51, no. 345, pp. 659–668, 2000.

[70] R. Gupta, R. Mehta, N. Kumar, and D. S. Dahiya, "Effect of chromium (VI) on phosphorus fractions in developing sunflower seeds," *Crop Research*, vol. 20, pp. 46–51, 2000.

[71] R. Gopal, A. H. Rizvi, and N. Nautiyal, "Chromium alters iron nutrition and water relations of Spinach," *Journal of Plant Nutrition*, vol. 32, no. 9, pp. 1551–1559, 2009.

[72] C. Prado, L. Rodríguez-Montelongo, J. A. González, E. A. Pagano, M. Hilal, and F. E. Prado, "Uptake of chromium by *Salvinia minima*: effect on plant growth, leaf respiration and carbohydrate metabolism," *Journal of Hazardous Materials*, vol. 177, no. 1–3, pp. 546–553, 2010.

[73] Z. Zaimoglu, N. Koksal, N. Basci, M. Kesici, H. Gulen, and F. Budak, "Antioxidative enzyme activities in *Brassica juncea* L. and *Brassica oleracea* L. plants under chromium stress," *International Journal of Food, Agriculture & Environment*, vol. 9, no. 1, pp. 676–679, 2011.

[74] M. Le Guédard, O. Faure, and J. J. Bessoule, "Soundness of in situ lipid biomarker analysis: early effect of heavy metals on leaf fatty acid composition of *Lactuca serriola*," *Environmental and Experimental Botany*, vol. 76, pp. 54–59, 2012.

[75] E. Rodriguez, R. Azevedo, P. Fernandes, and C. Santos, "Cr(VI) induces DNA damage, cell cycle arrest and polyploidization: a flow cytometric and comet assay study in *Pisum sativum*," *Chemical Research in Toxicology*, vol. 24, no. 7, pp. 1040–1047, 2011.

[76] M. Labra, F. Grassi, S. Imazio et al., "Genetic and DNA-methylation changes induced by potassium dichromate in *Brassica napus* L.," *Chemosphere*, vol. 54, no. 8, pp. 1049–1058, 2004.

[77] M. Labra, T. Di Fabio, F. Grassi et al., "AFLP analysis as biomarker of exposure to organic and inorganic genotoxic substances in plants," *Chemosphere*, vol. 52, no. 7, pp. 1183–1188, 2003.

[78] S. Knasmüller, E. Gottmann, H. Steinkellner et al., "Detection of genotoxic effects of heavy metal contaminated soils with plant bioassays," *Mutation Research*, vol. 420, no. 1–3, pp. 37–48, 1998.

[79] H. Wang, "Clastogenicity of chromium contaminated soil samples evaluated by *Vicia* root-micronucleus assay," *Mutation Research*, vol. 426, no. 2, pp. 147–149, 1999.

[80] C. Vannini, G. Domingo, M. Marsoni et al., "Proteomic changes and molecular effects associated with Cr(III) and Cr(VI) treatments on germinating kiwifruit pollen," *Phytochemistry*, vol. 72, no. 14-15, pp. 1786–1795, 2011.

11

Plant Fitness Assessment for Wild Relatives of Insect Resistant Bt-Crops

D. K. Letourneau and J. A. Hagen

Department of Environmental Studies, University of California, Santa Cruz, CA 95064, USA

Correspondence should be addressed to D. K. Letourneau, dletour@ucsc.edu

Academic Editor: William K. Smith

When field tests of transgenic plants are precluded by practical containment concerns, manipulative experiments can detect potential consequences of crop-wild gene flow. Using topical sprays of bacterial *Bacillus thuringiensis* larvicide (Bt) and larval additions, we measured fitness effects of reduced herbivory on *Brassica rapa* (wild mustard) and *Raphanus sativus* (wild radish). These species represent different life histories among the potential recipients of Bt transgenes from Bt cole crops in the US and Asia, for which rare spontaneous crosses are expected under high exposure. Protected wild radish and wild mustard seedlings had approximately half the herbivore damage of exposed plants and 55% lower seedling mortality, resulting in 27% greater reproductive success, 14-day longer life-spans, and 118% more seeds, on average. Seed addition experiments in microcosms and *in situ* indicated that wild radish was more likely to spread than wild mustard in coastal grasslands.

1. Introduction

Commercialized transgenic, insect resistant (IR) crops currently grown in the United States have virtually no wild relatives near production sites, thus ensuring that novel crop traits are unlikely to move into local wild gene pools. However, an assessment of the consequences of gene flow will be necessary in future deregulation decisions because most of the major and minor crops in the world either exist in the wild themselves or hybridize with wild relatives somewhere in their range [1–5]. Wild relatives of transformed plants that obtain IR traits through gene flow and introgression may be released from the pressure exerted by susceptible herbivores [6–14]. However, scant knowledge about the ecological factors that regulate the abundance, competitive ability, or geographic range of weeds limits our ability to predict whether novel plant defenses are likely to increase the weediness of wild crop relatives [14] or even whether herbivory has a negative or positive effect on plant growth and fitness [15–20]. Surprisingly, few tests have been conducted on the effects of herbivory on the spread of invasive plants [21, 22] or to quantify the effects of herbivory on plant vital rates [23].

Identifying and quantifying environmental risks associated with gene flow from transgenic crops is subject to methodological tradeoffs because of containment restrictions, especially for plant fitness effects, which require pollen production. Field tests with pollen-producing transgenic plants must be contained physically in cages or greenhouses or established at sites where wild relatives do not occur. Conditions in regions that have no natural populations of wild relatives may differ from areas of concern for hybrid formation in ways that affect the results, and therefore the relevance, of such field tests. An alternative method, used in this study, is to conduct tests *in situ*, where hybrids would be expected to occur spontaneously, by using herbivore protection/addition techniques on nontransgenic wild-type plants. This method could be adapted to test for fitness effects of IR transgenes in crop-wild hybrids worldwide, for example, the Bt-Brassica IPM programs in Asia and Africa (Grzywacz et al. [24]), and Bt-maize adoption in Mexico.

We present herbivore protection experiments using Bt sprays on two very different species in the Brassicaceae, wild radish (*Raphanus sativus* L., which occurs in California as a hybrid complex with *R. raphanistrum* L. [25]) and wild mustard (*Brassica rapa* L.) as a method to assess potential fitness differences should crop transgenes from Bt cole crops (such as Bt canola, Bt cabbage, or Bt broccoli) introgress into wild populations occurring in the local mosaic of agriculture and coastal wildlands. The flora surrounding cole crop fields includes naturalized cabbage plants (*B. oleracea* L.), exotic wild radish, and wild mustards (*B. rapa* L., *B. nigra* L., and others), as well as native relatives. Using the two local Brassicaceae species that differ the most in their habits and characteristics, we compared relative levels of plant mortality, longevity, reproductive success, and total seed production per seedling of Bt-protected versus Bt-exposed wild mustard (*B. rapa*) and wild radish (*R. sativus*) in the three main habitats where they occur naturally: cultivated (disked) agricultural fields, uncultivated agricultural field margins, and nearby coastal grasslands (Figure 1). These habitats differ in resources and vegetational quality and may differ in patterns of gene flow, so that they represent the range of conditions under which wild relatives of Bt cole crops occur locally. To explore the link between reproductive output and population size, one needs to quantify the relationship between seed production and recruitment [26, 27]. Therefore, we also compared plant recruitment with seed addition experiments in field plots and in microcosms of disked fields and grassland.

Our experiments on wild radish and wild mustard were designed to detect plant population responses in complex habitats and determine (1) if protection from Bt susceptible herbivore damage would result in increased survivorship, longevity, and/or fecundity compared to plants incurring damage within the natural range occurring on local plants, (2) if tolerance to herbivory varies between wild plant species or among habitats that differ categorically in terms of plant resources and vegetational background (disked agricultural soil versus agricultural field margins versus natural vegetation), and (3) if seed limitation is a likely regulatory mechanism for either species in different habitats. The advantage of simulating plant protection conferred by transgenic traits in wild plants is the ability to include multispecies interactions that can alter fitness effects in ways that differ fundamentally from outcomes predicted by experiments with isolated plants, caged or greenhouse trials, or with artificial herbivory (see [28]).

Drawbacks include the adequate matching of expression levels, persistence rates, and target insects with applications of insecticidal simulants, and any fundamental differences between actual transgenic hybrids and wild-type experimental plants. We compare the results of this simulation technique to our previous studies and to studies conducted with transgenic plants by other researchers to gauge the usefulness of a field simulation method in informing risk assessments and regulatory decision making more generally.

2. Materials and Methods

2.1. Study System. We compared individual fitness parameters of wild radish (Brassicaceae: *Raphanus sativus* L.) in 2003-2004 (year one) and wild radish and wild mustard

(Brassicaceae: *Brassica rapa* L.) in 2004-2005 (year two). In the California central coast region, both species are naturalized winter annuals, emerging as seedlings in October or November with the first rains. Wild mustard flowers earlier and produces seed in spring compared to radish, which flowers and produces seed through August or September. Occasionally, wild radish plants persist for two years. Both wild mustard and wild radish are self-incompatible, insect-pollinated, and belong to crop-weed-wild complexes [29, 30]. *R. sativus* and *B. rapa* have persistent seed banks and are common in agricultural fields, field margins, and coastal grasslands. These habitats differ in resources and vegetational quality. Agricultural habitats are commonly fertilized and irrigated, whereas adjacent grassland soils receive water only through seasonal rainfall. The highest disturbance levels are found in cultivated fields, which are disked to remove weeds and loosen soil before planting annual crops. Wild radish and wild mustard are treated with herbicides or controlled through machine or hand weeding by local growers and managers of nature reserves, railroad corridors, state and city parklands, and wetlands.

Wild radish and mustard host a variety of herbivores, including cabbage aphids (*Brevicoryne brassicae* L.), green peach aphids (*Myzus persicae* L.), flea beetles (*Phyllotreta cruciferae* L.), diamondback moths (*Plutella xylostella* L.), cabbage butterflies (*Pieris rapae* L.), cabbage loopers (*Trichoplusia ni* L.), slugs, and snails. We used weekly bacterial insecticide (Bt) spray to prevent tissue removal by susceptible lepidopterans such as *P. rapae* and thus create an herbivore exclusion treatment against some of the insect species that feed on these plants. All plants were sprayed weekly with either a suspension of 1 g Dipel Dry Flowable (DF) powder per 750 mL of deionized water (protected plants) or a suspension of the same concentration of deactivated Bt (exposed plants) as a control for added water and bacterial material. This type of Bt is an appropriate simulant for Cry1s, which are relevant for herbivores of *Brassica* spp. and their relatives. Frequent treatments are necessary to simulate persistent dosage levels of Bt transgene expression or to simply act as an herbivore exclusion treatment, because this microbial insecticide is broken down by UV radiation and washed off with rain; this method may underestimate the insecticidal effects of transgenes in geographic locations with high precipitation.

We added a single 1st or 2nd instar *Pieris rapae* L. (Family: Pieridae), reared from field-collected adults, at monthly intervals to each of the plants that were exposed to all herbivores. In coastal California, *P. rapae* adults are in flight year-round, ovipositing on host plants in the Brassicaceae (Letourneau pers. obs., [31]). Details of these treatments were based on data from three previous experiments. First, our laboratory trials showed that *P. rapae* larvae in Petri dishes on cabbage leaf disks sprayed with denatured Bt (control, $n = 10$) fed and survived, whereas movement of larvae on disks with active Bt ($n = 10$) rapidly ceased, followed by death. Product breakdown during drying and transport of sprayed foliage prevented us from determining and comparing Bt levels with the range of expression in transgenic plants. Second, cumulative samples in June–August, 1987,

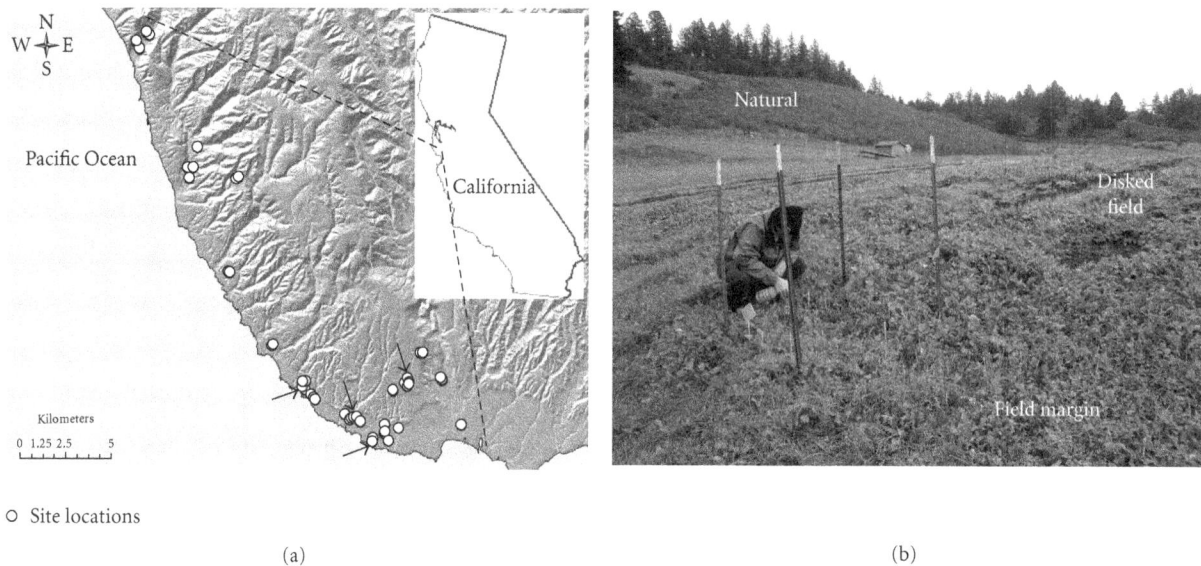

(a)   (b)

FIGURE 1: (a) Topographic map of California central coast, primarily Santa Cruz County, showing locations of four sites (Elkhorn Slough site was ~50 km south) used in year one (arrows) and of 90 1 m × 1 m plots established as 30 triplets of disked, field margin, and natural vegetation habitat plots in year two (GPS locations shown as white circles with black borders). (b) Example of disked field plot in year two, with T.R.F. Roubison spraying individual wild radish and wild mustard plants with Bt and denatured Bt amongst other forbs and grasses that emerged from soil seedbanks with the winter rains. Not shown are the associated field margin plot, which was within 5 m of the field, and the natural vegetation plot in uncultivated lands as shown in the distance.

a nonoutbreak year for lepidopterans, showed an average density of three lepidopteran eggs per *B. rapa* plant in unfertilized control plots within the study area for the current experiments [32, 33]. Finally, there were high levels of variability in ambient herbivory on *R. sativus* plants among sites and sample dates in preliminary field trials with 80–100 plants per habitat type at two sites in 2002 (0 to 100% per plant). The average number of damaged leaves in Bt-sprayed plots ranged from 8% to 55% of those damaged in ambient herbivory control plots (no Bt spray, no added larvae).

2.2. Study Sites. Experiments were conducted in the central coast region of California (36.974°N, −122.029°W), where 780 mm annual precipitation in this Mediterranean climate falls between October and May. In 2003-2004 (year one), sixty 1 m × 1 m field plots were established at each of five sites (Figure 1(a), arrows): the research farm at the University of California Santa Cruz Center for Agroecology and Sustainable Food Systems (UCSC CASFS), the UC Younger Lagoon Natural Reserve, east and west sites at Wilder Ranch State Park, and the Elkhorn Slough National Estuarine Research Reserve (not shown, 40 km south). These sites differ in historical land use, dominant vegetation, and local climate. In year two (2004-2005), three 1 m × 1 m experimental plots were established at each of 30 field sites along ~35 km of coastal Santa Cruz County, CA. In both years, one-third of the plots were designated for each of three habitat types (Figure 1(b)). Disked field habitat plots, located on the edge of an agricultural field (≤5 m from row crops), were dug with a shovel to turn over 10–20 cm of soil to simulate disking and disrupt any vegetation formerly

occurring on that plot. Field margin habitat plots were not disked or shoveled and had weedy vegetation, often on soil compacted by farm machinery. Both disked field and field margin plots were exposed to rain and irrigation run-off from fertilized fields (downhill placement from adjacent farm fields). Natural vegetation plots were established away from agriculture, had a mixture of naturally occurring native and exotic plants, and received no run-off or irrigation. All plots were established where dried plants of wild radish or wild mustard (or both) were already present within the plot or nearby (within seed-rain distance from plot).

2.3. Experimental Design. Treatment design was hierarchical in year one, with herbivory treatments (two levels) nested in habitat type (three levels—except at two locations where seedlings were scarce in disked plots), nested in location (five levels). Ten naturally occurring wild radish seedlings were selected in each plot (2,600 seedlings in total) and marked initially with numbered stakes and then with a numbered spiral binding ring around the plant stem. Paired plots were randomly designated to receive either weekly Bt sprays (protected plants) or denatured Bt sprays and larval additions (exposed plants). A single early instar *P. rapae* was placed on each experimental wild radish plant at three evenly spaced intervals between December and February, corresponding to plant stages from seedling to reproducing. Percent herbivore damage per plot was estimated once, 12 weeks after the first rains, by assigning a damage category to each leaf of each plant (0%, 1–10%, 11–25%, 26–50%, >50%) and converting the average category level per plant to the mid-point percentage of that category (e.g., 18% for

11–25%). The total number of viable radish seeds per small plant was counted after lightly crushing pods. For large plants, counts of viable radish seed for ≈ 15 pods were used to estimate total plant number of seeds by weight.

In year two, experimental plots (one set of three habitat types at each of 30 sites, Figures 1(a) and 1(b)) were divided into four 50 cm × 50 cm subplots to accommodate herbivore protection and herbivore exposure treatments for both wild radish and wild mustard. Before the first rain, to supplement any seedlings emerging naturally from the seed bank, eight radish seed pods collected from 25 mother plants in year one were added to each radish subplot and approximately 20 mustard seeds from 10 mother plants were added to mustard subplots. We reseeded subplots after 10 days if fewer than six seedlings of either plant species emerged. Three weeks after the first rains, four wild radish and four wild mustard seedlings per subplot (total of 1,260 plants) were numbered individually, assigned to receive either Bt sprays or inactivated Bt spray and larval additions, and followed weekly thereafter. *P. rapae* larvae were added at 3-4 week intervals, for a total of two to four larvae per plant over the season (up to four on long-lived plants). Percent herbivory per seedling was estimated (using categorical estimates of 0%, 1–5%, 6–60%, 61–90% and >90%, and their mid-points for calculating average estimates), and seedling mortality rates were determined eight weeks after the first rain. Percent herbivory was estimated again 21 weeks after the first rains. Although seed in disked plots were sown into bare soil, field margin plots usually had some standing vegetation and natural vegetation plots had very high cover rates of dried grasses and other annuals as well as some green perennial plants. At approximately 12 weeks after the first rain, as experimental plants and other species began to bolt and flower, bare soil cover and species richness of plants within each subplot was estimated. Wild mustard and radish plant status (rosette, bolting, budding, flowering, producing siliques or pods, or dead) was recorded on two of the four plants at week 15 after the rains. We continued to record the status of those plants every two weeks through week 39 when all plants had either died or finished producing seeds. Seed counts and estimates were done as in year one.

2.4. Seed Addition Experiments. For an increase in seed output to result in increased spread of the population, any production of additional seed has to result in additional plants surviving to reproduce in the habitat. To assess the potential fitness advantage of an increase in seed output by wild radish and wild mustard, we tested for seed limitation in a microcosm experiment with two of the habitat types (disked field and natural vegetation). Before the first rains (August/September 2006), large soil cores (25.4 cm diameter), with the existing dry vegetation and seed bank completely intact, were transferred to pots with care taken to minimize disturbance to the plant cover or soil profile. Two pairs of these large soil cores were taken from each of the 10 sites along the central coast with recently tilled soil, with one of each pair of the soil cores taken from a disked farm field before crops were sown and one taken from nearby areas

(same soil type) with natural vegetation intact. To test for any impact of increased seed yield, eight wild radish pods (lightly crushed) yielding 4–10 seeds/pod and 15 seeds of wild mustard were added to one of each pair of soil cores in three-gallon pots representing either disked agricultural fields or grassland. The number of seeds added approximated one-half the increase in number of seeds produced by a single wild radish plant (36.5 seeds) or wild mustard plant (26.5 seeds) when that mother plant has been protected from herbivory (average of years one and two from our field results). The resulting potted microcosms were used to ensure that runoff from heavy rains would not cause unaccountable seed losses. Pots were transported to the rooftop greenhouse at UCSC where they could be watered biweekly as needed and exposed to coastal weather for 10 weeks until wild mustard in the pots produced siliques and wild radish plants produced pods. A small scale field experiment at two of our sites (UCSC CASFS and the Homeless Garden Project) was used to further test these seed additions *in situ* in disked field and natural vegetation plots over six months in 2008-2009. Rings from pots were used to maintain added seeds in five replicate plots per habitat at each of the two sites.

2.5. Data Analyses. All analyses used PC-SAS v. 9.1 (SAS Institute 1990). In year one, both the average herbivore damage at 12 weeks after the first rain and the lifetime seed output for wild radish seedlings (rank transformed) were compared with respect to (1) herbivory treatment (protected versus exposed), (2) habitat type (disked field, field margin, and natural vegetation), and (3) interactions between herbivory treatment and habitat type. We used a General Linear Models (GLM) nested ANOVA to test for any significant effects, designating the error term as the type III mean square value for plot nested in the interaction term for site by treatment by habitat type. In year two, we used a logistic regression to compare categorical estimates of herbivory after eight weeks (0%, 2.5%, 25%, 75%, 95%), plant mortality (dead or alive), and lifetime production of siliques/pods (reproductive or not). Independent variables were plant species (wild radish versus wild mustard), herbivory treatment (protected versus exposed), and habitat type (disked field, field margin, and natural vegetation). Additionally, we used a GLM repeated measures ANOVA to test for changes in herbivory with time and to test for seasonal effects of habitat type, plant species, and interactions among these factors. Mean seed output per plant per plot, which could not be transformed to meet the assumptions of normality, was compared using a GLM ANOVA on ranks. Similarly, percent cover of bare soil after community development for 12 weeks was compared among subplots with ANOVA on ranks. All subplots for wild mustard and wild radish, and both herbivory treatments, were pooled to test for differences among habitat types because there were no significant differences in bare soil cover due to plant species or herbivore treatment.

The number of reproductive wild radish and wild mustard plants in the potted seed addition experiment was analyzed separately by plant species, with $n = 10$ microcosms for the factorial experiment with two levels of seed addition (seed added versus no seed added) and two levels of habitat type (disked field versus natural vegetation), using GLM ANOVA on ranks because the data could not be transformed to approximate a normal distribution. For seed addition experiments in the field, the total number of wild radish plants and wild mustard plants in each plot ($n = 5$ plots per plant per treatment per habitat type) were compared separately, six months after the first rain, using a nested GLM ANOVA with the error term for nesting within sites as a replicate (seed treatment by habitat by site) on ranks.

3. Results

3.1. Herbivory. In year one, 12 weeks after the first rains that caused seeds to germinate in experimental plots, juvenile wild radish plants protected with activated Bt sprays had an average of $5.8\% \pm 0.4\%$ herbivory caused by slugs, snails, beetles, and possibly lepidopterans that are not susceptible to Bt toxins [32]. This level of herbivore damage was significantly lower than the average $11.1\% \pm 0.8\%$ for unprotected wild radish plants exposed to all herbivores plus added *P. rapae* (ANOVA for treatment nested in site, $F_{5,218} = 8.8, P < 0.0001$). Herbivore damage was significantly greater in disked field plots ($9.7\% \pm 0.6\%$) than in field margin and natural vegetation plots ($7.8\% \pm 0.4\%$ and $8.0\% \pm 0.4\%$, resp.) (ANOVA for habitat type nested in site $F_{8,218} = 2.4, P = 0.0182$). Wild radish seedlings may have been more apparent to naturally occurring herbivores in disked plots free of other vegetation. Also, in the process of inoculating the plants in all habitats, we commonly observed spider predation on larvae; it is possible that such predation was more common when inoculated plants occurred in established vegetation. In all habitat types, however, protected plants were damaged significantly less than those exposed to all herbivores (no significant interaction between treatment and habitat type, $P = 0.1872$).

In year two, eight weeks after the first rains, wild radish and wild mustard seedlings protected with Bt sprays again received approximately one-half ($12.2\% \pm 1.7\%$ versus $27.4\% \pm 2.6\%$) the amount of herbivory received by unprotected plants exposed to all herbivores plus *P. rapae* larvae (Figure 2). Seedling damage was significantly lower in natural vegetation plots than in field margin habitat plots, with neither natural nor field margin habitats differing from disked plots ($23.1\% \pm 2.2\%$ in field margin versus $17.1\% \pm 1.8\%$ in natural vegetation and $20.8\% \pm 1.9\%$ in disked field plots), but this pattern was strongly expressed only in wild mustard and not in wild radish (Figure 2). The lack of a treatment by habitat type interaction indicates that protected plants suffered lower levels of herbivory than exposed plants in all habitat types. When most mustard and some radish plants were starting to flower, at 21 weeks after the first winter rain, mean damage levels per plant were $5.0\% \pm 1.2\%$ versus $9.9\% \pm 1.8\%$ for protected and exposed plants, respectively. Damage levels dropped significantly from eight to 21 weeks after the first rain, but only the herbivory treatment remained significant (no between subjects effect of species or habitat type, and

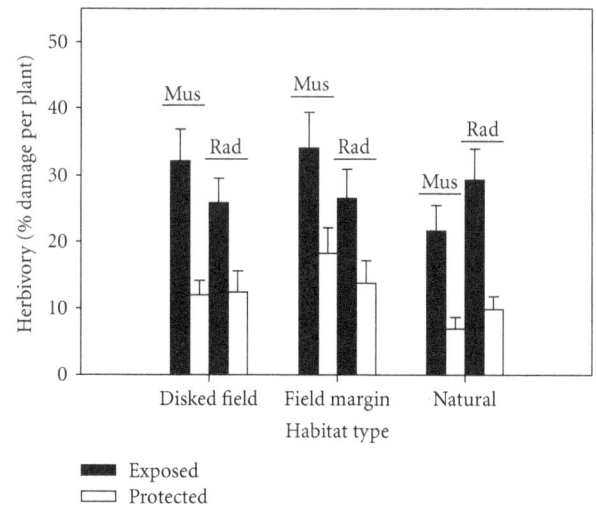

FIGURE 2: Year 2: mean (± 1 SE) percent damage per plant per plot 8 weeks after rains on exposed versus protected [tmt] wild mustard and wild radish [species] in disked field, field margin, and natural vegetation plots [hab]. Wild mustard bars, labeled "Mus," precede wild radish bars, labeled "Rad." Logistic regression (using $n = 30$ plot averages of 4 plants) showed significant effects for tmt ($df = 1$, $X^2 = 52.3, P = 0.0001$), hab ($df = 2, X^2 = 6.6, P = 0.0365$), and a nearly significant species*hab interaction ($df = 2, X^2 = 5.6$, $P = 0.0594$), but no significant effects for species, or species*tmt, tmt*hab, or species*tmt*hab interactions.

no within subject interactions among time, treatment, species, or habitat type) (repeated measures GLM: herbivory treatment $F_{1,178} = 4.47, P = 0.0359$). Lower herbivory levels later in the season indicated constant or lower herbivore pressure per unit tissue as plants grew larger or, especially for wild radish, an outcome of plants dropping damaged leaves.

3.2. Plant Mortality, Vegetation Cover, and Reproductive Success Rates. In year two, seedling mortality of wild mustard and wild radish was significantly lower for protected compared to exposed plants (Figure 3(a)). Exposed seedlings suffered 55% greater mortality than protected ones ($17.8\% \pm 1.8\%$ SE versus $11.9\% \pm 1.5\%$ SE). Seedling mortality was not significantly different among habitat types, though disked field plots tended to have lower seedling mortality rates overall (Figure 3(a)). There were no significant interactions involving habitat, herbivory, or species. The mortality rate of juvenile plants was 24% overall, with mortality rates not significantly different among plant species, treatments, or habitat types (Logistic regression, model value for Wald $X^2 = 13.7, P = 0.2528$), though mustard suffered higher juvenile mortality due to herbivory treatment than did wild radish (logistic regression for species by treatment interaction, Wald $X^2 = 4.2, P = 0.0413$).

Background vegetation had established fully by the time wild mustard and wild radish were reproductive, owing to the germination and growth of winter annuals after the fall rains began. By 12-13 weeks after the first rain, the average plant cover in all subplots was over 90%, including the disked field subplots, which initially had zero plant cover. The

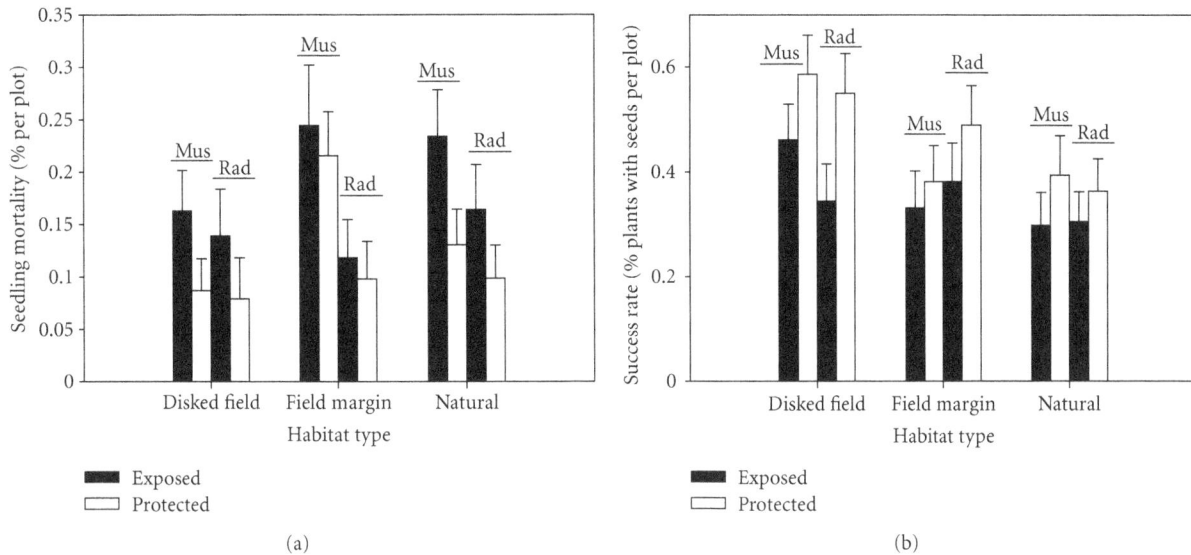

FIGURE 3: (a) Year 2: mean ($\pm$1 SE) percent seedling mortality per plot ($n = 30$ plots with 4 plants each) in year two on exposed versus protected [tmt] wild mustard and radish [species] in disked field, field margin, and natural vegetation plots [hab]. Wild mustard bars, labeled "Mus," precede wild radish bars, labeled "Rad." Logistic regression showed significant effects for species ($df = 1$, $X^2 = 9.7$, $P = 0.0019$) and tmt ($df = 1$, $X^2 = 6.2$, $P = 0.0127$), but not strongly for hab ($df = 2$, $X^2 = 5.0$, $P = 0.0826$), and none for any of the interactions: species*hab, species*tmt, tmt*hab, or species*tmt*hab. (b) Year 2: mean ($\pm$1 SE) percent of plants per plot ($n = 28$–30 plots with 2 plants/plot) producing pods/siliques on exposed versus protected [tmt] wild mustard and radish [species] in disked field, field margin, and natural vegetation plots [hab]. Wild mustard bars, labeled "Mus," precede wild radish bars, labeled "Rad." Logistic regression shows significant effects for tmt ($df = 1$, $X^2 = 7.0$, $P = 0.0082$) and hab ($df = 2$, $X^2 = 7.5$, $P = 0.0237$), but not for species, species*hab, species*tmt, tmt*hab, or species*tmt*hab interactions.

average plant species richness overall was 6.8 ± 0.1 SE in all subplots, with no significant differences among habitat types. The average percent cover of bare soil in disked plots was not significantly different from that of field margin habitats but remained significantly greater than in natural vegetation habitats ($10.5\% \pm 2.2\%$ SE in natural vegetation versus $9.4\% \pm 1.8\%$ SE in field margins versus $6.6\% \pm 1.7\%$ SE in disked field plots) (GLM ANOVA on rank percent bare soil cover $F = 3.15$, $P = 0.0450$).

A significantly greater percentage of protected wild radish and mustard plants produced siliques or pods ($46.3\% \pm 3.0\%$ SE) than did exposed plants ($36.5\% \pm 2.8\%$ SE), a 27% increase in reproductive success rate (Figure 3(b)). Plants in disked plots had significantly higher reproductive success rates than did plants in either field margin or natural vegetation plots, suggesting that greater plant competition or poorer resource conditions in the latter habitats made those habitats less favorable for wild radish or wild mustard reproductive success in general, without regard to herbivore pressure (Figure 3(b)). Wild radish is a longer-lived species than wild mustard (living for 26.2 weeks ± 0.7 SE versus 18.3 weeks ± 0.4 SE, ANOVA on ranked number of weeks, $F = 96.3$, $P < 0.001$), and plant longevity was increased overall by reducing herbivory (both species combined, no significant interaction between species and herbivory treatment). The average longevity for protected plants was 23.1 ± 0.7 SE weeks versus 21.4 ± 0.6 SE weeks for exposed plants (ANOVA on ranked number of weeks, $F = 4.9$, $P = 0.0276$). Neither habitat type nor any of the interactions among factors were

significant factors affecting plant longevity, and none of the interactions among herbivore pressure and species or habitat affected reproductive success (Figure 3(b)).

3.3. Seed Production. In year one, lifetime seed output of an average wild radish seedling, taking into account that early mortality results in zero seed production, was relatively low overall (Figure 4(a) compared to year two in Figure 4(b)). Nevertheless, protected wild radish plants in the reduced herbivory treatment produced significantly more seeds, showing an almost twofold increase in fitness in terms of average seeds per seedling (14.8 ± 2.3 SE seeds) compared to exposed plants (8.4 ± 1.4 SE seeds). Average seed output per wild radish seedling was also significantly greater in disked and field margin plots (13.1 ± 3.0 SE and 12.1 ± 2.2 SE, resp.) than in natural vegetation plots (10.5 ± 1.5 SE) (Figure 4(a)).

Similarly, in year two, protected wild radish and wild mustard seedlings produced significantly more seeds (100.6 seeds/seedling) on average than exposed plants with higher levels of herbivory (46.1 seeds/seedling) (Figure 4(b)), with no treatment by habitat or treatment by species interactions. Both wild mustard and wild radish responded similarly to the herbivory treatment, but wild mustard plants were more strongly responsible for a significant effect of habitat type (marginally significant interaction between species and habitat, $P = 0.0627$). Mustard plants produced fewer seeds per seedling in natural vegetation plots than in disked or field margin plots, even when protected from lepidopteran

(a)

(b)

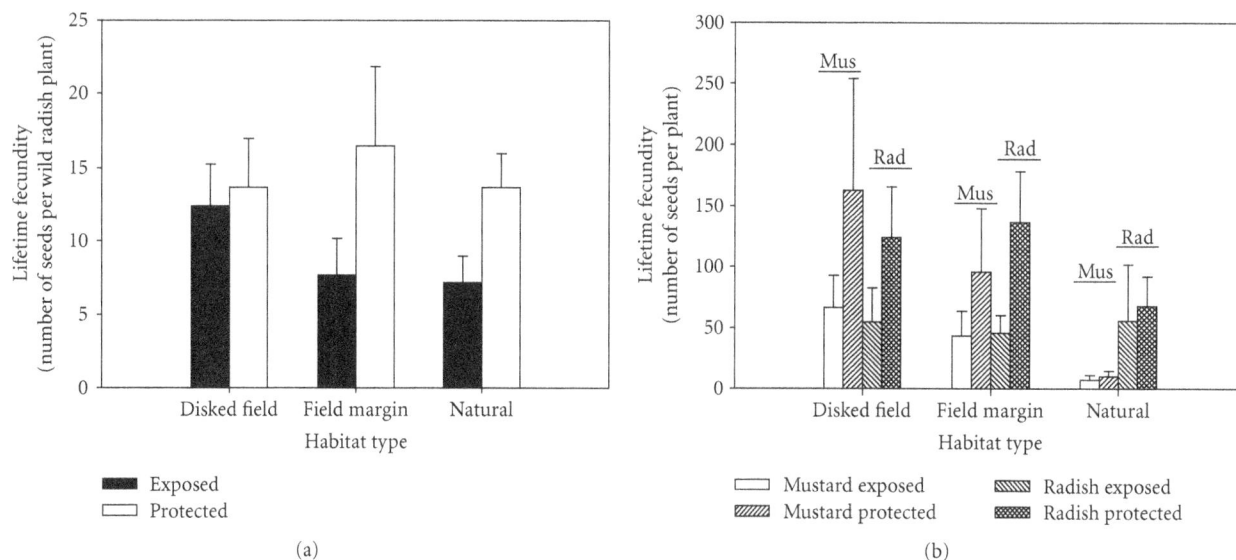

FIGURE 4: (a) Year 1: mean (± 1 SE) number of seeds per plant per plot ($n = 10$ plots with 10 plants each) produced by wild radish with two herbivory treatments (tmt: exposed to herbivores or protected from herbivores) in three habitat types (hab: disked field, field margin, or natural vegetation habitats); ANOVA on rank transformed levels of seed production per plant per plot showed significant effects for tmt (ANOVA for herbivory treatment nested in site, $F_{5,232} = 4.9$, $P = 0.0003$) and hab (ANOVA for habitat type nested in site, $F_{8,232} = 6.7$, $P < 0.0001$), with seed production greater on protected than exposed plants and in agricultural habitats than in natural vegetation, but no interaction between herbivory treatment and habitat type. Note that averages include plants that produced no seeds due to early mortality. (b) Year 2: mean (± 1 SE) number of seeds per plant per plot ($n = 28$–30 plots with 2 plants per plot) on exposed versus protected [tmt] wild mustard and radish [species] in disked field, field margin, and natural vegetation habitats [hab]. Wild mustard bars, labeled "Mus," precede wild radish bars, labeled "Rad." ANOVA on rank transformed levels of seed production per plant per plot showed significant effects for tmt ($F_{1,333} = 6.3$, $P = 0.0127$) and hab ($F_{2,333} = 3.7$, $P = 0.0246$), where seed set was significantly lower in natural habitats than in agricultural habitats, but the vulnerability of wild mustard in this regard was shown by a marginally significant effect of species*hab ($F_{2,333} = 2.8$, $P = 0.0627$), and no significant effects for species, species*tmt, tmt*hab, or species*tmt*hab interactions. Note that averages include plants that produced no seeds due to early mortality.

herbivory (Figure 4(b)). In a separate analysis of seed output *only* by those plants that lived to become reproductive and produce seed, radish produced significantly more seed than mustard (means of 215.2 ± 37.4 SE and 148.5 ± 28.7 SE, resp., ANOVA on ranks, $F_{1,392} = 14.0$, $P = 0.0002$), plants in disked and field margin habitats produced more than plants in natural vegetation habitats (means of 214.6 ± 40.5 SE, 210.7 ± 43.2 SE and 106.3 ± 36.0 SE, resp., ANOVA on ranks, $F_{1,392} = 8.0$, $P = 0.0004$) but only marginally significantly more seeds when protected from lepidopteran herbivory (202.8 ± 31.8 SE and 153.1 ± 34.5 SE per plant, resp., ANOVA on ranks, $F_{1,392} = 3.1$, $P = 0.0779$), with no treatment by habitat or treatment by species interactions.

3.4. Seed Limitation. Experimental addition of wild radish and wild mustard seeds resulted in higher recruitment, even when the number of seeds per microcosm or field plot was relatively low (estimated as 1/2 the additional seed produced by an average plant in the low herbivory treatment). Mean wild radish density was significantly greater when seeds were added to the soil compared to controls with no added seeds, whether in the disked soil microcosms (initially bare) or in the natural vegetation microcosms (with plant cover intact) (Table 1). Species richness of other plants emerging from the seed bank in disked soil microcosms averaged 4.5 (± 0.3 SE)

forbs and grasses, and natural vegetation microcosms had a mean of 5.2 (± 0.4 SE) other plant species, indicating that seed additions make a difference in recruitment of wild radish in the context of developing plant communities in these different habitat types. Seed addition also significantly increased mean recruitment of reproductive wild mustard plants (Table 1), but only in disked soil habitats. In contrast to wild radish, wild mustard density was significantly affected by habitat type (disked versus natural), such that few plants survived to flowering in natural vegetation microcosms. Species richness of other plants in these microcosms averaged 4.6 (± 0.3 SE) and 5.3 (± 0.3 SE) other forbs and grasses in disked and natural vegetation microcosms, respectively.

Seed additions carried out at two field sites showed that six months after the first rains, the number of wild radish plants established in field plots was (1) significantly greater when seeds were added than in control plots with no seeds added and (2) significantly greater in disked fields than in grasslands (Table 1). Seed addition produced the same result in disked fields as grasslands, as indicated by the lack of a treatment by habitat interaction. Similarly, the number of wild mustard plants was significantly increased by seed addition in disked field plots compared to controls, and disked plots had significantly more mustard plants than natural vegetation plots (grasslands). However, a significant

Table 1: Mean number of radish or mustard plants surviving to reproduce when the approximate half the mean seed output advantage per plant due to herbivore protection was added to microcosms ($n = 10$ per habitat type) and field plots ($n = 5$ per habitat per field site) compared to microcosms and field plots to which no seeds were added (control). Note that some seeds (whether added or present in the seed bank) exhibit dormancy.

Experiment	Disked soil habitat		Grassland habitat	
	Seeds added	Control	Seeds added	Control
Radish microcosms[1]	7.4 + 0.1	0.3 + 0.2	7.9 + 2.6	3.9 + 1.1
Mustard microcosms[2]	3.2 + 0.7	0.0 + 0.0	0.7 + 0.3	0.0 + 0.0
Radish field plots[3]	2.0 + 0.8	0.1 + 0.07	4.7 + 1.95	3.6 + 1.03
Mustard field plots[4]	7.0 + 1.96	1.1 + 0.68	0.0 + 0.0	0.0 + 0.0

[1] ANOVA on rank number of flowering radish plants per microcosm, with $F = 30.5$, $P = 0.0001$ for seed addition effects, $F = 1.4$, $P = 0.2452$ for disked versus natural soil microcosms, and $F = 6.7$, $P = 0.0136$ for seed addition by habitat interaction.

[2] ANOVA on rank number of flowering wild mustard plants per microcosm, with $F = 26.6$, $P < 0.0001$ for seed addition effects, $F = 7.4$, $P = 0.0101$ for disked versus natural soil, and $F = 4.2$, $P = 0.0485$ for seed addition by community interaction.

[3] Nested ANOVA on ranks for the number of wild radish plants after six months, seed addition treatment (site), $F_{2,56} = 4.72$, $P = 0.0165$; habitat (site) $F_{2,56} = 23.13$, $P \leq 0.0001$; treatment* habitat (site) $F_{2,56} = 1.63$, $P = 0.2125$.

[4] Nested ANOVA on ranks for the number of wild mustard plants after six months, seed addition treatment (site), $F_{2,56} = 8.0$, $P = 0.0123$; habitat (site) $F_{2,56} = 8.24$, $P = 0.0114$; treatment* habitat (site) $F_{2,56} = 8.0$, $P = 0.0123$.

interaction between seed addition treatment and habitat for wild mustard densities showed that seed addition effects were limited to disked plots, consistent with results of the microcosm experiment (Table 1). The impacts of seed addition on density of wild radish and wild mustard measured in these experiments are conservative, as a high proportion of seed for both species maintain dormancy so that seed addition can have a larger impact over time than that measured within a single season, such as in our experiments.

4. Discussion

4.1. Herbivory and Plant Fitness. Although there is general agreement that herbivores can have pronounced negative effects on plant fitness, plants also exhibit resistance to herbivores and tolerance to herbivore damage [34]. Thus, simple field tests have a place in directing further study and consideration of mitigation techniques toward the cases of expected transgene introgression which are most likely to increase the fitness and spread of weedy wild relatives in natural plant communities and agroecosystems (see Kwit et al. [35]). In our study, slight-to-moderate protection from herbivory, measured as the average difference between treatments on juvenile plants (5% on wild radish in year one, 17% on wild mustard in year two, and 16% on wild radish in year two), significantly increased lifetime seed output of an average plant to about double that measured for plants with greater mean herbivory. Typical fluctuations in herbivory,

then, that routinely reduce plant fitness in the field under present conditions, may be nullified to a significant degree with the introgression of a Bt trait for antiherbivore defense. The result could cause average seed output to shift to a higher level, depending on population levels of susceptible herbivores. A substantial proportion of plants protected in our treatments suffered some amount of foliar tissue removal (e.g., 45% of Bt sprayed wild mustard in 2003-2004) suggesting that both Bt susceptible and unsusceptible invertebrate species consumed plant tissue.

Lepidopteran resistant Bt-canola (*B. napus* L.), and possibly Bt broccoli or cabbage, would likely hybridize with *B. rapa* and other wild relatives if grown commercially in coastal California [36–40]. To test for potential fitness effects in two contrasting wild relatives in this study, we selected the two most abundant weedy relatives near cole crop fields, *B. rapa* and *R. sativus*, which differ in most life history traits. Although little is known about *R. sativus*, tests with *R. raphanistrum* show that it is unlikely to hybridize with *Brassica* crops (but see Gueritaine et al. [41]). Because crop-to-weed gene flow is highly idiosyncratic and difficult to preclude totally in many regions of the world and for many crop plants, research on the effects of transgenes on natural communities ([42] and references therein) and field tests where wild relatives of pest-resistant crops occur naturally are needed for risk assessment of transgenic crops [9, 12, 43–46]. Whether transgenes stably expressed in crop-wild hybrids [47] allow them to compete favorably [41, 48–51], persist, and spread more rapidly depends critically on the fitness advantage of the introduced resistance trait. Although wild mustard increased fitness overall when protected from the Bt susceptible subset of its herbivores [46], Bt-protected *B. rapa* was the lesser competitive Brassicaceae in nonagricultural areas with established vegetation. Both the overall fitness advantage of *B. rapa* plants protected with Bt *endotoxins*, and the weaker response in established vegetation are similar and complementary to the results of Stewart et al. [52] comparing transgenic Bt-*Brassica napus* and nontransgenic *B. napus* subjected to a defoliation episode that resulted in a 40%–60% decrease in percent herbivory, through the addition of larval diamondback moth in cultivated and natural vegetation plots. The similar increase in lifetime fitness in wild radish (*R. sativus*) with Bt sprays versus larval additions was more robust to habitat type differences than was *B. rapa*, suggesting that testing for fitness effects on wild relatives should involve representatives from more than one species under different field conditions.

Seed output of wild mustard that survived to maturity was not consistently affected by differential herbivory, with a significant increase in Bt-protected plants only in the last year of our study [46]. Therefore, our results and those of Stewart et al. [52] reflect lifetime fitness of *B. rapa*, including early mortality. This distinction may explain a lack of fecundity consequences from herbivory in some other potted plant studies of wild relatives of cole crops, which have shown strong compensation in terms of seed set ([53, 54] for *R. raphanistrum* and [55] for *B. nigra*). An average plant cover of ~90% grasses and forbs already established or building in the first months after germination of our test plants contrasted starkly with less crowded growing conditions in a majority of previous studies testing for tolerance to herbivory in pots (e.g., [20, 56–59]) or weeded field sites (e.g., [60, 61]). These studies, and the small scale field study of Strauss et al. [62] on wild radish are not inconsistent with our fitness results if plants in these protected conditions rarely suffered mortality or if seed output was calculated only from plants that lived to reproduce (rather than including zero seed production for plants that died as juveniles). Blatt et al. [63] suggested that an increase in stress through multiple attacks by herbivores, simultaneous damage by pathogens, or poor growing conditions may curb tolerance to herbivory and allow herbivores to exert measurable fitness effects on plants. In our study, early mortality resulting in zero seed set and aborted flowers resulting in plants with few seeds were important factors in explaining the overall very strong effects of herbivory (or release from herbivory) on average lifetime fitness per plant. The subset of those plants that survived to set seed in our experiments showed a less dramatic reduction in seed output owing to higher herbivory, which was only a marginally significant difference from protected plants. Surviving plants with higher herbivory produced 25% less seed than protected plants, compared to over 50% less when juvenile mortality was included. Clearly, the impact of Bt transgenes on mortality and fecundity of wild crop relatives is complicated by what kind of herbivory and how much herbivory the plant would have normally sustained, the developmental stage of the plant, plant species, and the surrounding habitat conditions. However, the potential for increased fecundity and rates of spread were upheld in field experiments under natural field conditions, whether using naturally occurring wild relatives with herbivore manipulations or transgenic Bt plants.

Researchers and regulators involved in risk assessment face data gaps in predicting the consequences of host plant IR transgenes in wild plant populations [9, 14, 43, 44]. In the process of constructing data sets to aid in these assessments, we expect limitations in scientific quality because demographic data on the size and rate of spread of plant populations derived from transgenic crop-wild hybrids cannot be collected without increasing the likelihood of transgene escape in the process. Containment restricts effective experimental designs with transgenic hybrids because they are conducted in cages, greenhouses, or at locations where the taxon does not occur, and so cannot form hybrids. In each of these cases, relevant biotic or abiotic factors are likely to be lacking. On the other hand, simulation experiments with nontransgenic plants are only as predictive as the simulation is accurate, at least to the degree necessary to represent relevant factors. Nevertheless, several experimental tests involving unintended effects of Bt insecticidal traits on plant fitness, and all with different experimental approaches, have come to similar conclusions about plant fitness effects of herbivore resistance traits. In addition to those mentioned previously, Snow et al. [44] found that significant reductions in seed predation by gelechiids and tortricid pests occurred when wild sunflower expressed Bt toxin; these insecticidal plants tended to have higher numbers of undamaged seeds

per plant, in part because of producing significantly more mature flowers.

We consider decreased female fitness due to naturally occurring levels of herbivory to be realistic outcomes for California agricultural and grassland habitats. Whereas individual fitness should be closely linked to seed production, the ecological population consequences of producing more seeds depend critically on whether those seeds will establish. Based on our seed addition experiments, an increase in seed production can allow more rapid spread of wild radish in local habitats. Results of our microcosm experiments are consistent with our casual observations of wild radish persisting in experimental plots that previously had no radish plants. Wild mustard, on the other hand, is not likely to invade grasslands by producing more seed but can increase numbers in disturbed sites such as agricultural fields.

Tiered frameworks for risk assessment (e.g., [64]) are designed to reach an informed prediction of the likelihood that harm will occur from an action, using the minimal amount of data. Using the framework proposed by Raybould and Cooper [45], tests for a fitness advantage associated with Bt-based resistance in wild plants constitute a 2nd tier hazard assessment. One advantage of herbivore exclusion tests at this step in a risk assessment process, designed to determine the likelihood of increased weediness in crop-weed hybrids with introgressed plant defense transgenes, is that the tests can be carried out in the regions and habitats where hybrids are likely to form. Another is that transgenic constructs of different wild relatives are not necessary. However, as with our study on wild mustard, factors such as hybrid fertility and vigor, hybrid seed emergence, differences between transgene expression and topical sprays (issues from genetic load to toxicity range on herbivores), and the competitiveness of the F1 and backcross generations (e.g., [48–51]) can partially mitigate or exacerbate any undesirable outcome of host plant resistance traits [45]. Yet assessing the potential for increased weediness of plant species with novel traits in different, realistic habitats is critical, given the range of potential economic and environmental consequences [65, 66]. Therefore, herbivore exclusion experiments are practical tools for predicting potential effects of a range of insect resistance traits on the population dynamics of crop relatives that may receive these traits through gene flow from transgenic crops.

Acknowledgments

This research was supported by USDA Biotechnology Risk Assessment Grant 2003-33120-13968, faculty research grants from the UCSC Academic Senate and Social Sciences Division, and graduate student fellowship and research grants from the National Science Foundation and UCSC Department of Environmental Studies. The authors thank the UCSC Center for Agroecology and Sustainable Food Systems, Elkhorn Slough Foundation, UCSC Natural Reserves system, California State Parks, J. Velzy, and local growers and land managers for greenhouse, field, and logistical assistance. T. Roubison assisted with all field and lab experiments. They also thank I. Parker for help with experimental design and logistical dilemmas. S. Bothwell, A. Zeilinger, R. Abarca, D. Barrantes, L. Barth, S. L. Bryan, C. Conlan, E. Encarnacion, A. Fintz, L. Funk, E. Hampson, E. Hariton, F. Hesse, C. Josephson, A. LeComte, J. Martin, S. Moskal, R. Muscutt, Y. Pellman, T. Rogers, A. H. Stroud, A. Warner, J. Wilson, M. B. Winston, and K. Wong assisted in conducting field and lab experiments. The paper was improved by anonymous reviewers, P. Barbosa, S. Bothwell, T. Cornellise, J. Jedlicka, T. Krupnik, C. Moreno, I. Parker, A. Racelis, and A. Zeilinger.

References

[1] T. Klinger, "Variability and uncertainty in crop-to-wild hybridization," in *Genetically Engineered Organisms: Assessing Environmental and Human Health Effects*, D. K. Letourneau and B. E. Burrows, Eds., CRC Press, Boca Raton, Fla, USA, 2002.

[2] A. A. Snow and P. M. Palma, "Commercialization of transgenic plants: potential ecological risks," *BioScience*, vol. 47, no. 2, pp. 86–96, 1997.

[3] A. A. Snow and P. M. Palma, "Commercialization of transgenic plants: potential ecological risks," *BioScience*, vol. 47, no. 4, p. 206, 1997.

[4] N. C. Ellstrand, "Current knowledge of gene flow in plants: implications for transgene flow," *Philosophical Transactions of the Royal Society B*, vol. 358, no. 1434, pp. 1163–1170, 2003.

[5] N. C. Ellstrand, H. C. Prentice, and J. F. Hancock, "Gene flow and introgression from domesticated plants into their wild relatives," *Annual Review of Ecology and Systematics*, vol. 30, pp. 539–563, 1999.

[6] H. Darmency, "The impact of hybrids between genetically modified crop plants and their related species: introgression and weediness," *Molecular Ecology*, vol. 3, pp. 37–40, 1994.

[7] R. K. Colwell, E. A. Norse, D. Pimentel, F. E. Sharples, and D. Simberloff, "Genetic engineering in agriculture," *Science*, vol. 229, no. 4709, pp. 111–112, 1985.

[8] J. M. Tiedje, R. K. Colwell, Y. L. Grossman et al., "The planned introduction of genetically engineered organisms: ecological considerations and recommendations," *Ecology*, vol. 70, p. 298, 1989.

[9] C. A. Hoffman, "Ecological risks of genetic engineering of crop plants: scientific and social analyses are critical to realize benefits of the new techniques," *BioScience*, vol. 40, p. 434, 1990.

[10] P. Kareiva, W. Morris, and C. M. Jacobi, "Studying and managing the risk of cross-fertilization between transgenic crops and wild relatives," *Molecular Ecology*, vol. 3, pp. 15–21, 1994.

[11] R. S. Hails, "Genetically modified plants—the debate continues," *Trends in Ecology and Evolution*, vol. 15, no. 1, pp. 14–18, 2000.

[12] L. L. Wolfenbarger and P. R. Phifer, "The ecological risks and benefits of genetically engineered plants," *Science*, vol. 290, no. 5499, pp. 2088–2093, 2000.

[13] R. S. Hails and K. Morley, "Genes invading new populations: a risk assessment perspective," *Trends in Ecology and Evolution*, vol. 20, no. 5, pp. 245–252, 2005.

[14] A. A. Snow, D. A. Andow, P. Gepts et al., "Genetically engineered organisms and the environment: current status and recommendations," *Ecological Applications*, vol. 15, no. 2, pp. 377–404, 2005.

[15] A. A. Agrawal, "Overcompensation of plants in response to herbivory and the by-product benefits of mutualism," *Trends in Plant Science*, vol. 5, no. 7, pp. 309–313, 2000.

[16] T. Jermy, "Evolution of insect/host plant relationships," *American Naturalist*, vol. 124, no. 5, pp. 609–630, 1984.

[17] M. Crawley, *Herbivory The Dynamics of Animal-Plant Interactions*, vol. 10, University of California Press, Berkeley, Calif, USA, 1983.

[18] A. J. Belsky, "Does herbivory benefit plants? A review of the evidence," *American Naturalist*, vol. 127, no. 6, pp. 870–892, 1986.

[19] M. J. Wise and W. G. Abrahamson, "Effects of resource availability on tolerance of herbivory: a review and assessment of three opposing models," *American Naturalist*, vol. 169, no. 4, pp. 443–454, 2007.

[20] D. J. Susko and B. Superfisky, "A comparison of artificial defoliation techniques using canola (*Brassica napus*)," *Plant Ecology*, vol. 202, no. 1, pp. 169–175, 2009.

[21] R. M. Keane and M. J. Crawley, "Exotic plant invasions and the enemy release hypothesis," *Trends in Ecology and Evolution*, vol. 17, no. 4, pp. 164–170, 2002.

[22] I. M. Parker and G. S. Gilbert, "When there is no escape: the effects of natural enemies on native, invasive, and noninvasive plants," *Ecology*, vol. 88, no. 5, pp. 1210–1224, 2007.

[23] H. Liu and P. Stiling, "Testing the enemy release hypothesis: a review and meta-analysis," *Biological Invasions*, vol. 8, no. 7, pp. 1535–1545, 2006.

[24] D. Grzywacz, A. Rossbach, A. Rauf, D. A. Russell, R. Srinivasan, and A. M. Shelton, "Current control methods for diamondback moth and other *brassica* insect pests and the prospects for improved management with lepidopteran-resistant Bt vegetable brassicas in Asia and Africa," *Crop Protection*, vol. 29, no. 1, pp. 68–79, 2010.

[25] S. G. Hegde, J. D. Nason, J. M. Clegg, and N. C. Ellstrand, "The evolution of California's wild radish has resulted in the extinction of its progenitors," *Evolution*, vol. 60, no. 6, pp. 1187–1197, 2006.

[26] M. J. Crawley, "Insect herbivores and plant population dynamics," *Annual Review of Entomology*, vol. 34, pp. 531–564, 1989.

[27] J. Bergelson, "Changes in fecundity do not predict invasiveness: a model study of transgenic plants," *Ecology*, vol. 75, no. 1, pp. 249–252, 1994.

[28] S. Y. Strauss and R. E. Irwin, "Ecological and evolutionary consequences of multispecies plant-animal interactions," *Annual Review of Ecology, Evolution, and Systematics*, vol. 35, pp. 435–466, 2004.

[29] S. Warwick and C. N. Stewart, "Crops come from wild plants—how domestication, transgenes, and linkage together shape ferality," in *Crop Ferality and Volunteerism*, J. Gressel, Ed., p. 422, Taylor & Francis/CRC Press, Boca Raton, Fla, USA, 2005.

[30] A. A. Snow and L. G. Campbell, "Can feral radishes become weeds?" in *Crop Ferality and Volunteerism*, J. Gressel, Ed., p. 422, Taylor & Francis, CRC Press, Boca Raton, Fla, USA, 2005.

[31] A. M. Shapiro, "Non-diapause overwintering by *Pieris rapae* (Lepidoptera: Pieridae) and *Papilio zelicaon* (Lepidoptera: Papilionidae) in California: adaptiveness of Type III diapause-induction curves," *Psyche*, vol. 91, pp. 161–169, 1984.

[32] D. K. Letourneau, J. A. Hagen, and G. S. Robinson, "*Bt*-crops evaluating benefits under cultivation and risks from escaped transgenes in the wild," in *Genetically Engineered Organisms:*

Assessing Environmental and Human Health Effects, D. K. Letourneau and B. E. Burrows, Eds., p. 33, CRC Press, Boca Raton, Fla, USA, 2002.

[33] D. K. Letourneau and L. R. Fox, "Effects of experimental design and nitrogen on cabbage butterfly oviposition," *Oecologia*, vol. 80, no. 2, pp. 211–214, 1989.

[34] R. S. Fritz and E. L. Simms, *Plant Resistance to Herbivores and Pathogens : Ecology, Evolution, and Genetics*, University of Chicago Press, Chicago, Ill, USA, 1992.

[35] C. Kwit, H. S. Moon, S. I. Warwick, and C. N. Stewart, "Transgene introgression in crop relatives: molecular evidence and mitigation strategies," *Trends in Biotechnology*, vol. 29, pp. 284–293, 2011.

[36] R. B. Jorgensen and B. Andersen, "Spontaneous hybridization between oilseed rape (*Brassica napus*) and weedy *Brassica campestris* (Brassicaceae): a risk of growing genetically modified oilseed rape," *American Journal of Botany*, vol. 81, no. 12, pp. 1620–1626, 1994.

[37] M. A. Rieger, T. D. Potter, C. Preston, and S. B. Powles, "Hybridisation between *Brassica napus* L. and *Raphanus raphanistrum* L. under agronomic field conditions," *Theoretical and Applied Genetics*, vol. 103, no. 4, pp. 555–560, 2001.

[38] S. I. Warwick, M. J. Simard, A. Legere et al., "Hybridization between transgenic *Brassica napus* L. and its wild relatives: *Brassica rapa* L., *Raphanus raphanistrum* L., *Sinapis arvensis* L., and *Erucastrum gallicum* (Willd.) O.E. Schulz," *Theoretical and Applied Genetics*, vol. 107, no. 3, pp. 528–539, 2003.

[39] H. Ammitzbøll and R. Bagger Jørgensen, "Hybridization between oilseed rape (*Brassica napus*) and different populations and species of *Raphanus*," *Environmental Biosafety Research*, vol. 5, no. 1, pp. 3–13, 2006.

[40] R. G. Fitzjohn, T. T. Armstrong, L. E. Newstrom-Lloyd, A. D. Wilton, and M. Cochrane, "Hybridisation within *Brassica* and allied genera: evaluation of potential for transgene escape," *Euphytica*, vol. 158, no. 1-2, pp. 209–230, 2007.

[41] G. Gueritaine, M. Sester, F. Eber, A.-M. Chevre, and H. Darmency, "Fitness of backcross six of hybrids between transgenic oilseed rape (*Brassica napus*) and wild radish (*Raphanus raphanistrum*)," *Molecular Ecology*, vol. 11, no. 8, pp. 1419–1426, 2002.

[42] E. Jenczewski, J. Ronfort, and A. M. Chèvre, "Crop-to-wild gene flow, introgression and possible fitness effects of transgenes," *Environmental Biosafety Research*, vol. 2, no. 1, pp. 9–24, 2003.

[43] M. Marvier and P. Kareiva, "Extrapolating from field experiments that remove herbivores to population-level effects of herbivore-resistant transgenes," in *Proceedings of the Workshop on Ecological Effects of Pest Resistance Genes in Managed Ecosystems*, pp. 57–64, Bethesda, Md, USA, 1999.

[44] A. A. Snow, D. Pilson, L. H. Rieseberg et al., "A Bt transgene reduces herbivory and enhances fecundity in wild sunflowers," *Ecological Applications*, vol. 13, no. 2, pp. 279–286, 2003.

[45] A. Raybould and I. Cooper, "Tiered tests to assess the environmental risk of fitness changes in hybrids between transgenic crops and wild relatives: the example of virus resistant *Brassica napus*," *Environmental Biosafety Research*, vol. 4, no. 3, pp. 127–140, 2005.

[46] D. K. Letourneau and J. A. Hagen, "Plant fitness assessment for wild relatives of insect resistant crops," *Environmental Biosafety Research*, vol. 8, no. 1, pp. 45–55, 2009.

[47] B. Zhu, J. R. Lawrence, S. I. Warwick et al., "Stable *Bacillus thuringiensis* (Bt) toxin content in interspecific F 1 and backcross populations of wild *Brassica rapa* after Bt gene transfer," *Molecular Ecology*, vol. 13, no. 1, pp. 237–241, 2004.

[48] M. D. Halfhill, J. P. Sutherland, H. S. Moon et al., "Growth, productivity, and competitiveness of introgressed weedy *Brassica rapa* hybrids selected for the presence of Bt cry1Ac and gfp transgenes," *Molecular Ecology*, vol. 14, no. 10, pp. 3177–3189, 2005.

[49] H. Ammitzboll, T. N. Mikkelsen, and R. B. Jorgensen, "Transgene expression and fitness of hybrids between GM oilseed rape and *Brassica rapa*," *Environmental Biosafety Research*, vol. 4, no. 1, pp. 3–12, 2005.

[50] J. Allainguillaume, M. Alexander, J. M. Bullock et al., "Fitness of hybrids between rapeseed (*Brassica napus*) and wild *Brassica rapa* in natural habitats," *Molecular Ecology*, vol. 15, no. 4, pp. 1175–1184, 2006.

[51] J. P. Londo, M. A. Bollman, C. L. Sagers, E. H. Lee, and L. S. Watrud, "Changes in fitness-associated traits due to the stacking of transgenic glyphosate resistance and insect resistance in *Brassica napus* L," *Heredity*, vol. 107, pp. 328–337, 2011.

[52] C. N. Stewart, J. N. All, P. L. Raymer, and S. Ramachandran, "Increased fitness of transgenic insecticidal rapeseed under insect selection pressure," *Molecular Ecology*, vol. 6, no. 8, pp. 773–779, 1997.

[53] K. Lehtila and S. Y. Strauss, "Effects of foliar herbivory on male and female reproductive traits of wild radish, *Raphanus raphanistrum*," *Ecology*, vol. 80, no. 1, pp. 116–124, 1999.

[54] A. A. Agrawal, S. Y. Strauss, and M. J. Stout, "Costs of induced responses and tolerance to herbivory in male and female fitness components of wild radish," *Evolution*, vol. 53, no. 4, pp. 1093–1104, 1999.

[55] G. A. Meyer, "Interactive effects of soil fertility and herbivory on *Brassica nigra*," *Oikos*, vol. 88, no. 2, pp. 433–441, 2000.

[56] C. Vacher, A. E. Weis, D. Hermann, T. Kossler, C. Young, and M. E. Hochberg, "Impact of ecological factors on the initial invasion of Bt transgenes into wild populations of birdseed rape (*Brassica rapa*)," *Theoretical and Applied Genetics*, vol. 109, no. 4, pp. 806–814, 2004.

[57] E. Boalt and K. Lehtilä, "Tolerance to apical and foliar damage: costs and mechanisms in *Raphanus raphanistrum*," *Oikos*, vol. 116, no. 12, pp. 2071–2081, 2007.

[58] M. Dicke, "Precise manipulation through a modeling study," *Journal of Chemical Ecology*, vol. 34, no. 7, pp. 943–944, 2008.

[59] C. B. Marshall, G. Avila-Sakar, and E. G. Reekie, "Effects of nutrient and CO_2 availability on tolerance to herbivory in *Brassica rapa*," *Plant Ecology*, vol. 196, no. 1, pp. 1–13, 2008.

[60] R. A. Lankau, "Specialist and generalist herbivores exert opposing selection on a chemical defense," *New Phytologist*, vol. 175, no. 1, pp. 176–184, 2007.

[61] M. G. Bidart-Bouzat and D. J. Kliebenstein, "Differential levels of insect herbivory in the field associated with genotypic variation in glucosinolates in *Arabidopsis thaliana*," *Journal of Chemical Ecology*, vol. 34, no. 8, pp. 1026–1037, 2008.

[62] S. Y. Strauss, J. K. Conner, and K. P. Lehtilä, "Effects of foliar herbivory by insects on the fitness of *Raphanus raphanistrum*: damage can increase male fitness," *American Naturalist*, vol. 158, no. 5, pp. 496–504, 2001.

[63] S. E. Blatt, R. C. Smallegange, L. Hess, J. A. Harvey, M. Dicke, and J. J. A. van Loon, "Tolerance of *Brassica nigra* to *Pieris brassicae* herbivory," *Botany*, vol. 86, no. 6, pp. 641–648, 2008.

[64] D. A. Andow and C. Zwahlen, "Assessing environmental risks of transgenic plants," *Ecology Letters*, vol. 9, no. 2, pp. 196–214, 2006.

[65] J. M. Randall, "Weed control for the preservation of biological diversity," *Weed Technology*, vol. 10, no. 2, pp. 370–383, 1996.

[66] A. Hilbeck, "Implications of transgenic, insecticidal plants for insect and plant biodiversity," *Perspectives in Plant Ecology, Evolution and Systematics*, vol. 4, no. 1, pp. 43–61, 2001.

12

Genome Mutation Revealed by Artificial Hybridization between *Chrysanthemum yoshinaganthum* and *Chrysanthemum vestitum* Assessed by FISH and GISH

Magdy Hussein Abd El-Twab[1] and Katsuhiko Kondo[2]

[1] Department of Botany and Microbiology, Faculty of Science, Minia University, El-Minia 61519, Egypt
[2] Laboratory of Plant Genetics and Breeding Research, Department of Agriculture, Faculty of Agriculture, Tokyo University of Agriculture, Funako, Kanagawa, Atsugi 1737, Japan

Correspondence should be addressed to Katsuhiko Kondo, k3kondo@nodai.ac.jp

Academic Editor: Jaume Pellicer

Present study has been done to investigate artificial interspecific crossability between Japanese *Chrysanthemum yoshinaganthum* ($2n = 36$) and Chinese *C. vestitum* ($2n = 54$), which were cultured *in vitro* and *in vivo* and characterization of their artificial hybrid chromosomes and type of changes assessed by FISH and GISH. GISH was applied by using biotin-labeled total genomic DNA probe of *C. vestitum*, which were mixed with blocking DNA of *C. yoshinaganthum*. Approximately 18 yellow-green colored chromosomes of *C. vestitum* were detected by the probe, approximately 18 yellow-red- mixed colored chromosomes could be common chromosomes of the two species, and nine chromosomes were relatively red of *Ch. yoshinaganthum* that were not detected by the probe. The expected average of FISH six signals of 5S rDNA sites and ten of 45S rDNA were observed on the chromosomes of three and six hybrid plants, respectively. Multicolor FISH signals showed unexpected average of seven and 14 yellow signals of 5S rDNA sites on seven and thirteen chromosomes simultaneous with ten and 11 red signals of 45S rDNA sites on ten and 11 chromosomes which were detected by the probes respectively. FISH mapping of the 5S rDNA at terminal sites was detected in hybrid chromosomes, for the first time. Yellow-color signals of the telomeres were detected by the biotin-labeled probe of the PCR-amplified telomeric probe in interphase and terminal sites in metaphase. All chromosomes that showed terminal signals except four chromosomes showed subterminal sites of telomeres indicating the presence of translocations.

1. Introduction

Chrysanthemum yoshinaganthum ($2n = 36$, tetraploid) is endemic to Japan while *C. vestitum* ($2n = 54$, hexaploid) is endemic to China. Studies on origin and chromosome constitutions of the native Japanese and Chinese species of *Chrysanthemum sensu lato* may contribute to satisfactory phylogenetic and taxonomic treatment of species relationships and the constitution and breeding cultivars. The Japanese and Chinese *Chrysanthemum* may be closely related to each other, make cross-hybrids, and perform a polyploid series, which plays an important role in chromosome evolution and would even perform introgressive hybridization in the nature [1–3]. In plant evolutionary studies and breeding, there is often a need to discriminate between the genomes of closely related taxa and to identify ancestors [4–9]. Since *Chrysanthemum* species has a self-incompatible breeding system, crosses between closely related or unrelated individuals would be unsuccessful, and, thus, in genetics, little is known in the species of the genus [10]. Therefore chromosome evidences of Chinese and Japanese species of *Chrysanthemum* should be compared to justify and clarify the cytogeographical disjunct and evolutionary concepts [1]. Some chromosome analyses in Japanese and Chinese *Chrysanthemum* and its allied genera have been carried out extensively (e.g., [1, 5, 11–16]). Several cross hybridizations between diploid and high-ploid species of *Chrysanthemum* were made by the artificial cross-pollination for chromosome and genome analysis (e.g., [5, 17, 18]).

In higher eukaryotes, rDNAs are organized into two distinct gene classes of the major rDNA cluster encoding 45S rDNAs, and the minor rDNA cluster encoding 5S rDNA. The minor rDNAs are usually found in loci that are separate from those of the major rDNAs and, unlike major rDNA, the minor rDNA is not involved in the nucleolus formation [19–22]. Another highly conserved sequence is found at the physical ends of chromosomes of nearly most of plant species. Multiple copies of a motif similar to the seven nucleotide unit TTTAGGG are added by an enzyme, telomerase, incorporating an RNA template, and function in stabilizing the chromosome ends and enabling semiconservative replication of the DNA up to the telomerase-added region [23].

As a pioneer among molecular cytogenetic techniques, FISH allows the identification of specific sequences in a structurally preserved cell, in metaphase or interphase. This technique, based on the complementary double-stranded nature of DNA, hybridizes labelled specific DNA (probe). The probe, bound to the target, will be developed into a fluorescent signal. FISH is a valuable method for studying the chromosomal distribution of DNA sequences and copy numbers at different sites, and to follow evolutionary changes in their physical organization in the genome [24]. FISH makes chromosomes of plant species providing the information of molecular characters of nucleolar organizing region (NOR) [25]. Several molecular evidences indicated that 45S rDNA is located on NOR-bearing chromosomes [4, 5, 26–28]. FISH mapping of rDNA would be a simple and effective way to characterize diverse collections of germplasm materials and breeding lines correctly. Investigating the number and location of rDNA loci is important to gain a better understanding of evolution and phylogeny of various species in higher plant [7, 20, 22, 28–30]. In plants, physical localization of the tandemly repeated genes as 5S rDNA, 45S rDNA, and *Arabidopsis* type of telomere repeats has provided a group of valuable chromosome landmark markers. The classes of rDNA and telomeres are a major type of repetitive sequence in the higher plant genome. The chromosomal localization of the 5S, 45S rDNA, and telomere repeats via FISH in the chromosome complements of several species in *Chrysanthemum sensu lato* was reported [3, 6, 7, 16, 20, 21, 26–28, 31–34].

The technique of genomic *in situ* hybridization (GISH) has been useful in identification of parental genomes and determination of levels and incorporation positions of alien chromatin in interspecific and intergeneric plant hybrids (e.g., [6, 18, 30, 32, 33, 35–43]).

This research is one of series that had the following objectives: (1) to investigate the interspecific cross-ability between the Japanese and Chinese species in *Chrysanthemum sensu lato*, (2) intergenomic characterization and the type of changes in the rDNA and telomere loci on the chromosome complement of the hybrids that resulted after hybridization of the species, (3) homology of the two parental chromosome complements, and (4) to determine whether the prevalence of genomic changes could start within the first generation. Therefore, FISH and GISH were applied on the chromosomes of F$_1$ hybrid between *C. yoshinaganthum* and *C. vestitum*.

2. Materials and Methods

2.1. Cross-hybridization. Artificial cross by hand pollination was made in the forenoon of a fine day between *Ch. yoshinaganthum* and *Ch. vestitum* during the flowering seasons of the species. The method followed Tanaka et al. [12]. The disc flowers of newly blooming capitula from the maternal plant were sprayed with distilled water from an atomizer. After drying the wet flowers, their stigma was artificially pollinated with pollen from the paternal plants. Each capitulum was covered with paper bag and labeled. Subsequently, the stigma of newly blooming disc flowers was constantly cut off to prevent further pollination.

2.2. Aseptic Culture of Seeds. Twelve best-ripen achenes were sown and germinated *in vitro* on 1/2 strength of Murashige and Skoog (MS; Murashige and Skoog 1962) solid medium supplemented with 1.5% sucrose and 0.3% gelrite at pH 5.5 under ca 2,000 lux illumination for 24 hours daily at 22°C after they were surface sterilized with 1% (v/v) benzalkonium chloride for five minutes, 1-2% (v/v) sodium hypochlorite solution for five minutes, and finally rinsed three times with distilled sterilized water. The other seeds were germinated *in vivo* in a compost of akadama, vermiculite, and decomposed leaves (3 : 2 : 1) in combination in beds in the greenhouse of the laboratory. The plant species and hybrids have been preserved in the Laboratory of Plant Genetics and Breeding Research, Department of Agriculture, Faculty of Agriculture, Tokyo University of Agriculture, Funako 1737, Atsugi City, Kanagawa Prefecture, Japan.

2.3. Chromosome Preparations. Following Kondo et al. [26] growing root tips were collected and pretreated in 0.002 M 8-hydroxyquinoline at 18°C for 1.5 hours. They were fixed in the 1 : 3 of glacial acetic acid and ethanol at 4°C for 2 hours. Fixed roots were excised and washed in distilled water many times to remove the fixative. Five to ten tips 2–5 mm long each from root tips were placed in 1.5 mL microcentrifuge tube containing the enzymatic buffer of 5% cellulase (Yakult), 2% pectolyase Y-23 (Kikkoman). They were incubated at 37°C for 20 mins, and, then, their soft meristematic tissues were washed in distilled water to remove the enzymatic solution and were squashed in 45% acetic acid. The coverslip was removed by the dry-ice freezing method, and the preparation was dried at room temperature.

2.4. Probes Preparation and Labeling. The probe 5S rDNA was produced by the polymerase chain reaction (PCR) method as described by Abd El-Twab and Kondo [20]. The probe of pTa71 (45S rDNA) consisted of a 9 kb Eco RI fragment of rDNA derived from *Triticum aestivum* L. [44], which was recloned into pUC19 plasmid. It was comprised of the coding sequence for the 18S, 5.8S, and 26S genes and the nontranscribed spacer sequences. In the probe of the Arabidopsis-type telomere sequence repeats was synthesized by using the polymerase chain reaction (PCR). PCR was carried out in the absence of template using primers (TTTAGGG)$_5$ and (CCCTAAA)$_5$ [7, 45]. The probe of total genomic DNA: DNA was isolated from 0.5–1.0 g of fresh young leaf material of the two parental species following the protocol Abd El-Twab et al. [32]. Total genomic DNA of *C. yoshinaganthum* and *C. vestitum* was mechanically sheared and labeled with biotin 14-dATP for use as the *in situ* probe.

2.4.1. Labeling of the Probes. The probes were labeled either with biotin-14-dATP nick translation kit (Invitrogen) or digoxigenin (DIG)-dUTP by Dig DNA labeling kit (Roche) labeling of each probe that carried out according to the manufacturer's protocols.

2.5. The Procedure of FISH and GISH. The probe hybridization mixture (20 μL per slide) contained about 50 ng of each DNA-labelled probe, 10% dextran sulphate, 2X SSC, 0.25% SDS (lauryl sulphate). The probe mixtures were denatured by boiling for 5 min, centrifuged briefly [21].

Chromosome Denaturation. The slides were denatured in a preheated formamide solution (70% in 2X SSC) at 72°C for 2 min, dehydrated in an ice-cold ethanol series (70, 80, and 100%, 3 min each), and then allowed to air-dry.

Hybridization and Washing Followed Abd El-Twab and Konodo [30]. The probe mixtures were applied to the appropriate slides and covered with glass coverslips without air bubbles. The slides were then incubated in a preheated humid chamber at 63°C for 2 h of hybridization. After gently removing the coverslips, the slides were washed for 2×5 min in 2X SSC at 63°C, transferred to an incubator at 37°C for 20 min, and allowed to cool to room temperature. Visualization of the hybridized probes followed Abd El-Twab et al. [43]. For double probes multicolor FISH, the yellow color was visualized by FITC after the hybridization with the biotin-labeled probes, and red color was visualized by antidigoxigenin rhodamine, and the counter stain was blue of DAPI, while the counter stain was red color of PI for single probe. The fluorescence signals were examined with an epifluorescence microscope with Nikon B-2A filter cassette, and microphotographs were taken on CCD camera (Pixera Pengium 600CL). Analysis of hybridization signals and superimposed images was produced using Adobe Photoshop 7.

3. Results and Discussion

We present here for the first time analytical chromosome study by FISH and GISH on the chromosome complements of interspecific F$_1$ hybrid between *Ch. yoshinaganthum* and *Ch. vestitum* to determine genome changes and identification of specific parental chromosomes in the artificial hybrid. Crosses between the maternal, Japanese *Ch. yoshinaganthum* ($2n = 36$, tetraploid), and the paternal, Chinese *Ch. vestitum* ($2n = 54$, hexaploid), produced fertile achenes after hand pollination which germinated and cultivated both *in vitro*,

TABLE 1: Crossability, seed germination, and seedling survival in F$_1$ hybrids between in this study.

Cross combination	Ch. Yoshinaganthum ($2n = 36$) X Ch. Vestitum ($2n = 54$)
Total number of heads	26
Total number of florets used	3696
Total number of fertile (aborted) achenes obtained	970 (340)
% fertile of achenes obtained (%aborted)	26.2 (9.2)
Total number of achenes sown *in vitro*	12
Total number of achenes sown *in vivo*	958
Total number of achenes germinated *in vitro*	14
Total number of achenes germinated *in vivo*	161
Total number of hybrids determined*/Total no. of live plants studied *in vitro* and (%)	12/12 (100%)
Total number of hybrids determined*/Total no. of live plants studied *in vivo* and (%)	74/74 (100%)

*Hybridity was determined by leaf morphology and/or chromosome numbers.

and *in vivo* under normal condition of culture (Table 1). Among 14 plants germinated *in vitro* and 74 *in vivo* (Table 1), all plants showed true hybrids based on the leaf morphological characters with 100% hybridity. Seven plants and two from those had been grown *in vivo* and *in vitro* respectively, studied and showed the expected chromosome number of $2n = 45$ (pentaploid), of which eight plants were used for GISH (ID: y-11-4-1-1 and -10) and FISH (ID: y-11-4-1-1, -2, -5, -10, -12, -38, -45, and -62). F$_1$ hybrids showed intermediate morphological characters of the two parents, but the hybrids resemble more closely to the hexaploid species since the leaf edge and blade inherited resemble to *Ch. vestitum*, thus the hybrids used to be more closely in habit to the higher polyploid parent.

GISH applied on somatic chromosomes in the artificial interspecific F$_1$ hybrid between *Ch. yoshinaganthum* and *Ch. vestitum*. Biotin-labeled total genomic DNA probe of *Ch. vestitum* was mixed with approximately 20–40 times blocking DNA of *Ch. yoshinaganthum* and applied on the somatic chromosomes. Yellow color was fluoresced and visualized by FITC and red color by PI. Approximately 18 yellow-green colored chromosomes of *Ch. vestitum* were hybridized with the probe, approximately 18 yellow-red-mixed colored chromosomes could be common chromosomes of the two species, and nine chromosomes were relatively red of *Ch. yoshinaganthum* that were not hybridized with the probe (Figures 1(a)–1(d); ID: y-11-4-1-10). Interphase and Prophase showed the parental chromosomes separated in two spatial domains without intermix (Figures 1(a)–1(c)).

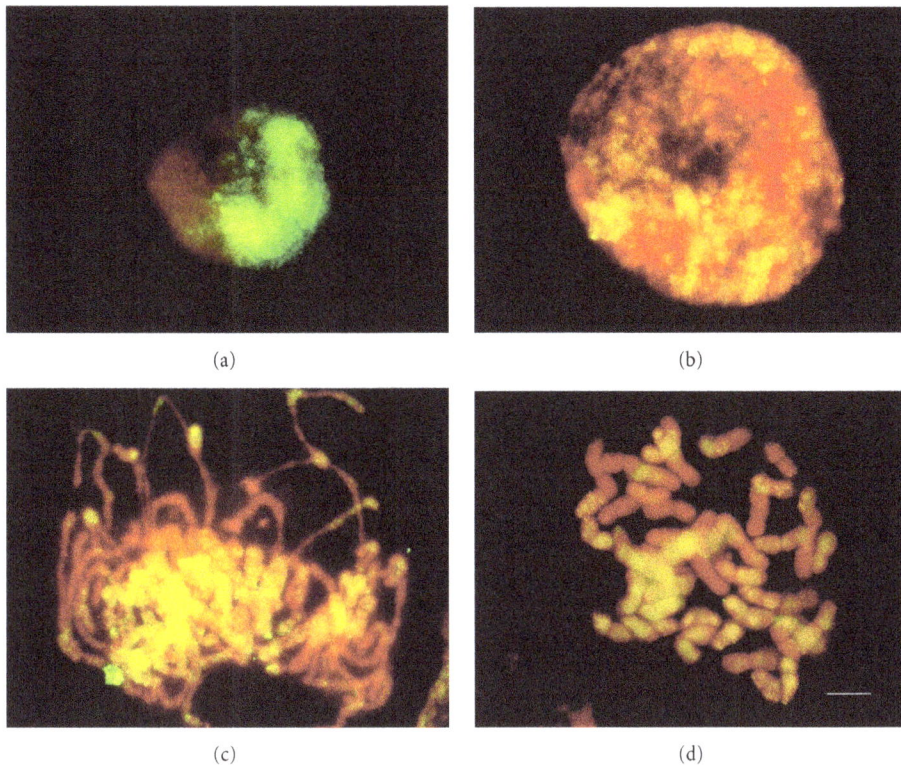

FIGURE 1: (a)–(d). GISH applied on somatic chromosomes in the artificial interspecific F_1 hybrid ($2n = 45$, pentaploid; ID: y-11-4-1-10) between *Ch. yoshinaganthum* and *Ch. vestitum*. Biotin-labeled total genomic DNA probe of *Ch. vestitum* ((a)–(d)) was mixed with approximately 20–40 times blocking DNA of *Ch. yoshinaganthum* and used. Yellow-colored chromosomes of *Ch. vestitum* were hybridized with the probe and visualized by FITC, and approximately nine red-colored chromosomes of *Ch. yoshinaganthum* were not and stained partially red-color by PI. Interphase (a) and prophase ((b)-(c)) show the parental chromosomes separated in two spatial domains without intermix. Bar = 10 μm.

In the present FISH study, three probes of 5S rDNA, pTa71, and telomere sequence repeats were used to investigate the genome characterization of the F_1 hybrids. Multicolor FISH by applying the biotin-labeled probe of 5S rDNA and digoxigenin-labeled pTa71 (Figures 2(a), 2(c), 2(e) and 2(f)) and bicolor FISH by applying biotin labeled of the telomere sequence repeats probe (Figures 2(b) and 2(d); ID: y-11-4-1-10) were applied on somatic chromosomes in the artificial interspecific F_1 hybrid (Figures 2(c) and 2(f); IDs: y-11-4-1-2 and y-11-4-1-5) between *Ch. yoshinaganthum* and *Ch. vestitum*. Yellow-green color of the 5S rDNA loci was fluoresced and visualized by FITC, red color of the 45S rDNA loci by rhodamine and whole chromosomes with blue color by DAPI. Superimposed image locates the distribution of the seven yellow-green-color signals (Figures 2(c) and 2(f) I) and 14 of the 5S rDNA sites (Figures 2(e) and 2(f) II) on seven and thirteen chromosomes. Ten red signals (Figures 2(c) and 2(f) I) and 11 of the 45S sites (Figures 2(e) and 2(f) II) on ten and eleven chromosomes were hybridized with the probes, respectively. The expected average of FISH six signals of 5S rDNA sites and ten of 45S rDNA were observed in the chromosomes of three and six plants, respectively (data not shown). Yellow-color signals of the telomeres that hybridized with the biotin-labeled telomeric probe in interphase (Figure 2(b)). All chromosomes showed terminal

signals except four chromosomes showed sub-terminal sites of telomeres (Figure 2(d)).

Allopolyploidization is viewed as a highly dynamic process and a major force in the evolution of higher plants [46] and might be a source of genomic stress that facilitates rapid genome rearrangement and evolution [6, 47]. Several analyses of the constitution and interaction among the polyploid genomes in *Chrysanthemum* species by means of hybridization among taxa with various chromosomes numbered or high ploidy species were made (e.g. [16, 48]). Genome rearrangement seems to be a common attribute of polyploids and still much remains unknown about polyploid plant species, including their general mode of formation [49]. The typical form of *Ch. yoshinaganthum* was found to have variation in chromosome number of $2n = 36$ [50] and of $2n = 72$ [51] therefore; it was proposed to have chromosome complement with more than one diploid progenitor [5, 28]. *Ch. vestitum* was similar in morphology to hexaploid Japanese species regarding flower head and chromosome number, while their karyotypes were different [13] even after using fluorescent banding [52]. The autopolyploid of *Ch. vestitum* [13] could not be supported after fluorescent banding and FISH [28, 53]. Natural interspecific hybrids were found in Japan and China [14, 54]. Thus, certain introgressive hybridizations among

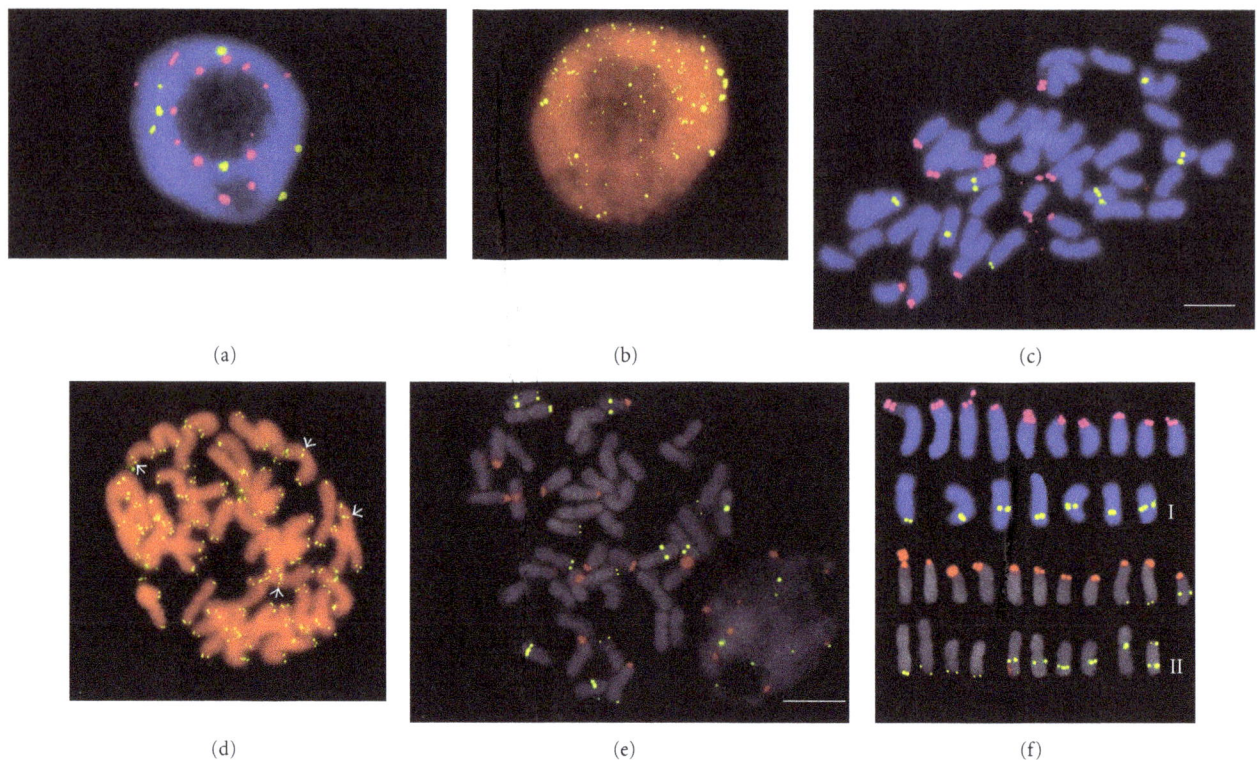

FIGURE 2: Tricolor FISH ((a), (c), (e), and (f).) and bicolor ((b) and (d)) applied on mitotic chromosomes in the artificial interspecific F$_1$ hybrid between *Ch. yoshinaganthum* and *Ch. vestitum* (2n = 45, pentaploid, ID: y-11-4-1-10 ((b) and (d)), y-11-4-1-2 ((c) and (f) I), and y-11-4-1-5 ((e) and (f) II). (a), (c), (e), and (f). Superimposed image locates the distribution of the yellow signals of seven yellow-green-color signals ((c) and (f) I) of 5S rDNA sites and 14 ((e) and (f) II) on seven and thirteen chromosomes and ten red signals ((c) and (f) I) and 11 ((e) and (f) II) of 45S rDNA sites in ten and eleven chromosomes that were detected by the probes. (b) and (d). Yellow-color signals of the telomeres that were detected by the biotin-labeled telomeric probe in interphase (b) and terminal sites in metaphase (d). All chromosomes showed terminal signals except four chromosomes that showed subterminal sites of telomeres (pointed by four arrows). Bar = 10 μm.

species of *Chrysanthemum* might involve with differentiation and speciation among both Chinese and Japanese *Chrysanthemum* species. Moreover, artificial F$_1$ hybrid between *Ch. yoshinaganthum* and *Ch. vestitum* could be produced to perform a heptaploid plant when *Ch. yoshinaganthum* was used as the female parent [13]. In the present study, the chromosome number of the artificial hybrids was the average of the two parents when *Ch. yoshinaganthum* was used as the female parent.

The fluoresced color degree of GISH could be used for interpreting homology between the genomes and species relationships after hybridization of the labeled total genomic DNA of one of the parental probes mixed with/without blocking DNA, with the parental chromosomes of the hybrids in *Chrysanthemum sensu lato* [6, 17, 18, 21, 30, 42]. GISH applied on metaphase chromosomes of hybrid combination after using blocking DNA considered closely related [32], while other combinations considered distantly related after GISH without blocking DNA [18, 33, 41, 42]. GISH could differentiate at least 18 chromosomes of *C. vestitum* different to *C. boreale* among chromosomes of their hybrid [8]. In this study GISH coloration could support the speculation of that *C. vestitum* might have at least 18 chromosomes that have a different genome structure to

C. yoshinaganthum, and the other chromosomes have the common genome homology. Accordingly, *C. vestitum* that is geographically isolated might have different ancestor from *C. yoshinaganthum*, which could be color differentiated by GISH after using blocking DNA.

FISH hybridization signals of the 5S and 45S rDNA sites in the chromosomes of *Ch. yoshinaganthum* were four interstitial sites of 5S rDNA [20], while it was ten terminal sites of 45S rDNA [27] and four [31]. *Ch. vestitum* were eight interstitial signals of the 5S rDNA sites [20] and ten signals of the 45S rDNA [28]. Chromosome mutations could be detected in different sitespecific of 45S rDNA in the chromosome complement of hybrids in *Chrysanthemum sensu lato*, which were resulted due to translocation and recombination indicating rapid genome changes after the artificial hybridization between *C. indicum* and *C. vestitum* as well as *C. remotipinum* and *Ch. chanetii* [6, 16]. Extra interstitial sites of 45S rDNA by FISH were detected in metaphase chromosomes of the artificial intergeneric hybrid between *Ch. horaimontanum* and *Tanacetum vulgare* [9], while deletion and/or translocation of the 5SrDNA sites were detected by FISH in the chromosomes of the artificial interspecific hybrids between *Ch. remotipinum* and *Ch. chanetii* [16]. The present FISH data on the chromosome complement of the hybrid had extra one 45S rDNA site and from one to eight extra 5S rDNA sites more than the mean averages of the parental chromosomal sites, as well as four interstitial translocations of the telomeric sequence repeats. The FISH physical mapping of the 5S rDNA loci was detected at terminal sites for the first time. The results imply that rapid genome changes of F_1 hybrid produced by chromosomal breakage followed by inversion recombination or translocation and inversion recombination, which might be considered an essential adjustment and required for the harmonious coexistence of the two or more different genomes in the nucleus of *Chrysanthemum* allopolyploids [6]. In the present study, the genome changes of chromosomal rearrangement of translocation and inversion of the present allopolyploids might be resulted soon after cross-fertilization of the two gametes which were germinated and grown to healthy mature plant. Therefore, the polyploid nature of *Chrysanthemum* is now considered very complex. Translocation and/or inversion may not involve a loss or an addition of chromosome material, but they do become frequently associated with differences, duplications, and unbalanced combinations of genetic units [55]. Therefore, breakage inversion, translocation, recombination, and so forth, might be a reason behind the high variation in the chromosome constitution after production of allopolyploids in the artificial hybrid combinations as was observed in FISH patterns, which might be important of variation through the evolution of the species in *Chrysanthemum sensu lato*.

Although mutations in the chromosomes of several intergeneric combinations of the artificial hybrids produced by hand pollination [6, 7, 33], and cell fusion [43] could be detected after the germination of the fertilized embryos in the *in vitro* cultures-free hormones, but also the in interspecific hybrids that germinated *in vivo*. Changing the number of NORs in the chromosome complements in some taxa might be due to the NORs of some chromosomes that are free to jump at least between some preferential chromosomal sites, either by means of adjacent transposable elements or due to recombination hot spots that find homologous sequences inside or proximal to the NORs [56]. Present study and others [6, 16, 21] could detect transposition of NORs as well as 5S rDNA sites to new chromosome sites through translocation of rDNA loci between homologous or nonhomologous chromosomes, or after duplication of those loci during/after direct artificial hybridization of the F_1 hybrids. Such chromosome rearrangements may be facilitated by transposable elements that can provide the substrates for recombination.

Present results and others [6, 16, 42] imply that rapid genome changes in artificial F_1 hybrids are produced by translocation/deletion and recombination, which is an essential adjustment and required for the harmonious coexistence of the two or more different genomes in the nucleus of chrysanthemum allopolyploids. Therefore, the genome changes of allopolyploids might be resulted soon after hybridization and considered as an important additional source of genetic diversity in polyploid species. The discovery of many artificial F_1 hybrid combinations that have rapid genome changes in synthetic polyploids in *Chrysanthemum*

sensu lato provides a new avenue of investigation into the molecular cytogenetics events, which shape the outcome of allopolyploidy. It seems that the synthetic polyploid has the adaptive function of stabilizing genotype through mutations, which might possess an adaptive advantage to various new environments.

Acknowledgments

This study was supported by the Grant-in-Aid for Scientific Research Program (A) no. 10044209 of Japan Society for the Promotion of Science (Representative: Katsuhiko Kondo) and the National Bioresource Project "*Chrysanthemum sensu lato*" of the Japanese Ministry of Education, Science, Sports and Culture (Representative: Katsuhiko Kondo). "Memory of chromosome researches in the Anthemideae, Asteraceae and retirement to Prof. Joan Valles, Barcerona University, Spain."

References

[1] K. Kondo, R. Tanaka, D. Hong, M. Hizume, Q. Yang, and M. Nkata, "A chromosome study of *Ajania* and its allied genera in the Chrysantheminae, the Anthemideae, the Compositae in Chinese highlands," in *Karyomorphological and Cytogenetical Studies in Plants Common to Japan and China*, R. Tanaka, Ed., pp. 1–13, Hiroshima University, 1994.

[2] K. Kondo, R. Tanaka, S. Ge, D. Hong, and M. Nakata, "Chromosome studies of *Ajania* and its alied genera," in *Compositae: Systematics. Prooceeding of the International Compositae Conference*, D. J. N. Hind and H. J. Beentje, Eds., vol. 1, pp. 425–434, Royal Botanic Gardens, Kew, NY, USA, 1994.

[3] M. H. Abd El-Twab and K. Kondo, "Physical mapping of 5S, 45S, Arabidopsis-type telomere sequence repeats and AT-rich regions in *Achillea millefolium* showing intrachromosomal variation by FISH and DAPI," *Chromosome Botany*, vol. 4, pp. 37–45, 2009.

[4] M. Ørgaard and J. S. Heslop-Harrison, "Investigations of genome relationships between *Leymus*, *Psathyrostachys* and *Hordeum* inferred by genomic DNA:DNA *in situ* hybridization," *Annals of Botany*, vol. 73, no. 2, pp. 195–203, 1994.

[5] K. Kondo and M. H. Abd El-Twab, "Analysis of intera-generic relationships *Sensu stricto* among the members of *Chrysanthemum sensu lato* by using fluorescence *in situ* hybridization and genomic *in situ* hybridization," *Chromosome Science*, vol. 6, pp. 87–100, 2002.

[6] M. H. Abd El-Twab and K. Kondo, "Rapid genome reshuffling induced by allopolyploidy in F_1 hybrid in *Chrysanthemum remotipinum* (formerly *Ajania remotipinna*) and *Chrysanthemum chanetii* (formerly *Dendranthema chanetii*)," *Chromosome Botany*, vol. 2, pp. 1–9, 2007.

[7] M. H. Abd El-Twab and K. Kondo, "FISH physical mapping of 5S rDNA and telomere sequence repeats identified a peculiar chromosome mapping and mutation in *Leucanthemella linearis* and *Nipponanthemum nipponicum* in *Chrysanthemum sensu lato*," *Chromosome Botany*, vol. 2, pp. 11–17, 2007.

[8] M. H. Abd El-Twab and K. Kondo, "Isolation of chromosomes and mutation in the interspecific hybrid between *Chrysanthemum boreale* and *Ch. vestitum* using fluorescence *in situ* hybridization and genomic *in situ* hybridization," *Chromosome Botany*, vol. 2, pp. 19–24, 2007.

[9] M. H. Abd El-Twab and K. Kondo, "Identification of parental chromosomes, intra-chromosomal changes and relationship of the artificial intergeneric hybrid between *Chrysanthemum horaimontanum* and *Tanacetum vulgare* by single color and simultaneous bicolor of FISH and GISH," *Chromosome Botany*, vol. 2, pp. 113–119, 2007.

[10] K. Wolff and J. Peters-van Rijn, "Rapid detection of genetic variability in *chrysanthemum* (*Dendranthema grandiflora* Tzvelev) using random primers," *Heredity*, vol. 71, p. 4, 1993.

[11] R. Tanaka, M. Nakata, and M. Aoyama, "Cytogenetic studies on wild *Chrysanthemum* from China IV. Karyotype of *Ch. morii*," *Chromosome Information Service*, vol. 43, pp. 18–19, 1987.

[12] R. Tanaka, S. Kawasaki, J. Yonezawa, K. Taniguchi, and H. Ikeda, "Cytogenetic studies on wild *Chrysanthemum* from China V. F1-hybrids of *Chrysanthemum lavandulifolium* var. *sianense* X *Ch. boreale*," *Cytologia*, vol. 54, pp. 365–372, 1989.

[13] M. Nakata, D. Hong, J. Qiu, H. Uchiyama, R. Tanaka, and S. Chen, "Cytogenetic studies on wild *Chrysanthemum sensu lato* in China. I. Karyotype of *Dendranthema vestitum*," *Japanese Journal of Botany*, vol. 66, pp. 199–204, 1991.

[14] M. Nakata, D. Hong, J. Qiu, H. Uchiyama, R. Tanaka, and S. Chen, "Cytogenetic studies on wild *Chrysanthemum sensu lato* in China. II. A natural hybrid between *Dendranthema indicum* ($2n = 36$) and *D. vestitum* ($2n = 54$) from Hbei Province," *Japanese Journal of Botany*, vol. 67, pp. 92–100, 1992.

[15] K. Kondo, R. Tanaka, S. Ge, D. Hong, and M. Nakata, "Cytogenetic studies on wild *Chrysanthemum sensu lato* in China. IV. Karyomorphological characteristics of three species of Ajania," *Japanese Journal of Botany*, vol. 67, pp. 324–329, 1992.

[16] M. H. Abd El-Twab and K. Kondo, "Rapid genome changes after inter specific hybridization between *Dendranthema indica X D. vestita* identified by fluorescent *in situ* hybridization and 4, 6-diamidino-2-phenylindole," *Chromosome Science*, vol. 7, pp. 77–81, 2003.

[17] M. H. Abd El-Twab and K. Kondo, "Discrimination and isolation of terminal chromosomal regions of *Dendranthema occidentali-japonense* in the chromosomes of F_1 hybrid of *Dendranthema boreale* by using GISH," *Chromosome Science*, vol. 4, pp. 87–93, 2000.

[18] M. H. Abd El-Twab and K. Kondo, "Molecular cytogenetic identification of the parental genomes in the intergeneric hybrid between *Leucanthemella linearis* and *Nipponanthemum nipponicum* during meiosis and mitosis," *Caryologia*, vol. 54, no. 2, pp. 109–114, 2001.

[19] J. Inafuku, M. Nabeyama, Y. Kikuma, J. Saitoh, S. Kubota, and S. I. Kohno, "Chromosomal location and nucleotide sequences of 5S ribosomal DNA of two cyprinid species (Osteichthyes, Pisces)," *Chromosome Research*, vol. 8, no. 3, pp. 193–199, 2000.

[20] M. H. Abd El-Twab and K. Kondo, "Physical mapping of 5S rDNA in chromosomes of *Dendranthema* by fluorescence *in situ* hybridization," *Chromosome Science*, vol. 6, pp. 13–16, 2002.

[21] M. H. Abd El-Twab and K. Kondo, "Identification of mutation and homologous chromosomes in four cultivars of *Dendranthema grandiflora* by physical mapping of 5S and 45S rDNA using fluorescence genomic *in situ* hybridization," *Chromosome Science*, vol. 8, pp. 81–68, 2004.

[22] M. H. Abd El-Twab and K. Kondo, "FISH physical mapping of 5S, 45S and *Arabidopsis*-type telomere sequence repeats in *Chrysanthemum zawadskii* showing intra-chromosomal variation and complexity in nature," *Chromosome Botany*, vol. 1, pp. 1–5, 2006.

[23] J. Fuchs and I. Schubert, "Localization of seed protein genes on metaphase chromosomes of *Vicia faba* via fluorescence *in situ* hybridization," *Chromosome Research*, vol. 3, no. 2, pp. 94–100, 1995.

[24] G. E. Harrison and J. S. Heslop Harrison, "Centromeric repetitive DNA sequences in the genus *Brassica*," *Theoretical and Applied Genetics*, vol. 90, no. 2, pp. 157–165, 1995.

[25] J. S. Heslop-Harrison, T. Schwarzacher, K. Anamthawat-Jonson, A. R. Leich, M. Shi, and I. J. Leich, "In situ hybridization with automated chromosome denaturation," *Technique*, vol. 3, pp. 109–116, 1991.

[26] K. Kondo, Y. Honda, and R. Tanaka, "Chromosome marking in *Dendranthema japonica* var. *wakasaense* and its closely related species by fluorescence *in situ* hybridization using rDNA probe.," *La Kromosomo*, vol. 81, pp. 2785–2791, 1996.

[27] K. Khaung, K. Kondo, and R. Tanaka, "Physical mapping of rDNA by fluorescent *in situ* hybridization using pTa71 probe in three tetraploid species of *Dendranthema*," *Chromosome Science*, vol. 1, pp. 25–30, 1997.

[28] M. H. Abd El-Twab and K. Kondo, "Physical mapping of 45S rDNA loci by fluorescent *in situ* hybridization and Evolution among polyploid *Dendranthema* species," *Chromosome Science*, vol. 7, pp. 71–76, 2003.

[29] S. P. Adams, I. J. Leitch, M. D. Bennett, M. W. Chase, and A. R. Leitch, "Ribosomal DNA evolution and phylogeny in Aloe (Asphodelaceae)," *American Journal of Botany*, vol. 87, no. 11, pp. 1578–1583, 2000.

[30] M. H. Abd El-Twab and K. Kondo, "Fluorescence *in situ* hybridization and genomic *in situ* hybridization to identify the parental genomes in the intergeneric hybrid between *Chrysanthemum japonicum* and *Nipponanthemum nipponicum*," *Chromosome Botany*, vol. 1, pp. 7–11, 2006.

[31] Y. Honda, M. H. Abd El-Twab, H. Ogura, K. Kondo, R. Tanaka, and T. Shidahara, "Counting sat-chromosome numbers and species characterization in wild species of *Chrysanthemum sensu lato* by fluorescence *in situ* hybridization using pTA71 probe," *Chromosome Science*, vol. 1, pp. 77–81, 1997.

[32] M. H. Abd El-Twab, K. Kondo, and D. Hong, "Isolation of a particular chromosome of *Ajania remotipinna* in a chromosome complement of an artificial F_1 hybrid of *Dendranthema lavandulifolia* X *Ajania remotipinna* by use of genomic *in situ* hybridization," *Chromosome Science*, vol. 3, pp. 21–28, 1999.

[33] M. H. Abd El-Twab and K. Kondo, "Identification of nucleolar organizing regions and parental chromosomes in F_1 hybrid of *Dendranthema japonica* and *Tanacetum vulgare* simultaneously by fluorescence *in situ* hybridization," *Chromosome Science*, vol. 3, pp. 59–62, 1999.

[34] M. H. Abd El-Twab and K. Kondo, "Visualization of genomic relationships in allotetraploid hybrids between *Chrysanthemum lavandulifolium* X *Ch. chanetii* by fluorescence *in situ* hybridization and genomic *in situ* hybridizarion," *Chromosome Botany*, vol. 3, pp. 19–25, 2008.

[35] T. Schwarzacher, A. R. Leitch, M. D. Bennett, and J. S. Heslop-harrison, "*In Situ* localization of parental genomes in a wide hybrid," *Annals of Botany*, vol. 64, no. 3, pp. 315–324, 1989.

[36] K. Anamthawat-Jónsson, T. Schwarzacher, A. R. Leitch, M. D. Bennett, and J. S. Heslop-Harrison, "Discrimination between closely related Triticeae species using genomic DNA as a probe," *Theoretical and Applied Genetics*, vol. 79, no. 6, pp. 721–728, 1990.

[37] S. T. Bennett, A. Y. Kenton, and M. D. Bennett, "Genomic *in situ* hybridization reveals the allopolyploid nature of *Milium montianum* (Gramineae)," *Chromosoma*, vol. 101, no. 7, pp. 420–424, 1992.

Genome Mutation Revealed by Artificial Hybridization between Chrysanthemum yoshinaganthum and Chrysanthemum
vestitum Assessed by FISH and GISH

115

[38] A. S. Parokonny, A. Kenton, Y. Y. Gleba, and M. D. Bennett, "The fate of recombinant chromosomes and genome interaction in *Nicotiana asymmetric* somatic hybrids and their sexual progeny," *Theoretical and Applied Genetics*, vol. 89, no. 4, pp. 488–497, 1994.

[39] Chen Qin, R. L. Conner, and A. Laroche, "Identification of the parental chromosomes of the wheat-alien amphiploid *Agrotana* by genomic *in situ* hybridization," *Genome*, vol. 38, no. 6, pp. 1163–1169, 1995.

[40] C. Takahashi, I. J. Leitch, A. Ryan, M. D. Bennett, and P. E. Brandham, "The use of genomic *in situ* hybridization (GISH) to show transmission of recombinant chromosomes by a partially fertile bigeneric hybrid, *Gasteria lutzii* X *Aloe aristata* (Aloaceae), to its progeny," *Chromosoma*, vol. 105, no. 6, pp. 342–348, 1996.

[41] H. Ogura and K. Kondo, "Application of genomic *in situ* hybridization to the chromosome complement of the intergeneric hybrid between *Leucanthemella linearis* (Matsum.) Tzuvelev and *Nipponanthemum nipponicum* (Franch.et Maxim.) Kitamura," *Chromosome Science*, vol. 2, pp. 91–93, 1998.

[42] K. Kondo, M. H. Abd El-Twab, and R. Tanaka, "Fluorescence *in situ* hybridization identifies reciprocal translocation of somatic chromosomes and origin of extra chromosome by an artificial, intergeneric hybrid between *Dendranthema japonica* X *Tanacetum vulgare*," *Chromosome Science*, vol. 3, pp. 15–19, 1999.

[43] M. H. Abd El-Twab, H. Shinoyama, and K. Kondo, "Evidences of intergeneric somatic-hybrids between *Dendranthema grandiflora* cv. Shuho-no-chikara and *Artemisia sieversiana* and their chromosomal mutations by using fluorescence *in situ* hybridization and genomic *in situ* hybridization," *Chromosome Science*, vol. 8, pp. 29–34, 2004.

[44] W. L. Gerlach and J. R. Bedbrook, "Cloning and characterization of ribosomal RNA genes from wheat and barley," *Nucleic Acids Research*, vol. 7, no. 7, pp. 1869–1885, 1979.

[45] A. V. Cox, S. T. Bennett, A. S. Parokonny, A. Kenton, M. A. Callimassia, and M. D. Bennett, "Comparison of plant telomere locations using a PCR-generated synthetic probe," *Annals of Botany*, vol. 72, no. 3, pp. 239–247, 1993.

[46] P. S. Soltis and D. E. Soltis, "The role of genetic and genomic attributes in the success of polyploids," *Proceedings of the National Academy of Sciences of the United States of America*, vol. 97, no. 13, pp. 7051–7057, 2000.

[47] X. P. Zhao, Y. Si, R. E. Hanson et al., "Dispersed repetitive DNA has spread to new genomes since polyploid formation in cotton," *Genome Research*, vol. 8, no. 5, pp. 479–492, 1998.

[48] N. Shimotomai, "Zurkaryogentik der gattung *Chrysanthemum*," *Journal of Science, Hiroshima University, Series B*, vol. 2, pp. 1–100, 1933.

[49] B. S. Gaut, M. L. T. D'Ennequin, A. S. Peek, and M. C. Sawkins, "Maize as a model for the evolution of plant nuclear genomes," *Proceedings of the National Academy of Sciences of the United States of America*, vol. 97, no. 13, pp. 7008–7015, 2000.

[50] R. Tanaka, "On the speciation and karyotypes in diploid and tetraploid species of *Chrysanthemum yoshinaganthum* ($2n = 36$)," *Cytologia*, vol. 25, pp. 43–58, 1960.

[51] S. Fukai, Y. Kamigaichi, N. Yamasaki, W. Zhang, and M. Goi, "Distribution, morphological variations and cpDNA PCR-RFLP analysis of *Dendranthema yoshinaganthum*," *Journal of the Japanese Society for Horticultural Science*, vol. 71, no. 1, pp. 114–122, 2002.

[52] K. Kondo, K. K. Khaung, R. Tanaka, and M. Nakata, "Fluorescent banding patterns in hexaploid *Dendranthema occidentali-japonense* and *D. vestitum*," *La Kromosomo*, vol. 70-80, pp. 2739–2745, 1995.

[53] K. K. Khaung, *Species relationships of the polyploid Dendranthema, sec. Dendranthema (the Chrystheminae, the Anthemideae, the Compositae)*, A thesis, School of Science, Hiroshima University, 1997.

[54] M. Nakata, "Possible natural hybrids with chromosome $2n = 62$ between *Chrysanthemum wakasaense* and *C. morifolium*," *Bulletin National of Science Museum, Tokyo, Series B*, vol. 15, pp. 143–149, 1989.

[55] E. J. Gardner, M. J. Simmons, and D. P. Snustad, "Variations in Chromosome Structure," in *Principles of Genetics*, chapter 18, pp. 488–510, John Wiley & Sons, New York, NY, USA, 8th edition, 1984.

[56] I. Schubert and U. Wobus, "*In situ* hybridization confirms jumping nucleolus organizing regions in Allium," *Chromosoma*, vol. 92, no. 2, pp. 143–148, 1985.

13

Phytochemical Composition and Antioxidant Potential of *Ruta graveolens* L. *In Vitro* Culture Lines

Renuka Diwan, Amit Shinde, and Nutan Malpathak

Department of Botany, University of Pune, Pune Maharashtra 411007, India

Correspondence should be addressed to Nutan Malpathak, mpthak@unipune.ac.in

Academic Editor: Gaoming Jiang

Ruta graveolens L. is a medicinal plant used in traditional systems of medicine for treatment of psoriasis, vitiligo, leucoderma, and lymphomas with well-known anti-inflammatory and anticancer properties. Therefore antioxidant potential of *R. graveolens* (*in planta* and *in vitro*) was investigated. As antioxidants present in plant extracts are multifunctional, their activity and mechanism depends on the composition and conditions of the test system. Therefore, the total antioxidant capacity was evaluated using assays that detect different antioxidants: free radical scavenging (DPPH and ABTS), transition metal ion reduction (phosphomolybdenum assay), reducing power, and nitric oxide reduction. Content of furanocoumarin-bergapten in the extracts showed good corelation with free radical scavenging, transition metal reduction and reducing power, while total phenolic content showed good corelation with nitric oxide reduction potential. Antioxidant activity of *in vitro* cultures was significantly higher compared to *in vivo* plant material. The present study is the first report on comprehensive study of antioxidant activity of *R. graveolens* and its *in vitro* cultures.

1. Introduction

Free radicals, together with secondarily formed radicals, are known to play an important role in the pathogenesis of many chronic conditions like atherosclerosis, arthritis, diabetes, ischemia, reperfusion injuries, central nervous system injury, and cancer [1, 2]. Hence, the study of antioxidant status during a free radical challenge can be used as an index of protection against the development of these degenerative processes in experimental condition for therapeutic measures.

Ruta graveolens is used in homeopathic, ayurvedic, and unani preparations [3] because this herb is so efficacious in various diseases (*Ruta* derived from Greek "reuo" means to set free). It has been extensively used in treatment of leucoderma, vitiligo, psoriasis, multiple sclerosis, cutaneous lymphomas, rheumatic arthritis and recently reported to possess anti-inflammatory and anticancer activity [4, 5]. Antioxidants in plants are affected by area, climatic conditions, and pest attack [6, 7]; therefore *in vitro* cultures are being investigated as alternate source of natural antioxidants [8]. For estimation of total antioxidant potential many

authors have stressed the need to perform more than one type of antioxidant activity measurement to take into account the various mechanisms of antioxidant action [9]. With this perspective the present study investigates the total antioxidant activity evaluated using DPPH, ABTS, phospho-molybdenum complex, reducing power, and Nitric oxide reduction assay of six selected culture lines of *R. graveolens*. Correlation between activity and phytochemical composition of extract (for total phenolics, flavonoids, flavonols, and furanocoumarin) was determined. To our knowledge, the present study is the first report on comprehensive study of antioxidant activity of *R. graveolens* and its *in vitro* cultures.

2. Methodology

2.1. Plant Material. Fruits, shoots, and roots were obtained from *Ruta graveolens* L. (Rutaceae) plants grown in Botanic garden, Department of Botany, University of Pune.

2.2. In Vitro Cultures. Three cell culture lines with varying degree of differentiation (dispersed cell line RC1, aggregated cell line RC3 and differentiated cell line RC6) [10] were selected for evaluating antioxidant potential. Shoot line RS2 [11] and genetically transformed clone Ia3 (Manuscript under revision by Acta Physiologiae Plantarum) were selected due to their lower doubling time and high furanocoumarin productivity.

2.3. Preparation of Extracts.

In vivo fruits, shoots, and roots were used as reference. Finely pulverized plant material (100 mg) was cold-extracted in ultrapure methanol overnight and centrifuged at 10,000 rpm for 20 min and supernatant was filtered. Supernatant was evaporated to dryness at room temperature and dissolved in methanol (ultrapure) to achieve final concentration of 1 mg/mL.

2.4. Phytochemical Composition.

Estimation of total phenolic, flavonoids, flavanols, and furanocoumarins. Total phenols were estimated as gallic acid equivalents (GAEs), expressed as mg gallic acid/g extract by Singleton et al. [12]. The content of flavonoids was determined by method described by Miliauskas et al. [6] using rutin as a reference compounds. The amount of flavonoids in plant extracts was expressed in rutin equivalent (RE). The flavonols content was determined by Miliauskas et al. [6] method. The content of flavanols was expressed in rutin equivalents (REs) as described above. Furanocoumarins psoralen, bergapten, and xanthotoxin were extracted and estimated according to method described elsewhere [11].

2.5. Antioxidant Activity

(1) DPPH. Free radical scavenging estimated according to Szabo et al. [13].

(2) ABTS. The free radical scavenging was measured by method of Teow et al. [14].

(3) Phosphomolybdenum Reduction. The total antioxidant capacity of extracts was evaluated by method of Prieto et al. [15] and expressed as equivalents of ascorbic acid (μmol/g of extract).

(4) Reducing Power. It was determined using method described by Oyaizu [16].

(5) Nitric Oxide Scavenging Activity. The nitric oxide scavenging activity was measured according to the method described by Marcocci et al. [17].

2.6. Statistical Analysis.

The influence of various treatments on growth and phytochemical content was analyzed by one-way analysis of variance (ANOVA). Free radical scavenging activity (DPPH and ABTS), reduction of transition metal ions by phosphomolybdenum complex, reducing power, nitric oxide reduction assay, and total phenolics, flavonoids and flavonols were determined in triplicates.

3. Results and Discussions

3.1. Phytochemical Composition of In Vitro Lines.

Plant phenolics constitute one of the major groups of compounds acting as primary antioxidants or free radical scavengers. Flavonoids are one of the most diverse and widespread group of natural compounds and are one of the most important natural phenolics. These compounds possess a broad spectrum of activities including radical scavenging properties [7]. Therefore it was necessary to determine total amounts of phenolics, flavonoids, and flavanols in the selected extracts.

In vivo root extracts has the highest phenolic content (80 mg GAE/g dry wt). *In vivo* shoot extracts showed lower phenolic content (37 mg GAE/g dry wt) as compared to *in vitro* shoot line RS2 (41 mg GAE/g dry wt) and transformed shoot clone Ia3 (50 mg GAE/g dry wt) (Figure 1(a)). The total flavanoid content in the *in vivo* plant extract ranged from 0.4 to 0.8 mg/g plant extract in RE, whereas total flavanoid content in the *in vitro* cultures ranged from 1 to 17.7 mg/g plant extract (Figure 1(b)). *In vitro* shoot cultures showed 2.18-fold increase in flavanoid content as compared to *in vivo* shoots. The total flavanol content in the *in vivo* plant extract ranged from 0.36 to 0.68 mg/g plant extract in RE, whereas total flavanol content in the *in vitro* cultures ranged from 0.5 to 2.4 mg/g plant extract (Figure 1(b)).

Furanocoumarins are one of the main active constituents of *Ruta graveolens* and are also reported to be potent antioxidants [18, 19]. Potent antioxidant activity (DPPH, APPH) of *Angelica dahuricae* extracts was attributed to the furanocoumarins [20]. Therefore amount of psoralen, bergapten, and xanthotoxin in the extracts were determined (Figure 1(c)). Fruits showed highest furanocoumarins content *in planta*. Amongst *in vitro* cultures differentiation dependent accumulation of bergapten and xanthotoxin was seen [11, 12] with high bergapten and xanthotoxin accumulation in culture lines RC5, RS2, and Ia3, and high psoralen accumulation in RC1 and RC3 (Figure 1(c)).

3.2. Antioxidant Activity

3.2.1. DPPH Radical Scavenging Activity.

The radical scavenging activity was expressed as percentage of reduction of initial DPPH absorbance by the extracts at different concentrations (Figure 2). *In vitro* culture extracts, shoot line RS2, and transformed line Ia3 showed highest radical scavenging activity of 86.4% and 89.1% (EC$_{50}$: 33 and 60 μg/mL), respectively, as compared to 50% and 64% (EC$_{50}$: 400, 130 μg/mL) inhibition by roots and shoots. The percentages can be considered as full absorption inhibition of DPPH, because after completion of reaction the final solution always possesses some yellowish colour and therefore its absorption inhibition compared to colourless methanol solutions cannot reach 100% as shown by Miliauskas et al. [6].

3.2.2. ABTS$^+$ Radical Scavenging Activity.

Amongst all extracts tested, highest antioxidant potential was shown by

(a) Total phenolic content in *in vivo* and *in vitro* cultures of *R. graveolens*

(b) Total flavanoids and flavanols in *in vivo* and *in vitro* cultures of *R. graveolens*

(c) furanocoumarins in *in vivo* and *in vitro* cultures of *R. graveolens*

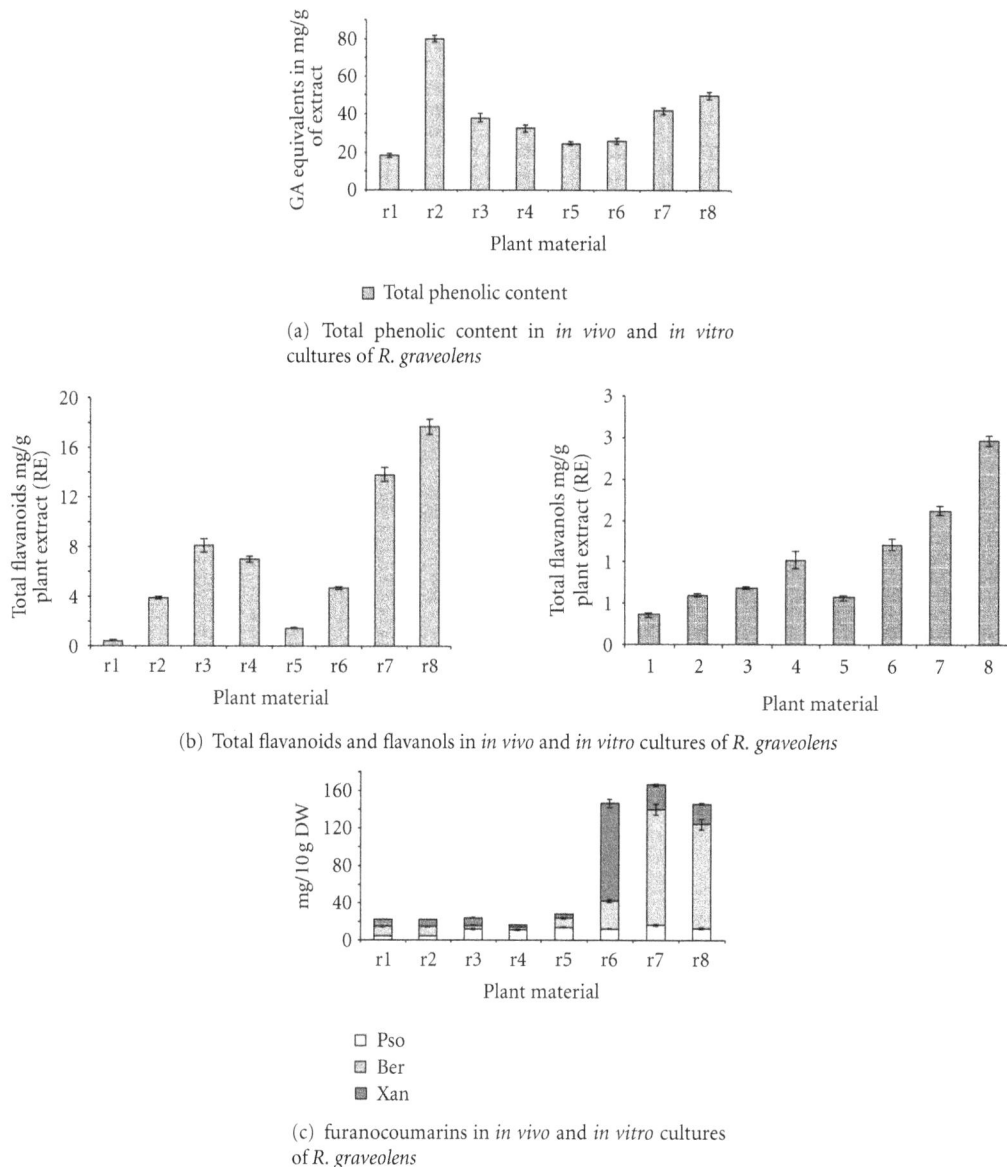

FIGURE 1: Values are mean of three replicates ± std. dev. values significant at $P \leq 0.095$ as calculated by two-way Anova (VassarStats), r1 = *in vivo* fruits, r2 = *in vivo* roots, r3 = *in vivo* shoots, r4 = dispersed suspension RC1, r5 = aggregated suspension RC3, r6 = differentiated suspension RC5, r7 = shoot line RS2, r8 = transformed clone Ia3.

TABLE 1: Correlation analysis.

Antioxidant activity	Phenolics	Flavanols	Flavanoids	Psoralen	Bergapten	Xanthotoxin
DPPH	0.433	0.089	0.124	0.407	**0.779**	0.150
ABTS	0.248	0.105	0.047	0.571	**0.865**	0.111
Phosphomolybdenum assay	0.221	0.388	0	0.221	**0.704**	0.384
Reducing power	0.163	**0.518**	0.399	0.090	**0.578**	0.151
Nitric oxide reduction	**0.83**	0.34	0.23	0.42	0.213	0.287

shoot line RS2 and transformed line Ia3 (EC50; 65 and 115 μg/mL), respectively (Figure 3).

Comparing results of the two radical scavenging tests (DPPH : ABTS) showed good correlation between their activity ($r^2 = 0.98$).

3.2.3. Phosphomolybdenum Assay. The different extracts at a concentration of (50 μg/mL) were assayed for their antioxidant potency by the formation of green phosphomolybdenum complex (Figure 4). *In vitro* shoots line RS2 and transformed clone Ia3 extract had strongest effects on

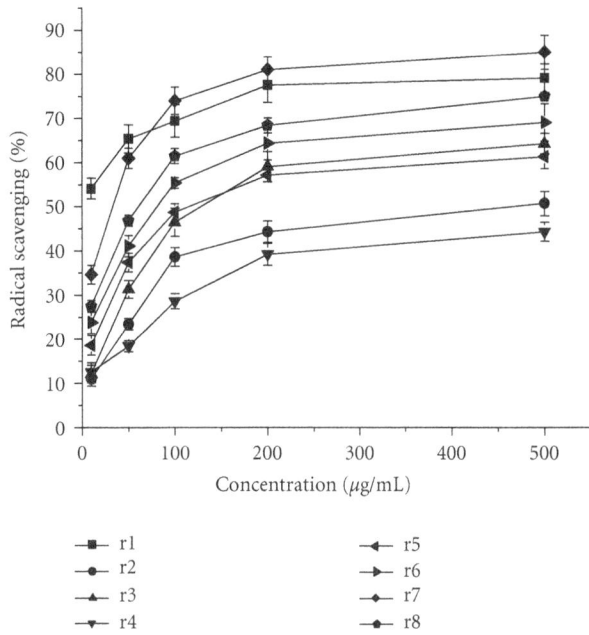

FIGURE 2: DPPH radical scavenging activity of *R. graveolens in vivo* and *in vitro* cultures. Values are mean of three replicates ± std. dev. values significant at $P \leq 0.095$ as calculated by two-way Anova (VassarStats), r1 = *in vivo* fruits, r2 = *in vivo* roots, r3 = *in vivo* shoots, r4 = dispersed suspension RC1, r5 = aggregated suspension RC3, r6 = differentiated suspension RC5, r7 = shoot line RS2, r8 = transformed clone Ia3.

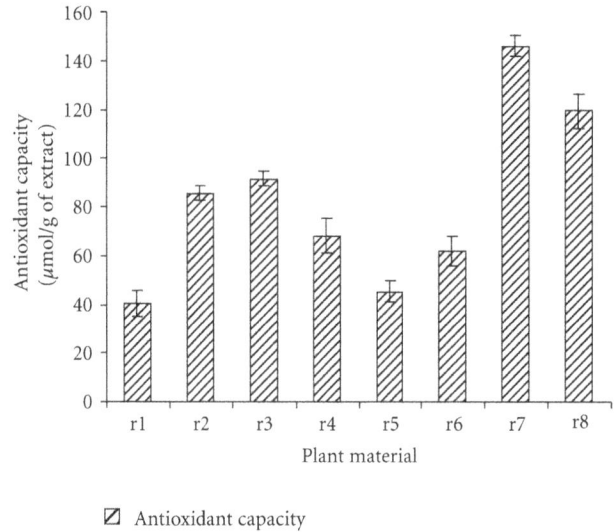

FIGURE 4: Phosphomolybdenum assay of *in vivo* and *in vitro* culture extracts of *R. graveolens.* Values are mean of three replicates ± std. dev. values significant at $P \leq 0.095$ as calculated by two way Anova (VassarStats), r1 = *in vivo* fruits, r2 = *in vivo* roots, r3 = *in vivo* shoots, r4 = dispersed suspension RC1, r5 = aggregated suspension RC3, r6 = differentiated suspension RC5, r7 = shoot line RS2, r8 = transformed clone Ia3.

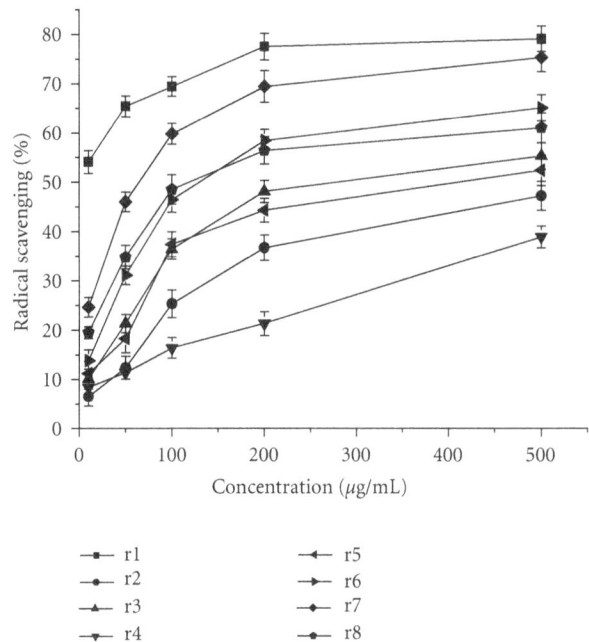

FIGURE 3: ABTS radical scavenging activity of *R. graveolens in vivo* and *in vitro* cultures. Values are mean of three replicates ± std. dev. values significant at $P \leq 0.095$ as calculated by two way Anova (VassarStats), r1 = *in vivo* fruits, r2 = *in vivo* roots, r3 = *in vivo* shoots, r4 = dispersed suspension RC1, r5 = aggregated suspension RC3, r6 = differentiated suspension RC5, r7 = shoot line RS2, r8 = transformed clone Ia3.

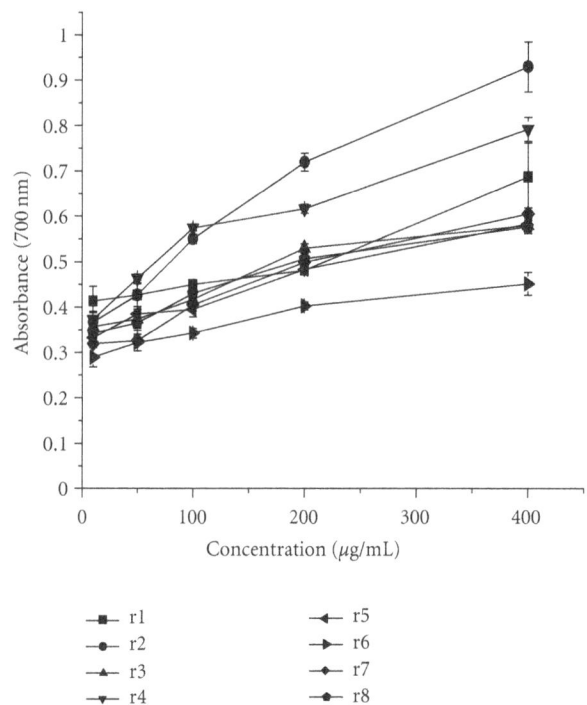

FIGURE 5: Reducing power *in vivo* and *in vitro* culture extracts of *R. graveolens.* Values are mean of three replicates ± std. dev. values significant at $P \leq 0.095$ as calculated by two way Anova (VassarStats), r1 = *in vivo* fruits, r2 = *in vivo* roots, r3 = *in vivo* shoots, r4 = dispersed suspension RC1, r5 = aggregated suspension RC3, r6 = differentiated suspension RC5, r7 = shoot line RS2, r8 = transformed clone Ia3.

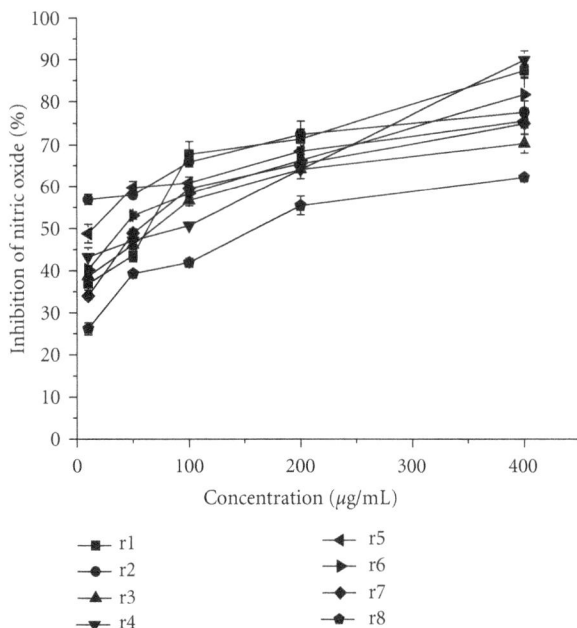

FIGURE 6: Percentage inhibition of nitric oxide radicals by *R. graveolens in vivo* and *in vitro* cultures extracts. Values are mean of three replicates ± std. dev. values significant at $P \leq 0.095$ as calculated by two-way Anova (VassarStats), r1 = *in vivo* fruits, r2 = *in vivo* roots, r3 = *in vivo* shoots, r4 = dispersed suspension RC1, r5 = aggregated suspension RC3, r6 = differentiated suspension RC5, r7 = shoot line RS2, r8 = transformed clone Ia3.

reducing Mo radical (146, 119 equivalents of ascorbic acid μmol/g of extract, resp.) which was 1.5-fold higher than *in vivo* shoots (91 μmol ascorbic acid/g).

It is known that hydrogen and electron transfer from antioxidant analytes to $DPPH^-$, $ABTS^+$ and Mo(VI) complex occur in the DPPH, $ABTS^+$ and phosphomolybdenum assay methods. The transfers occur at different redox potentials in the two assays and also depend on the structure of the antioxidant. $DPPH^-$ and $ABTS^+$ scavenging assays detect antioxidants such as flavonoids and polyphenols, whereas the phosphomolybdenum method usually detects antioxidants such as ascorbic acid, some phenolics, a-tocopherol, and carotenoids [15]. Ascorbic acid, glutathione, cysteine, tocopherols, polyphenols, and aromatic amines can be detected by the two assay models [21]. Thus the antioxidant potential of the extracts differed in these two types of assays. Fruit extracts which showed highest $DPPH^-$ and $ABTS^+$ scavenging activity (EC$_{50}$: 9 μg/mL) showed lowest antioxidant potential in the phosphomolybdenum assay (50 equivalents of ascorbic acid μmol/g of extract). From correlation analysis it was seen that bergapten content showed good correlation with DPPH ($r^2 = 078$), ABTS ($r^2 = 0.86$) and phosphomolybdenum assay ($r^2 = 0.7$) (Table 1).

3.2.4. Reducing Power Assay. Dose-dependant increase in reducing power of Fe^{3+} to Fe^{2+} was observed in the extracts (Figure 5). Amongst all extracts tested, highest activity was observed for root extract (absorbance of 0.93 at 400 μg/mL).

Dispersed suspension (RC1) extract (400 μg/mL) showed maximum absorbance (0.81) amongst all the *in vitro* cultures tested. No significant difference was observed between *in vitro* shoot cultures and *in vivo* shoots. Reducing power of BHT was observed to be 0.8 at 1 mg/mL. It was seen that reducing power showed partial correlation with total flavanol ($r^2 = 0.51$) and bergapten ($r^2 = 0.58$) content.

3.2.5. Nitric Oxide Scavenging Activity. Dose-dependant increase in nitric-oxide radical scavenging activity was observed at studied concentrations (Figure 6). Highest activity was showed by *in vivo* root extracts (EC$_{50}$: 10 μg/mL). Amongst *in vitro* cultures, aggregated cell suspension culture RC3 showed highest activity (EC$_{50}$: 13.1 μg/mL) followed by differentiated cell suspensions RC5 and dispersed cell suspensions RC1 (EC$_{50}$: 41.8 and 93.9 μg/mL), respectively. EC$_{50}$ of Butylated hydroxyl toluene (BHT) was >300 μg/mL, which is considerably higher than that of *R. graveolens* extracts. Thus *in vivo* and *in vitro* extracts of *R. graveolens* are potent scavengers of NO radicals. Nitric oxide reducing ability showed good correlation with total phenolic content ($r^2 = 0.83$) of the extracts.

Results obtained here showed that extracts from *in vitro* cultures showed strong antioxidant activity (DPPH, ABTS, reducing power, phosphomolybdenum assay) as compared to *in vivo* plant material. It has been reported that dedifferentiated cultures (callus and suspension) have less antioxidant potential [22]. However, our results indicated good antioxidant potential (reducing power and nitric oxide scavenging) of *R. graveolens* cell suspension. The observed elevated activity of *in vitro* cultures can be attributed to furanocoumarins especially bergapten and phenolics (Table 1) [23].

Production of antioxidants through plant tissue cultures is an effective strategy which also offers additional advantage of optimization of production of these agents by changing culture conditions and production of active antioxidant principles throughout year [9]. Present study highlighted the use of *in vitro* cultures as a source of antioxidants. Collectively, furanocoumarin content and antioxidant activities of *R. graveolens* will help to select suitable culture type as a source of natural antioxidants and nutraceuticals to enhance health benefits.

Acknowledgment

The financial support provided by University Grants Commission (UGC), New Delhi, India is duly acknowledged. All authors contributed equally to this work.

References

[1] N. C. Cook and S. Samman, "Flavonoids—Chemistry, metabolism, cardioprotective effects, and dietary sources," *Journal of Nutritional Biochemistry*, vol. 7, no. 2, pp. 66–76, 1996.

[2] J. T. Kumpulainen and J. T. Salonen, *Natural Antioxidants and Anticarcinogens in Nutrition, Health and Disease*, Royal Society of Chemistry, Cambridge, UK, 1999.

[3] K. C. Preethi, C. K. K. Nair, and R. Kuttan, "Clastogenic potential of *Ruta graveolens* extract and a homeopathic preparation in mouse bone marrow cells," *Asian Pacific Journal of Cancer Prevention*, vol. 9, no. 4, pp. 763–769, 2008.

[4] S. Pathak, A. Multani, P. Banerji, and P. Banerji, "Ruta 6 selectively induces cell death in brain cancer cells but proliferation in normal peripheral blood lymphocytes: a novel treatment for human brain cancer," *International Journal of Oncology*, vol. 23, no. 4, pp. 975–982, 2003.

[5] K. Preethi, G. Kuttan, and R. Kuttan, "Anti-tumour activity of *Ruta graveolens* extract," *Asian Pacific Journal of Cancer Prevention*, vol. 7, no. 3, pp. 439–443, 2006.

[6] G. Miliauskas, P. R. Venskutonis, and T. A. van Beek, "Screening of radical scavenging activity of some medicinal and aromatic plant extracts," *Food Chemistry*, vol. 85, no. 2, pp. 231–237, 2004.

[7] X. Liu, M. Zhao, J. Wang, B. Yang, and Y. Jiang, "Antioxidant activity of methanolic extract of emblica fruit (*Phyllanthus emblica* L.) from six regions in China," *Journal of Food Composition and Analysis*, vol. 21, no. 3, pp. 219–228, 2008.

[8] M. Gulluce, M. Sokmen, D. Daferera et al., "In vitro antibacterial, antifungal, and antioxidant activities of the essential oil and methanol extracts of herbal parts and callus cultures of *Satureja hortensis* L.," *Journal of Agricultural and Food Chemistry*, vol. 51, no. 14, pp. 3958–3965, 2003.

[9] E. N. Frankel and A. S. Meyer, "The problems of using one-dimensional methods to evaluate multifunctional food and biological antioxidants," *Journal of the Science of Food and Agriculture*, vol. 80, no. 13, pp. 1925–1941, 2000.

[10] R. Diwan and N. Malpathak, "Histochemical localization in *Ruta graveolens* cell cultures: elucidating the relationship between cellular differentiation and furanocoumarin production," *In Vitro Cellular and Developmental Biology*, vol. 46, no. 1, pp. 108–116, 2010.

[11] R. Diwan and N. Malpathak, "Novel technique for scaling up of micropropagated *Ruta graveolens* shoots using liquid culture systems: a step towards commercialization," *New Biotechnology*, vol. 25, no. 1, pp. 85–91, 2008.

[12] V. Singleton, R. Orthofer, and R. Lamuela-Raventos, "Analysis of total phenols and other oxidation substrates and antioxidants by means of Folin-Ciocalteu reagent," in *Methods in Enzymology*, L. Packer, Ed., vol. 299, pp. 152–315, Academic Press, San Diego, Calif, USA, 1999.

[13] M. R. Szabo, C. Idiţoiu, D. Chambre, and A. X. Lupea, "Improved DPPH determination for antioxidant activity spectrophotometric assay," *Chemical Papers*, vol. 61, no. 3, pp. 214–216, 2007.

[14] C. Teow, V. Truong, R. McFeeters, R. Thompson, K. Pecota, and C. Yencho, "Antioxidant activities, phenolic and β-carotene contents of sweet potato genotypes with varying flesh colours," *Food Chemistry*, vol. 103, no. 3, pp. 829–838, 2007.

[15] P. Prieto, M. Pineda, and M. Aguilar, "Spectrophotometric quantitation of antioxidant capacity through the formation of a phosphomolybdenum complex: specific application to the determination of vitamin E," *Analytical Biochemistry*, vol. 269, no. 2, pp. 337–341, 1999.

[16] M. Oyaizu, "Studies on products of the browning reaction. Antioxidative activities of browning reaction products prepared from glucosamine," *Japanese Journal of Nutrition*, vol. 44, no. 6, pp. 307–315, 1986.

[17] L. Marcocci, J. Maguire, M. Droy-Lefaix, and L. Packer, "The nitric oxide-scavenging properties of *Ginkgo biloba* extract EGb 761," *Biochemical and Biophysical Research Communications*, vol. 201, no. 2, pp. 748–755, 1994.

[18] A. Karagozler, B. Erdag, Y. Emek, and D. Uygun, "Antioxidant activity and proline content of leaf extracts from *Dorystoechas hastate*," *Food Chemistry*, vol. 111, pp. 400–447, 2008.

[19] X. Piao, I. Park, S. Baek, H. Kim, M. Park, and J. Park, "Antioxidative activity of furanocoumarins isolated from *Angelicae dahuricae*," *Journal of Ethnopharmacology*, vol. 93, no. 2, pp. 243–246, 2004.

[20] I. Grzegorczyk, A. Matkowski, and H. Wysokinska, "Antioxidant activity of extracts from *in vitro* cultures of *Salvia officinalis* L.," *Food Chemistry*, vol. 104, no. 2, pp. 536–541, 2007.

[21] F. Borges, F. Roleira, N. Milhazes, L. Santana, and E. Uriarte, "Simple coumarins and analogues in medicinal chemistry: occurrence, synthesis and biological activity," *Current Medicinal Chemistry*, vol. 12, no. 8, pp. 887–916, 2005.

[22] I. Kostova, "Synthetic and natural coumarins as cytotoxic agents," *Current Medicinal Chemistry*, vol. 5, no. 1, pp. 29–46, 2005.

[23] R. Marwah, M. Fatope, R. Al Mahrooqi, G. B. Varma, H. Al Abadi, and S. Al-Burtamani, "Antioxidant capacity of some edible and wound healing plants in Oman," *Food Chemistry*, vol. 101, no. 2, pp. 465–470, 2007.

14

Observations on the Morphology, Pollination and Cultivation of Coco de Mer (*Lodoicea maldivica* (J F Gmel.) Pers., Palmae)

Stephen Blackmore,[1] **See-Chung Chin,**[2] **Lindsay Chong Seng,**[3] **Frieda Christie,**[1]
Fiona Inches,[1] **Putri Winda Utami,**[4] **Neil Watherston,**[1] **and Alexandra H. Wortley**[1]

[1] *Royal Botanic Garden Edinburgh, 20a Inverleith Row, Edinburgh EH3 5LR, UK*
[2] *Singapore Botanic Gardens, 1 Cluny Road, Singapore 259569, Singapore*
[3] *Plant Coversation Action Group, P.O. Box 392, Victoria, Mahé, Seychelles*
[4] *Center for Plant Conservation-Bogor Botanical Gardens, Kebun Raya Indonesia, Indonesian Institute of Sciences,*
  Jl. Ir. H. Juanda No. 13 P.O. BOX 309, Bogor 16003, Indonesia

Correspondence should be addressed to Stephen Blackmore, s.blackmore@rbge.org.uk

Academic Editor: Hiroshi Tobe

We present a range of observations on the reproductive morphology, pollination biology and cultivation of *Lodoicea maldivica* (coco de mer), an endangered species with great ecological, economic and cultural importance. We review the history of study of this charismatic species. Morphological studies of the male inflorescence indicate its importance as a year-round food source to the Seychelles fauna. *In situ* observations suggest a number of potential biotic and abiotic pollination mechanisms including bees, flies, slugs, and geckos; trigonid bees are identified as the most likely potential natural pollinator. We outline a successful programme for *ex situ* pollination, germination, and cultivation of the coco de mer, highlighting the importance of temperature, humidity and light levels as well as maintaining an undisturbed environment. In combination with continued protection and monitoring, this advice may aid the future *in situ* and *ex situ* conservation of the coco de mer.

1. Introduction

Lodoicea maldivica (J F Gmel.) Pers. (Figure 1) is endemic to the Seychelles and is a remarkable species not only in biology, but also in ecological, economic and cultural significance. A tall, straight-trunked, dioecious palm with large fan-shaped leaves, it holds world records for the heaviest fruit of any palm (up to 45 kg [1]; Figure 1(e)), the heaviest seed in the plant kingdom (up to 25 kg [1]; Figure 1(f)), and the largest female flowers of any palm [2] (see Figures 1(b) and 1(c)).

As described in detail by Lionnet [3] and Fischer et al. [4], *L. maldivica* was first recorded by Garcia de Orta in 1563 as *coco das Maldivas* in *Colóquios dos Simples e Drogas he Cousas Medicinais da Índia e assi Dalgũas Frutas Achadas Nella Onde se Tratam Algũas Cousas Tocantes a Medicina, Pratica, e Outras Cousas Boas Pera Saber* [5]. At this time, its large, highly valued nuts (Figure 1(f)) could be found all around the Indian Ocean, including the Maldives, India, and

Sri Lanka, and were traded as far afield as China (where they were renowned for their supposed medicinal properties), Japan and Indonesia. Since there was no evidence regarding the source of the nuts, they were initially believed to be produced by a submarine plant species: hence the common name, *coco de mer* [4, 6]. The origin of the nuts was finally traced to Praslin Island, Republic of Seychelles, by the expedition of Chevalier Marion Dufresne in 1768 [3, 4]. In 1769, Dufresne's second-in-command, Duchemin, returned to Praslin Island and exported such a large quantity of coco de mer nuts that he flooded and practically destroyed their market [3].

The ecological importance of the coco de mer is considerable. Palm forests are a key feature of the natural and secondary vegetation of the Seychelles [7] and stands of *L. maldivica*, in particular, support a unique endemic fauna including the Seychelles black parrot (*Coracopsis nigra* L. ssp. *barclayi* Newton), Seychelles bulbul (*Hypsipetes crassirostris*

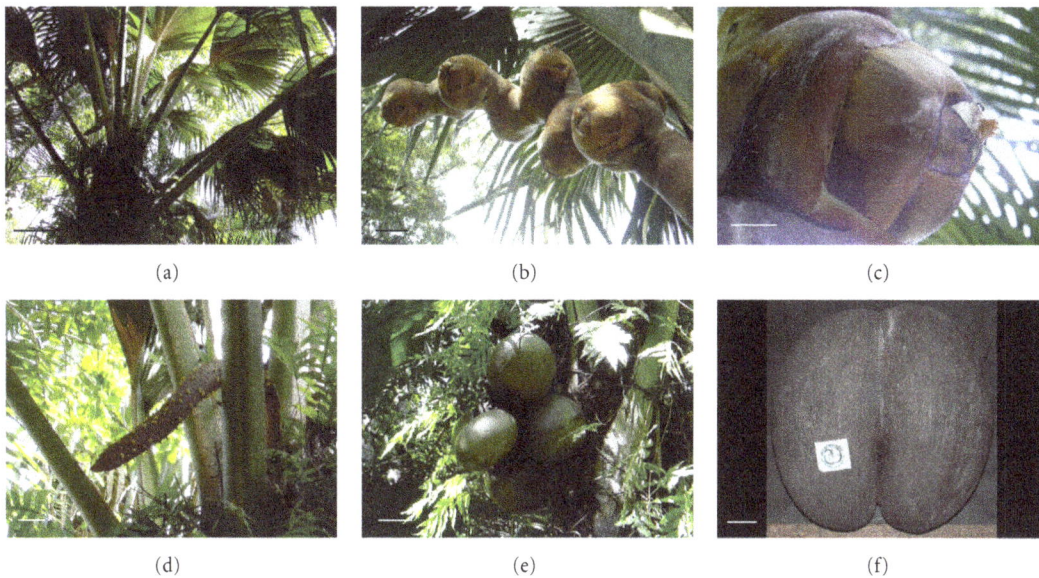

FIGURE 1: Morphology of *Lodoicea maldivica* (coco de mer). (a) Mature male tree showing axillary inflorescence, Vallée de Mai, Praslin, Republic of Seychelles, (b) detail of female inflorescence with developing fruits, Kebun Raya Indonesia, (c) female flower at anthesis, (d) male inflorescence with flowers at anthesis, (e) developing fruits, (f) mature nut sent from Vallée de Mai to the Royal Botanic Garden Edinburgh for propagation. Scale bars: (a) = approx. 1 m; (b) = approx. 10 cm; (c) = 2 cm; (d) = 10 cm (e) = 10 cm; (f) = 10 cm.

Newton), five species of gecko (*Ailuronyx seychellensis* Duméril and Bibron, *A. tachyscopaeus* Gerlach and Canning, *A. trachygaster* Duméril, *Phelsuma astriata* Tornier, and *P. sundbergi* Rendahl), a chameleon (*Archaius tigris* (Kuhl) Townsend et al.) and a snail (*Pachnodus praslinus* Gerlach) [6].

Economically, coco de mer nuts are the most valuable in the world (making them a prime target for poachers; see below); a value recently exemplified when one was given as a wedding gift by the government of the Seychelles to the UK heir to the throne and his wife, the Duke and Duchess of Cambridge. Long considered an antidote to various poisons [7], the suggestive form of the male inflorescence (Figure 1(d)), as well as the nuts themselves (Figure 1(f)), has spawned a wide range of commercial products such as beverages and perfumes, loosely based on the "doctrine of signatures" (and most of which have no true *L. maldivica* content). Since its designation as a UNESCO World Heritage Site in 1983, the *L. maldivica* forest of the 18 Ha Vallée de Mai Nature Reserve on Praslin Island has also become significant to the Seychelles' economy through admission fees—so much so that the funds raised are sufficient to maintain a second Seychelles World Heritage Site at Aldabra Atoll, primarily for the protection of the world's largest population of the giant tortoise, *Geochelone gigantea* Schweigger.

The coco de mer was traditionally useful to the inhabitants of Praslin Island, its leaves being made into thatch (sewn also with thread made from the veins of the leaves), baskets, hats and mats, trunk into furniture, crates and walking sticks, husk into rope, and nuts into utensils and vessels for water storage or liquor manufacture [8]. The indumentum of the young leaves was used for wound dressing and stuffing pillows [8], while the jelly found inside immature nuts was considered a delicacy [7, 8]. The ripe fruits contain a tough endocarp, used to make the spots on dominoes [8]. *L. maldivica* remains culturally important and is known by numerous common names including the *coco jumeau* (twin coconut) and the highly suggestive *coco indecent coco fesse* and *cul de negresse* [3]. Seychelles legend suggests that the trees reproduce on stormy nights, when the males become mobile and "walk" to the waiting females; witnessing such an event results in death [9]. In the nineteenth century, General Gordon of Khartoum's work *Eden and its Two Sacramental Trees* depicted Praslin Island as the Garden of Eden and the coco de mer as its "tree of knowledge." On a more practical note, Gordon also made some of the first herbarium specimens of the species, along with a number of pertinent observations including the presence of a red, gelatinous "pollination droplet" at the female flowers and the paired arrangement of the fruits. Furthermore, he was the first to discuss its method of pollination and details of development: on an annotated drawing held the Royal Botanic Gardens, Kew (see Figure 2), he noted "fecundation takes place by bees, or by placing male flowers from baba on apex of immature nut … it bears in its 40–50 year [sic] … the fruit takes 7 years to ripen." Gordon was also the first person to note the dangers of over-harvesting, lobbying the British government to purchase and protect parts Praslin Island and its *L. maldivica* forests [4].

Lodoicea maldivica is the sole species in its genus (Figure 3) and is one of six monotypic, endemic palm genera found in the archipelago [7, 10]. First formally described as a species of *Cocos* (*C. maldivica* J F Gmel.) in 1791, it was last monographed by Fauvel in 1915 [11]; his 138-page work remains the most complete. The genus was named for Louis (Lat: Lodoicus) XV of France. *Lodoicea* is classified, with

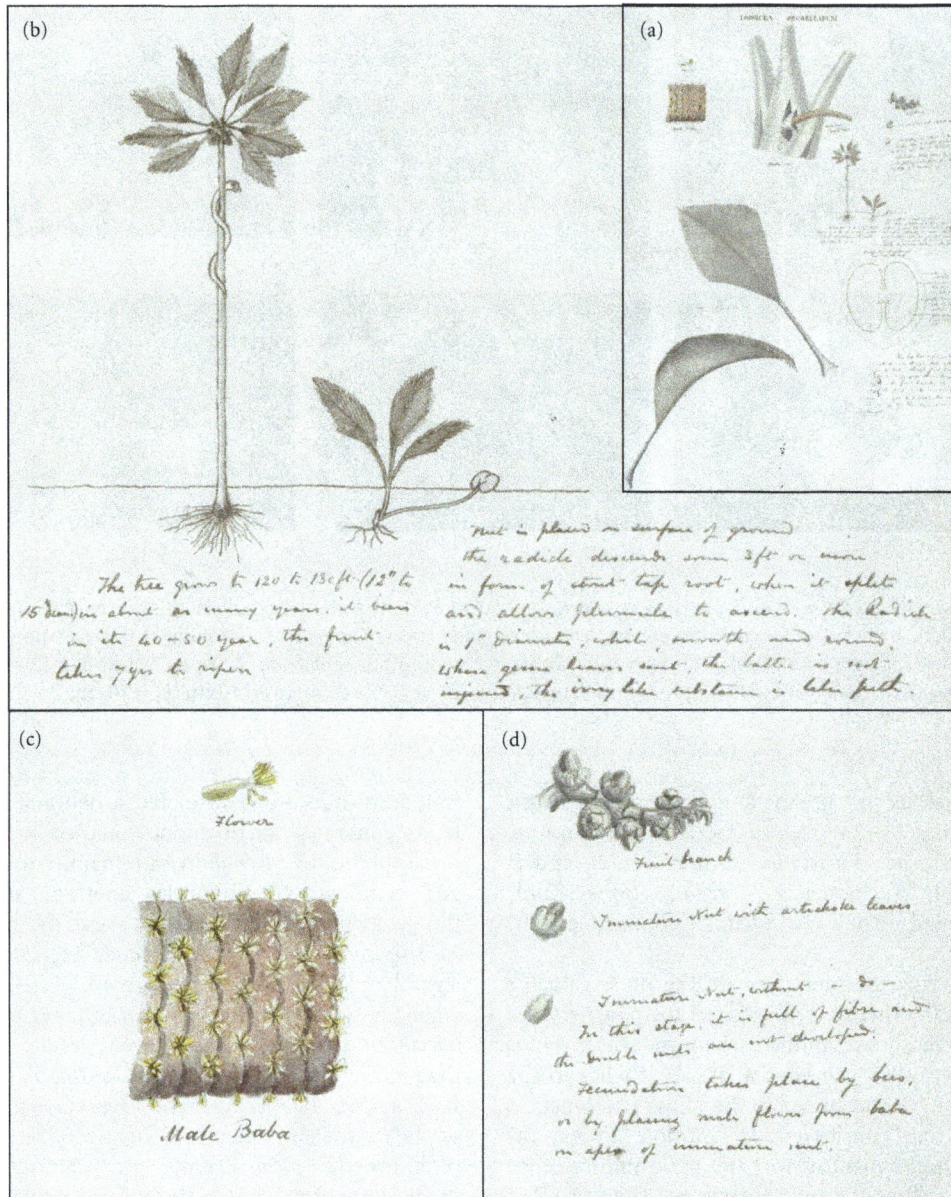

FIGURE 2: General Gordon of Khartoum's annotated drawings of the coco de mer, reproduced with the kind permission of the Director and the Board of Trustees, Royal Botanic Gardens, Kew. (a) Archive sheet in its entirety (the sheet measures approximately 54 cm by 75 cm), showing leaves and male inflorescence "baba" of *Lodoicea maldivica* (here called "*Lodoicea sechellarum*"), (b) enlarged detail of his diagram and notes on the tree's stature and germination: note the snake which is thought to represent Gordon's idea that Praslin Island was the biblical Garden of Eden. Gordon's annotations read "The tree grows to 120 to 130 ft (12″ to 15″ diam) in about as many years. It bears in its 40–50 year, the fruit takes 7 years to ripen," and "Nut is placed on surface of ground. The radicle descends some 3 ft or more in form of stout tap-root, when it splits and allows plumule to ascend. The radicle is 1″ diameter, white, smooth, and round. When green leaves emit, the latter is not injured, the ivory substance is like pith", (c) enlarged detail of male flower and inflorescence ("male baba"), (d) enlarged detail of fruiting branch and immature nuts. Annotations read, "Immature nut with artichoke leaves," and "Immature nut without [artichoke leaves]. In this stage it is full of fibre, and the double nuts are not developed. Fecundation takes place by bees, or by placing male flowers from baba on apex of immature nut."

three other genera (*Latania* Comm. ex Juss., *Borassodendron* Becc. and *Borassus* L.), in subtribe Lataniinae of tribe Borasseae in palm subfamily Coryphoideae [2, 12]. Although phylogenetic relationships within Lataniinae are neither well resolved nor supported [12, 13], it is considered most likely sister to *Borassus* and *Borassodendron*, with *Latania* forming the basally-branching lineage of the subtribe [12, 13]. All Lataniinae are dioecious [2]. Morphologically, *Lodoicea* is perhaps the closest to *Borassus*, the most widespread of the four genera [14], and the only other taxon in the subtribe to produce a bilobed nut [2, 15], with some authors suggesting that it evolved from a *Borassus*-like ancestor [16].

(a)

(b)

(c)

□ Approximate previous distribution
□ Approximate present distribution

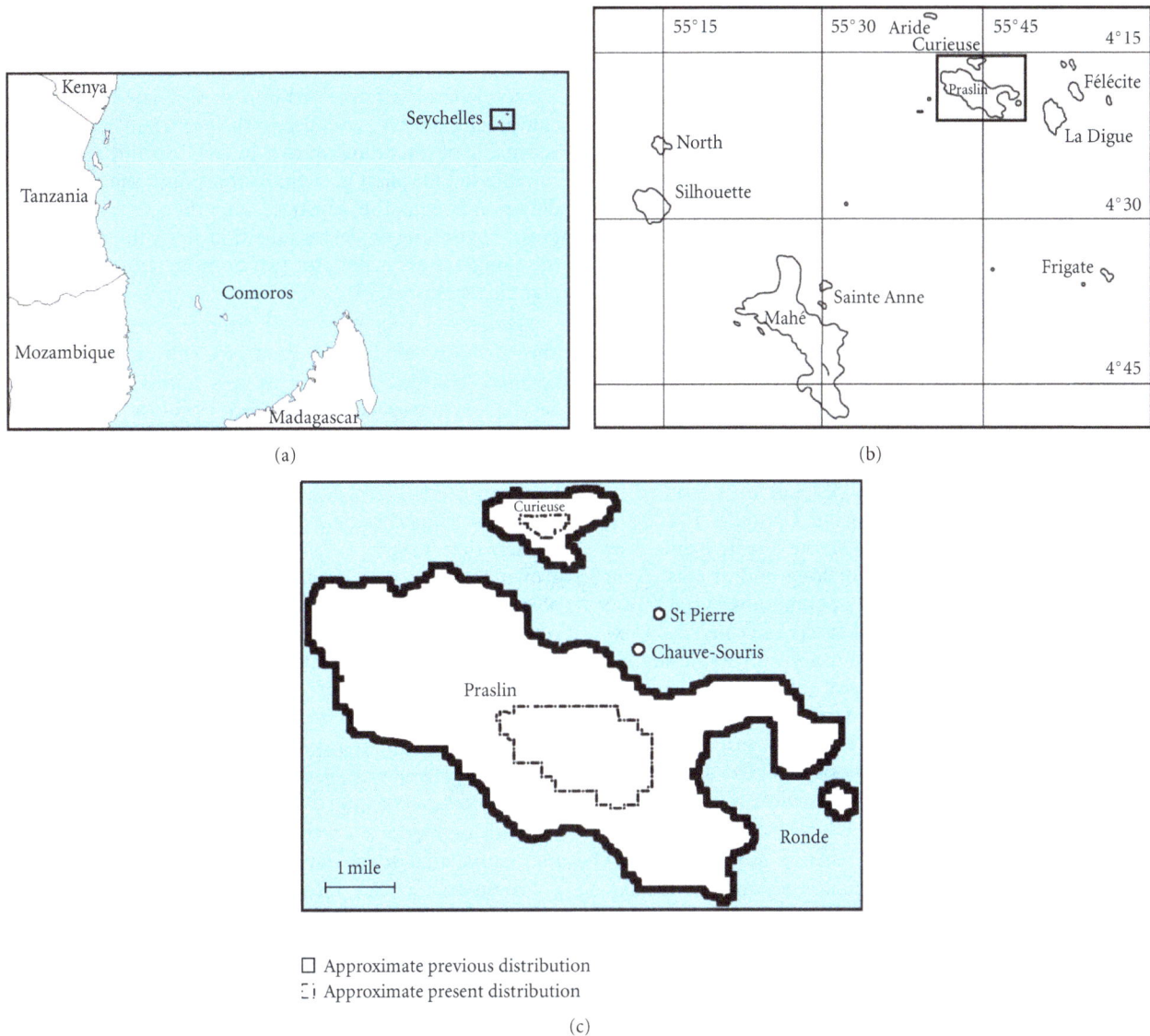

FIGURE 3: Location of coco de mer populations. (a) location of Seychelles (boxed) in the Indian Ocean, (b) location of Praslin and associated islands within the Seychelles, (c) previous and present distribution of coco de mer populations.

Lodoicea maldivica is confined in the wild to the islands of Praslin and Curieuse, which together have an area of less than 5,000 Ha [7, 17]. The species was once found also on St. Pierre, Chauve-Souris, and Île Ronde (Figure 3), and is, therefore, considered to be a relict species [6, 7]. Previously, *L. maldivica* could be found as dense stands in valleys and within mixed-species forest on slopes and ridges; its distribution is now concentrated only in the valleys, with the slopes and ridges dominated by introduced species [6].

The largest population of *L. maldivica* is found on Praslin Island, where it remains locally dominant in the Vallée de Mai [10]. Two reports in the past 60 years [3, 17] have described a population of 4,000 trees including approximately equal numbers of staminate and pistillate individuals [3]. However, Savage and Ashton [18] found that male trees made up 64.3% of the flowering population, which they attributed to differences in life expectancy of male and female trees (and

considered consistent with the widespread hypothesis of wind pollination, which has also recently been suggested by genetic studies [19]). By 2002, the Vallée de Mai population had been reduced to 1,162 mature trees, of which c. 55% were male [6]. Recent studies suggest totals of c. 21,000 trees on Praslin Island and a further 3,800 trees on Curieuse [4], or around 24,500 to 27,000 in the Seychelles as a whole [19, 20].

Lodoicea maldivica is categorised in the IUCN *Red List of Threatened Species* as Endangered (categories B1ab [ii, iii, v] + 2ab [ii, iii, v]) [21], a status to which it was recently upgraded. The major threat to the species is long-term over-exploitation of the nuts, which has had a significant detrimental effect upon natural recruitment and regeneration, thought to be affecting the demographic structure of whole stands [20, 22]. The trees, which may live for up to 350 years [2], are estimated to take a century to reach full size and 25 years to reach reproductive age [7, 17], so this effect may be slow both to develop and to counteract. The nuts (Figure 1(f)) are harvested and sold as souvenirs, fetching prices of up to $400 for a polished specimen [6, 20], or $65 per kg for the kernels alone [20]. The species has been protected and the nut trade legally controlled since 1995, but poaching continues to represent a severe constraint upon regeneration in the wild [20, 23]. Fire is another major threat which has repeatedly impacted on one small Praslin Island population, at Fond Ferdinand. These threats are compounded by the highly restricted distribution of the species, due largely to the fact that the nuts are too heavy to roll uphill, and also that they sink in water, rendering successful dispersal limited [4, 6].

The vegetative morphology and anatomy of *L. maldivica* has been described by Tomlinson [24] and Seubert [25]. Its record-breaking stature may be explained as a case of island gigantism [10], a phenomenon that has been widely discussed in terms of fauna [26–33] but little studied in plants (one notable exception being Carlquist [34, 35]). In terms of reproductive structures, the female palms bear flowers on axillary spadices 1-2 m long and at least 5 cm thick. The female flowers (five to 13 per inflorescence [3]) are borne singly within a pair of broad bracts and comprise a six-lobed perianth sheathing a conical ovary with sessile stigma. The male spadices are similar in size (Lionnet [3] describes them as "the size and thickness of a man's arm") with flowers in cincinni [2], partially covered by bracts, from which the 15–18 stamens are exserted at anthesis [2, 7]. Fertilised fruits may be up to 50 cm long, ovoid, and usually contain a single-seeded, bilobed (occasionally 3–6-lobed) nut [7].

Although there has recently been a growth of interest in pollination biology, particularly among the smaller palms (e.g., [36–39]), relatively little is known about that of canopy-layer palms such as *L. maldivica*. The most widespread assumption is that the species is predominantly anemophilous but as-yet unproven theories, both biotic and abiotic, on the pollination of the coco de mer abound. For example, Good [17] suggested the flowers are pollinated by wind and also visited by insects. Edwards et al. [16] also considered wind, and also rain, to be important in pollination, while Savage & Ashton [18] regarded wind as the main vector, and also observed bees visiting staminate inflorescences. Fleischer-Dogley et al. [19] considered that both biotic and abiotic pollen dispersal may be important in maintaining genetic diversity. Corner [15] suggested pollination may be effected by animals such as geckos, based on the honey-like aroma of the inflorescences. Fischer et al. [4] also suggested that geckos may be involved, and indeed they have been observed visiting both male and female inflorescences. *Phelsuma sundbergi*, in particular, has been noted to show a strong association with male *L. maldivica* inflorescences [40]. Finally, Gerlach [41] reported a small dolichopodid fly as the pollinator.

Once fertilised, the fruits spend some ten months enlarging, followed by around five years ripening, during which time the endosperm solidifies and the exocarp dries and thins [7]. Fallen fruits typically do not germinate for a further six months, during which time the husk disintegrates. The process of germination, which is similar to that of related palm genera such as *Latania* [7], has been well documented by Bailey [7], Purocher [42] and Lionnet [3]: the base of the cotyledon itself develops into an elongated shoot, its tip comprising a cavity containing the embryo [7]. This shoot is capable of extending up to 4 m [43], until the tip reaches suitable soil for plant growth. At this point, the shoot curves downwards into the substrate and the embryo begins to grow, continuing to derive nutrition from the nut, through the elongated cotyledon, for two or more years until the new plant is established [9].

Despite a long history of attempts, relatively little is known about the ideal conditions for cultivation of the coco de mer. In 1890, *L. maldivica* was found to be the only Seychelles palm species that was impossible to cultivate at the Royal Botanic Gardens, Kew and elsewhere [44], although it was successfully cultivated at tropical localities such as Peradeniya (Sri Lanka) and Zanzibar, and is now exhibited at several tropical botanic gardens. Bailey ([7] p.18) stated vaguely that it requires "a full tropical climate ... and good growing conditions." He also noted that the *ex situ* success of germination likely depends on the maturity of the nut when planted. In their native habitat, *L. maldivica* plants are cultivated simply by lying them on a patch of moist soil [7]. It can take as long as four years before the resultant seedling is established and its connection with the nut is lost [7].

Comparable in stature and cultural significance to the cedar of Lebanon, *Cedrus libani* A.Rich. and the giant sequoia, *Sequoiadendron giganteum* (Lindl.) J. Buchholz, [3], the coco de mer is a worthy counterpart to the charismatic megafauna that today dominate conservation biology. As a "keystone" species forming dense forests on which a great variety of species depend, its conservation is crucial and depends upon a detailed understanding of the species' biology (particularly reproductive biology), ecology and cultivation. In this paper, we report a range of observations on the *in situ* and *ex situ* pollination biology, *ex situ* germination and cultivation of the coco de mer, data which are frequently lacking for large, Old World palms. Although they remain incomplete due to the enigmatic nature of the coco de mer, these notes may aid future *in situ* and *ex situ* conservation efforts for the species and the unique community that it sustains.

2. Materials and Methods

2.1. Study Sites. Materials were collected and pollinator observations made in the Vallée de Mai, Praslin Island, with supplementary collections made on Curieuse Island, both in the Republic of Seychelles (Figure 3). The Seychelles have a humid tropical climate with a strong maritime influence. Year-round average temperatures range from 27°C in summer to 30°C in spring and mean monthly rainfall ranges from 75 mm (summer) to 355 mm (January) (http://www.metoffice.gov.uk/weather/africa/eafrica_past .html). The main seasonal differences in climate are due to the direction of the prevailing winds: north western from November to April and south eastern from May to October.

2.2. Morphology of Male Inflorescence and Flower. The morphological studies described below focus on the male inflorescence. Material of male inflorescences collected on Curieuse Island was stored in Copenhagen Solution (70% industrial methylated spirit, 28% distilled water, 2% glycerol). Inflorescences were dissected under a Leica dissecting microscope and observed using both this and a Zeiss Axioskop compound microscope fitted with a digital camera; images were captured using Zeiss Axiovision software and edited using Adobe Photoshop Elements 5.0.

2.3. In Situ Pollination Studies. Field work to observe and capture the animal species associated with fertile male and female inflorescences took place in the Vallée de Mai, Praslin Island. This 19.5 Ha site contains over 7,000 individuals [19], which were monitored by regular observation for signs of flower maturation (male and female plants) and the presence and behaviour of potential pollinators (all observed visitors to male and female trees noted). All female trees in the Vallée de Mai were also monitored for the appearance of a pollination droplet. These observations were made by a team of *in situ* staff from 2003–2010, during all seasons of the year, primarily during the day with additional observations made at night. Captured animals were examined under LM and SEM for traces of pollen which would indicate their potential as pollen vectors. Any pollen grains observed on the vectors were compared with SEM images of pollen extracted directly from *L. maldivica* anthers. For larger animals, gut contents and faecal deposits were also analysed. SEM stubs were either uncoated (for specimens examined using variable pressure technology) or pulse sputter coated with a gold or gold/palladium (60/40) target and examined with JEOL 880, Leica Stereoscan 440, Hitachi S800, and Zeiss Supra 55VP scanning electron microscopes.

2.4. Ex Situ Pollination Studies. Artificial pollination experiments were conducted on a mature female *L. maldivica* tree at Kebun Raya Indonesia, with pollen from male trees at Singapore Botanic Gardens. Male flowers were harvested just prior to anthesis, packaged in a Petri dish, and sent by high-speed courier to Kebun Raya Indonesia, where mature female flowers were pollinated using a small brush. Three rounds of artificial pollination took place, in April 2004, January 2005, and April 2007, using fresh pollen each time. A further pollination took place in August 2007, using pollen stored from the April attempt.

2.5. Germination and Cultivation Studies. *Ex situ* germination and cultivation of *L. maldivica* outside of the tropics have proven notoriously difficult in the past (e.g., [44]). Three nuts were received at the Royal Botanic Garden Edinburgh from the Vallée de Mai, Seychelles, in 2003 (courtesy of the Seychelles Island Foundation), with a further nut sent in January 2005. Germination and cultivation conditions mimicked the natural environment of the Seychelles forests as closely as possible, particularly in terms of temperature, humidity and light levels.

Germination was carried out in a propagation house, in a purpose-made crate (Figure 4(a)), lined with black polythene to ensure total darkness, placed over heated pipes, and filled with *Sphagnum* L. moss. The nuts were placed on the surface of the *Sphagnum* and left untouched. Temperatures in the house were maintained at between 19 and 33°C (usually 20–30°C), and in the germination case between 19 and 39°C (usually 20–35°C); relative humidity between 60 and 100% (usually 90–100%).

Following production of the cotyledonary shoot, the nut was gradually raised above the surface of the soil on pots, to ensure that its tip did not come into contact with the substrate and therefore the shoot continued to extend (Figure 4(c)). When the shoot was about 20 cm long, the nut and shoot were transferred to a purpose-built container, comprising three bottomless pots placed on top of one another to provide a depth of at least 1 m for root growth (Figure 4(d)). The lowest level comprised a double layer of pots to provide extra strength. The container was filled with an equal mix of peat-free compost, fine bark chippings and medium bark chippings to maintain long-term fertility, mixed with perlite, vermiculite, and charcoal. The nut and shoot were planted into a hole within the substrate, so as not to put too much pressure upon the shoot, to a depth such that the majority of the nut was above the surface, then packed around with more of the compost mix. After watering in, the nut was mulched with *Sphagnum* to maintain humidity, and the plant was placed in the sunniest part of the glasshouse (lowest night temperature 18°C, daytime temperature c. 25°C, humidity 90–100%), and left undisturbed as far as possible. It was watered frequently and the *Sphagnum* replenished as necessary.

3. Results

3.1. Morphology of Male Inflorescence and Flower. Male trees of *L. maldivica* in the Vallée de Mai were observed to produce flowers and pollen throughout the year, with a peak of new inflorescence production around November, coincident with the onset of the monsoon season. Male individuals were found to bear up to six long-lasting inflorescences at a time, of which two to three were actively fertile (female inflorescences, while also produced year-round, were observed to be much less frequent). Our observations of one male inflorescence (Figure 5) indicate that a single rachilla contains at least 35 flowers, separated by bract scales, and that each flower contains at least 25 anthers, significantly more than reported in recent works [2]. Given that there may be as many as 200 inflorescence branches in an inflorescence, and two or more inflorescences on a tree, this suggests that each tree may produce at least 350,000 stamens in a single season. With thousands of pollen grains in each anther, as observed under the light microscope, each tree is capable of producing over a billion (US) pollen grains per year.

Each rachilla of the male inflorescence contained flowers at a range of stages of development (Figures 5(a) and 5(b)): when some flowers were at anthesis, others contained anthers entirely undifferentiated (Figure 5(c)). Likewise, pollen from this range of flowers comprised all developmental stages, from premeiosis (no microspores apparent) through the tetrad stage, to mature pollen (Figures 5(e) and 5(f)).

FIGURE 4: Propagation and cultivation of *L. maldivica* at the Royal Botanic Garden Edinburgh. (a) nuts prepared for germination in purpose-made heated crate, lined with *Sphagnum* moss, (b) nut with cotyledonary shoot, (c) care of germinating nut by positioning on raised pots, (d) germinating nut transferred to purpose-built container, (e) radicle at c. two months after emergence, (f) young tree with two leaves, August 2009. Scale bars: (a) = 10 cm; (b) = 10 cm; (c) = 10 cm; (d) = 20 cm; (e) = 5 cm; (f) = 50 cm.

3.2. In Situ Pollination Studies. Whilst the males of *L. maldivica* produce pollen throughout the year (Figures 1(d), 5(a)), the female inflorescences (Figures 1(b) and 1(c)) were found to be much rarer. In the Vallée de Mai, the maturation of the female flowers on an inflorescence was found to be usually sequential although on cultivated specimens at Kebun Raya Indonesia (Bogor Botanic Gardens), all female flowers observed were roughly concurrent in development. When receptive, all female flowers monitored (Figure 1(c)) were observed to exude yellowish-brown nectar. They are also reported to produce a reddish, gelatinous liquid, considered to act as a pollination droplet, although this is extremely transitory and difficult to observe. Pollination drops are most frequently associated with gymnosperms, where they may act as an attractant to pollinators or a pollen trap, which also draws the grains into the female flower.

Pollen of the coco de mer is dehisced poricidally (at least initially, but c.f. Dransfield et al. [2]) from the tips of the anther thecae (Figure 5(f)). It is coated with pollen-kitt (Figure 5(f)), by which many grains are often adhered together. The pollen grains (Figures 5(e) and 5(f)) were found to be typical of the palm family: elliptic, monosulcate, c. $40\,\mu m \times 50\,\mu m$ in diameter, with a reticulate surface, in agreement with the observations of Ferguson et al. [45]. All pollen observed on the various potential vectors captured agreed with this description. In total, eight animal species (in six genera) were observed (and, in the case of invertebrates, captured) visiting the male inflorescence of *L. maldivica* in the Vallée de Mai. These species, all of which depend to some extent upon the coco de mer as a resource, may also all be considered potential pollinators of the coco de mer (Figure 6).

(i) Small stingless bees likely of the genus *Trigona* Jurine (Figure 6(a)). These bees were found to have large numbers of *L. maldivica* pollen grains all over their bodies, concentrated on the hind legs and the rear part of the abdomen (Figure 6(b)), where they were frequently trapped by dense hairs forming scopae. *Trigona* bees were also frequently seen to be attracted to the male inflorescences of trees at Singapore Botanic Gardens.

(ii) Introduced honeybees (*Apis mellifera* L.; Figure 6(c)). These bees also displayed dense masses of *L. maldivica* pollen in the pollen sacs (scopae) of their hind legs, also held in place by hairs (Figure 6(d)).

(iii) Dolichopodid flies of the genus *Cyrturella* Collin. When examined under SEM (Figure 6(e)), these appeared very similar to those identified as *Ethiosciapus* cf. *bilobatus* by Gerlach [41]. No pollen grains were found on the bodies of these flies.

(iv) White slugs (*Vaginula seychellensis* Fischer; Figure 6(f)), often observed to be feeding on male inflorescences—indeed this is by far the most frequent place in which they are observed. The behaviour of

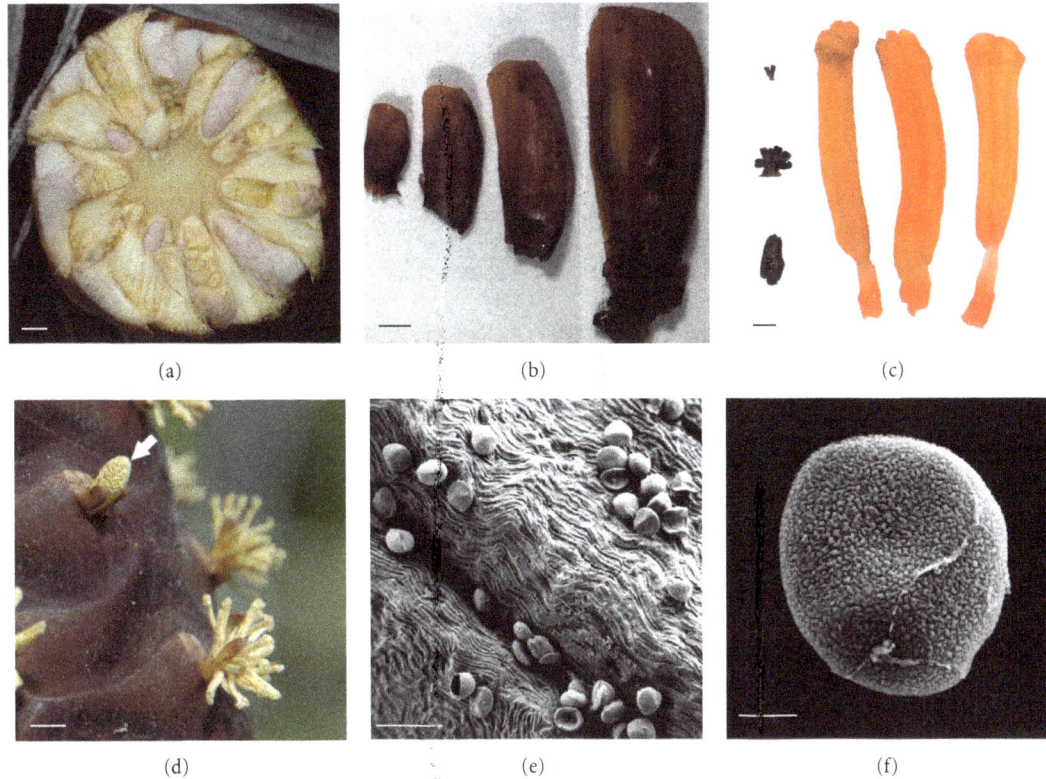

FIGURE 5: Inflorescence structure of male *L. maldivica*. (a) Transverse section of fresh male inflorescence *in situ* on Curieuse Island, (b) four flowers at different stages of development, dissected from the same inflorescence branch, viewed with a dissecting microscope, (c) four anthers at different stages of development, roughly approximating to the development of the flowers in Figure 5(b), viewed with dissecting and compound microscope, (d) male flowers at anthesis *in situ*, showing anthers with apical pore for pollen release (these are more easily seen in the slightly less open flower, arrowed), (e) pollen grains on anther surface viewed under SEM, (f) single pollen grain viewed under SEM showing reticulate surface with pollen-kitt. Scale bars: (a) = 5 mm; (b) = 1 mm; (c) = 500 μm; (d) = 5 mm; (e) = 100 μm; (f) = 10 μm.

this species was observed to be as follows: they spend the majority of time concealed within the vegetation, emerging to feed when no other macrofauna are present, when they position themselves atop a male flower in order to consume it.

(v) Three species of brown or "bronze" geckos (*Ailuronyx tachyscopaeus, A. trachygaster,* and *A. seychellensis,* shown in Figure 6(g)). These endemic geckos were frequently observed in association with the trees and are believed to feed on nectar and pollen from the inflorescences. No specimens were taken, but droppings (most likely from *A. seychellensis*) examined under SEM, consisted almost entirely of coco de mer pollen (Figure 6(h)).

(vi) Green "giant day" geckos (*Phelsuma sundbergi*; Figure 6(i)). This endemic, CITES-listed [46] species was frequently observed feeding both on insects visiting the male inflorescences, and on the inflorescences themselves [40, 47] and sleeping hanging from the leaves of the coco de mer; no specimens were taken and no faeces were found. *Phelsuma sundbergi* are highly territorial and appear to represent the dominant terrestrial macrofauna species on *L. maldivica*

species during daylight hours. Other species, such as *V. seychellensis,* as well as other individuals of *P. sundbergi,* will avoid visiting an inflorescence when one *P. sundbergi* individual is already present, and niche partitioning between gecko species has been proposed as an important ecological factor in Seychelles palm forests [40].

Visitors to the female inflorescences included two gecko species, *Ailuronyx sp.* and *Phelsuma sp.,* which were observed feeding on nectar, and the white slug *Vaginula seychellensis.* Numerous ants, tentatively identified as *Solenopsis mameti* Donisthorpe, also visited female inflorescences and consumed the nectar of plants in the Vallée de Mai, as well as at Kebun Raya Indonesia (Figure 6(j)), but it is unlikely that they are pollinators as they have never been seen foraging for pollen. Other, unidentified flying insects have also been observed around receptive female flowers (identified by the presence of a red pollination droplet) in the Vallée de Mai. Wasp species (Figure 6(k)) were also observed to be attracted to female flowers at Kebun Raya Indonesia; these are not native to the Vallée de Mai so could not be a natural pollinator, but their presence indicates that the female flowers are attractive to species of Hymenoptera. The Seychelles' endemic brown snail (*Stylodonta studeriana*

FIGURE 6: Candidate pollinators of coco de mer. (a) Probable *Trigona sp.* (stingless bee), captured in the Vallée de Mai, (b) *Trigona sp.*, detail of abdomen and hind legs showing pollen grains of coco de mer, (c) *Apis mellifera* (European honeybee), visiting male inflorescence on Curieuse (note full pollen sacs on hind legs), (d) *Apis mellifera*, captured in the Vallée de Mai: detail of hind leg pollen sac showing pollen grains of coco de mer enclosed by hairs (e) *Cyrturella sp.* (dolichopodid fly), captured in the Vallée de Mai, (f) *Vaginula seychellensis* (white slug) feeding on male flowers of coco de mer, Vallée de Mai (g) *Ailuronyx seychellensis* (bronze gecko), Vallée de Mai, (h) *Ailuronyx seychellensis*, contents of droppings (pollen of *L. maldivica*) viewed under SEM, (i) *Phelsuma sundbergi* (giant day gecko) feeding on male flowers of coco de mer, Vallée de Mai, (j) ant, possibly *Solenopsis mameti*, visiting female flower of *L. maldivica*, Vallée de Mai, (k) wasp visiting female inflorescence of *L. maldivica*, Singapore Botanic Gardens. Scale bars: (a) = 500 μm; (b) = 100 μm; (c) = 2 mm; (d) = 200 μm; (e) = 500 μm; (f) = 2 cm; (g) = 2 cm; (h) = 50 μm; (i) = 5 mm; (j) = 5 mm; (k) = 10 cm.

Férussac) was also observed by General Gordon to feed on the tree but has not been observed to visit the inflorescences.

3.3. Ex Situ Pollination Studies.

The first attempt at artificial pollination at Kebun Raya Indonesia involved eight flowers, of which two produced fruits. The second, involving seven flowers, also resulted in two fruits. The third pollination of eight flowers produced one fruit, and the final pollination of seven flowers (using stored pollen) was unsuccessful. These figures are comparable with the rate of successful fertilisation seen in nature: female inflorescences typically produce only one or two (occasionally up to four) fertilised fruits. The two fruits from the second pollination, in 2005, failed to survive. The remaining three fruits, two from the 2004 pollination

and one from the 2007 pollination, remain on the trees in an immature state.

3.4. Germination and Cultivation Studies.

Of the four nuts received at the Royal Botanic Garden Edinburgh, two successfully germinated: one received in 2003, and the sole nut received in 2005. The latter (Figure 4(b)) germinated rapidly, within six months of reception, indicating that it was a relatively mature nut which had lain on the ground *in situ* for some time.

The strength of the two cotyledonary shoots was markedly different, seen by their very different girth, with the 2005 shoot being approximately twice as thick as the 2003 shoot. After receiving a great deal of media attention (e.g., [48]), the 2003 shoot did not survive to produce a radicle or hypocotyl. A radicle emerged from the 2005 germination in September 2006, after approximately 16 months' growth (Figure 4(e)). The shoot then grew at a rate of roughly 1 cm per day, attaining a height of 40 cm by the time the leaf was beginning to unfurl in early November 2006. Since that time, the plant has grown at a rate of c. 5 mm increase in height per day, producing one leaf each year. At present, the plant has reached a height of c. 5 m, with three fully emerged leaves and a fourth just emergent at soil level (Figure 4(f)). The second leaf was observed to be larger and stronger than the first, attaining a length of c. 3 m, which is probably a result of the increasing maturity of the plant. A third leaf has fanned out but is not yet fully extended, at almost 2 m in length. Both the second and third leaves were notably slower to develop than the first, which may be due to duller weather conditions during their development. Approximately six years after germination, the original seed now appears to be hollow, indicating that the juvenile plant is now producing all its own sustenance through photosynthesis, rather than from the seed.

4. Discussion

An archetypal "keystone" species [49], the coco de mer lies at the centre of a complex network of interacting and interlinked organisms, which is not yet fully understood. In this paper, we discuss the fauna associated with the inflorescences, in particular the male inflorescence and pollen. The architecture of *L. maldivica* inflorescences has already been well described [2, 7, 43], and instead the morphological studies presented here focus upon aspects of male floral biology and pollen production related to pollination. The morphological studies described above indicate the high reproductive potential of the male plants of *L. maldivica*: each mature plant produces billions of pollen grains per year. Despite this, the status of the species remains fragile and reproduction limited to two wild populations.

4.1. In Situ and Ex Situ Pollination Studies. Despite the general assumption that it is wind-pollinated, at least half a dozen animal species have directly been observed visiting the male flowers of *L. maldivica*. Due to the dominance of *L. maldivica* in the Vallée de Mai, and the year-round flowering of the male trees, the pollen is clearly an important food resource for the variety of vertebrate and invertebrate species know to feed upon it. Pollen is widely known to be used as a food source by beetles, bees and flies [50]; however, in the Vallée de Mai, it appears also to be the primary food source for the white slug, *Vaginula seychellensis* (Figure 6(f)) and bronze geckos, *Ailuronyx spp.* (Figures 6(g) and 6(h)). All three species of gecko found in association with the inflorescences are known to feed (although not exclusively) on *L. maldivica* pollen. The difficulty in observing the female pollination droplet may suggest that this is also a highly prized and rapidly consumed food source and, indeed, *Phelsuma spp.* have been observed licking nectar from the surface of female flowers.

Because pollination is vital to the reproduction, and, therefore, *in situ* conservation, of this rare and endangered species, our studies focused on determining which species (or abiotic mechanisms) are the most important natural pollinators of *L. maldivica*. No conclusive pollinators have so far been identified for any species of Lataniinae (although honeybees such as *A. mellifera* have been observed to visit the male flowers of *Borassus madagascariensis* Bojer and *B. sambiranensis* Jum. & H.Perrier, the latter now treated as a synonym of *B. aethiopum* [51]).

To provide conclusive proof of pollination in *L. maldivica*, it would be necessary to observe an individual visiting the male flowers of a palm in the natural population, actively or passively collecting pollen, and transferring this to a receptive female flower. However, despite the numerous observations visitations to male flowers described above, this has not been achieved. The only organisms so far observed visiting female inflorescences (*in situ*) are geckos (*Ailuronyx sp.* and *Phelsuma sp.*) and ants. Geckos are, by their territorial nature, unlikely regularly to effect transfer of pollen between male and female flowers. Ferguson et al. [45] agree that vertebrates such as geckos are unlikely to act as pollinators. These authors also note the negative effect these creatures may have upon pollination by feeding upon bees, which may be the most successful pollinators of the coco de mer [45]. Ants have never been observed foraging on pollen at male flowers and, despite their abundance, are rarely effective pollinators of plant species [52], with any instances tending to be limited to small herbs [53].

Pollination by any of the biotic and abiotic means discussed here likely occurs occasionally. Wind pollination is known to be common in dioecious species [54]. For *L. maldivica*, most of the support for a theory of wind pollination derives from *ex situ* pollination experiments in which fertilisation is successfully achieved by placing male inflorescences above receptive female flowers, indicating effective pollen transfer through the air, as well as circumstantial genetic evidence [19]. However, such transfer differs from what would be required under natural circumstances in several ways: transfer is vertical rather than horizontal, occurs over much shorter distances than would be found between male and female trees in a forest situation, and relies upon the experimenter timing the transfer to coincide with the fertile period of the female flower. Furthermore, the pollen of *L. maldivica* (Figures 5(e) and 5(f)) is not typical of wind-pollinated species, being sticky with abundant pollen-kitt on the surface. In addition, the female flowers do not bear the exposed, feathery or sticky stigmas commonly associated with anemophilous species, although the pollination droplet could fulfil a similar role. However, unlike gymnosperms, in which a pollination droplet does facilitate wind pollination, pollen of the coco de mer is not released into the air in large quantities, but is instead retained on the anthers for collection by foraging animals. Finally, the strong scent produced by both male and female flowers suggests adaptations for attracting pollinating insects. Thus, we consider wind to be at most a secondary means of pollination in *L. maldivica*.

We consider it most likely that the successful pollinator of *L. maldivica* would be a flying vector. Although several species of vertebrates and invertebrates have been observed visiting the inflorescences of male trees, the likelihood that sufficient numbers of animals, carrying sufficient pollen for successful pollination, could move from the inflorescence of the male tree to the much rarer female inflorescence of a different tree is highest for flying species (i.e., flying insects; birds and bats have never been observed visiting *L. maldivica* trees), especially given the height of the palms involved.

The two native animal species most likely to represent candidate pollinators of the coco de mer are, therefore, *Trigona sp.* bees and *Cyrturella sp.* flies. No species of beetle (Coleoptera) were observed in association with the inflorescences, either at night or during the day. During many years of observation, no flies of the genus *Ethiosciapus* have been found in association with coco de mer inflorescences (c.f. [41]). As far as is known, all dolichopodid flies (including both *Cyrturella* and *Ethioscapus*) are carnivorous, feeding on other invertebrates rather than foraging for pollen, and, therefore, are very unlikely to act as pollinators. Indeed, the relative hairlessness of *Cyrturella* (Figure 6(e)) compared to both species of bees observed here (Figures 6(a)–6(d)), suggests it is not adapted for pollen transfer. The large number of *L. maldivica* pollen grains adhering to the bodies of *Trigona* individuals (Figures 6(a) and 6(b)), whilst none have been found on the *Cyrturella* specimens, confirms this, and suggests that the bees are more likely to be effective pollinators—as suggested by General Gordon as long ago as the nineteenth century. The large numbers of grains also found on the introduced honeybee, *Apis mellifera*, indicate the possibility that this species nowadays plays a role in pollination of the coco de mer.

Previous authors have also observed bees visiting the male inflorescence of the coco de mer [45]. Although trigonid bees have not yet been observed visiting the female inflorescences, the presence of other hymenoptera (e.g., wasps observed at Singapore Botanic Gardens; Figure 6(k)) attracted to the female flowers indicates the secretion of chemicals attractive to these insects. The *L. maldivica* gynoecium is syncarpous [2], a state that is frequently related to the presence of a septal nectary, which would also indicate the likelihood of insect attractants. Corner's observation that the female flowers of the coco de mer smell of honey [15] might also indicate that they, like the male flowers, are attractive to bees. The male flowers are also reported to have a distinct scent [47].

The hypothesis of trigonid bee-pollination is in line with observations of other species in the Lataniinae: *Trigona sp.* have been observed visiting the male inflorescences of *Borassodendron* and *Hyphaene* [45] and have also been seen to visit the female inflorescence of *B. borneense* J. Dransf., apparently feeding on nectar [45]. To provide conclusive evidence, it would be necessary to observe a bee visiting male flowers of an *L. maldivica* palm in the natural population, actively or passively collecting pollen, and transferring this to a receptive female flower. This demonstration remains, at present, elusive. (Note that on Curieuse Island, the trees exist in a much more open habitat than that of the Vallée de Mai, and male and female individuals are generally at a greater

distance from one another. Nevertheless, natural regeneration does occur, suggesting that successful pollination takes place and, therefore, that pollinators (or abiotic pollination mechanisms) must be present here also. *Phelsuma sundbergi* geckos are present on Curieuse, as are various species of bee. In this more open habitat is possible that wind also plays a part in pollination).

Outside of its natural range, artificial pollination has proven successful in fertilising *L. maldivica*, with five fruits produced from 23 pollinated flowers using fresh pollen, of which three survived. The numbers of fruits produced per inflorescence (one or two) were slightly fewer than is normal in natural pollination in the Seychelles (three to five per inflorescence [47]), but fertilisation was not attempted for all flowers on each inflorescence. The fertility of these fruits remains to be seen; when they are mature, attempts will be made to germinate them. Preliminary results suggest that successful *ex situ* pollination relies upon fresh pollen (pollen stored for four months proved unable to fertilise female flowers of coco de mer at Kebun Raya Indonesia).

4.2. Germination and Cultivation Studies. Despite initial difficulties [44], germination and cultivation of *L. maldivica* in botanic gardens around the world, both tropical and temperate, is now proving a successful, if time-consuming, endeavour, and a great deal has been learned about the best methods to achieve results *ex situ*. For instance, the critical importance of temperature, humidity and light levels for germination and survival is now understood (with slower growth observed during years with lower light availability). As observed by Bailey as long ago as 1942, the condition and maturity of the nut is also of great importance and may be difficult to ascertain when specimens are received from a distant source—hence the variable success rate seen in botanic gardens around the world. Furthermore, it is crucial to avoid tampering with germinating nuts, as the attachment of the shoot to the nut is perhaps the most fragile part of the plant at this stage and remains so until all the nutrients from the seed are expended. It will shortly be necessary to move the successfully germinated coco de mer at the Royal Botanic Garden Edinburgh into a larger pot, as its roots are becoming cramped and its largest leaf is touching the glasshouse roof and becoming scorched. Such a transfer is not without risk, in particular, the risk of detaching the plant from the nut before it is completely hollow when it may still be providing nutrients. Thus, this repotting has been delayed as long as possible and will be undertaken with the utmost care.

In the Seychelles, the traditional methodology for the cultivation of coco de mer is to plant the nut in a hole one meter deep filled with topsoil, manure (or other organic matter), and dried leaves, covered with a thin layer of soil and dry grass (to prevent the nut from being stolen) and kept moist. However, this has met with only c. 50% success. Germination in a small depression on the surface of the substrate, similar to the method used in Edinburgh, has met with almost 100% success but runs the risk of the valuable nuts being stolen. It is said that if the soil is too sandy the cotyledonary shoot will grow downwards until it rots, never returning to the surface to produce a radicle, and that this can be prevented by placing a flat rock in the bottom of the planting hole. An alternative, practised by the Seychelles Forestry Department, is to germinate the nuts in a cool, damp, shady spot then transfer them to a well-lit site for production of the shoot; this is most analogous to the procedure carried out at the Edinburgh Garden. Cultivation is most successful in well-drained, well-lit sites although if planted in shady sites the cotyledonary shoot is able to extend underground and produce a plant some distance from the planting site, in warmer and better-lit conditions. The more exposed the site, the more attention must be paid to watering.

The specimen at the Royal Botanic Garden Edinburgh has now survived six years from the emergence of the first shoot, has three large leaves and a fourth emergent, and is growing at an expected rate. Future growth (particularly once the seed is no longer providing nutrients) remains uncertain since it is not known, for instance, whether the species requires any mycorrhizal symbionts to survive. The sex of the Edinburgh plant has not yet been determined; and without genetic analyses, this will not be possible to determine until the first flowering, which may take 30 years or more. The plant has the potential to provide a new centrepiece for the Royal Botanic Garden Edinburgh Tropical Palm House, when the existing palm (a *Sabal bermudana* L. H. Bailey specimen more than 200 years old) becomes too tall. Being a slow-growing species, the coco de mer has the potential to fill this role for many decades.

Acknowledgments

The authors thank the Seychelles Islands Foundation, its Chairman, Maurice Loustau-Lalanne and Chief Executive, Frauke Fleischer-Dogley, wardens of the Vallée Atterville Cedras, Victorin Laboudallon and Marc Jean-Baptiste, and their rangers. They are grateful to Ross Bayton, Katy Beaver, Peter Edwards, Christopher Kaiser-Bunbury, Bärbel Koch and Pat Matyot for discussions. Insect identifications were undertaken by Graham Rotheray at the National Museums of Scotland. They also thank the anonymous reviewers for their helpful comments on the manuscript. Archival materials from the National Botanic Gardens, Glasnevin, Ireland, and the Royal Botanic Gardens, Kew, were provided by Matthew Jebb and Julia Buckley.

References

[1] P. B. Tomlinson, "The uniqueness of palms," *Botanical Journal of the Linnean Society*, vol. 151, no. 1, pp. 5–14, 2006.

[2] J. Dransfield, N. W. Uhl, C. B. Asmussen, W. J. Baker, M. M. Harley, and C. E. Lewis, *Genera Palmarum: The Evolution and Classification of Palms*, Royal Botanic Gardens, Kew, Richmond, UK, 2008.

[3] G. Lionnet, "The double coconut of the Seychelles," *West Australian Nut and Tree Crops Association Yearbook*, vol. 2, pp. 6–20, 1976.

[4] B. E. Fischer, F. Fleischer-Dogley, and A. Fischer, *Coco de Mer: Myths and Eros of the Sea Coconut*, A. B. Fischer, Berlin, Germany, 2008.

[5] H. J. Schlieben, "Coco de mer—die romantische Geschichte einer Palme," *Natur und Museum*, vol. 102, no. 8, pp. 281–291, 1972.

[6] F. Fleisher-Dogley and T. Kendle, *The Conservation Status of the Coco de Mer, Lodoicea maldivica (Gmelin) Persoon: A Flagship Species*, Royal Botanic Gardens, Kew, London, UK, 2002.

[7] L. H. Bailey, "Palms of the Seychelles," *Gentes Herbarum*, vol. 6, pp. 1–48, 1942.

[8] W. Andre and K. Beaver, "Traditional uses of endemic palms and pandans," *Kapisen*, no. 4, pp. 13–14, 2005.

[9] G. Lionnet, *Coco de Mer: The Romance of a Palm*, L'Ile Aux Images, Pailles, Mauritius, 1986.

[10] J. Procter, *Vegetation of The Granitic Islands of the Seychelles*, Kluwer Academic, Boston, Mass, USA, 1984.

[11] A. A. Fauvel, "Lo cocotier de mer des Iles Seychelles," *Annales due Musee Colonial de Marseille, Series 3*, vol. 1, pp. 169–307, 1915.

[12] C. B. Asmussen, J. Dransfield, V. Deickmann, A. S. Barfod, J. C. Pintaud, and W. J. Baker, "A new subfamily classification of the palm family (Arecaceae): evidence from plastid DNA phylogeny," *Botanical Journal of the Linnean Society*, vol. 151, no. 1, pp. 15–38, 2006.

[13] W. J. Baker, V. Savolainen, C. B. Asmussen-Lange et al., "Complete generic-level phylogenetic analyses of palms (Arecaceae) with comparisons of supertree and supermatrix approaches," *Systematic Biology*, vol. 58, no. 2, pp. 240–256, 2009.

[14] R. P. Bayton, "A revision of *Borassus* L. (Arecaceae)," *Kew Bulletin*, vol. 62, no. 4, pp. 561–586, 2007.

[15] E. J. H. Corner, *The Natural History of Palms*, University of California Press, Berkeley, Calif, USA, 1966.

[16] P. J. Edwards, J. Kollmann, and K. Fleischmann, "Life history evolution in *Lodoicea maldivica* (Arecaceae)," *Nordic Journal of Botany*, vol. 22, no. 2, pp. 227–237, 2002.

[17] R. Good, "The coco-de-mer of the Seychelles," *Nature*, vol. 167, no. 4248, pp. 518–519, 1951.

[18] A. J. P. Savage and P. S. Ashton, "The population structure of the double coconut and some other Seychelles palms," *Biotropica*, vol. 15, no. 1, pp. 15–25, 1983.

[19] F. Fleischer-Dogley, C. J. Kettle, P. J. Edwards, J. Ghazoul, K. Määttänen, and C. N. Kaiser-Bunbury, "Morphological and genetic differentiation in populations of the dispersal-limited coco de mer (*Lodoicea maldivica*): implications for management and conservation," *Diversity and Distributions*, vol. 17, no. 2, pp. 235–243, 2011.

[20] L. Rist, C. N. Kaiser-Bunbury, F. Fleischer-Dogley, P. J. Edwards, N. Bunbury, and J. Ghazoul, "Sustainable harvesting of coco de mer, *Lodoicea maldivica*, in the Vallée de Mai, Seychelles," *Forest Ecology and Management*, vol. 260, no. 12, pp. 2224–2231, 2010.

[21] IUCN, "IUCN Red List of Threatened Species. Version 2011.2," 2011.

[22] A. J. P. Savage and P. S. Ashton, "Tourism is affecting the stand structure of the coco-de-mer," *Principes*, vol. 35, no. 1, pp. 47–48, 1991.

[23] K. Fleischmann, P. Heritier, C. Meuwly, C. Kuffer, and P. J. Edwards, "Virtual gallery of the vegetation and flora of the Seychelles," *Bulletin of the Geobotanical Institute ETH*, vol. 69, pp. 57–64, 2003.

[24] P. B. Tomlinson, "Palmae," in *Anatomy of the Monocotyledons II*, C. R. Metcalfe, Ed., Clarendon Press, Oxford, UK, 1961.

[25] E. Seubert, "Root anatomy of palms—I. Coryphoideae," *Flora*, vol. 192, no. 1, pp. 81–103, 1997.

[26] J. B. Foster, "Evolution of mammals on islands," *Nature*, vol. 202, no. 4929, pp. 234–235, 1964.

[27] J. de Vos, L. W. van der Hoek Ostende, and G. D. van den Bergh, *Patterns in Insular Evolution of Mammals: A key to Island Palaeogeography*, Springer-Verlag, Dordrecht, The Netherlands, 2007.

[28] M. V. Lomolino, "Body size of mammals on islands: the Island rule reexamined," *The American Naturalist*, vol. 125, no. 2, pp. 310–316, 1985.

[29] M. V. Lomolino, "Body size evolution in insular vertebrates: generality of the island rule," *Journal of Biogeography*, vol. 32, no. 10, pp. 1683–1699, 2005.

[30] C. R. McClain, A. G. Boyer, and G. Rosenberg, "The island rule and the evolution of body size in the deep sea," *Journal of Biogeography*, vol. 33, no. 9, pp. 1578–1584, 2006.

[31] S. Meiri, N. Cooper, and A. Purvis, "The island rule: made to be broken?" *Proceedings of the Royal Society B*, vol. 275, no. 1631, pp. 141–148, 2008.

[32] S. Meiri, T. Dayan, and D. Simberloff, "The generality of the island rule reexamined," *Journal of Biogeography*, vol. 33, no. 9, pp. 1571–1577, 2006.

[33] J. B. Losos and R. E. Ricklefs, "Adaptation and diversification on islands," *Nature*, vol. 457, no. 7231, pp. 830–836, 2009.

[34] S. Carlquist, "Island biology: we've only just begun," *BioScience*, vol. 22, no. 4, pp. 221–225, 1972.

[35] S. Carlquist, *Island Biology*, Columbia University Press, New York, NY, USA, 1974.

[36] E. J. Berry and D. L. Gorchov, "Reproductive biology of the dioecious understorey palm *Chamaedorea radicalis* in a Mexican cloud forest: pollination vector, flowering phenology and female fecundity," *Journal of Tropical Ecology*, vol. 20, no. 4, pp. 369–376, 2004.

[37] A. Otero-Arnaiz and K. Oyama, "Reproductive phenology, seed-set and pollination in *Chamaedorea alternans*, an understorey dioecious palm in a rain forest in Mexico," *Journal of Tropical Ecology*, vol. 17, no. 5, pp. 745–754, 2001.

[38] R. K. Rosa and S. Koptur, "Preliminary observations and analyses of pollination in *Coccothrinax argentata*: do insects play a role?" *Palms*, vol. 53, no. 2, pp. 75–83, 2009.

[39] L. A. Núñez-Avellaneda and R. Rojas-Robles, "Reproductive biology and pollination ecology of the milpesos palm *Oenocarpus batana* in the Colombian Andes," *Caldasia*, vol. 30, no. 1, pp. 101–125, 2008.

[40] T. Noble, N. Bunbury, C. N. Kaiser-Bunbury, and D. J. Bell, "Ecology and co-existence of two endemic day gecko (*Phelsuma*) species in Seychelles native palm forest," *Journal of Zoology*, vol. 283, no. 1, pp. 73–80, 2011.

[41] J. Gerlach, "Pollination in the coco-de-mer, *Lodoicea maldivica*," *Palms*, vol. 47, no. 3, pp. 135–138, 2003.

[42] Y. F. Purocher, "Seychelles botanical treasure: the "coco-de-mer" palm (*Lodoicea maldivica* Pers.)," *La Revue Agricole del' Ile Maurice*, vol. 26, no. 2, pp. 69–87, 1947.

[43] P. B. Tomlinson, *The Structural Biology of Palms*, Clarendon Press, Oxford, UK, 1990.

[44] W. Watson, "The coco-de-mer in cultivation," *Nature*, vol. 43, no. 1097, pp. 19–20, 1890.

[45] D. K. Ferguson, A. J. Havard, and J. Dransfield, "The pollen morphology of the tribe *Borasseae* (*Palmae: Coryphoideae*)," *Kew Bulletin*, vol. 42, no. 2, pp. 405–422, 1987.

[46] UNEP-WCMC, "UNEP-WCMC Species Database: CITES-Listed Species," 2009.

[47] K. Beaver and L. Chong Seng, *Vallée de Mai*, SPACE, Mahé, Seychelles, 1992.

[48] C. Holden, "Random samples: the big seed," *Science*, vol. 301, pp. 1180–1181, 2003.

[49] R. T. Paine, "A note on trophic complexity and community stability," *The American Naturalist*, vol. 103, no. 929, pp. 91–93, 1969.

[50] P. Bernhardt, "Anther adaptation in animal pollination," in *The Anther: Form, Function and Phylogeny*, W. G. D'Arcy and R. C. Keating, Eds., chapter 9, Cambridge University Press, Cambridge, UK, 1996.

[51] R. P. Bayton, C. Obunyali, and R. Ranaivojaona, "A re-examination of *Borassus* in Madagascar," *Palms*, vol. 47, no. 4, pp. 206–219, 2003.

[52] R. Peakall, S. N. Handel, and A. J. Beattie, "The evidence for, and importance of, ant pollination," in *Ant-Plant Interactions*, C. R. Huxley and D. F. Cutler, Eds., Oxford University Press, Oxford, UK, 1991.

[53] J. C. Hickman, "Pollination by ants: a low-energy system," *Science*, vol. 184, no. 4143, pp. 1290–1292, 1974.

[54] W. G. D'Arcy, *Anthers and Stamens and What they Do*, Cambridge University Press, Cambridge, UK, 1996.

15

A Comparative Analysis of the Mechanical Role of Leaf Sheaths of Poaceae, Juncaceae, and Cyperaceae

Andreas Kempe, Martin Sommer, and Christoph Neinhuis

Department of Biology, Institute of Botany, Faculty of Science, Technische Universität Dresden, 01062 Dresden, Germany

Correspondence should be addressed to Andreas Kempe; andreas.kempe@tu-dresden.de

Academic Editor: Zed Rengel

Similarities in structural organization of the culm in Poaceae, Juncaceae, and Cyperaceae such as leaf sheaths and the presence of intercalary meristems at every node suggest the same mechanical properties and, accordingly, the same functionality. Meristems are zones of tissue formation, which constitute areas of weakness along the entire culm and provide the basis for rapid shoot elongation. Leaf sheaths clasp the culm preventing the shoot from breaking, ensuring the rigidity to grow erectly and to avoid damage of the meristematic tissue. The mechanical influence of leaf sheaths was investigated in members of Poaceae, Juncaceae, and Cyperaceae in the flowering stage. Mechanical properties of *Poa araratica*, *Bromus erectus*, *Arrhenatherum elatius* (Poaceae), *Luzula nivea* (Juncaceae), and *Carex arctata* (Cyperaceae) were determined in three-point bending before and after the removal of leaf sheaths. The presence of leaf sheaths results in smoothing the distribution of flexural rigidity and therefore avoids stress peaks. The achieved maxima of relative contribution of leaf sheaths to entire flexural rigidity ranged from 55% up to 81% for Poaceae, 72% for *C. arctata*, and 40% for *L. nivea*. Across the investigated families, the mechanical role of leaf sheaths could be verified as essential for culm stability during development and beyond.

1. Introduction

One characteristic of the morphological feature of grass culms is the presence of intercalary meristems above every node. The growth is not restricted to one apical meristem; rather, the culm is able to grow at every internode [1, 2]. Therefore, the culm is showing frequent alternations of weak, nonreinforced and stiff, fully developed tissues below and above the nodes particularly during development. The increased vertical growth speed, enabled through many areas of growth, comes at a cost. Additional stabilizing structures to reinforce the less stable meristems have to be provided. The leaf sheaths envelope the culm in the most fragile section avoiding possible damage by providing the crucial stiffness [3–5].

Culms of Poaceae without regard to their leaf sheaths are studied concerning flexural rigidity, Young's modulus, the relation between bending and torsional stiffness, and with regard to anatomy [2, 6–10]. The mechanical significance of leaf sheaths especially for necessity in development of culms was shown for *Arundo donax* [11] and *Miscanthus* [12].

Zebrowski [5] paid special attention on this aspect in *Triticale* as well as Niklas [3, 4] for *Arundinaria tecta* and *Avena sativa*. The contribution of leaf sheaths to the entire rigidity of the culm of *Arundinaria tecta* reached 33% on average [4]. Previous work, in which Young's modulus and flexural rigidity of two cultivars of *Avena sativa* were measured, revealed an increasing Young's modulus of the culm over a period of two weeks before reaching the state of anthesis [3]. Interestingly, even after anthesis the leaf sheaths continue to play a significant mechanical function according to their higher second moment of area due to its geometric contribution to flexural rigidity. Similar results were obtained from *Tiricale* [5], in which even three weeks after anthesis the leaf sheaths still contribute significantly to flexural rigidity of the entire culm.

Although structurally very similar to grasses, mechanical properties of rushes (Juncaceae) or sedges (Cyperaceae) have not been studied in detail. Ennos [13] and Etnier [14] studied Cyperaceae, pointing out a characteristic triangular cross-section of culms. Neither biomechanical properties of Juncaceae nor a contribution of leaf sheaths to bending stiffness of Juncaceae and Cyperaceae have been characterized so far.

Our study aims at comparing the function of leaf sheaths in several grasses and grass-like plants from different families. In particular, the stabilizing role of leaf sheaths along the culm, as described for other Poaceae, will be addressed.

2. Material and Methods

For our study, five species from three families of the order of Poales were chosen to cover different plant heights: *Poa araratica* Trautv., *Bromus erectus* Huds., *Arrhenatherum elatius* (L.) P. Beauv. ex J. Presl & K. Presl (Poaceae); *Luzula nivea* L. (DC) (Juncaceae); and *Carex arctata* Boott. from Cyperaceae. *Luzula nivea*, *Carex arctata*, and *Poa araratica* were taken from the Botanic Garden Dresden. *Bromus erectus* and *Arrhenatherum elatius* were available directly at the meadow in front of the Institute of Botany of the TU Dresden. Plants were harvested at full stage of anthesis, since shoot elongation stops at this point. *B. erectus*, *C. arctata*, and *A. elatius* represent tall grasses with culm lengths up to ca. 1.5 m, while those of *P. araratica* in contrast reach up to 0.5 m in length and *L. nivea* up to 0.8 m, respectively.

Flexural stiffnesses of stems were analysed in three-point bending tests using a standard testing machine (Zwick/Roell Z 2.5). Even though the risk of ovalisation and local buckling of the hollow culms at the supports of testing device is obvious, this method proved to be the most favourable compared to cantilever bending or Euler buckling tests [11]. In particular, for avoiding dehydration of fresh tissue while preparing and measuring, the fast and simple three-point bending method is best suited. In comparison to cantilever bending, it operates without embedding or clamping one end of the culm. Further, damage may occur directly at the meristematic zones when applying Euler buckling due to the highly anisotropic structure of culms and leaf sheaths, particularly considering open or overlapping leaf sheaths, may cause damage directly at the meristematic zones when applying Euler buckling.

The influence of leaf sheaths on flexural rigidity was investigated by measuring culm segments before and after removal of leaf sheaths in three-point bending. Only the leaf blades were removed before testing. In order to minimize the influence of shear, tests were carried out with a minimum span-to-depth ratio of 20 [15]. The specimens length was set between 40 and 70 mm, and accordingly the distance between the supports ranged depending on species from 60 mm for *A. elatius* and *B. erectus*, 40 mm for *L. nivea* and *C. arctata* and 30 mm for *P. araratica*, respectively. The entire culm was divided in as much segments as possible. All stems were tested in fresh condition and used for testing within two hours after harvesting. The flexural rigidity was calculated from the slope of the linear regression of the applied bending force versus deflection. For further evaluation, the relative contribution of leaf sheaths to entire flexural rigidity is calculated by dividing flexural rigidity of the culm without leaf sheath through flexural rigidity of the culm with leaf sheath. The tests were performed quasistatically at a testing speed of 1 mm/min,

reducing speed-dependent influences. Sources of errors are reduced by constant work conditions. Keeping the material in water prior to measurement reduced the influence of dehydration. However, biological materials characteristically show a broad variation of mechanical properties with respect to environmental conditions, for example, temperature, wind exposition, or soil nutrient content. The obtained data were evaluated by Kruskal-Wallis test with regard to differences of segments with and without leaf sheaths. P values less than 0.05 were considered significant.

3. Results

For each species, six individuals were tested. All representatives of Poaceae possess four to five internodes. The lengths of the internodes increase while the overlap of leaf sheaths decreases towards the apex. The basal leaf sheaths clasped the culms of Poaceae over the whole internode length (Table 1). For all these plants, the absolute maximum leaf sheath lengths are found at the apical internode. In contrast, *C. arctata* has three to four and *L. nivea* six to seven internodes. Compared to the grasses, internode and leaf sheath lengths of *L. nivea* are rather constant, while in *C. arctata* the longest internode and leaf sheath are found above the basal one with a subsequent decrease towards the apex. The leaf sheaths of *C. arctata* extend over the entire basal internode. *L. nivea* does not show a leaf sheaths overlap at any internode as the other species do. *B. erectus*, *C. arctata*, and *L. nivea* have closed leaf sheaths, whereas the others two species have open leaf sheaths.

The values for flexural rigidity obtained for Poaceae are very similar. The leaf sheath's function as stiffening structure has been lost or is only minimal in the basal internodes (leaf sheaths occasionally are dried and fall off), while it increases towards the apical internode. There, the relative contribution of leaf sheaths to entire flexural rigidity is at a maximum and contributes significantly to the entire rigidity of the internodes (Table 1). The relative contributions range on average from 55 (±10)% for *A. elatius* and about 76 (±8)% for *P. araratica* to 81 (±18)% for *B. erectus* (Table 1).

In contrast, the leaf sheaths of *C. arctata* account for 66 (±14)% to 72 (±26)% of the flexural rigidity at the meristematic zones of all internodes (Table 1). The leaf sheath's contribution to flexural rigidity of *L. nivea* equally reaches ca. 40 (±11)% in the three apical internodes while decreasing towards the base (Table 1). All species exhibit a high natural variability, which most probably arises from environmental influences.

Representative gradients of flexural rigidity for each species depending on distance from base and position within internodes are shown in Figure 1. The rigidity of culms including leaf sheaths (black dots) is always higher than that of the culms lacking leaf sheaths (circles); that is, leaf sheaths increase the rigidity of the culm segment which they cover (Table 2). The limited length of leaf sheaths did not always allow the measuring for more than one segment per internode (*B. erectus*, *P. araratica*, and *L. nivea*). In these cases, it was not possible to detect a gradient (Figures 1(b), 1(c), and 1(d)). In *A. elatius* and *C. arctata*, however,

TABLE 1: Internode length, overlap of leaf sheaths, and relative contribution of leaf sheath to flexural rigidity of entire culm of all species. Standard deviation is bracketed and expressed as a percentage. Asterisks denote significant differences (Kruskal-Wallis test) between the segment below the node and the first segment above the meristem with and without leaf sheaths.

	A. elatius	B. erectus	P. araratica	L. nivea	C. arctata
Culm length [mm]	1398 (126)	1053 (109)	467 (29)	683 (54)	980 (74)
Number of internodes	5	4-5	5	6-7	3-4
Internode length (% of entire length)					
Internode					
1 (base)	6.3 (3.3)	3.2 (1.1)	5.0 (2.3)	9.8 (5.9)	10.4 (5.7)
2	12.0 (3.2)	8.5 (3.8)	10.7 (1.4)	14.1 (3.3)	41.1 (3.0)
3	16.1 (0.9)	17.2 (4.9)	14.5 (0.6)	16.5 (1.8)	32.4 (4.5)
4	21.3 (2.0)	30.3 (5.9)	21.5 (1.3)	16.5 (1.8)	16.2 (1.8)
5	44.3 (6.9)	40.8 (9.1)	44.0 (1.5)	16.0 (1.3)	
6				15.9 (2.5)	
7				11.1 (7.8)	
Overlap of leaf sheaths (%)					
Internode					
1 (base)	111.1	181.5 (38.5)	122.0 (23.3)	66.9 (10.4)	143.3 (90.3)
2	84.4 (15.9)	111.1 (53.4)	78.2 (4.5)	56.6 (13.1)	27.5 (1.7)
3	72.8 (9.4)	62.4 (15.5)	63.9 (3.6)	41.9 (7.0)	24.7 (6.5)
4	66.7 (8.3)	41.9 (5.0)	44.4 (3.4)	38.0 (4.5)	41.4 (13.5)
5	45.2 (7.1)	34.1 (4.3)	22.6 (1.3)	36.2 (4.4)	
6				33.2 (5.8)	
7				49.8 (25.6)	
Relative contribution of leaf sheath to flexural rigidity of entire culm (%)					
Internode					
1 (base)	—	—	10.3	5.64	71.9 (2.8)
2	—	—	12.2 (9.9)	6.2 (4.7)	72.3 (26.0)*
3	11.1 (9.5)	33.6 (8.1)	21.8 (7.3)	11.0 (9.8)*	71.7 (19.4)*
4	26.8 (16.4)	55.7 (24.8)*	30.2 (9.2)*	21.3 (10.8)*	65.9 (14.2)
5	54.9 (10.5)*	81.3 (18.1)*	76.3 (7.6)*	39.9 (10.6)*	
6				40.0 (5.9)*	
7				37.2 (11.9)*	

the internode could be subdivided revealing that the contribution of leaf sheaths to flexural rigidity decreases with increasing distance from the meristem. Further on, A. elatius, B. erectus, and C. arctata lose the basal leaf sheaths, and the rigidity of the culms exhibits higher values compared to the apex (e.g., A. elatius 1.9 Nm^2 ($\pm$0.4) at base and 0.5 Nm^2 ($\pm$0.2) at apex, see Table 2). Despite high rigidity of basal internodes in P. araratica and L. nivea, the leaf sheaths are still present and further increase the entire rigidity, although only to a low amount. A steep increase of the culm's rigidity along the internode in apical direction (2nd and 3rd internodes of B. erectus and C. arctata) shows the development from fresh unreinforced tissue in active meristems into fully differentiated tissue. Small or hardly recognisable gradients consequently are observed in mature culms (from 1st up to 4th internode of A. elatius and P. araratica).

4. Discussion

Culms of grasses and grass-like plants possess an extraordinary slenderness ratio. The evolutionary benefit is to gain a maximum height of the inflorescences and thus of the infructescences, which guarantees a greater distance for seeds to be dispersed [10, 13]. However, that architecture exhibits some risks with regard to mechanical stability. With a high slenderness ratio, the grass culms are at risk of buckling due to wind load and own weight reducing the success of dispersal [3, 4]. Thus, the plants have evolved certain strategies in order to survive successfully. One method to handle height and stability is supported growth in a dense stand, for example, described for cereals [7, 10]. In absence of supporting plants, the height is restricted by the response of plants to environmental influences [7]. Another characteristic is the multiple growth in each of the intercalaryy meristems that account for rapid shoot elongation. For this reason grasses, and grass-like plants typically exhibit abrupt transitions from fully developed stiff to weak meristematic tissues within a short distance at every node. In agreement with Niklas [3, 4], Spatz et al. [11], and Zebrowski [5], we found that the young culms have a low flexural rigidity near the meristem above the node, whereas the culm below the node reaches full rigidity. This mechanical weak point is compensated by leaf

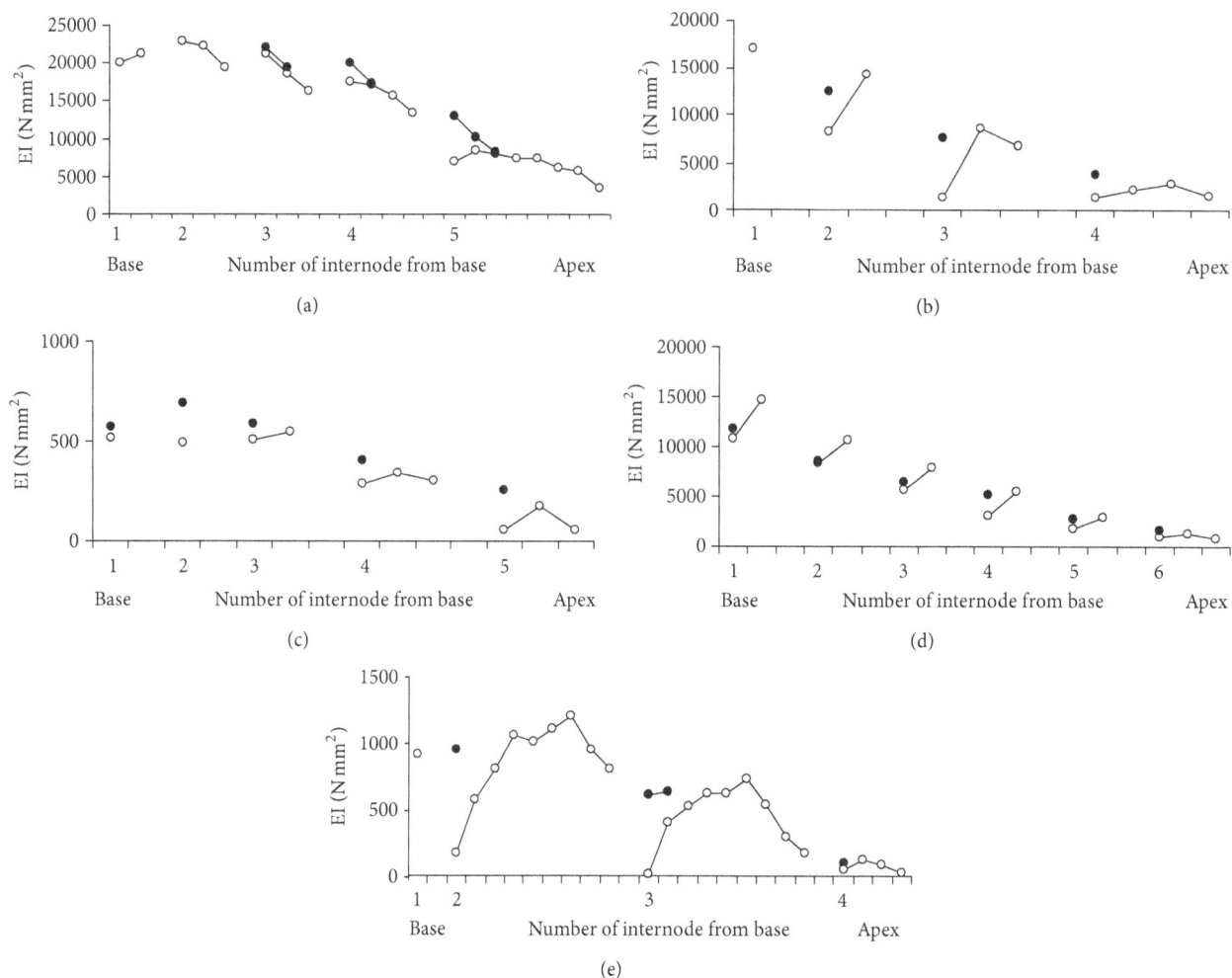

FIGURE 1: Representative gradients of flexural rigidity for individuals of (a) *A. elatius* (plant height h = 1473 mm), (b) *B. erectus* (h = 1071 mm), (c) *P. araratica* (h = 509 mm), (d) *L. nivea* (h = 662 mm), and (e) *C. arctata* (h = 1046 mm) with (•) and without (○) leaf sheath. Lines are interrupted where nodes are located.

sheaths, which are able to provide additional rigidity. In our investigations, the achieved maxima of relative contribution of leaf sheaths to entire flexural rigidity ranged from about 81% for *B. erectus*, 76% for *P. araratica*, 72% for *C. arctata*, and 55% for *A. elatius* to 40% for *L. nivea*. These data are in a same range as, for example, from Niklas [4] (33% relative contribution) or Zebrowski [5] (almost 100% relative contribution next to the meristem). Consequently, risk of failure due to local weakness is reduced by the presence of stiff leaf sheaths. The highly varying flexural rigidity along the plant axis of the leafless culm gets smoothed through the stabilization by leaf sheaths. We found this relationship to be ubiquitous, independent of plant heights, internode, and leaf sheath lengths, or differing in meristematic activity. Once the culm development is completed, high overall rigidity depends to a lesser extent on leaf sheaths, as shown for some basal internodes. For example, the achieved minima of relative contribution of leaf sheaths ranged from 6% for *L. nivea.* to 10% for *P. araratica*. *A. elatius* and *B. erectus* shed their basal leaf sheaths. Our studies have shown that all investigated species depend on leaf sheaths for protection and

supporting of the mechanical weak points that occur during development of the culm. The relative contribution of leaf sheaths to entire flexural rigidity not only differs between the species but also depends on the position of the internodes within the culm as well as on the growth stages. For Poaceae, these contributions are distinctly higher towards the apex than towards the base. We also found that the zones of high relative contribution to flexural rigidity represent zones of greater meristematic activity. But the contribution may also remain on fully developed culms and still support the fully developed plant (e.g., in base of *P. araratica*). Differences due to open or closed leaf sheaths are not identifiable.

Which is new now relating to the representatives of rushes and sedges? *C. arctata* exhibits a flexural rigidity of nearly zero next to the meristem of every node, and the contribution of the leaf sheath is high at these positions. Consequently, the meristems may be active along the entire stem at anthesis, in contrast to the Poaceae. Anatomical differences between *C. arctata* and the grasses are a smaller overlap of leaf sheaths, without influence on contribution to entire stability close to the meristem. But the gradient of flexural rigidity along

TABLE 2: Mean of flexural rigidity of entire internodes with and without leaf sheaths. Standard deviation is bracketed and expressed as a percentage.

	A. elatius	B. erectus	P. araratica	L. nivea	C. arctata
	Flexural rigidity of entire internodes without leaf sheaths (Nm^2)				
Internode					
1 (base)	1.92 (19.2)	1.31 (14.5)	0.05	1.43	0.08 (56.0)
2	1.70 (25.9)	2.33 (67.5)	0.05 (10.3)	1.20 (29.4)	0.08 (39.7)
3	1.26 (34.1)	1.16 (61.2)	0.05 (11.0)	0.78 (37.1)	0.03 (60.0)
4	0.92 (43.7)	0.62 (75.9)	0.03 (11.8)	0.50 (39.3)	0.01 (52.0)
5	0.49 (42.9)	0.28 (34.9)	0.01 (67.5)	0.28 (50.0)	
6				0.17 (50.1)	
7				0.07 (45.3)	
	Flexural rigidity of entire internodes with leaf sheaths (Nm^2)				
Internode					
1 (base)	—	—	0.06	1.35	0.09 (32.6)
2	—	—	0.06 (15.3)	1.01 (30.8)	0.08 (41.1)
3	1.27 (42.2)	1.28 (41.2)	0.06 (5.5)	0.77 (32.2)	0.06 (32.3)
4	1.01 (43.3)	0.79 (32.2)	0.04 (6.5)	0.51 (38.7)	0.03 (45.8)
5	0.65 (41.2)	0.42 (8.1)	0.02 (18.2)	0.34 (45.2)	
6				0.20 (45.8)	
7				0.10 (44.5)	

the stem seems less smoothed (Figure 1). Pointing out the overall rigidity of C. arctata is about one order of magnitude lower than the Poaceae of same height (resp. in the order of the much smaller P. araratica); the smoothing character is appropriate.

Every species has its individual gradient. L. nivea with more short internodes, relative short leaf sheaths, and small leaf sheath overlap shows a lower contribution of leaf sheaths to the entire flexural rigidity than the other investigated species. Particularly, in basal internodes the smoothing character of leaf sheaths is low. Nevertheless, leaf sheaths of L. nivea persist still at the base and contribute to all internodes rigidity during anthesis such as P. araratica. These results confirm an equal functional morphology of Cyperaceae and Juncaceae as was found for Poaceae and are in agreement with earlier studies.

However, all these results denote only a snapshot during development of fast-changing material properties. Therefore, the demonstrated differences between the species may be a result of different systematic groups (Cyperaceae, Juncaceae, and Poaceae) or due to slightly different ontogenetic stages.

Summing up, whereas the leaf sheaths exhibit full flexural rigidity at the meristem, the rigidity of the culm shows minimal flexural rigidity at the meristem which then increases towards the following node. The maximum contribution of leaf sheaths to flexural rigidity is supposed to occur during the growth of the culm, when the activity of the meristems is high.

Furthermore, failure of the culm is reduced by the leaf sheath construction, which reduces ovalisation [3]. Inspired by the mechanical role of leaf sheaths in rattan palms [16], another great benefit of grasses may be obvious. Under excessive bending, leaf sheaths may break at first. Assuming the enveloped culm still has the required flexibility, it bends easily and remains intact. Its flexibility ensures the survival of the culm and provides the possibility to reorientate for further growth. It could be observed that, before anthesis, lodged cereals are capable to reerect the upper internodes [17, 18]. The loss of the leaf sheath is a programmed and calculable investment for the plant. After full development of culm and reduced contribution of leaf sheaths to flexural rigidity, the stability contribution by leaf sheaths is negligible and the necessity of fail-safe properties becomes dispensable when the maturity ends, for example, in the basal internodes of Poaceae. This leaf-sheath-culm construction contains a fail-safe ability to survive break damage and consequently the possibility to reerect itself during elongation.

In summary, we showed the following.

(1) Leaf sheaths contribute significantly to the flexural rigidity of the entire culm at flowering stage.

(2) Alternating rigidity gradients within the whole culm length were compensated by the leaf sheaths.

(3) Leaf sheaths' contribution to flexural rigidity were found to be crucial for stability in the species of Poaceae as well as to the studied species of Juncaceae and Cyperaceae.

Acknowledgments

This work was financed by the European Union and the Free State of Saxony (ECEMP E2 - 13927/2379). The authors thank the Botanic Garden Dresden for providing them with plants. They acknowledge Sandrine Isnard, Lena Frenzke, Marko Storch, and Hanns-Christof Spatz for helpful discussions and constructive advices.

References

[1] K. J. Niklas, "Plant height and the properties of some herbaceous stems," *Annals of Botany*, vol. 75, no. 2, pp. 133–142, 1995.

[2] H. Prat, "Recherches sur la structure et le mode de croissance des chaumes," *Annales des Sciences Naturelles Botanique*, vol. 10, pp. 81–145, 1935.

[3] K. J. Niklas, "The mechanical significance of clasping leaf sheaths in grasses: evidence from two cultivars of *Avena sativa*," *Annals of Botany*, vol. 65, no. 5, pp. 505–512, 1990.

[4] K. J. Niklas, "The mechanical roles of clasping leaf sheaths: evidence from *Arundinaria tecta* (Poacea) shoots subjected to bending and twisting forces," *Annals of Botany*, vol. 81, no. 1, pp. 23–34, 1998.

[5] J. Zebrowski, "Complementary patterns of stiffness in stem and leaf sheaths of *Triticale*—measurements of ultrasound velocity," *Planta*, vol. 187, no. 3, pp. 301–305, 1992.

[6] L. S. Evans, Z. Kahn-Jetter, C. Marks, and K. R. Harmoney, "Mechanical properties and anatomical components of stems of 42 grass species," *Journal of the Torrey Botanical Society*, vol. 134, no. 4, pp. 458–467, 2007.

[7] M. J. Crook and A. R. Ennos, "Mechanical differences between free-standing and supported wheat plants, *Triticum aestivum* L," *Annals of Botany*, vol. 77, no. 3, pp. 197–202, 1996.

[8] G. H. Dunn and S. M. Dabney, "Modulus of elasticity and moment of inertia of grass hedge stems," *Transactions of the American Society of Agricultural Engineers*, vol. 39, no. 3, pp. 947–952, 1996.

[9] J. Grace and G. Russell, "The effect of wind on grasses: III. Influence of continuous drought or wind on anatomy and water relations in festuca arundinacea schreb," *Journal of Experimental Botany*, vol. 28, no. 2, pp. 268–278, 1977.

[10] J. Zebrowski, "Dynamic behaviour of inflorescence-bearing *Triticale* and *Triticum* stems," *Planta*, vol. 207, no. 3, pp. 410–417, 1999.

[11] H. C. Spatz, H. Beismann, F. Brüchert, A. Emanns, and T. Speck, "Biomechanics of the giant reed *Arundo donax*," *Philosophical Transactions of the Royal Society B*, vol. 352, no. 1349, pp. 1–10, 1997.

[12] K. Kaack and K. U. Schwarz, "Morphological and mechanical properties of *Miscanthus* in relation to harvesting, lodging, and growth conditions," *Industrial Crops and Products*, vol. 14, no. 2, pp. 145–154, 2001.

[13] A. R. Ennos, "The mechanics of the flower stem of the sedge *Carex acutiformis*," *Annals of Botany*, vol. 72, no. 2, pp. 123–127, 1993.

[14] S. A. Etnier, "Twisting and bending of biological beams: distribution of biological beams in a stiffness mechanospace," *The Biological Bulletin*, vol. 205, no. 1, pp. 36–46, 2003.

[15] J. F. V. Vincent, *Biomechanics—Materials: A Practical Approach*, IRL Press at Oxford University Press, Oxford, UK, 1992.

[16] S. Isnard and N. P. Rowe, "Mechanical role of the leaf sheath in rattans," *New Phytologist*, vol. 177, no. 3, pp. 643–652, 2008.

[17] R. A. Fischer and M. Stapper, "Lodging effects on high-yielding crops of irrigated semidwarf wheat," *Field Crops Research*, vol. 17, no. 3-4, pp. 245–258, 1987.

[18] P. M. Berry, M. Sterling, J. H. Spink et al., "Understanding and reducing lodging in cereals," *Advances in Agronomy*, vol. 84, pp. 217–271, 2004.

16

Responses of Green Leaves and Green Pseudobulbs of CAM Orchid *Cattleya laeliocattleya* Aloha Case to Drought Stress

Jie He, Hazelman Norhafis, and Lin Qin

Natural Sciences and Science Education Academic Group, National Institute of Education, Nanyang Technological University, 1 Nanyang Walk, Singapore 637 616

Correspondence should be addressed to Jie He; jie.he@nie.edu.sg

Academic Editor: Akira Nagatani

This study examined the responses of green leaves (GL) and green pseudobulbs (GPSB) of CAM orchid *Cattleya laeliocattleya* Aloha Case to drought stress. After being subjected to drought stress, the decrease in water content (WC) was much greater in GPSB than in GL, indicating that GPSB facilitated a slow reduction in the WC of GL. This finding was further supported by the result of relative water content (RWC) of GL, which started to decrease only after 3 weeks of drought stress. Decreases of midday F_v/F_m ratios of GL occurred in all plants. However, the degrees of decrease were much greater in drought-stressed GL than in well-watered GL. Reduced F_v/F_m ratio (<0.8) at early morning was observed in drought-stressed GL after 3 weeks of treatments. Decreases in total chlorophyll (Chl) content, electron transport rate (ETR), photochemical quenching, qP, and nonphotochemical quenching, qN, were severer in GPSB than in GL after drought treatment. CAM acidity was significantly lower in both GL and GPSB after 2 weeks of drought treatment compared to well-watered plants. However, decrease of CAM acidity was smaller in GL than in GPSB. These results suggest that both GL and GPSB of CAM orchid *Cattleya* plantswere susceptible to drought stress.

1. Introduction

Epiphytic orchids found in tropical environments are directly or indirectly exposed to natural air currents and solar radiation and receive only intermittent rains. In addition to coping with rapid changes in natural air current and light intensity, there is also the need for epiphytic orchids to adapt to periodic drought [1]. Many epiphytic orchids develop morphological features to conserve and/or store water to cope with drought. A morphological feature includes the presence of swollen stems called pseudobulbs that serve as a reserve for water and carbohydrates [2].

Among the many abiotic factors involved in the survival of epiphytes, water availability is probably the most important environmental factor limiting growth and survival of epiphytes [3]. Thus, tolerance to water deficit is a decisive factor in their survival. The presence of the pseudobulb may facilitate a slow reduction in the leaf water content and decline in water potential during a period of drought [1, 4, 5]. The pseudobulb is characterized by the presence of very thick cuticle, absence of stomata, and the abundance of water-storing cells [6]. This makes the pseudobulb an integral organ in the survival and growth of orchids [7, 8]. This is even more so for green pseudobulbs (GPSB) which can photosynthesise and hence contribute positively to carbon balance.

Although large amounts of water and carbohydrates are stored in the GPSB, many epiphytic orchids are sensitive to prolonged water deficit [9, 10]. Under conditions of water deficit, photosynthetic capacities in terms of utilization of radiant energy and carbon fixation of GL and GPSB may be reduced under drought stress. However, photosynthetic responses of GL and GPSB of CAM orchid plants to drought stress are poorly understood. The main objective of this study was first to investigate the effects of drought stress on the water status of both GL and GPSB of CAM orchid *Cattleya laeliocattleya* Aloha Case. To investigate the water status of plants, WC was determined for both GPSB and GL after the plants were subjected to drought stress. The massive reduction of WC in plants can have a negative impact on cell expansion and cell growth, resulting in overall growth reduction [11]. Compared to WC, RWC is considered a better indicator of plant water status, which measures the absolute content of water in fresh plant tissues relative to the maximum WC in the tissues at full turgidity [11]. It has been reported that well-watered plants have RWC values between 85% and 95% whereas RWC can decrease up to 40% in severely drought-stressed plants [12]. Therefore, RWC was determined but only for GL as it is not feasible to do this measurement for GPSB. In this study, photosynthetic utilization of radiant energy was also addressed through the measurements of photosynthetic pigments and the various chlorophyll fluorescence parameters [13]. The decrease in photosynthetic light use efficiency caused by drought could affect the carbon fixation measured by CAM acidity, and ultimately, the growth and development of plants [14, 15]. Hence, CAM acidity was also determined after the plants were subjected to drought stress.

2. Materials and Methods

2.1. Plant Material. Mature plants of CAM orchid *Cattleya laeliocattleya* Aloha Case were obtained from a commercial nursery. Each plant consisted of a single succulent green leaf (GL), a green pseudobulb (GPSB), and varying numbers of roots. The plants were repotted with around 2-3 plants per pot and allowed to acclimatize for four weeks in the plant physiology house at the National Institute of Education, Singapore, with a maximal photosynthetic photon flux density (PPFD) of 700 to 800 μmol m^{-2} s^{-1} and a daily temperature which ranged from 24°C to 33°C. During the four weeks of acclimatization, all plants were watered twice daily at 0800 h and 1700 h, respectively. At the end of the week, fertilizers and pesticides were sprayed to the leaves.

2.2. Drought Stress Treatment. The experiment was designed to study the course of changes in physiology occurring in both GL and GPSB during drought stress. After four weeks of acclimatization, one half of the plants were subjected to drought stress (no watering), while the other half of the plants were kept well watered (control) twice daily as described above. The period of drought stress was four weeks.

2.3. Measurement of WC. Fresh weight (FW) of plant tissues was recorded using a weighing balance (Sartorius, Fisher General Scientific Private Limited, Singapore) before being wrapped in aluminium foil and dried in an oven at 105°C for 48 h. Samples were then weighed to determine dry weight (DW). WC was expressed as (FW − DW)/FW.

2.4. Measurement of RWC. A small portion of GL was cut and immediately weighed with an analytical balance to determine FW. The sample was floated on water in the dark for 24 h prior to measurement of its saturated weight (SW). The sample was then dried in the oven at 105°C for 48 h to obtain its DW. RWC was calculated as RWC = (FW − DW)/(SW − DW) × 100%.

2.5. Measurement of Chl Fluorescence F_v/F_m Ratio. Readings were taken every two hours from 0800 to 1800 h after 1, 2,

3, and 4 weeks of drought stress treatment, respectively. All measurements were made with Plant Efficiency Analyzer (PEA, Hansatech Instruments Ltd., England). The GL were predarkened with clips for 15 min prior to measurement and then placed under the light pipe and irradiated with the pulsed lower intensity-measuring beam to measure F_0, initial Chl fluorescence. F_m, maximum Chl fluorescence, was assessed by 0.8 s of saturated pulse (>3000 μmol photon m^{-2} s^{-1}). The variable fluorescence yield, F_v, was determined by $F_m − F_0$. The efficiency of excitation energy captured by open photosystem II (PSII) reaction centers in dark adapted plant samples was estimated by the fluorescence (F_v/F_m ratio).

2.6. Measurement of Electron Transport Rate (ETR), Photochemical Quenching (qP), and Nonphotochemical Quenching (qN). For these measurements, GL and GPSB were harvested at 0730 h and were predarkened under a black cloth for 15 min prior to measurements. ETR, qP, and qN were determined using the Imaging Pam Chl Fluorometer (Walz, Effeltrich, Germany) at 25°C under different PPFDs in the laboratory as described by He et al. [13].

2.7. Measurement of CAM Acidity (Titratable Acidity, TA). TA was determined at both 0800 h and 1800 h. 1 g of GL and GPSB was transferred into heat tolerant tubes containing 1 ml of distilled water (neutral pH) and immersed into a boiling water bath for 15 min. The tubes were allowed to cool and were subsequently titrated against 0.01 M sodium hydroxide solution, NaOH(aq), using three drops of phenolphthalein as an indicator. The volume of NaOH(aq) needed to reach the end point was recorded. The samples were then dried at 80°C to obtain the weight of dry matter (DM). TA was calculated using the formula TA = (0.01 × volume)/DM. CAM acidity, which was the dawn/dusk fluctuation of TA, was calculated from the difference between TA at 0800 h and 1800 h.

2.8. Measurements of Chl Content. 0.05 g of each GL and GPSB sample was weighed and cut into smaller pieces. Chl content was extracted from these samples using dimethylformamide and quantified spectrophotometrically at wavelengths of 647 and 664 nm [16].

2.9. Statistical Analysis. A t-test was used to test for the differences between well-watered and drought-stressed GL or GPSB (MINITAB, Inc., Release 15, 2007).

3. Results

3.1. WC and RWC of GL and GPSB after Drought Stress Treatment. During a 4-week period of drought stress treatment, decrease in WC was much greater in GPBS than in GL of drought-stressed plants. For instance, compared to well-watered plants, significant decrease of WC was observed in GPSB of drought-stressed plants after two weeks of treatment ($P < 0.05$) while reduced WC was observed in GL only after 3 weeks ($P < 0.05$) of drought stress (Figure 1(a)). It was also shown that RWC of GL was still about 80% after 2 weeks of

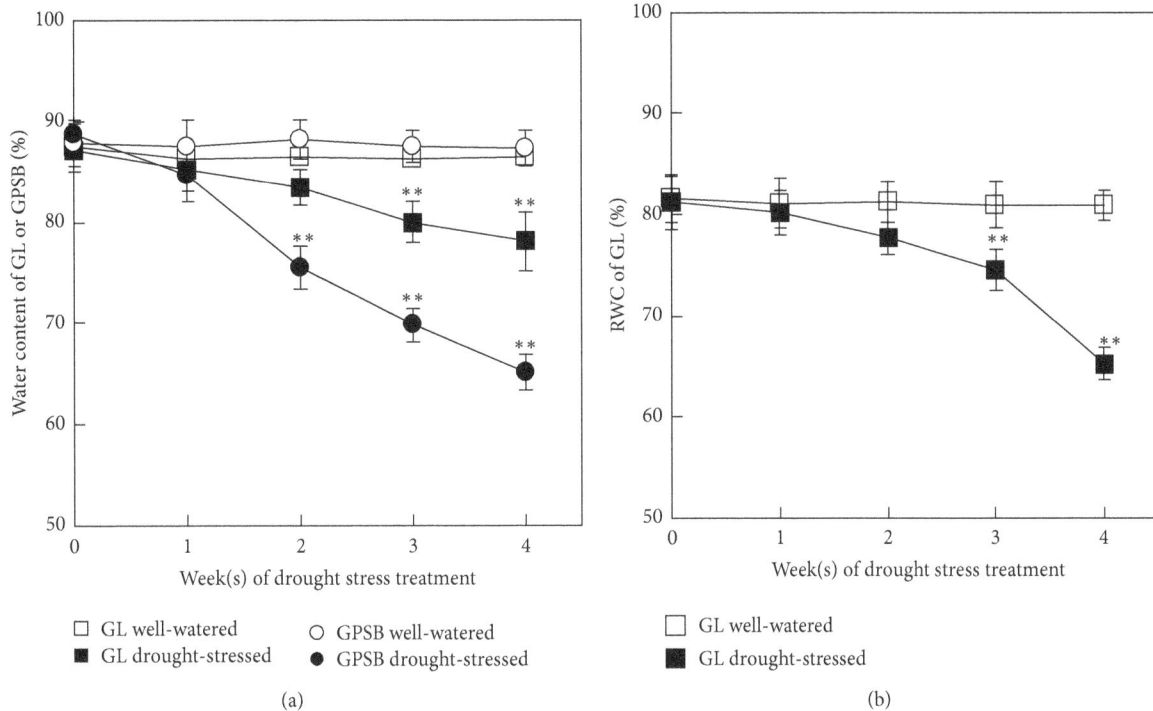

FIGURE 1: Changes in WC of GL and GPBS (a) and RWC (b) of GL during a drought stress period in well-watered and drought-stressed CAM orchid *Cattleya laeliocattleya* Aloha Case plants. Each point is the mean of 6 measurements from 6 different plants. Vertical bars represent standard errors. When the standard error bars cannot be seen, they are smaller than the symbols. ** Above the means of drought-stressed GL and GPSB are statistically different from their well-watered control plants, respectively ($P < 0.05$).

drought stress, and it only started to decrease after 3 weeks of drought treatment (Figure 1(b)).

3.2. Diurnal Changes of Chl Fluorescence F_v/F_m Ratio of GL after Drought Stress Treatment.

F_v/F_m ratio assesses whether the stress experienced by a plant affects photosynthetic light utilization [13]. As it was not feasible to measure F_v/F_m ratio from intact GPSB, diurnal changes of F_v/F_m ratio were only monitored from the same position of GL after drought stress treatment for 1, 2, 3, and 4 weeks, respectively. Although midday depression of F_v/F_m ratios was recorded in GL with lower values in drought-stressed GL than in well-watered GL after 1 and 2 weeks of treatment ($P < 0.05$), there were no significant differences in all F_v/F_m ratios during early morning from the same GL (Figures 2(a) and 2(b)). However, significant difference in F_v/F_m ratios between GL and GPSB was observed throughout the day ($P < 0.05$) after 3 and 4 weeks of treatment (Figures 2(c) and 2(d)). The differences in F_v/F_m ratios between well-watered and drought-stressed GL were much greater after 4 weeks (Figure 2(d)) than after 3 weeks (Figure 2(c)) of treatment.

3.3. ETR, qP, and qN of GL and GPSB after Drought Stress Treatment.

Differences in photosynthetic light utilization between GL and GPSB after drought stress treatments were studied by measuring ETR, qP, and qN under different PPFDs in the laboratory. The measurements were carried out after 2 and 3 weeks as samples of GPSB were severely damaged

after 4 weeks of treatment. For well-watered plants, ETR of GL increased with increasing PPFD from 25 to 600 μmol m^{-2} s^{-1} and then decreased with further increasing PPFD when they were measured under PPFDs > 600 μmol m^{-2} s^{-1} (Figures 3(a) and 3(c)). For GPSB of well-watered plants, although their light response curves were similar to those of GL, their highest ETR values were recorded under PPFD of about 400 μmol m^{-2} s^{-1} (Figures 3(b) and 3(d)). There were no significant differences in ETR in GL between well-watered and drought-stressed plants ($P > 0.05$) after 2 weeks of drought stress (Figure 3(a)) while a significant lower ETR was measured under PPFDs > 250 μmol m^{-2} s^{-1} in drought-stressed GL than in well-watered GL after 3 weeks of treatment (Figure 3(c)). For GPSB, a significant lower ETR was measured in drought-stressed plants than in well-watered plants when they were measured under PPFDs greater than 200 and 400 μmol m^{-2} s^{-1}, respectively (Figures 3(b) and 3(d)). For both GL and GPSB, qP decreased with increasing PPFD for all plants (Figure 4). For GL, there were no differences in qP between well-watered and drought-stressed plants ($P > 0.05$) after 2 weeks of treatment (Figure 4(a)). However, qP values were significantly lower ($P < 0.05$) in GL of drought-stressed than that of well-watered plants (Figure 4(c)). For GPSB, qP values of drought-stressed plants were significantly lower ($P < 0.05$) in drought-stressed plants than in well-watered plants after 2 and 3 weeks of treatment (Figures 4(b) and 4(d)). These results show that GPSB have a lower light utilization compared to that of GL,

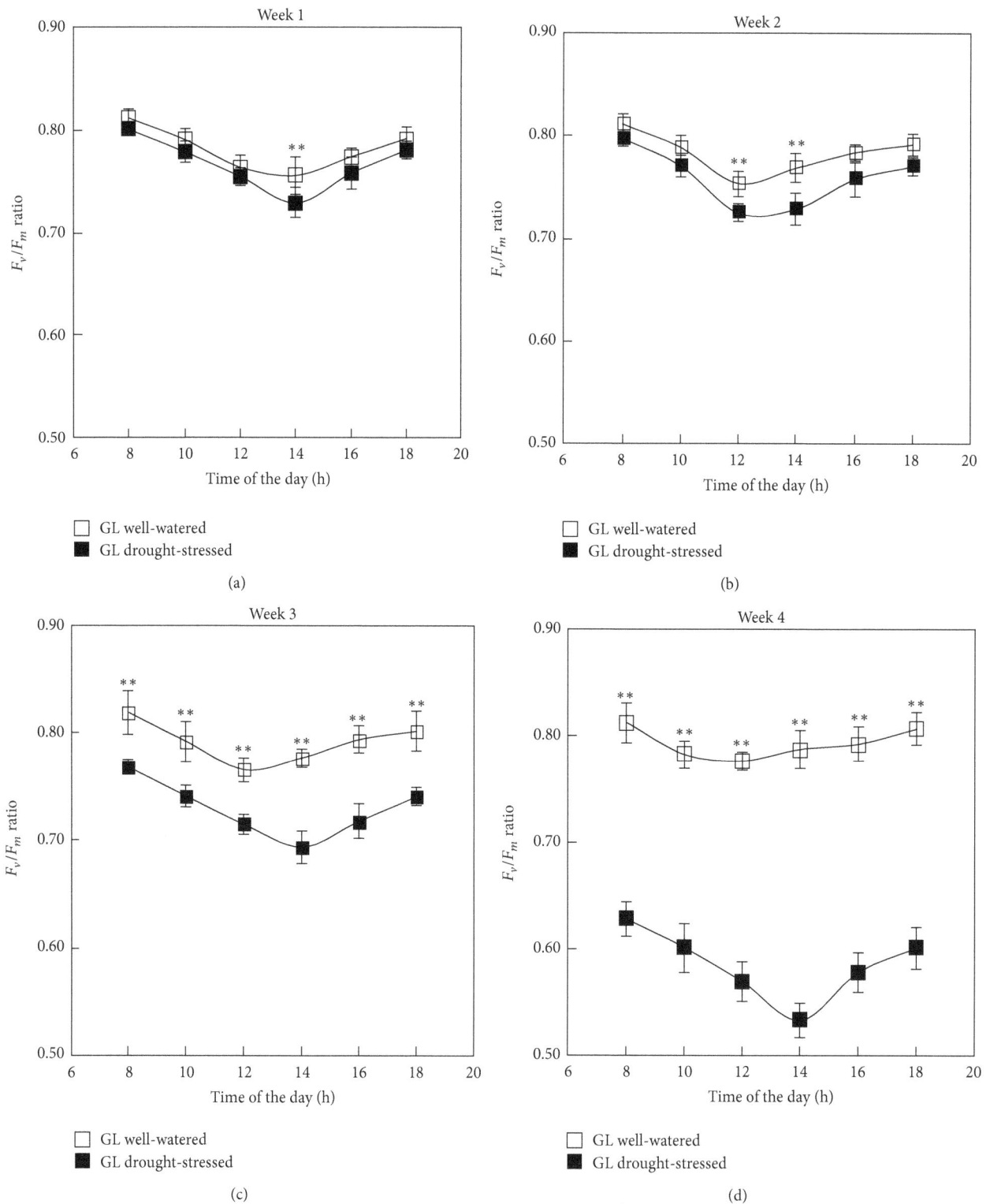

FIGURE 2: Diurnal changes of Chl fluorescence F_v/F_m ratio after CAM orchid *Cattleya laeliocattleya* Aloha Case plants were subjected to drought stress for 1, 2, 3, and 4 weeks. Means of 8 measurements from 8 different plants. Vertical bars represent standard errors. When the standard error bars cannot be seen, they are smaller than the symbols. **Above the means of GL and GPSB of well watered plants are statistically different from the drought-stressed plants ($P < 0.05$).

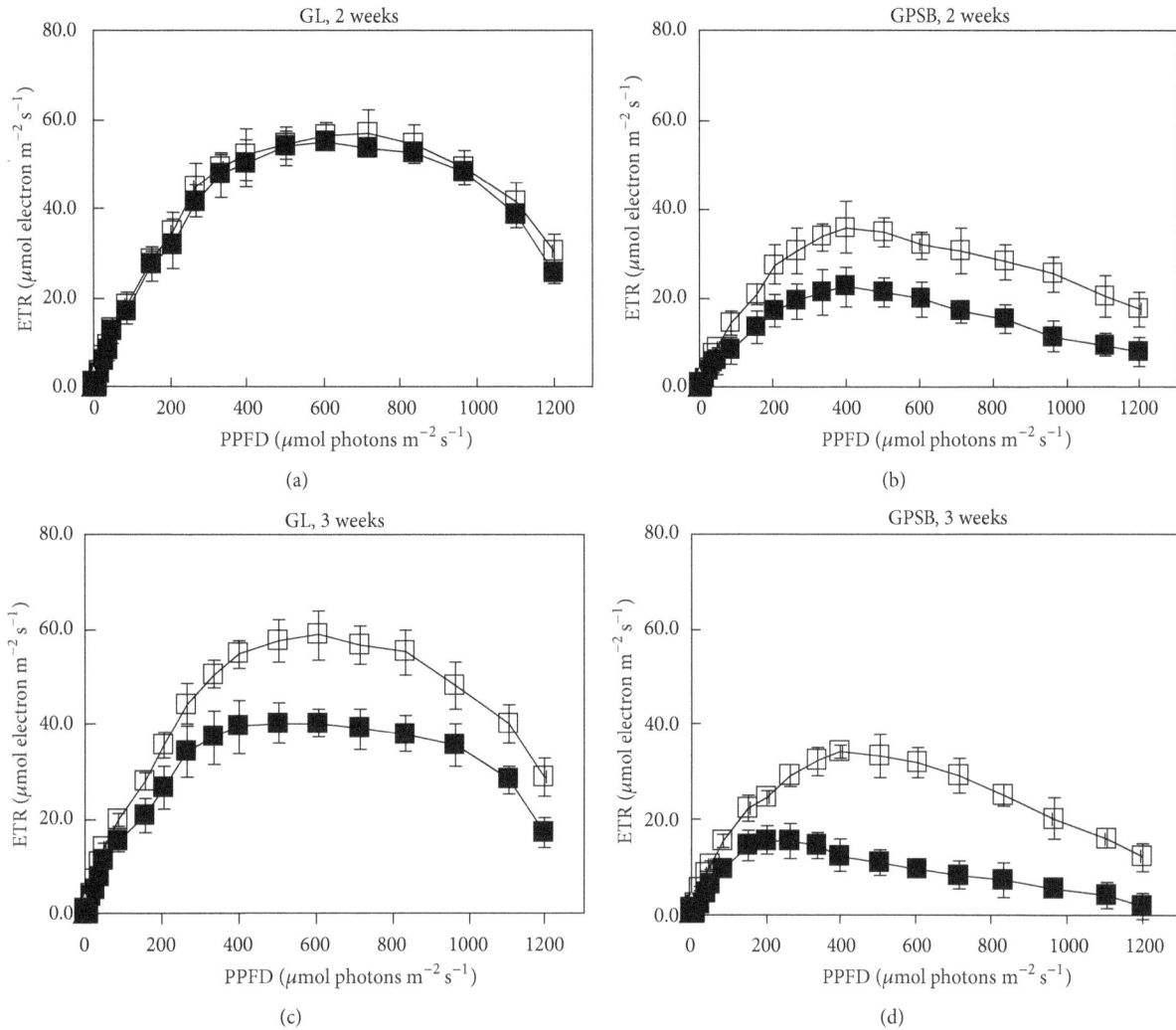

FIGURE 3: ETR of GL ((a), (c)) and GPSB ((b), (d)) under different PPFDs after CAM orchid *Cattleya laeliocattleya* Aloha Case plants were subjected to drought stress for 2 ((a), (b)) and 3 weeks ((c), (d)). Means of 5 measurements were measured from 5 different plants. Vertical bars represent standard errors. When the standard error bars cannot be seen, they are smaller than the symbols.

and the capability of using light energy was further decreased after being subjected to drought stress. For heat dissipation measured as qN, both GL and GPSB of all plants had similar treads. Values of qN increased with increasing PPFDs from 25 to 400 μmol m^{-2} s^{-1} (Figure 4) and after that they plateaued until the highest PPFD 6 of 1200 μmol m^{-2} s^{-1}. However, qN values were lower in GL and GPSB of drought-stressed plants compared to well-watered plants after 2 and 3 weeks of treatment under PPFDs greater than 400 μmol m^{-2} s^{-1}. The decreases were greater in drought-stressed GPSB than in drought-stressed GL, indicating that drought stress resulted in decreases of heat dissipation in both GL and GPSB with greater impact on GPSB than on GL.

3.4. Total Chl Content and CAM Acidity after Drought Stress Treatment. Two weeks after drought stress, there was no significant difference in total Chl content in GL between drought-stressed and well-watered plants ($P > 0.05$). However, a significant lower total Chl content was found in GL

of drought-stress plants compared to well-watered plants ($P < 0.05$) after 3 weeks of treatment (Figure 5(a)). For GPSB, drought-stressed plants had a significant lower total Chl content than well-watered plants after 2 and 3 weeks of treatment with greater decrease after 3 weeks of treatment (Figure 5(b)). It was interesting to see that CAM acidity of GL significantly decreased compared to well-watered plants after 2 weeks of treatment ($P < 0.05$). This decrease was much greater after 3 weeks of drought stress (Figure 5(c)). For GPSB, significant decreases of CAM acidity in drought-stressed plants were also recorded after 2 and 3 weeks of treatment (Figure 5(d)). Decreases of CAM acidity in drought-stressed GPSB were greater than in drought-stressed GL (Figures 5(c) and 5(d)).

4. Discussion

The results of this study showed that there were significant decreases in WC of GL and GPSB (Figure 1(a)) and RWC

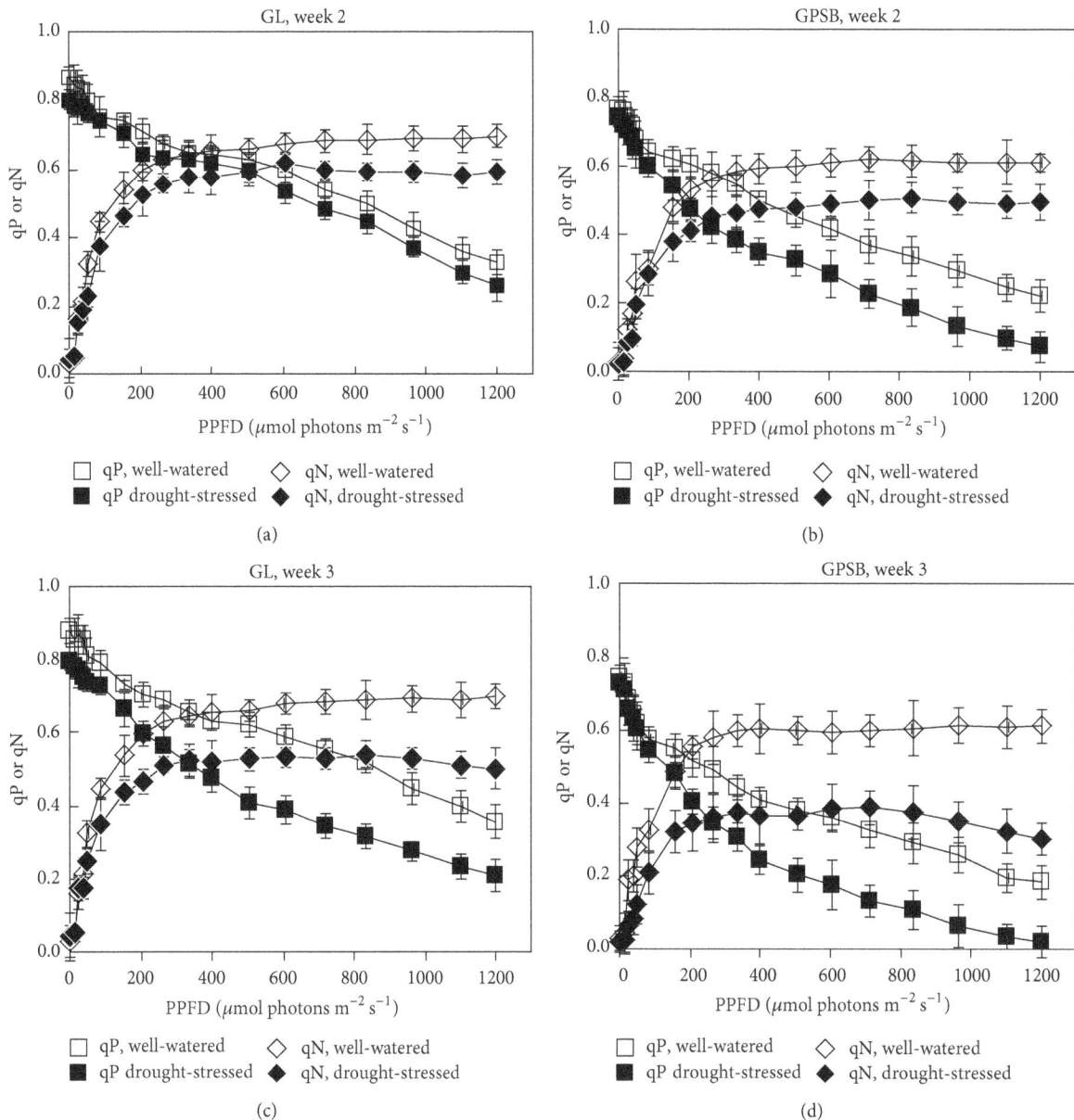

FIGURE 4: qP and qN of GL ((a), (c)) and GPSB ((b), (d)) under different PPFDs after CAM orchid *Cattleya laeliocattleya* Aloha Case plants were subjected to drought stress for 2 ((a), (b)) and 3 weeks ((c), (d)). Means of 5 measurements were from 5 different plants. Vertical bars represent standard errors. When the standard error bars cannot be seen, they are smaller than the symbols.

of GL (Figure 1(b)) after CAM orchid *Cattleya laeliocattleya* Aloha Case plants were subjected to drought stress for 3 weeks. Stancato et al. [1] also observed that RWC decreased continuously in GL and GPSB of epiphytic CAM orchid (*Cattleya forbesii* Lindl. × *Laelia tenebrosa* Rolfe) after they were subjected to up 7 to 45 days of drought stress. It was also found that the decrease in WC of drought-stressed plants was greater in GPSB than in GL (Figure 1(a)). However, the decrease in WC of GPSB was one week before that of GL, indicating that GPSB facilitated a slow reduction in the leaf WC. It was further supported by the result of relative WC of GL, which was still about 80% after 2 weeks of drought stress (Figure 1(b)). In the present study, stomatal conductance was not measured in the CAM *Cattleya* orchid plant due to

the closure of stomata even at night after being subjected to drought stress. However, the leaf RWC is a common indicator for plant water status. CAM *Cattleya* orchid plants under drought stress for 3 weeks were able to maintain leaf RWC above 70%. This may not be possible without the presence of GPSB, which were capable of delaying excessive water loss in times of drought. As an epiphytic orchid, the GPSB are important water storage organs to provide a continual source of water in times of drought. In our previous study, we reported that the constant level of GL soluble sugar could be maintained through the mobilization of soluble and insoluble sugars of GPSB [13]. However, there were no distinct patterns in the change of sugar during drought stress treatment in the present study (data not shown). The accumulation of other

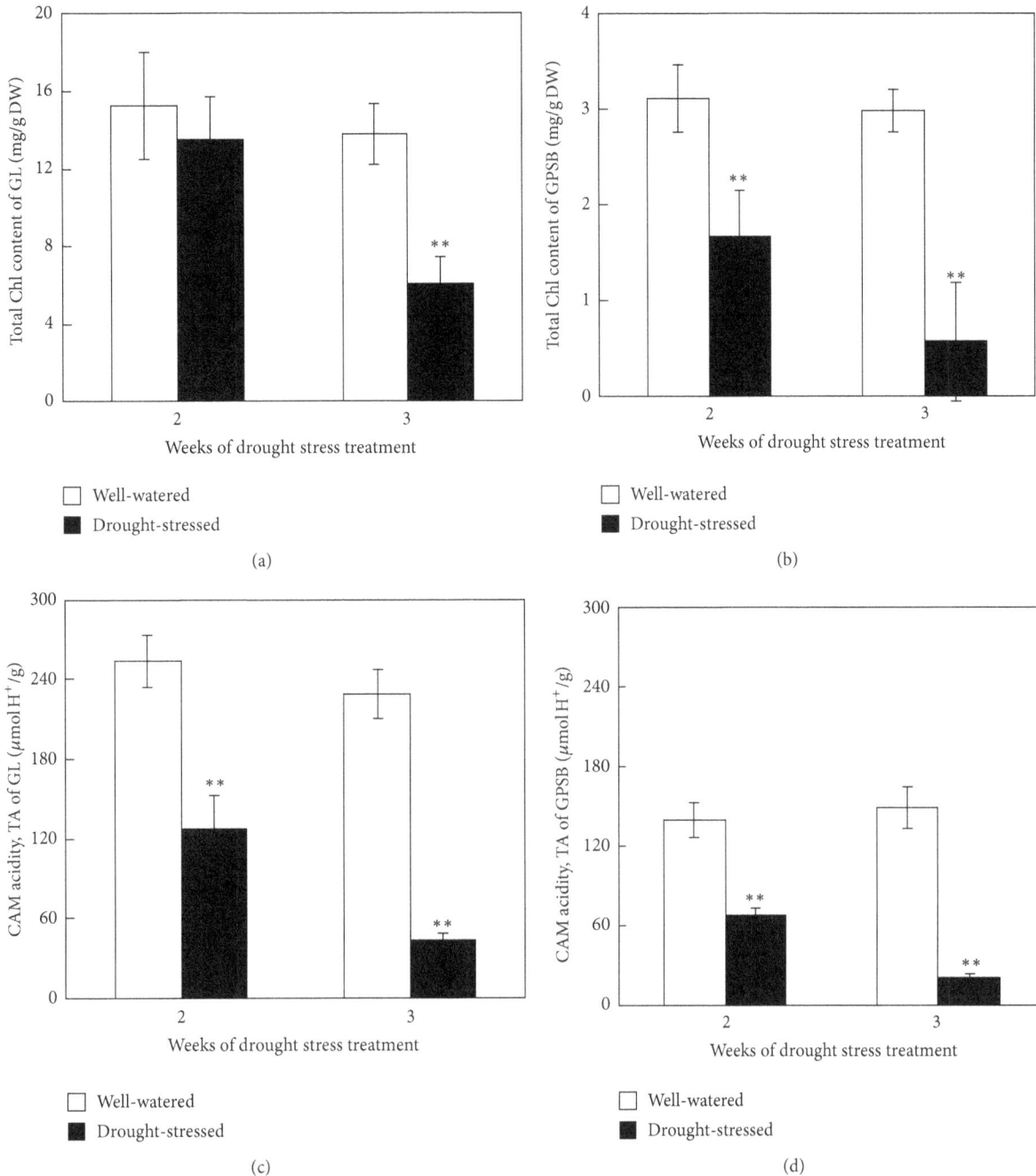

FIGURE 5: Total Chl content ((a), (b)) and CAM acidity ((c), (d)) and TA of GL ((a), (c)) and GPSB ((b), (d)) under different PPFDs after CAM orchid *Cattleya laeliocattleya* Aloha Case plants were subjected to drought stress for 2 and 3 weeks. Means of 5 measurements were measured from 5 different plants. Vertical bars represent standard errors. When the standard error bars cannot be seen, they are smaller than the symbols. ** Above the means of GL and GPSB of drought-stressed plants are statistically different from the well-watered plants ($P < 0.05$).

osmolytes, such as proline for osmotic adjustment between GL and GPSB, merits our future studies.

Studying with epiphytic CAM orchid plants, Stancato et al. [1] found that F_v/F_m ratio remained constant in well-watered plants but decreased in plants under water deficit up to 45 days. Lower F_v/F_m ratio (<0.8) could be an indication of photoinhibition that resulted from excessive light [17]. Photoinhibition can be dynamic or chronic. Dynamic photoinhibition is the protective downregulation

of photosynthesis by diversion of electrons away from PSII by increasing nonphotochemical quenching [18]. This is common amongst sun plants and is normally a short-term reversible reduction of the photosynthetic efficiency [19]. Chronic photoinhibition, on the other hand, damages the PSII reaction centres and causes the photodestruction of the photosynthetic pigments [17]. Shade plants are especially susceptible to chronic photoinhibition following exposure to high light. It is also possible for plants growing under low light to suffer from chronic photoinhibition if they were subjected to drought [20]. Results of the present study showed that dynamic photoinhibition occurred in all plants. However, the effects of photoinhibition during midday in GL were more severe in drought-stressed plants than in well-watered plants (Figure 2). Chronic photoinhibition characterized by the decrease in F_v/F_m ratio measured at 0800 h was observed in GL of drought-stressed plants after 3 and 4 weeks of treatment (Figures 2(c) and 2(d)).

With increasing drought stress, the plants suffer lower quantum yield potential of photosynthesis as a result of lesser reaction centers in PSII and result in a lower F_v/F_m value throughout the day [21, 22]. In addition, the reductions of ETR (Figure 3), qP, and qN measured under higher PPFD (Figure 4) for both GL and GPSB after drought stress suggested that GPSB were more susceptible to both drought and light stresses than GL. This is supported by the fact that the decreases in these parameters in GPSB were one week before those in GL. The greater sensitivity of GPSB to drought stress compared to GL could be due to its faster rate of water decrease after subjecting to drought stress. Studied with *Triticum aestivum* L plants, Hassan [23] also found that drought stress treatment, where irrigation was withheld till soil moisture of 13%, reduced the levels of qP and qN significantly in leaves. As stated earlier, GPSB of *Cattleya* orchid plants also acted as a water storage source which supplied water to its GL [7]. The importance of maintaining the turgidity of GL could be due to, as shown by He et al. [13], the higher photosynthetic capability of GL as compared to other green parts. Decreases in Chl fluorescence parameters (Figures 2, 3, and 4) could be due to the changes in photosynthetic pigments under environmental stress [17, 24]. For example, Paknejad et al. [24] worked on wheat cultivars and reported that total Chl decreased with increasing drought severity of 80% soil moisture reduction. In the present study, decrease in total Chl content was much earlier, and the degree of reduced total Chl content was much greater in GL than in GPSB (Figures 5(a) and 5(b)). Decreases of total content in GL of drought-stressed plants after 3 weeks of treatment further confirmed that chronic photoinhibition (lower F_v/F_m ratio at 0800 h, Figure 2(c)) occurred in GL only after 3 weeks of drought stress treatment.

It was reported that the organic acids fixed in GL enter the GPSB at night for storage and GPSB could fix the internal respiratory CO_2 into organic acid [8]. Winter et al. [14] found that under conditions of drought stress, nocturnal acidification was greatly depressed. Drought stress may cause stomata closure in GL due to water conservation. However, closure of stomata limits gaseous exchange. This reduces the amount of fixed CO_2 which causes a reduction of TA content. This could explain the significantly lower dawn/dusk fluctuation of TA of GL and GPSB of drought-stressed plants as compared to well-watered plants as earlier as one week of drought stress in both GL and GPSB (Figures 5(c) and 5(d)). While CAM plants are excellent at conserving water to survive in extremely arid environments, they do not photosynthesise as efficiently and, as a result, suffer a decrease in growth [15].

5. Conclusion

CAM orchid *Cattleya laeliocattleya* Aloha Case plants were susceptible to drought stress. Drought stress had a negative impact on water content, RWC, Chl fluorescence F_v/F_m ratio, ETR, qN, qP, and dawn/dusk fluctuation of TA of GL and GPSB. Compared to GL, although GPSB were more sensitive to drought stress, to a certain extent GPSB played an important role in maintaining the water level of GL. The accumulation of organic molecules such as proline for osmotic adjustment between GL and GPSB merits our future studies.

Abbreviations

Chl:	Chlorophyll
DW:	Dry weight
ETR:	Electron transport rate
F_m and F_v:	Maximal and variable fluorescence yields obtained from a dark-adapted sample upon application of a saturation pulse of radiation, respectively
FW:	Fresh weight
GL:	Green leaves
GPSB:	Green pseudobulbs
PPFD:	Photosynthetic photon flux density
PSII:	Photosystem II
qN:	Nonphotochemical quenching
qP:	Photochemical quenching
RWC:	Relative water content
SW:	Saturated weight
TA:	Titratable acidity
WC:	Water content.

Acknowledgments

This project was supported by teaching materials' vote of the National Institute of Education, Nanyang Technological University, Singapore, and Academic Research Fund (RI 5/07HJ), Ministry of Education, Singapore.

References

[1] G. C. Stancato, P. Mazzafera, and M. S. Buckeridge, "Effect of a drought period on the mobilisation of non-structural carbohydrates, photosynthetic efficiency and water status in an epiphytic orchid," *Plant Physiology and Biochemistry*, vol. 39, no. 11, pp. 1009–1016, 2001.

[2] C. J. Goh and M. Kluge, "Gas exchange and water relations in epiphytic orchids," in *VAscular Plants as Epiphytes, Evolution and Ecophysiology*, U. Lüttge, Ed., pp. 139–166, Springer, Heidelberg, Germany, 1989.

[3] G. Zotz and M. T. Tyree, "Water stress in the epiphytic orchid *Dimerandra emarginata* (G. Meyer) hoehne," *Oecologia*, vol. 107, no. 2, pp. 151–159, 1996.

[4] J. B. Ertelt, "Horticultural aspects of growing and displaying a wide variety of epiphytes," *Selbyana*, vol. 13, pp. 95–98, 1992.

[5] X. N. Zheng, Z. Q. Wen, and C. S. Hew, "Response of *Cymbidium sinense* to drought stress," *Journal of Hoticultural Scienc*, vol. 67, pp. 295–299, 1992.

[6] J. Arditti, "Fundamentals of orchid biology," *Fundamentals of Orchid Biology*, 1992.

[7] C. K. Y. Ng and C. S. Hew, "Orchid pseudobulbs "False" bulbs with a genuine importance in orchid growth and survival!," *Scientia Horticulturae*, vol. 83, no. 3-4, pp. 165–172, 2000.

[8] C. S. Hew and J. W. H. Yong, *The Physiology of Tropical Orchids in Relation to the Industry*, World Scientific Publishing, Singapore, 2nd edition, 2004.

[9] R. Sinclair, "Water relations of tropical epiphytes.I: relationships between stomatal resistance, relative water content and the components of water potential," *Journal of Experimental Botany*, vol. 34, no. 12, pp. 1652–1663, 1983.

[10] R. Sinclair, "Water relations of tropical epiphytes.II: performance during droughting," *Journal of Experimental Botany*, vol. 34, no. 12, pp. 1664–1675, 1983.

[11] S. A. Anjum, X.-Y. Xie, L.-C. Wang, M. F. Saleem, C. Man, and W. Lei, "Morphological, physiological and biochemical responses of plants to drought stress," *African Journal of Agricultural Research*, vol. 6, no. 9, pp. 2026–2032, 2011.

[12] S. T. Cockerham and B. Leinauer, *Turfgrass Water Conservation*, University of California, Agriculture and Natural Resources, Davis, Calif, USA, 2011.

[13] J. He, B. H. G. Tan, and L. Qin, "Source-to-sink relationship between green leaves and green pseudobulbs of C_3 orchid in regulation of photosynthesis," *Photosynthetica*, vol. 49, no. 2, pp. 209–218, 2011.

[14] K. Winter, G. S. Meier, and M. M. Caldwell, "Respiratory CO_2 as carbon source for nocturnal acid synthesis at high temperatures in three species exhibiting crassulacean acid metabolism," *Plant Physiology*, vol. 81, pp. 390–394, 1986.

[15] L. Berg, *Introductory Botany: Plants, People, and the Environment*, Thomson Brooks/Cole, Belmont, Calif, USA, 2nd edition, 2008.

[16] A. R. Wellburn, "The spectral determination of chlorophylls a and b, as well as total carotenoids, using various solvents with spectrophotometers of different resolution," *Journal of Plant Physiology*, vol. 144, no. 3, pp. 307–313, 1994.

[17] S. B. Powles, "Photoinhibition of photosynthesis induced by visible light," *Annual Review of Plant Physiology*, vol. 35, pp. 15–44, 1984.

[18] F. Ghetti, G. Checcucci, and J. F. Bornman, *Environmental UV Radiation: Impact on Ecosystems and Human Health and Predictive Models*, Springer, Amsterdam, The Netherlands, 2001.

[19] C. B. Osmond, "What is photoinhibition? Some insights from comparisons of sun and shade plants," in *Photoinhibition of Photosynthesis. From Molecular Mechanisms to the Field*, N. R. Baker and J. R. Bowyer, Eds., pp. 1–24, BIOS, Oxford, UK, 1994.

[20] D. O. Hall and K. K. Rao, *Photosynthesis*, Cambridge University Press, UK, 6th edition, 2001.

[21] G. H. Krause and S. Somersalo, "Fluorescence as a tool in photosynthesis research: application in studies of photoinhibition, cold acclimation and freezing stress," *Philosophical Transactions of the Royal Society B*, vol. 323, pp. 281–293, 1989.

[22] B. Venkateswarlu, A. K. Shanker, C. Shanker, and M. Maheswari, *Crop Stress and Its Management: Perspectives and Strategies*, Springer, Amsterdam, The Netherlands, 2012.

[23] I. A. Hassan, "Effects of water stress and high temperature on gas exchange and chlorophyll fluorescence in *Triticum aestivum* L.," *Photosynthetica*, vol. 44, no. 2, pp. 312–315, 2006.

[24] F. Paknejad, M. Nasri, H. R. T. Moghadam, H. Zahedi, and M. J. Alahmadi, "Effects of drought stress on chlorophyll fluorescence parameters, chlorophyll content and grain yield of wheat cultivars," *Journal of Biological Sciences*, vol. 7, no. 6, pp. 841–847, 2007.

17

The Evolutionary Dynamics of Apomixis in Ferns: A Case Study from Polystichoid Ferns

Hong-Mei Liu,[1] Robert J. Dyer,[2,3] Zhi-You Guo,[4] Zhen Meng,[5] Jian-Hui Li,[5] and Harald Schneider[2,6]

[1] *Key Laboratory of Southern Subtropical Plant Diversity, Fairylake Botanical Garden, Shenzhen & Chinese Academy of Sciences, Shenzhen 518004, China*

[2] *Department of Life Sciences, Natural History Museum, London SW7 5BD, UK*

[3] *Research Group in Biodiversity Genomics and Environmental Sciences, Imperial College London, Silwood Park Campus, Ascot, Berkshire SL5 7PY, UK*

[4] *Department of Biological Sciences, Qiannan Normal College for Nationalities, Duyun 558000, China*

[5] *Computer Network Information Center, Chinese Academy of Sciences, Beijing 100190, China*

[6] *State Key Laboratory of Systematic and Evolutionary Botany, Institute of Botany, Chinese Academy of Sciences, Beijing 100093, China*

Correspondence should be addressed to Hong-Mei Liu, sorolepidium@gmail.com and Harald Schneider, h.schneider@nhm.ac.uk

Academic Editor: Karl Joseph Niklas

The disparate distribution of apomixis between the major plant lineages is arguably one of the most paradoxical phenomena in plant evolution. Ferns are particularly interesting for addressing this issue because apomixis is more frequent than in any other group of plants. Here, we use a phylogenetic framework to explore some aspects of the evolution of apomixis in ferns and in particular in the polystichoid ferns. Our findings indicate that apomixis evolved several times independently in three different clades of polystichoid ferns. A lineage-wide perspective across ferns indicates a correlation between apomixis and the species richness of lineages; however BiSSE tests did not recover evidence for a correlation of apomixis and diversification rates. Instead, evidence was recovered supporting an association between the establishment of apomixis and reticulate evolution, especially in the establishment of triploid hybrids. Diversification time estimates supported the hypothesis of short living apomictic lineages and indicated a link between the establishment of apomixis and the strengthening of the monsoons caused by the lifting of the Qinghai-Tibetan plateau. In general our results supported the hypothesis for the rare establishment of apomictic lineages, high extinction risks, and low speciation rates.

1. Introduction

The evolution of asexual reproduction in multicellular eukaryote lineages such as animals, fungi, and plants continues to attract the interests of biologists [1–3]. The prevalence of sexual reproduction in these lineages is paradoxical considering the theoretical advantages associated to asexual reproduction [3, 4]. Hypotheses to explain this paradox have concerned lower rates of speciation and higher rates of extinction in asexual lineages. Essentially, the short-term ecological advantages of asexual reproduction are balanced by the long-term effect of lower diversification rates. This hypothesis is consistent with reported evidence for short-lived apomictic lineages, for example, the derived fern genus *Astrolepis* [5], and the developmental feature changes required for the loss of sexuality in plants [3]. However, it is inconsistent with other studies that recover evidence for species rich lineages that lack any reported evidence for at least rare sexual events [6, 7], and a recent study on the eudicot genus *Oenothera*, which shows "increasing diversification associated with loss of sexual recombination and segregation" [4]. This conflicting evidence questions our general understanding of the evolution of asexual reproduction, and the causes of its uneven distribution among animals and land plants.

Here, we investigate the association of apomixis and changes of diversification rates in a derived group of ferns. Apomixis in ferns is a distinctive form of asexual reproduction that combines the production of unreduced spores (diplospory) and the formation of sporophytes from somatic cells of the prothallium (apogamy) [8–11]. Apomixis is known from different lineages of land plants but is reported to be more common in ferns [12, 13]. Some reviews suggest estimates for the occurrence of apomixis is up to 10% of extant fern species globally [12], and up to 17% in some local fern floras [13]. The high frequencies in ferns contrast strongly with the much lower frequencies of apomictic taxa in angiosperms and gymnosperms [2, 3, 14, 15]. However, apomixis is not only unevenly distributed among the major lineages of land plants but also among fern lineages (Figure 1; Table 1). Within the derived ferns, two of the most species-rich families, Dryopteridaceae and Pteridaceae, comprise approximately 70% of reported apomictic fern species (Figure 1; Table 1). This pattern suggests that apomixis may be linked to species richness [16]. However, apomixis is rare in several other species-rich families of derived ferns such as Polypodiaceae (Figure 1; Table 1). The conflicting evidence suggests at least one missing factor, such as environmental conditions or karyological traits, for example, hybridization and polyploidy [3]. This highlights the need to consider additional properties, some of which are unique to ferns, when trying to understand the evolution of apomixis in ferns.

Polyploidy and reticulate evolution are strongly linked with the occurrence and establishment of apomixis in both ferns and angiosperms [5, 17–23]. This link is especially true for triploid and pentaploid taxa [13, 19, 23]. However, several cases of diploid apomictic ferns are known; for example, about 11% of apomictic species of *Pteris* are diploids [19, 23]. It is also significant that the vast majority of apomictic ferns are reported to be obligate, lacking the ability to reproduce sexually [9], whereas in angiosperms many "apomictic" taxa are facultative, exhibiting a mixed mating system that combines the advantages of sexual and asexual reproduction. This has a large impact on the establishment of apomixis, and the various transition pathways possible.

We also need to consider the similarity in the ecological-evolutionary potential of sexual, obligate selfing (intraga-metophytic) ferns. Obligate intragametophytic selfing of polyploids results in fixed heterozygote genotypes that are comparable to apomictic lineages. However, obligate inbreeding appears to be uncommon in ferns, with the majority of ferns reproducing either via obligate outcrossing or mixed mating systems [24, 25].

Finally, the high frequency of apomixis in Japanese ferns [13] indicates a possible geographic influence on the occurrence of apomictic taxa. In angiosperms, apomictic species tend to occupy often more northern distribution ranges than their sexual relatives [26]. But such a northern range effect has not been reported for ferns.

In this study, we address the following issues: (1) the uneven distribution of apomixis in fern lineages, (2) the association of apomixis with polyploidy and hybridization, (3) the association of apomixis and odd ploidy levels, (4) the connection between apomixis and changes in diversification rates, and (5) the link between the establishment of apomixis and geographic distribution and/or geological events. To address these questions, we assessed the reported occurrence of apomixis across ferns and reconstructed the evolution of apomixis in polystichoid ferns. The latter are an outstanding example given our knowledge on the phylogeny of this species-rich lineage of derived ferns [27–36], including the various reports on reticulate evolution and polyploidy [37], and reports on the reproductive biology of these ferns showing a range of modes from obligate outcrossing to apomixis [34, 38, 39]. It has been suggested that apomixis may be restricted to three lineages of polystichoid ferns which all have their putative origin in SE Asia [34]. Thus, we also discuss evidence supporting the hypothesis linking the evolution of apomixis with the strengthening of the monsoon climate due to the rise of the Qinghai-Tibetan plateau from the mid Miocene to Pleistocene [40].

2. Materials and Methods

To investigate the occurrence of apomixis and poly-ploidy in ferns and in particular polystichoid ferns, we carefully reviewed all existing reports on embryo development [39, 41, 42] and chromosome numbers (see http://darwintree.cn/special_topic/fern/home.jsp). The latter were obtained mainly from exploring data accessible in Tropicos (http://www.tropicos.org/), published chromosome number lists [43], and from pertinent publications [37, 44]. In the context of polystichoid ferns, we also used the information provided at David Barrington's *Polystichum* website (http://www.uvm.edu/~dbarring/polystichum.htm). Where possible, we traced back the original reports to confirm the taxon identity and reported results. However, confirmation of species identity through the study of vouchers was not applicable to the vast majority of the records due to the lack of adequate vouchers, and/or the amount of time and funding required to carry out this practice. This limitation introduces some taxonomic uncertainty given the difficulties of species identification and changes of taxonomic concept over the last 100 years. This problem must be considered especially for single reports of apomixis and/or chromosome numbers per species. In addition, single observations are also insufficient to explore the fixation of reproductive modes and ploidy levels in species. These problems cannot be avoided but we tried to reduce this putative bias as much as possible by scrutinizing the evidence and by integrating different sources of information. In general, apomixis was only accepted for studies that are unlikely to report induced apomixis, meaning apomixis created under laboratory conditions [39, 45]. A significant problem may be the assumption that no record of apomixis for taxa is the same as the absence of apomixis. To compensate for this shortcoming, we carried out investigations into

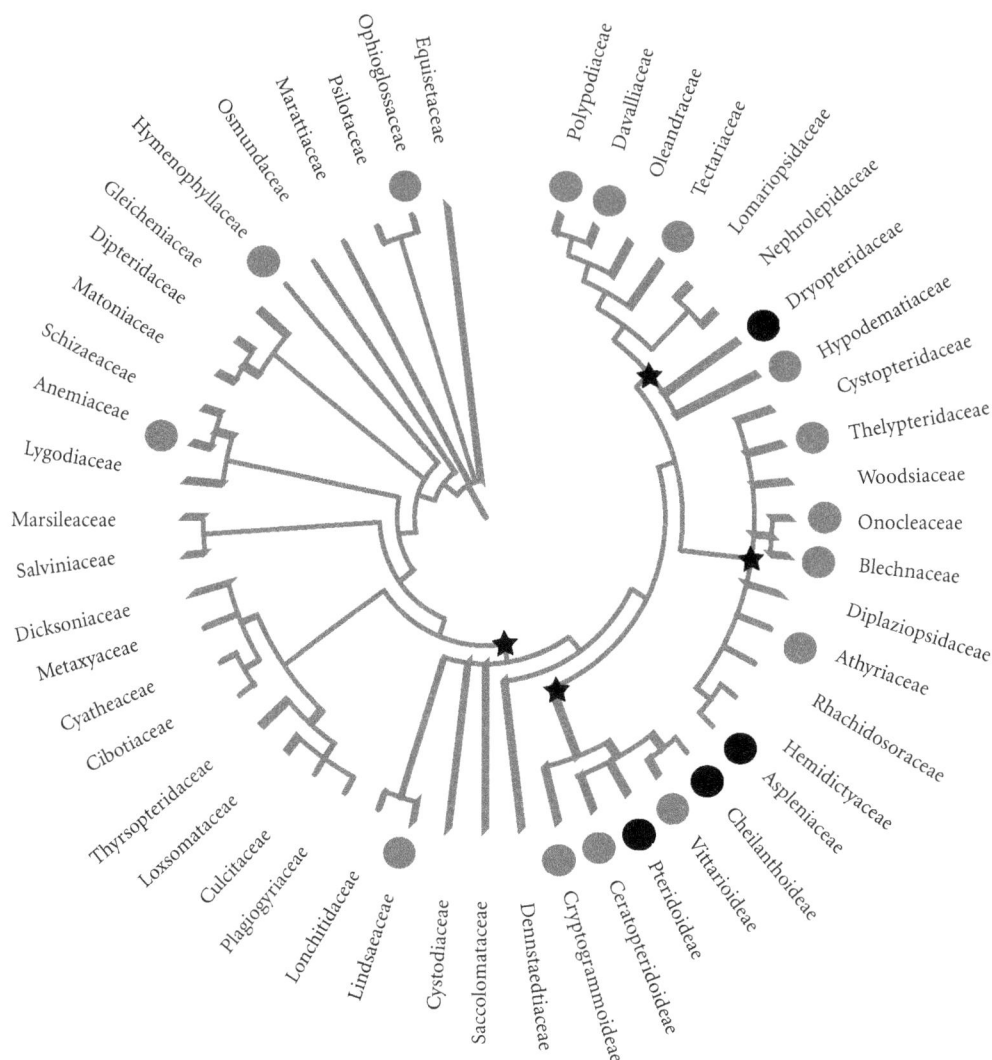

FIGURE 1: Global phylogeny of fern families as recovered in recent phylogenetic analyses [53, 61] with some poorly resolved nodes collapsed. Circles indicate occurrence of apomixis in this family. Grey circle: <5% of known apomictic fern taxa belong to this family; black circles: >5% of known apomictic fern taxa belong to this family. Pteridaceae are replaced by five subfamilies to illustrate the uneven distribution of apomixis in the highly diverse family. Fern families with more than 400 extant species are printed in bold. Stars indicate lineages of importance: 1 = Polypodiales, 2 = Pteridaceae, 3 = Eupolypods 1, and 4 = Eupolypods 2.

the reproductive biology of polystichoid ferns, particularly *Cyrtogonellum* (*Cyrtogonellum caducum*). This was achieved by growing gametophytes from spores to obtain direct observation on the formation of sexual organs and the formation of sporophytic outgrowth with or without fertilization. The results of these experiments were compared with previous reports on the occurrences of apomictic reproduction in these ferns [42].

Cyrtogonellum caducum sporophytes were collected from Guizhou Province, China. Fertile pinnae were kept in clean paper bags under dry conditions until spores were shed. Spores were cultured in plastic containers filled with humus collected together with the sporophytes that was sterilized at 120–130°C for thirty minutes. The containers were covered with transparent plastic film, on which two to three small holes were made in order to reduce the risk of contamination

and desiccation. The containers were kept in the laboratory and exposed to natural light conditions throughout the experiments. Cultures were moistened with tap water to prevent desiccation and were periodically examined to record the onset of spore germination, the growth of gametophytes, the formation of antheridia and archegonia, and the onset of the development of sporophytes. Theses conditions were not significantly different from conditions used to grow sexual taxa. We can not exclude the possibility of induction of apomixis via cultivation condition and this may need to be addressed in future studies. So far, all reported evidence suggested this taxon to reproduce apomictically in both cultural experiments as well as natural conditions.

We have compiled all data on apomixis in ferns and lycophytes and presented it in a new website "*Apomixis in Ferns*" (http://darwintree.cn/special_topic/fern/home.jsp).

TABLE 1: Summary of the recorded cases of apomixis in fern families. Natural occurring apomixis (excluding induced occurrences) was reported for about 242 species of ferns. These reports suggest apomictic occurrence of at least 3% of the currently estimated 10,000 to 11,000 fern species. Species rich fern families (>400 species) are marked in bold. The sequence of families follows Figure 1, families defined as in the most recent classifications of ferns [56, 57].

Family	NR APO	ASSO POLY	MULT ORG
Equisetaceae	0	NA	NA
Psilotaceae	0	NA	NA
Ophioglossaceae	1	++	NA
Marattiaceae	0	NA	NA
Osmundaceae	0	NA	NA
Hymenophyllaceae	8	++	++
Dipteridaceae	0	NA	NA
Gleicheniaceae	0	NA	NA
Matoniaceae	0	NA	NA
Anemiaceae	1	?	NA
Lygodiaceae	0	NA	NA
Schizaeaceae	0	NA	NA
Marsileaceae	0	NA	NA
Salviniaceae	0	NA	NA
Cibotiaceae	0	NA	NA
Culcitaceae	0	NA	NA
Cyatheaceae	0	NA	NA
Dicksoniaceae	0	NA	NA
Loxsomataceae	0	NA	NA
Metaxyaceae	0	NA	NA
Plagiogyriaceae	0	NA	NA
Cystodiaceae	0	NA	NA
Lindsaeaceae	9	++	++
Lonchitidaceae	0	NA	NA
Saccolomataceae	0	NA	NA
Dennstaedtiaceae	0	NA	NA
Pteridaceae	84	++	++
Aspleniaceae	20	++	++
Athyriaceae	16	++	++
Blechnaceae	1	?	NA
Cystopteridaceae	0	NA	NA
Diplaziopsidaceae	0	NA	NA
Hemidictyaceae	0	NA	NA
Onocleaceae	1	NA	NA
Rhachidosoraceae	0	NA	NA
Thelypteridaceae	5	++	++
Dryopteridaceae	87	++	++
Hypodematiaceae	1	?	NA
Lomariopsidaceae	0	NA	NA
Nephrolepidaceae	0	NA	NA
Tectariaceae	1	?	NA
Oleandraceae	0	NA	NA
Davalliaceae	2	++	++
Polypodiaceae	6	++	++

NR APO: number of species per family recorded to be apomictic; ASSO POLY: evidence for association of apomixis and polyploidy; MULT ORG: evidence for multiple origins of apomixis in this family considering existing phylogenetic evidence. Symbols: ++: yes, ?: unknown as a result of lack of evidence, NA: not applicable as an absence of apomixis or a single taxon with apomixis.

The website provides basic information about occurrence of apomixis in ferns and lycophytes, including a list of all recorded apomictic taxa as well as the corresponding references.

To enable us to explore the evolution of apomixis in polystichoid ferns, we developed a phylogenetic hypothesis based on *rbcL* sequence data that were obtained from Gen-Bank in March 2012. We downloaded all sequences available for the genera *Cyrtogonellum*, *Cyrtomidictyum*, *Cyrtomium*, *Phanerophlebia*, and *Polystichum*. In total 215 accessions were acquired and were manually aligned and visually inspected using Mesquite 2.75 [46]. Sequences of poor quality were excluded if more than one accession per taxon was available. Phylogenetic reconstruction of this data was compared with published phylogenies of polystichoid ferns [27–36] to highlight any conflicts in the phylogenetic relationships observed. Comparisons were specifically made between our results and the consensus tree provided at the *Polystichum* website (http://www.uvm.edu/~dbarring/polystichum.htm). Several species were found to be polyphyletic and we deleted sequences that were likely due to the result of misidentification. Where possible we checked vouchers but in many cases we needed to trust the identification provided by the colleagues generating these data. In general, the probability of incorrectly identified material was considered to be limited in this dataset because the majority of DNA sequences used was generated by researcher groups that comprised one or several experts in the taxonomy of polystichoids ferns. Finally, we reduced the dataset to have one accession per taxon. This dataset of 143 taxa was then used to reconstruct phylogenetic hypotheses used in this study. To reduce the impact of missing karyological data, we further reduced the dataset to 82 taxa for which the number of chromosomes was known.

Standard methods were used to obtain phylogenetic hypotheses based on the 143-taxon set and 82-taxon set. First, we used jModeltest [47] to determine the best-fit model for both data sets. Then, we carried out maximum parsimony analyses in PAUP 4.0 [48] using heuristic searches with 100 random starting point and default options, and maximum likelihood analyses in PHYML 3.0 [49] as implemented in the plugin for Geneious 5.4. [50]. The latter approach was also used to obtain bootstrap values for the maximum likelihood tree using 1000 bootstrap replicates. Bayesian inference of phylogeny was carried out using MrBayes 3.2.1 [51] and BEAST 1.7.1 [52]. Both Bayesian analyses were carried out using a single partition and the model selected in jModeltest, but BEAST analyses required further parameter choices. These included the assumption of a relaxed molecular clock, a calibration of the onset of the diversification of polystichoid ferns in the form of a log-normal distribution with a standard deviation of 1 and a shift from 0 by 34.5 ma as found in [53], and a Death-Birth speciation process. All Bayesian results were inspected using TRACER 1.5 [54] and FIGTREE 1.3.1 [55] before further usage.

We scored the reproductive biology of each species as sexual (0) or apomictic (1) for all taxa included. The chromosome numbers were used to obtain two characters. First, we scored the taxon to be either diploid (0) or a polyploid (1) without any further differentiation of the ploidy level. Secondly, we scored each taxon as having an even (0) or odd (1) ploidy level, that is, contrasting diploid, tetraploid, hexaploid, and octoploid taxa against triploid, pentaploid, and heptaploid taxa. A few taxa were scored as polymorphic for one or all of these characters. Limitations of the available data were taken into consideration, such as the lack of evidence about the reproductive biology of

several diploid taxa nested within apomictic clades. Diploid apomicts were recorded for several derived fern genera [19, 23, 58] and may also occur in polystichoid ferns. However, we did not have sufficient evidence to reject the alternative that diploids represent sexual taxa. In this study, we explore two assumptions: diploid taxa always represent sexuals versus diploids in apomictic clades reproduce via apomixis. So far, diploid apomictic taxa have not been recorded for polystichoid ferns but we cannot reject their putative occurrence. The link between apomixis and biogeography was inferred by scoring the distribution occurrence either as: within SE Asia (including China, Japan, Korea, and Indochina) or as outside of SE Asia. Taxa occurring in both areas were scored as within SE Asia, assuming this as the area of origin. Information concerning distributions was obtained from GBIF, Tropicos, and publications on polystichoid ferns referred to throughout this study.

These scores were then plotted onto the phylogenetic hypotheses obtained in the analyses of the 143 and 82 taxon datasets. We used maximum parsimony reconstruction with unordered and/or ordered characters as well as maximum likelihood reconstruction with the MK1 model to infer the evolution of these characters. Pagel's test for correlation of characters over a phylogeny [59] was used to infer correlations between the three characters scored: absence/presence of apomixis, ploidy level as diploidy or polyploidy, and absence/presence of odd ploidy levels (e.g., 2x, 4x, 6x versus 3x, 5x). This test was carried out using the implementation in Mesquite and with the 82-taxon dataset because these analyses cannot be carried out with incompletely scored taxa.

The association of apomixis and diversification rates of polystichoid ferns was investigated using the BiSSE method as implemented in Mesquite [60]. The 0 hypothesis (independence) was compared with the tested hypothesis (dependence) using likelihood ratio tests. Significance of the results was further investigated by comparing the likelihood values of the tested hypothesis with the likelihood distribution based on 1,000 simulated trees. First, we explored the association of the three characters scored with changes in speciation rate, extinction rate, frequency of character state changes, or combinations of these three parameters. Secondly, we inferred the hypothesis of unidirectional change from sexual to apomixis in these ferns by comparing the probabilities of character state changes $q01$ (sexual to apomictic) and $q10$ (apomictic to sexual). The results of estimated and fixed (one close to 0 and one close to 1) values of $q01$ and $q10$ were compared. In the absence of direct reports on the reproductive biology, the putative presence of sexual reproduction or apomictic reproduction was inferred via two assumptions: (1) 32 spores per sporangium indicated apomictic reproduction whereas 64 spores per sporangium indicated sexual reproduction; (2) diploids were assumed to be sexually reproducing despite evidence for diploid apomictic ferns in other fern genera [19, 23, 58]. The reduction of the spore number from 64 to 32 spores was considered as an indicator of apomixis in ferns belonging to the order Polypodiales. This criterion was introduced based on the documented conservation of the spore number in this fern order and the discovered mechanisms of apomixis in ferns. This criterion can not be used in other ferns such as tree ferns. Theoretically, some other mechanism may result in the reduction of spores in Polypodiales but not a single case has been convincingly reported. The vast number of observation on the reproductive biology of diploid ferns supports these assumptions but we took care to consider the potential error introduced. The second assumption was expected to increase the signal for putative reversals from apomictic to sexual reproduction. Thus, the effect of this value was inferred by carrying out the analyses by scoring diploids nested within apomictic clades either as apomictic or sexual.

3. Results

Phylogenetic analyses and character reconstruction for apomixis and polyploidy showed that apomixis was restricted to three clades in the polystichoid ferns. These three clades were (1) the core *Cyrtomium* clade, (2) the *Cyrtogonellum* clade, and (3) the *Xiphopolystichum* clade (Figure 2). In contrast, polyploidy was found to occur in all clades of polystichoid ferns. However, the occurrence of triploid taxa was restricted to the clades with apomictic reproduction. Apomictic reproduction was confirmed by gametophyte cultures in *Cyrtogonellum caducum*, based on the observation of sporophytic outgrowths and the lack of archegonia (Figure 3). However, functional antheridia were observed (Figure 3).

Maximum likelihood reconstruction supported a minimum number of three independent origins of apomixis (Figure 2). Apomixis was supported as the ancestral state for the *Cyrtogonellum* clade ($P > 0.95$), but not for the core *Cyrtomium* clade ($P = 0.05$) or the *Xiphopolystichum* ($P = 0.5$). However, the subclades of *Xiphopolystichum* had probabilities $P = 0.50$ and 0.97, and the subclade of core *Cyrtomium* had probabilities of $P = 0.50$ and 0.03 (Figure 2).

Pagel's test for the correlated evolution of discrete characters on phylogenies provided evidence for significant correlations between apomixis and the occurrence of triploids ($\ln(0) = -53.719693$, $\ln(1) = -26.701372$, Diff $= 27.018321$, $P < 0.01$), apomixis and polyploidy ($\ln(0) = -74.733399$, $\ln(1) = -66.158126$, Diff $= 8.575273$, $P < 0.01$), and polyploidy and occurrence of triploids ($\ln(0) = -74.733399$, $\ln(1) = -66.162897$, Diff $= 0.570502$, $P = 0.01$). Apomixis was also found to be associated with biogeography ($\ln(0) = -57.204702/\ln(1) = -52.350905$, Diff $= 4.845666$, $P = 0.02$).

BiSSE analyses for association of character evolution and diversification rate recovered small differences between log likelihoods of analyses assuming character-independent rates to those assuming character-dependent rates with one or all three parameters (speciation rate λ, extinction rate μ, and character transition rate q) considered. None of these values were significant considering a $P < 0.05$ criterion: $\ln(\text{independent}) = -0.164.15245282$, $\ln(\text{dependent}) = -0.163.6640405$, Diff $= 0.4860$.

BiSSE analyses on the frequency of character changes q01 versus q10 did not support the two extreme models with q01 $\ll$ q10 and q01 $\gg$ q10 while the model q01 $<$ q10 was found to have the best fit.

4. Discussion

Both the global phylogeny of ferns as well as the detailed reconstruction of the evolution of apomixis in polystichoid ferns recovered evidence for multiple origins of apomixis (Figures 1 and 2; Table 1). Examination of previous studies supports the hypothesis that apomixis evolved many times in ferns, and that it occurs with high frequency in some clades and not at all in others (Table 1). Species richness appears to be one factor associated to apomixis at the global level (Table 1), but this association is not supported in the polystichoid fern lineage. In this lineage we did not find a correlation between speciation/extinction rates and apomixis (see results of BiSSE analyses). Nevertheless, the analyses of the polystichoid dataset did support the hypothesised association of apomixis and reticulate evolution involving polyploidy [5, 17, 20–23]. The main caveat here is that polyploidy alone is insufficient to confirm reticulation because chromosome numbers do not provide sufficient evidence to distinguish between auto- and allopolyploidy. However, we suggest that polyploidy is a strong proxy for reticulate evolution and expect similar results to be recovered in more detailed studies of the evolution of apomixis in other fern families.

The polystichoid dataset also showed clear evidence for a correlation between apomixis and the occurrence of triploid and pentaploid taxa. This link may be caused by the capacity for tetraploid and hexaploid hybrids to reproduce via intragametophytic selfing, which results in fixed heterozygosity [62]. This capacity may lead to a reduction in the potential establishment of such apomictic hybrids. In contrast, apomixis is more likely to be established in triploid and pentaploid hybrids, because meiotic incompatibilities at these uneven ploidy levels would result in female sterility. A transition to apomixis in such hybrids would overcome female sterility. Despite the clear evidence for this correlation, it does not answer all questions concerning the origin of apomixis in ferns.

To address these questions we need to consider the limitation of our current knowledge on the occurrence of apomixis in ferns. For example, it is very difficult to estimate the minimum number of times in which apomixis has originated due to ambiguities in reports on the absence/presence of apomixis. The reproductive biology of many species is poorly or completely unknown. Thus, the absence in reports of apomixis may not always indicate the absence of apomixis in natural populations. Similarly, single reports of apomictic reproduction may also be misleading if apomixis is not a fixed character state of a single recognized taxon. Further reports using cultivated material must be examined carefully because it is possible to confuse experimentally induced apomixis and naturally occurring apomixis [39, 45]. Apomixis can be easily induced in many fern taxa and spontaneous apomixis may happen frequently in nature. For example, we considered the recent report of apomixis

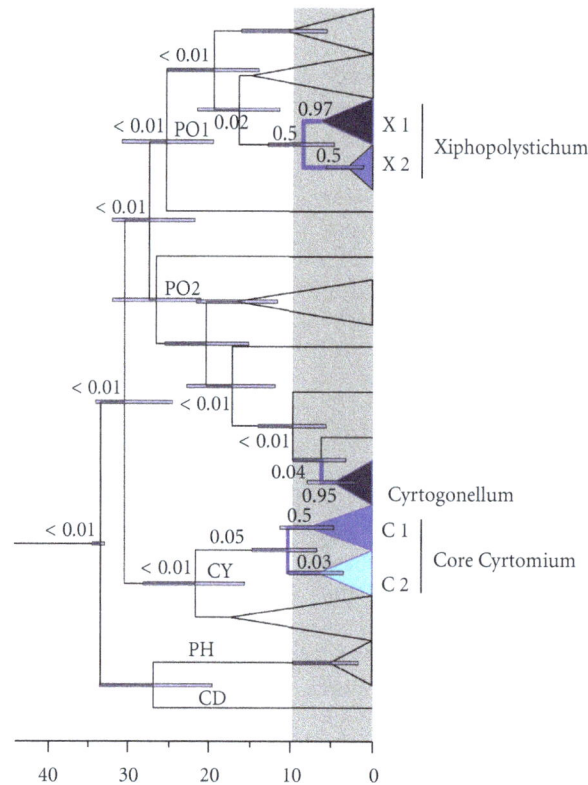

FIGURE 2: Diversification of polystichoid ferns through time as inferred using *rbcL* chloroplast DNA sequences (obtained from GenBank in March 2012) and Bayesian estimates of divergence times using a relaxed molecular clock model. The shown topology is consistent with the results of published phylogenetic analyses of polystichoid ferns, see also David Barrington's *Polystichum* website (http://www.uvm.edu/~dbarring/polystichum.htm). The three groups comprising apomictic taxa are indicated as follows: dark color: all or nearly all taxa are reported to be apomictic, middle blue: half or most taxa are reported to be apomictic, and bright blue: minority of apomictic taxa. Thick blue branches: earliest putative origin of apomixis indicated in maximum parsimony reconstructions under the assumption of accelerated gain and putative secondary lost of apomixis. Values above branches indicate probability for ancestral apomixis in the clades depicted as triangles and branches leading to these clades (with <0.01 as the minimum value shown). The grey vertical lines indicate known for enhancement of the monsoon regimes in SE Asia. Abbreviations: CD: *Cyrtomidictyum*, CY: *Cyrtomium*, PH: *Phanerophlebia*, PO1: *Polystichum* clade 1, and PO2: *Polystichum* clade 2. C1 includes *Cyrtomium* species with mainly apomictic reproduction such as *C. falcatum*, *C. fortunei*, and *C. macrophyllum*, whereas C2 includes *Cyrtomium* known to show sexual reproduction such as *C. grossum* and *C. nephrolepioides*, as well as the apomictic *C. hemionitis*. X1 includes the apomictic taxa *Polystichum luctuosum*, *P. tsus-simense*, and *P. xiphophyllum*, whereas X2 includes the apomictic *P. neolobatum* and the nonapomictic *P. hillebrandii*.

in *Polystichum polyblepharum* [39] to be an example of laboratory-induced apomixis. However, we are only interested in cases in which apomictic taxa have originated spontaneously in nature.

In many cases, diploid chromosome number can be an indicator of sexual reproduction but diploid apomictic ferns have been discovered in several groups [19, 23, 58]. In such cases, it is not always certain that all individuals of a morphological defined species reproduce either sexually or asexually. The results may also be biased by the putative ecological advantages of apomictic taxa when compared to their sexual relatives. They may contribute to unbalanced sampling probabilities, that is, a high probability of sampling the common apomictic taxa versus a low probability of collecting the rare sexual taxa. The inferences made in this study do not account for these uncertainties. They may therefore result in a higher frequency of observed transitions from apomixis to sexual reproduction than expected.

Only unidirectional transition from sexual reproduction to apomixis has been recorded in ferns, suggesting that observations of the reverse transitions in the results of this study are artefacts of our methodology.

Incomplete taxon sampling may have influenced our findings that apomixis has little to no effect on speciation and extinction rates. Neither the large nor the small taxa set is close to representing the species of these genera. Our results therefore need to be tested on a more comprehensive sampling effort in the future. This problem is particularly important under the model of postorigin differentiation of apomictic taxa because the phylogenies used here are likely to underestimate the number of biological entities present. Evidence for this process was reported for the *Cyrtomium fortunei* complex [63]. Despite these limitations, our results are consistent with the hypotheses for the rare establishment of asexual lineages and a generally high extinction risk in these lineages [3].

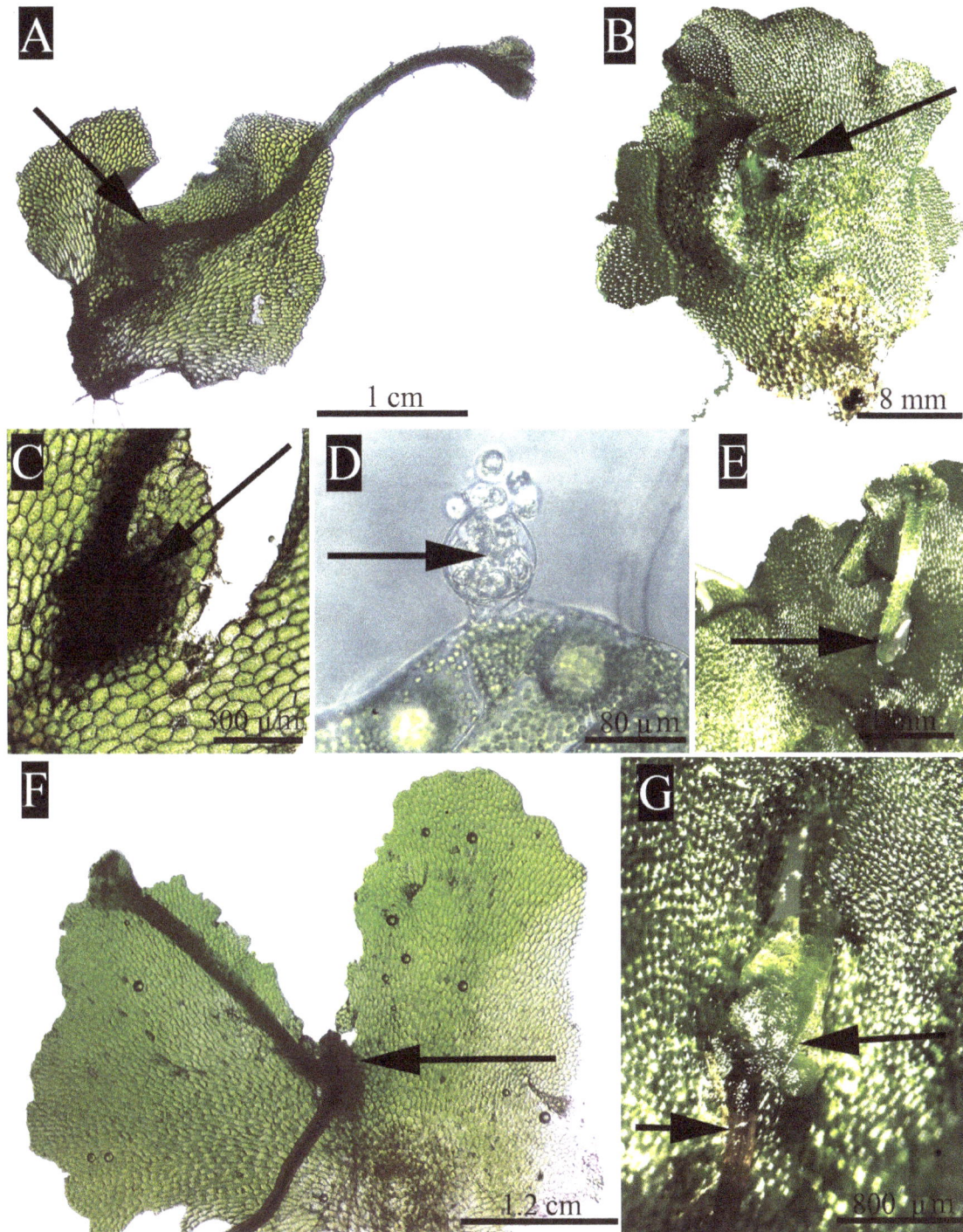

FIGURE 3: Apomictic reproduction in *Cyrtogonellum caducum*, observation from gametophyte cultures obtained from spores of GZY-LB. (A) Heart-shaped prothallium with rhizoids (bottom of the image) and sporophytic outgrowth (arrow). (B) Close up of the early stage sporophytic outgrowth without differentiated organs (arrow), no indication for any archegonium was recovered. (C) Early stage of sporophytic outgrowth developed from somatic cells (arrow). (D) Close up of a mature antheridium (arrow) with spermatozoids. (E) Close up of the sporophytic outgrowth (arrow) with the differentiation of the first leaf. (F) Close up of the prothallium showing a slightly older sporophytic outgrowth (arrow) with one leave and one root located closely to the notch. (G) Sporophytic outgrowth at the notch of the gametophyte (upper arrow); the first root is already developed (lower arrow).

Our results are also consistent with the hypothesis of young (in geological time), recurrently formed apomictic taxa. The divergence estimates did not recover evidence for apomictic lineages that originated more than 8 million years before the present (Figure 2). These results are consistent with other published results [5] and theoretical expectations [3]. The individual age of each apomictic lineage may be still much younger given that apomixis is likely to have been established multiple times in each of these clades.

All polystichoid clades comprising apomictic taxa occur either exclusively or predominantly in SE Asia [27, 30, 31, 34], which was recognized as the center of diversity and putative area of origin of polystichoid ferns [28]. Thus, it is likely that apomixis originated in SE Asia and has not been established during the colonization of other regions. The time interval for the establishment of apomictic lineages coincides with estimates for the rise of the Qinghai-Tibetan plateau and changes in the monsoon regimes in SE Asia [40]. The establishment of apomixis in polystichoid ferns may therefore be associated to the strengthened monsoon regimes, and thus associated to the increased seasonality of precipitation. The SE Asian monsoon hypothesis is consistent with some other observations of apomictic ferns. Apomixis is especially common among the xerophytic cheilanthoid ferns [17, 20–22]. Apomictic species of Aspleniaceae occur preferably in areas with season rainfalls such as parts of Africa, for example, *Asplenium aethiopicum* complex, and northern Central America, for example, *Asplenium monanthes* complex [9, 10]. However, further detailed investigations are required to establish a close link between the evolution of apomixis in ferns and the strengthening of the monsoon climates in SE Asia. However, further investigation into the effect of selection on the establishment of apomixis requires caution because the causality may be much less clear than anticipated [64].

Although there is strong support for multiple origins of apomixis in polystichoid ferns, it does not sufficiently explain the assembly of the observed taxonomic and morphological variation in the core *Cyrtomium* and *Cyrtogonellum* clades. Segregation in apomictic lineages was recently documented for *Cyrtomium fortunei* [63]. This process may also contribute to the diversity of the *Cyrtogonellum* clade because our results suggest the possibility of a single origin of apomixis in this lineage. However, this could be also the result of unsampled sexual taxa, due to their rarity. Despite uncertainties, the segregation hypothesis currently appears to be the best explanation for this complex. Future studies should take into account the evidence for antheridia in apomictic reproducing gametophytes of *Cyrtogonellum caducum*. Sexual organs were reported to be absent in most apomictic ferns, especially those formed by the Döpp-Manton pathway [9–12]. The formation of antheridia may result in the fertilization of sexual polystichoid ferns by spermatozoids formed by apomictic taxa. This process has not been convincingly investigated in ferns yet.

The documented phylogenetic pattern introduces new questions concerning the uneven distribution of apomixis in these ferns. Above, we explored various factors such as ecological spatial aspects and hybridization. However, not all lineages of polystichoids occurring in SE Asia contain apomicts. Thus, other factors require to be considered in addressing the observed clustering of apomixis in the phylogeny of ferns. Considering evidence in angiosperms [65–67], the study of genome structure and gene expression may result in the discovery of preadaptations to apomixis present in some ferns. However, the current evidence does not allow rejecting the hypothesis of a random evolution.

5. Conclusion

Our study supports a scenario in which apomixis was repeatedly established in fern lineages that experienced frequent reticulate evolution combined with polyploidization. The apomictic lineages showed no increase in speciation rate, instead all apomictic lineages appeared to be short lived despite some evidence for postorigin diversification. In general, the results support the hypothesis that apomictic ferns are evolutionary dead ends in the long term but maintain the short-term potential to be highly successful in particular ecological conditions such as climates with strong seasonality. Finally, the study adds to the rapidly growing number of studies showing the power of the comparative approaches using robust phylogenetic frameworks to infer key questions in macroecological and macroevolutionary research. So far, the waste majority of studies using phylogenetic evidence to study apomixis in ferns are focused on species complexes [5, 17–21]. As shown here and in a study on the genus *Pteris* [23], studies focusing on species-rich genera, families, or the whole phylogenetic tree of ferns will likely recover the answers to the core question: "why is apomixis so common in ferns?"

Acknowledgments

The authors thank several colleagues for discussions of various aspects addressed in this study in particular Tim Barraclough (London, UK), David Barrington (Vermont, USA), Vincent Savolainen (London, UK), and Xian-Chun Zhang (Beijing, China). The authors acknowledge also the helpful comments of one anonymous reviewer. The State Key Laboratory of Systematic and Evolutionary Botany, Institute of Botany, the Chinese Academy of Sciences financially supported this study (to H. M. Liu). H. Schneider acknowledges the Senior Visiting Professorship awarded by the Chinese Academy of Sciences.

References

[1] D. R. Marshall and A. H. D. Brown, "The evolution of apomixis," *Heredity*, vol. 47, pp. 1–15, 1981.

[2] M. Mogie, *The Evolution of Asexual Reproduction in Plants*, 1992.

[3] E. Hörandl, "A combinational theory for maintenance of sex," *Heredity*, vol. 103, no. 6, pp. 445–457, 2009.

[4] M. T. J. Johnson, R. G. FitzJohn, S. D. Smith, M. D. Rausher, and S. P. Otto, "Loss of sexual recombination and segregation is associated with increased diversification in evening primroses," *Evolution*, vol. 65, pp. 3230–3240, 2011.

[5] J. B. Beck, M. D. Windham, and K. M. Pryer, "Do asexual polyploid lineages led short lives? A case study from the fern genus *Astrolepis*," *Evolution*, vol. 65, pp. 3217–3229, 2011.

[6] D. Fontaneto, E. A. Herniou, C. Boschetti et al., "Independently evolving species in asexual bdelloid rotifers," *PLoS Biology*, vol. 5, no. 4, pp. 914–921, 2007.

[7] T. Schwander, L. Henry, and B. J. Crespi, "Molecular evidence for ancient asexuality in timema stick insects," *Current Biology*, vol. 21, no. 13, pp. 1129–1134, 2011.

[8] W. Döpp, "Cytologische und genetische Untersuchungen innerhalb der Gattung *Dryopteris*," *Planta*, vol. 29, no. 4, pp. 481–533, 1939.

[9] I. Manton, *Problems of Cytology and Evolution in the Pteridophyta*, Cambridge University Press, New York, NY, USA, 1950.

[10] A. F. Braithwaite, "A new type of apogamy in ferns," *New Phytologist*, vol. 63, pp. 293–305, 1964.

[11] J. D. Lovis, "Evolutionary patterns and processes in ferns," *Advances in Botanical Research*, vol. 4, pp. 229–415, 1978.

[12] T. G. Walker, "Apomixis and vegetative reproduction in ferns," in *Reproductive Biology and Taxonomy of Vascular Plants*, J. G. Hawkes, Ed., pp. 152–161, Botanical Society of the British Isles, Middlesex, UK, 1966.

[13] C. H. Park and M. Kato, "Apomixis in the interspecific triploid hybrid fern *Cornopteris christenseniana* (Woodsiaceae)," *Journal of Plant Research*, vol. 116, no. 2, pp. 93–103, 2003.

[14] U. Grossniklaus, G. A. Nogler, and P. J. Van Dijk, "How to avoid sex: the genetic control of gametophytic apomixis," *Plant Cell*, vol. 13, no. 7, pp. 1491–1497, 2001.

[15] R. A. Bicknell and A. M. Koltunow, "Understanding apomixis: recent advances and remaining conundrums," *Plant Cell*, vol. 16, pp. S228–S245, 2004.

[16] H. Schneider, E. Schuetipelz, K. M. Pryer, R. Cranfill, S. Magallón, and R. Lupia, "Ferns diversified in the shadow of angiosperms," *Nature*, vol. 428, no. 6982, pp. 553–557, 2004.

[17] A. L. Grusz, M. D. Windham, and K. M. Pryer, "Deciphering the origins of apomictic polyploids in the *Cheilanthes yavapensis* complex (Pteridaceae)," *American Journal of Botany*, vol. 96, no. 9, pp. 1636–1645, 2009.

[18] Y. M. Huang, S. Y. Hsu, T. H. Hsieh, H. M. Chou, and W. L. Chiou, "Three *Pteris* species (Pteridaceae: Pteridophyta) reproduce by apogamy," *Botanical Studies*, vol. 52, no. 1, pp. 79–87, 2011.

[19] J. Schneller and K. Krattinger, "Genetic composition of Swiss and Austrian members of the apogamous *Dryopteris affinis* complex (Dryopteridaceae, Polypodiopsida) based on ISSR markers," *Plant Systematics and Evolution*, vol. 286, no. 1-2, pp. 1–6, 2010.

[20] J. B. Beck, P. J. Alexander, L. Alphin et al., "Does hybridization drive the transition to asexuality in diploid *Boechaera*?" *Evolution*, vol. 66, no. 4, pp. 985–995, 2012.

[21] J. B. Beck, M. D. Windham, G. Yatskievych, and K. M. Pryer, "A diploids-first approach to species delimitation and interpreting polyploid evolution in the fern genus *Astrolepis* (pteridaceae)," *Systematic Botany*, vol. 35, no. 2, pp. 223–234, 2010.

[22] E. M. Sigel, M. D. Windham, L. Huiet, G. Yatskievych, and K. M. Pryer, "Species relationships and farina evolution in the cheilanthoid fern genus *Argyrochosma* (Peridaceae)," *Systematic Botany*, vol. 36, no. 3, pp. 554–564, 2011.

[23] Y. S. Chao, H. Y. Liu, Y. C. Ching, and W. L. Chiou, "Polyploidy and speciation in *Pteris* (Pteridaceae)," *Journal of Botany*, vol. 2012, Article ID 817920, 7 pages, 2012.

[24] P. S. Soltis and D. E. Sotlis, "Evolution of inbreeding and outcrossing in ferns and fern-allies," *Plant Species Biology*, vol. 5, pp. 1–11, 1990.

[25] E. R. J. Wubs, G. A. De, H. J. During et al., "Mixed mating system in the fern *Asplenium scolopendrium*: implications for colonization potential," *Annals of Botany*, vol. 106, no. 4, pp. 583–590, 2010.

[26] E. Hörandl, "Evolution and biogeography of alpine apomictic plants," *Taxon*, vol. 60, no. 2, pp. 390–402, 2011.

[27] D. P. Little and D. S. Barrington, "Major evolutionary events in the origin and diversification of the fern genus *Polystichum* (Dryopteridaceae)," *American Journal of Botany*, vol. 90, no. 3, pp. 508–514, 2003.

[28] C. X. Li, S. G. Lu, and Q. Yang, "Asian origin for *Polystichum* (Dryopteridaceae) based on rbcL sequences," *Chinese Science Bulletin*, vol. 49, no. 11, pp. 1146–1150, 2004.

[29] C. X. Li, S. G. Lu, and D. S. Barrington, "Phylogeny of Chinese *Polystichum* (Dryopteridaceae) based on chloroplast DNA sequence data (*trnL-F* and *rps4-trnS*)," *Journal of Plant Research*, vol. 121, no. 1, pp. 19–26, 2008.

[30] H. M. Liu, X. C. Zhang, Z. D. Chen, and Y. L. Qiu, "Inclusion of the eastern Asia endemic genus *Sorolepidium* in *Polystichum* (Dryopteridacae): evidence from the chloroplast *rbcL* and *atpB* genes," *International Journal of Plant Sciences*, vol. 689, pp. 1311–1323, 2007.

[31] H. M. Liu, X. C. Zhang, W. Wang, and H. Zeng, "Molecular phylogeny of the endemic fern genera *Cyrtomidictyum* and *Cyrtogonellum* (Dryopteridaceae) from East Asia," *Organisms Diversity and Evolution*, vol. 10, no. 1, pp. 57–68, 2010.

[32] J. M. Lu, D. Z. Li, L. M. Gao, X. Cheng, and D. Wu, "Paraphyly of *Cyrtomium* (Dryopteridaceae): evidence from *rbcL* and *trnL-F* sequence data," *Journal of Plant Research*, vol. 118, no. 2, pp. 129–135, 2005.

[33] J. M. Lu, D. S. Barrington, and D. Z. Li, "Molecular phylogeny of the polystichoid ferns in Asia based on *rbcL* sequences," *Systematic Botany*, vol. 32, no. 1, pp. 26–33, 2007.

[34] H. E. Driscoll and D. S. Barrington, "Origin of Hawaiian *Polystichum* (Dryopteridaceae) in the context of a world phylogeny," *American Journal of Botany*, vol. 94, no. 8, pp. 1413–1424, 2007.

[35] L. R. Perrie, P. J. Brownsey, P. J. Lockhart, and M. F. Large, "Evidence for an allopolyploid complex in New Zealand *Polystichum* (Dryopteridaceae)," *New Zealand Journal of Botany*, vol. 41, no. 2, pp. 189–215, 2003.

[36] L. B. Zhang and H. He, "*Polystichum speluncicola* sp. nov. (sect. *Haplopolystichum*, Dryopteridaceae) based on morphological, palynological, and molecular evidence with reference to the non-monophyly of *Cyrtogonellum*," *Systematic Botany*, vol. 35, no. 1, pp. 13–19, 2010.

[37] J. M. Lu, X. Cheng, D. Wu, and D. Z. Li, "Chromosome study of the fern genus *Cyrtomium* (Dryopteridaceae)," *Botanical Journal of the Linnean Society*, vol. 150, no. 2, pp. 221–228, 2006.

[38] P. S. Soltis and D. E. Soltis, "Population structure and estimates of gene flow in the homosporous fern Polystichum munitum," *Evolution*, vol. 41, pp. 620–629, 1987.

[39] G. Migliaro, Y. Gabriel, and J. M. Galan, "Gametophyte development and reproduction of the Asian fern *Polystichum polyblepharum* (Roem. Ex Kunze) C. Presl, (Dryopteridaceae, Polypodiopsida)," *Plant Biosystems*, vol. 146, no. 2, pp. 368–373, 2012.

[40] Z. S. An, J. E. Kutzbach, W. L. Prell, and S. C. Porter, "Evolution of Asian monsoons and phased uplift of the Himalaya—Tibetan plateau since Late Miocene times," *Nature*, vol. 411, no. 6833, pp. 62–66, 2001.

[41] G. Yatskievych, "Antheridiogen response in *Phanerophlebia* and related fern genera," *American Fern Journal*, vol. 83, pp. 30–36, 1993.

[42] W. M. Bao, Q. X. Wang, and S. J. Dai, "Morphology and ontogeny of the gametophyte of *Cyrtogonellum inaequale* Ching," in *Ching Memorial Volume*, Beijing, China, 1999.

[43] X. Cheng and S. Z. Zhang, *Index to Chromosome Numbers of Chinese Pteridophyta*, 2010.

[44] M. Kato, N. Nakato, X. Cheng, and K. Iwatsuki, "Cytotaxonomic study of ferns of Yunnan, southwestern China," *The Botanical Magazine Tokyo*, vol. 105, no. 1, pp. 105–124, 1992.

[45] V. Raghavan, *Developmental Biology of Fern Gametophytes*, 1989.

[46] W. P. Maddison and D. R. Maddison, "Mesquite 2. 75: a modular system for evolutionary analysis," 2012, http://mesquite-project.org/mesquite/mesquite.html.

[47] D. Posada, "jModelTest: phylogenetic model averaging," *Molecular Biology and Evolution*, vol. 25, no. 7, pp. 1253–1256, 2008.

[48] D. L. Swofford, *PAUP*. Phylogenetic Analysis Using Parsimony (* and Other Methods). Version 4*, Sinauer Associates, Sunderland, Mass, USA, 2003.

[49] S. Guindon, J. F. Dufayard, V. Lefort, M. Anisimova, W. Hordijk, and O. Gascuel, "New algorithms and methods to estimate maximum-likelihood phylogenies: assessing the performance of PhyML 3.0," *Systematic Biology*, vol. 59, no. 3, pp. 307–321, 2010.

[50] A. J. Drummond, B. Ashton, S. Buxton et al., "Geneious v. 5. 4.," 2011, http://www.geneious.com/.

[51] F. Ronquist and J. P. Huelsenbeck, "MrBayes 3: bayesian phylogenetic inference under mixed models," *Bioinformatics*, vol. 19, no. 12, pp. 1572–1574, 2003.

[52] A. J. Drummond, M. A. Suchard, D. Xie, and A. Rambaut, "Bayesian phylogenetics with BEAUti and the BEAST 1. 7.," *Molecular Biology and Evolution*, vol. 29, no. 8, pp. 1969–1973, 2012.

[53] E. Schuettpelz and K. M. Pryer, "Evidence for a Cenozoic radiation of ferns in an angiosperm-dominated canopy," *Proceedings of the National Academy of Sciences of the United States of America*, vol. 106, no. 27, pp. 11200–11205, 2009.

[54] A. Rambaut and A. J. Drummond, "TRACER: MCMC Trace Analysis Tool v. 1. 5.," 2009, http://beast.bio.ed.ac.uk/.

[55] A. Rambaut, "FigTree 1. 31.," 2009, http://tree.bio.ed.ac.uk/.

[56] M. Christenhusz, H. Schneider, and X. C. Zhang, "A linear sequence of extant families and genera of lycophytes and ferns," *Phytotaxa*, vol. 19, pp. 7–54, 2011.

[57] A. R. Smith, K. M. Pryer, E. Schuettpelz, P. Korall, H. Schneider, and P. G. Wolf, "A classification for extant ferns," *Taxon*, vol. 55, no. 3, pp. 705–731, 2006.

[58] S. J. Lin, M. Kato, and K. Iwatsuki, "Diploid and triploid offspring of triploid agamosporous fern *Dryopteris pacifica*," *The Botanical Magazine*, vol. 105, no. 3, pp. 443–452, 1992.

[59] M. Pagel, "Detecting correlated evolution on phylogenies: a general method for the comparative analysis of discrete characters," *Proceedings of the Royal Society B*, vol. 255, no. 1342, pp. 37–45, 1994.

[60] W. P. Maddison, P. E. Midford, and S. P. Otto, "Estimating a binary character's effect on speciation and extinction," *Systematic Biology*, vol. 56, no. 5, pp. 701–710, 2007.

[61] S. Lehtonen, "Towards resolving the complete fern tree of life," *PLoS ONE*, vol. 6, no. 10, Article ID e24851, 2011.

[62] H. V. Hunt, S. W. Ansell, S. J. Russell, H. Schneider, and J. C. Vogel, "Dynamics of polyploid formation and establishment in the allotetraploid rock fern *Asplenium majoricum*," *Annals of Botany*, vol. 108, no. 1, pp. 143–157, 2011.

[63] R. Ootsuki, H. Sato, N. Nakato, and N. Murakami, "Evidence for genetic segregation in the apogamous fern *Cyrtomium fortunei* (Dryopteridaceae)," *Journal of Plant Research*, vol. 125, no. 5, pp. 605–612, 2012.

[64] E. Hörandl, C. Dobeš, J. Suda et al., "Apomixis is not prevalent in subnival to nival plants of the European Alps," *Annals of Botany*, vol. 108, no. 2, pp. 381–390, 2011.

[65] P. J. van Dijk and K. Vijverberg, "The significance of apomixis in the evolution of the angiosperms: a reappraisal," *Regnum Vegetabile*, vol. 143, pp. 101–116, 2005.

[66] J. G. Carman, "Asynchonous expression of duplicate genes in angiosperms may cause apomixis, bispory, tetraspory, and polyembryony," *Biological Journal of the Linnean Society*, vol. 61, no. 1, pp. 51–94, 1997.

[67] D. Haig and M. Westoby, "Genomic imprinting in endosperm: its effect on seed devlopment in crosses between species, and between different ploidies of the same species, and its implications for the evolution of apomixis," *Philosophical Transactions of the Royal Society London Series B*, vol. 333, pp. 1–13, 1991.

18

Role of Ascorbate in the Regulation of the *Arabidopsis thaliana* Root Growth by Phosphate Availability

Jarosław Tyburski, Kamila Dunajska-Ordak, Monika Skorupa, and Andrzej Tretyn

Chair of Plant Physiology and Biotechnology, Nicolaus Copernicus University, Gagarina 9, 87-100 Toruń, Poland

Correspondence should be addressed to Jarosław Tyburski, tybr@umk.pl

Academic Editor: Philip White

Arabidopsis root system responds to phosphorus (P) deficiency by decreasing primary root elongation and developing abundant lateral roots. Feeding plants with ascorbic acid (ASC) stimulated primary root elongation in seedlings grown under limiting P concentration. However, at high P, ASC inhibited root growth. Seedlings of ascorbate-deficient mutant *(vtc1)* formed short roots irrespective of P availability. P-starved plants accumulated less ascorbate in primary root tips than those grown under high P. ASC-treatment stimulated cell divisions in root tips of seedlings grown at low P. At high P concentrations ASC decreased the number of mitotic cells in the root tips. The lateral root density in seedlings grown under P deficiency was decreased by ASC treatments. At high P, this parameter was not affected by ASC-supplementation. *vtc1* mutant exhibited increased lateral root formation on either, P-deficient or P-sufficient medium. Irrespective of P availability, high ASC concentrations reduced density and growth of root hairs. These results suggest that ascorbate may participate in the regulation of primary root elongation at different phosphate availability via its effect on mitotic activity in the root tips.

1. Introduction

It has been reported that phosphorus (P) availability affects root architecture in *Arabidopsis*. Seedlings of this plant, exposed to low P concentration ($1\,\mu$M), formed a highly branched root system with abundant lateral roots and a short primary root. Under these growth conditions primary, and secondary roots had an abundance of long root hairs. Under high P concentrations (1 mM), the root system was composed of a long primary root with few lateral roots and short root hairs [1–3].

P deficiency responses of the root system are dependent on changes in cell proliferation. Under low P conditions, the reduction in primary root growth is due to inhibition of cell division and cell differentiation within the primary root meristem. The mitotic activity is relocated to the sites of lateral root formation, resulting in increased lateral root density. However, similar to the primary root tip, cell differentiation in older lateral roots occurs within the apical root meristem, which results in the arrest of lateral root elongation [4]. Besides the reduction in cell division rate, low P treatment inhibits cell elongation and reduces the number of cells in root elongation zone [4, 5].

Little is known about the mechanisms underlying the alteration in growth and development of *Arabidopsis* roots after plants are exposed to limiting P supply. It has been demonstrated that an arrest of root growth under low P concentration is dependent on the direct contact of growing root tips with the medium where P sensing occurs. This effect is mediated by two multicopper oxidases, LPR1 and LPR2, expressed in the root tip including the meristem and the root cap [6]. Recently, an important role in phosphate sensing and in root response to low P was attributed to *PHOSPHATE DEFICIENCY RESPONSE 2* gene (*PDR2*) encoding a P5-type ATPase that is required for proper expression of the *SCARECROW* gene (*SCR*). SCR is a key regulator in tissue patterning and stem-cell maintenance in apical root meristem [7]. Both, LPR1 and PDR2 are associated with the endoplasmic reticulum (ER). This finding points at the role of an ER-resident pathway in adjusting root meristem activity in response to external phosphate [8].

Published data suggests that the reduced elongation rate of primary roots of P-starved plants results from increased auxin accumulation in the root apical meristem [9]. However, other authors have proposed an auxin-independent process responsible for the primary root growth arrest under low P conditions. Increased lateral root formation under low P availability is related to auxin transport and signaling [3].

Besides the hormones, other endogenous factors mediate root developmental responses to nutrient deprivation. Shin and Schatchman [10] have demonstrated that K^+ deprivation is followed by an increase in H_2O_2 production in specific regions of the root. Accumulation of reactive oxygen species (ROS) in the roots of K^+-starved plants was localized just behind the elongation zone (where K^+ active transport and translocation take place). It was also shown that H_2O_2 plays a role in controlling the expression of some genes in response to K^+ deprivation [10]. ROS accumulation in P-deprived roots has also been reported. Under low P conditions, increased ROS production was observed in the cortex. In contrast to P-starved roots, ROS accumulation in K^+- and NO_3^--deprived organs occurred in the epidermis rather than in the cortex [11].

Another component of cellular redox systems—ascorbate, is directly involved in the regulation of two processes that mediate morphogenic responses of root systems to nutrient availability: cell division and elongation. It has been shown that high ascorbic acid (ASC) concentrations are required for normal progression of the cell cycle in meristematic tissues [12–14]. ASC was identified as a factor necessary for G1-S transition. ASC addition to the cells of the root quiescent center induces these normally nondividing cells to pass from G1 into the S phase [13, 15]. Besides its effect on cell proliferation, ASC stimulates cell elongation by increasing cell wall extensibility [16, 17].

In the present study, we address the role of ascorbate in the regulation of root system architecture under different P availability. Wild-type *Arabidopsis* seedlings grown on media containing low or high P concentrations supplemented with various ASC concentration as well as ascorbate-deficient *vtc1* mutants [18], were used in the study. Ascorbate concentration, and its effects on the primary root length, lateral root density, and length were analyzed. We also demonstrate the effect of ASC on the mitotic activity in the primary root meristem of seedlings grown under different phosphate regimes. Finally, the role of ASC in root hair development is assessed.

2. Materials and Methods

2.1. Plant Material. *Arabidopsis* wild-type plants (ecotype *Col 0*) and vitamin C-1 (*vtc1*) mutants were used. Seeds of wild-type and mutant plants were purchased from Nottingham *Arabidopsis* Stock Centre (University of Nottingham, UK). Seeds were soaked in sterile distilled water for 30 min. Surface sterilization of the seeds was performed with 95% (v/v) ethanol for 5 min followed by 20% (v/v) bleach for 7 min. Subsequently, the seeds were washed several times in sterile water and sown onto culture media in Petri dishes. Seedlings were grown on the Murashige and Skoog

[19] medium modified according to López-Bucio et al. [2]. In order to provide phosphate deficiency or phosphate sufficiency (control) conditions two basal media were applied. Phosphate deficient media contained $1 \mu M$ NaH_2PO_4, and P sufficient media contained $1 mM$ Na_2HPO_4 [2]. Seedlings were grown on unsupplemented media or on media supplemented with 100, 200, 300, 400, or $500 \mu M$ ASCNa. ASCNa solution was filter-sterilized before adding to the autoclaved medium. Before the culture was initiated, the dishes were placed in dark at 4°C for 48 hours to promote and synchronize germination [2]. Seedlings were grown at 25°C under continuous white light with standard irradiation ($431 \mu mol\ m^{-2}\ s^{-1}$) provided by Osram 30 W/11-860 "Daylight" fluorescent tubes (Osram, Berlin, Germany).

2.2. The Analysis of the Root System Architecture. Control and ascorbate-treated seedlings were photographed and the images were analyzed using Image Gauge software (Fujifilm, Japan). Length of the primary roots, number and length of the lateral roots, and the number and length of root hairs were determined. Density of lateral roots, that is, number of lateral roots/cm of primary root, and root hair density, that is, number of root hairs/mm of primary root, were calculated. Length of the primary roots, lateral root density, and lateral root length were determined after 12 days of culture on basal- or ASC-supplemented media. Root hair density and length were analyzed using 5-day-old seedlings.

2.3. Determination of Cell Division Frequency. Cell division frequency was determined microscopically in root tip squash preparations by inspecting at least 1000 cells and expressed as a percentage of mitotic cells in a single-root-tip squash. Roots were fixed in a 3 : 1 ethanol : acetic acid (v/v), hydrolyzed in 1 N HCl for 1 min, washed in a 50 mM phosphate buffer (pH 7.5), and stained with 4,6-diamidino-2-phenylindole (DAPI; $0.1\ mg\ L^{-1}$) for 10 min. Segments, 1 mm in length, starting from the root tip were dissected, squashed in a drop of 50 mM phosphate buffer (pH 7.5), and observed under fluorescent microscope at an excitation wavelength of 365 nm. Mitotic activity in the root tips was analyzed after 8 and 12 days of culture.

2.4. Ascorbate and Dehydroascorbate Determination. Concentrations of the reduced and oxidized form of ascorbate were determined in the apices of primary roots (samples consisted of approximately final 0.5 cm of the primary root) of *Arabidopsis* seedlings. The samples were homogenized in liquid nitrogen and extracted for 15 min with 6% trichloroacetic acid (TCA) at 0°C. Subsequently, 6% TCA was added to the samples up to 0.5 cm^3. The homogenate was centrifuged at $15\ 000 \times g$ for 5 min at 4°C. ASC and DHA were determined after the reduction of dehydroascorbate to ascorbate with dithiotreitol (DTT) and measured as described by Kampfenkel et al. [20]. The assay is based on the reduction of Fe^{3+} to Fe^{2+} by ASC and the spectrophotometric detection of Fe^{2+} complexed with $2.2'$-dipirydyl. DHA content ($\mu g/g\ fw$) in extracts was calculated from the difference between ASC+DHA and ASC ($\mu g/g\ fw$). ASC and DHA content in root apices were determined after

12 days of culture. Ascorbate redox state was expressed as $([ASC]/[ASC+DHA]) \times 100$.

2.5. H₂O₂ Localization in the Root-Hair-Forming Zone of the Primary Root.

2.5. H₂O₂ Localization in the Root-Hair-Forming Zone of the Primary Root. Patterns of H_2O_2 accumulation were studied with 2′,7′-dichlorodihydrofluorescein (DCFH) diacetate. DCFH-diacetate can cross the plasma membrane and, after being deacetylated by endogenous esterase, liberates DCFH in the cytoplasm, where it is oxidized in the reaction with H_2O_2 to highly fluorescent 2′,7′-dichlorofluorescein (DCF) [21]. In order for the dye to infiltrate the cells, roots were incubated for 15 min in 50 mM phosphate buffer (pH 7.5) containing 50 μM DCFH-diacetate and subsequently rinsed with phosphate buffer and imaged employing the Eclipse (Nicon) confocal microscope using 488 nm excitation and 525 nm emission spectra [22]. Optical sections were collected with a z focus increment of 5 μm. H_2O_2 distribution patterns were analyzed in 5-day-long seedling of wild-type control plants, plants treated with 500 μM ASC, and *vtc1* mutant plants. The experiment was repeated three times. The photographs show the representative view chosen from at least 15 plants analyzed in each experiment.

2.6. Statistics.

2.6. Statistics. Student's *t*-test was applied to determine the statistical significance of the results as compared to the control. The data for ASC/DHA determination represent the mean and standard deviation (SD) of at least three independent experiments. At least 30 seedlings were used in each replicate to obtain the ascorbate extracts. Data for root number and root length represents the mean and SD of three independent experiments with at least 25 seedlings in each replicate.

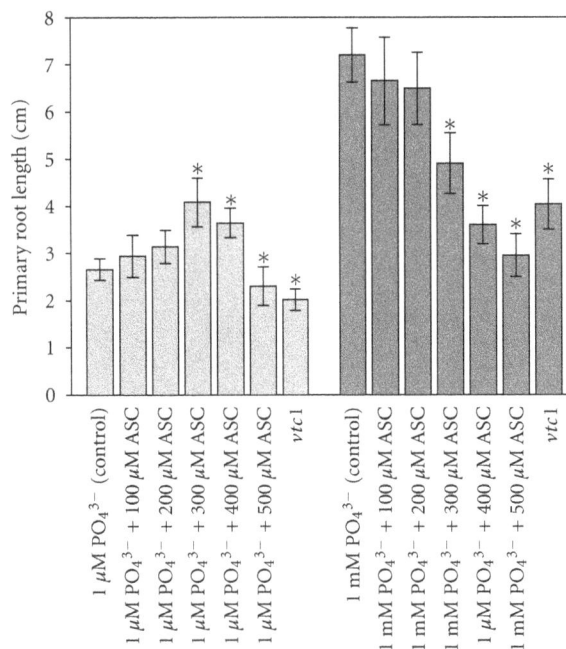

FIGURE 1: Effect of phosphate availability and ascorbate concentration on the length of the primary root of the *Arabidopsis* seedlings. Wild-type (Col 0) and *vtc1* mutant seedlings were grown 12 days on vertically oriented agar plates containing low (1 μM) P or high (1 mM) P medium, after that, primary root length was determined. Wild-type seedlings were cultured on media supplemented with varying concentrations of ASC or on ascorbate—free media (control). Seedlings of *vtc1* mutant were cultured on ascorbate—free media. Values shown represent the mean of at least 50 seedlings ±SD. An asterisk denotes significant differences from the control with $P < 0.05$.

3. Results

3.1. The Effect of ASC on Root System Architecture in the Arabidopsis Seedlings Grown at Low (1 μM PO₄³⁻) and High (1 mM PO₄³⁻) Phosphate Concentration.

3.1. The Effect of ASC on Root System Architecture in the Arabidopsis Seedlings Grown at Low (1 μM PO₄³⁻) and High (1 mM PO₄³⁻) Phosphate Concentration. Control seedlings grown under low P concentration produced shorter primary root when compared to those cultured on the medium supplemented with high P (Figures 1, 2(a), and 2(f)). In contrast to the primary root, lateral roots of plants grown at low P conditions were longer, when compared to those, formed by seedlings grown at high P availability (Table 1(a)). Supplementing ASC to the medium affected the growth of both, the primary root and the lateral roots of seedlings cultured under P-deficient or P-sufficient medium (Figure 2, Table 1(a)). When plants were grown under low P, ASC in concentrations of 300 and 400 μM significantly stimulated primary root elongation. Highest stimulatory effect was observed when 300 μM ASC was added to the medium. ASC concentrations higher than 400 μM did not result in further stimulation of primary root growth. On the contrary, seedlings cultured in the presence of 500 μM ASC formed slightly shorter roots than untreated controls, however, the difference was not statistically significant (Figure 1). Further increase in ASC concentration in the medium resulted in significant inhibition of primary root growth (data not shown). As opposed to plants grown under P deficiency, when plants

were cultured under high P, ASC concentrations applied did not stimulate primary root growth. The lowest ASC concentrations, that is, 100 and 200 μM did not significantly affect primary root growth. Higher concentrations, in a dose-dependent manner, inhibited primary root elongation (Figure 1). Seedlings of *vtc1* mutants cultured on both, P-deficient or P-sufficient medium, formed significantly shorter primary roots than wild-type plants (Figures 1, 2(a), 2(d), 2(f), and 2(i)). When *vtc1* mutants were fed with 300 μM ASC, they formed roots of the length comparable with wild-type plants (Figures 2(e) and 2(j)).

The length of lateral roots was negatively affected by exogenous ASC under both phosphate regimes (Table 1(a)). When plants were grown under low P availability, all ASC concentrations applied significantly decreased lateral root length (Table 1(a)). When plants were grown on medium containing 1 mM P, 100 and 200 μM ASC did not affect lateral root elongation, however, this process was strongly inhibited by higher ASC concentrations (Table 1(a)). Lateral roots of *vtc1* seedlings did not differ in length from the wild-type seedlings when plants were grown on either the P-deficient or P-sufficient medium (Table 1(a)).

The density of lateral roots was calculated to normalize the effects of P availability and ASC application on primary

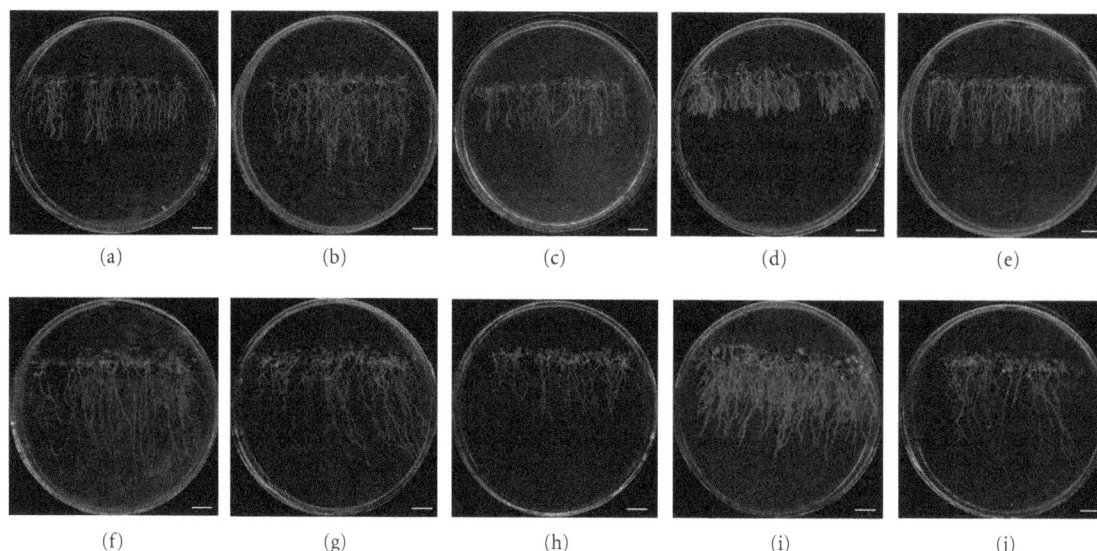

FIGURE 2: Effect of phosphate availability and ascorbate concentration on *Arabidopsis* root architecture. Wild-type (Col 0) seedlings were grown in the presence of the low ($1\,\mu$M) P concentration (a) or on the same medium supplemented with $300\,\mu$M ASC (b) or $500\,\mu$M ASC (c). *vtc1* seedlings were grown in the presence of the low ($1\,\mu$M) P concentration (d) or on the same medium supplemented with $300\,\mu$M ASC (e). Wild-type Col 0 seedlings were grown in the presence of high (1 mM) P (f) or on the same medium supplemented with $300\,\mu$M ASC (g) or $500\,\mu$M ASC (h). *vtc1* seedlings were grown in the presence of high (1 mM) P (i) or on the same medium supplemented with $300\,\mu$M ASC (j). Seedlings were photographed 12 days after germination. Bar = 1 cm.

TABLE 1: Effect of ASC on the lateral root number, lateral root length, and lateral root density.

(a) Lateral root length (mm)

$1\,\mu$M PO_4^{3-}						
Control	$100\,\mu$M ASC	$200\,\mu$M ASC	$300\,\mu$M ASC	$400\,\mu$M ASC	$500\,\mu$M ASC	*vtc1*
$8,36\pm0,29$	$6,30\pm0,49^*$	$5,55\pm0,43^*$	$5,12\pm0,37^*$	$4,52\pm0,36^*$	$3,51\pm0,30^*$	$7,46\pm0,24$
1 mM PO_4^{3-}						
Control	$100\,\mu$M ASC	$200\,\mu$M ASC	$300\,\mu$M ASC	$400\,\mu$M ASC	$500\,\mu$M ASC	*vtc1*
$5,29\pm0,52$	$5,69\pm0,58$	$4,94\pm0,61$	$3,64\pm0,40^*$	$3,62\pm0,36^*$	$2,71\pm0,32^*$	$5,55\pm0,15$

(b) Lateral root density (number of lateral roots/cm of primary root)

$1\,\mu$M PO_4^{3-}						
Control	$100\,\mu$M ASC	$200\,\mu$M ASC	$300\,\mu$M ASC	$400\,\mu$M ASC	$500\,\mu$M ASC	*vtc1*
$3,22\pm0,47$	$2,76\pm0,56$	$2,55\pm0,45^*$	$1,96\pm0,43^*$	$1,68\pm0,40^*$	$1,87\pm0,38^*$	$5,30\pm1,05^*$
1 mM PO_4^{3-}						
Control	$100\,\mu$M ASC	$200\,\mu$M ASC	$300\,\mu$M ASC	$400\,\mu$M ASC	$500\,\mu$M ASC	*vtc1*
$1,17\pm0,28$	$1,31\pm0,27$	$1,03\pm0,17$	$1,25\pm0,18$	$1,20\pm0,19$	$0,91\pm0,32$	$3,39\pm0,27^*$

Lateral root number and lateral root density were determined after 12 days of culture. *Arabidopsis* seedlings were grown on media containing $1\,\mu$M PO_4^{3-} or 1 mM PO_4^{3-} or on the same media supplemented with 100–$500\,\mu$M ASC. Seedlings of *vtc1* mutant were grown on medium containing either $1\,\mu$M PO_4^{3-} or 1 mM PO_4^{3-}. Data represent mean and standard deviation. *Significant differences from control with $P < 0.05$.

root length. This parameter decreased approximately 2-fold in plants grown under high P conditions, when compared to plants grown under low P conditions (Table 1(b)). Treatment with ascorbate concentration of $200\,\mu$M or higher resulted in a significant decrease of lateral root density in plants cultured under low P. When plants were grown at 1 mM P, ASC concentration did not change the lateral root density. Irrespective of the P concentration in medium, seedlings of

vtc1 mutant were characterized by a significantly increased lateral root density when compared to wild-type plants (Table 1(b)).

3.2. *ASC Concentration and Redox State in Arabidopsis Seedlings Grown at Low ($1\,\mu$M PO_4^{3-}) and High (1 mM PO_4^{3-}) Phosphate Concentration.* The concentration of endogenous ascorbate (ASC) and dehydroascorbate

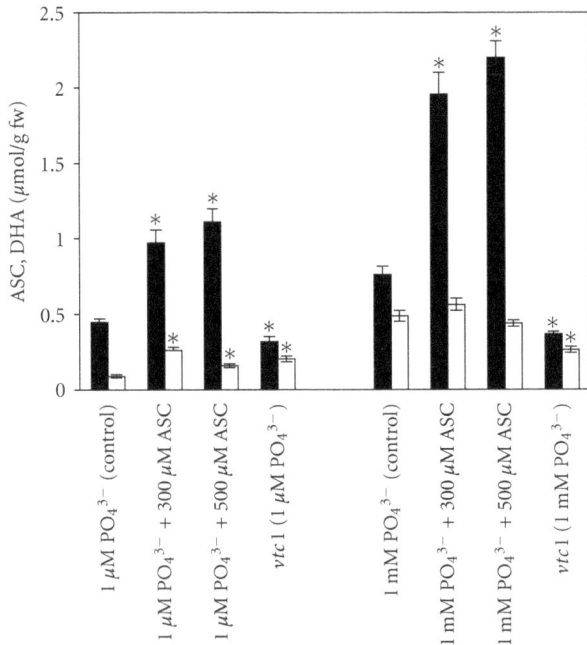

FIGURE 3: Ascorbate (black bars) and dehydroascorbate (white bars) content in root apices of the seedlings of wild-type *Arabidopsis thaliana* (Col 0) and *vtc1* mutant grown 12 days on vertically oriented agar plates on the low ($1 \mu M$) P or high (1 mM) P medium. Another set of wild-type seedlings was grown on low P or high P media supplemented with 300 or $500 \mu M$ ASC. Mean and $\pm SD$ is shown. An asterisk denotes significant differences from the control with $P < 0.05$.

TABLE 2: Effect of phosphate availability and exogenous ASC on ascorbate redox state $(([ASC]/[ASC+DHA]) \times 100)$ in root apices of *Arabidopsis* seedlings.

$1 \mu M$ PO$_4^{3-}$			
control	$300 \mu M$ ASC	$500 \mu M$ ASC	*vtc1*
$83,3 \pm 5,0$	$78,9 \pm 7,9$	$87,5 \pm 7,8$	$61,1 \pm 8,9^*$
1 mM PO$_4^{3-}$			
control	$300 \mu M$ ASC	$500 \mu M$ ASC	*vtc1*
$60,9 \pm 7,2$	$77,6 \pm 7,2^*$	$83,3 \pm 5,1^*$	$58,2 \pm 4,7$

Ascorbate redox state $(([ASC]/[ASC+DHA]) \times 100)$ in root tips of *Arabidopsis* seedlings was determined after 12 days of culture. *Arabidopsis* seedlings were grown on media containing $1 \mu M$ PO$_4^{3-}$ or 1 mM PO$_4^{3-}$ or on the same media supplemented with 100 or $500 \mu M$ ASC. Seedlings of *vtc1* mutant were grown on medium containing either $1 \mu M$ PO$_4^{3-}$ or 1 mM PO$_4^{3-}$. *Significant differences from control with $P < 0.05$.

(DHA) was determined in the primary root apices (Figure 3). Additionally, ASC concentration in the media was measured during the culture, in order to detect possible ASC breakdown in the media. However, no significant decrease in ASC concentration in the media was detected (data not shown).

Root tips of seedlings grown under high P concentration accumulated more ASC and DHA than those grown under P deficiency. And under high P concentration, the participation of DHA in total ascorbate pool was higher when compared to low P conditions (Figure 3, Table 2). Root tips of *vtc1* mutants cultured on phosphate-sufficient medium contained significantly less amounts of both, ASC and DHA than the wild-type plants. However, when mutant plants where grown on P-deficient medium, only ASC content was reduced in comparison to the wild-type plants, whereas DHA concentration was slightly, but significantly, increased (Figure 3). Total ascorbate pool in the root tips of *vtc1* was more oxidized when compared to wild-type plants if seedlings were grown under P deficiency. In contrast, ascorbate redox state did not differ significantly between *vtc1* and wild-type, if plants were grown on phosphate-sufficient medium (Table 2).

Adding ASC to either P-deficient or P-sufficient medium resulted in a strong increase in ascorbate content in the root tips. It should be noted that seedlings, grown under low P conditions, reacted to ASC treatments with a smaller increase in ascorbate concentration in the root apices when compared

to the roots of plants grown under high P conditions (Figure 3). Under low P amounts, an increase in ASC content was accompanied by an elevation in DHA concentration. However, due to a strong rise in ASC concentration, an ascorbate redox state remained unchanged when compared to untreated control (Table 2). Contrary to P-deficient plants, DHA concentration was unchanged, by ASC treatment, in root tips of plants cultured on high P medium (Figure 3). Consequently, an increase in ASC content in root apices of ASC-treated plants shifted ascorbate/dehydroascorbate ratio towards more reduced redox state (Table 2).

3.3. Effects of ASC on Cell Divisions in the Primary Root Tips of the Arabidopsis Seedlings Grown at Low ($1 \mu M$ PO$_4^{3-}$) and High (1 mM PO$_4^{3-}$) Phosphate Concentration.

To study whether the effects of ASC concentration on root elongation in seedlings grown under low P or high P concentration are mediated by changes in the cell division rate in the root meristem, we determined the frequency of dividing cells in the primary root tips of wild-type control plants, ASC-treated plants and *vtc1* mutant plants. Cell division activity was assayed twice during the culture period: 8 and 12 days after germination. Mitosis frequency in root tips of wild-type plants grown under low P was approximately six times lower than the corresponding part of the plants' root cultured at high P (Figure 4). At low P conditions, increasing ASC concentrations in the medium resulted in a gradual stimulation of mitotic activity. Highest increase in the rate of cell divisions was observed in seedlings cultured in the presence of $300 \mu M$ ASC. With higher ASC concentrations, its stimulatory effect was decreased proportionally to the ASC concentration (Figure 4). In contrast to seedlings grown at $1 \mu M$ P, treatments with a series of ASC concentrations led to a gradual decrease in the mitotic activity in root tips of high P-grown plants. Mitotic activity in the root tips of *vtc1* was strongly reduced when compared to wild-type controls irrespective of P concentration in the medium (Figure 4).

3.4. Effects of ASC on Root Hair Density and Elongation.

Root hair development was studied in the area encompassing the final 5 mm of the primary root where root hairs are formed

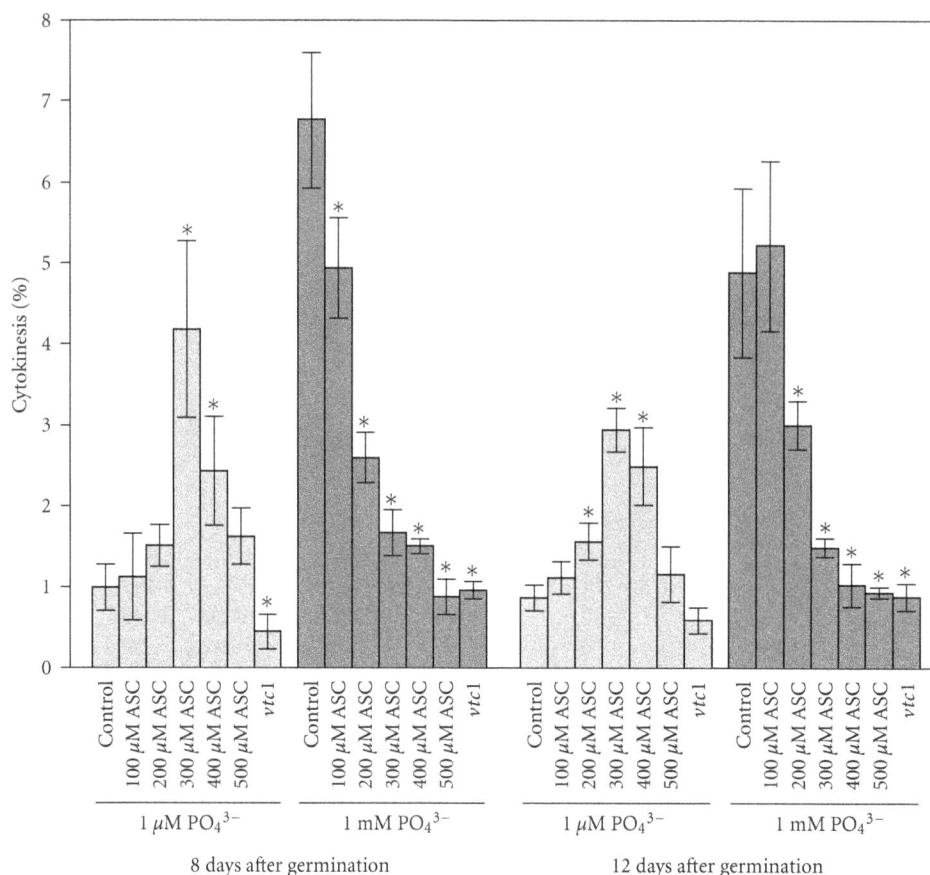

FIGURE 4: Effect of phosphate availability and ascorbate concentration on the cytokinesis frequency in the root tips of *Arabidopsis thaliana*. Mitotic activity in the root tips of wild-type (Col 0) and *vtc1* seedlings was analyzed after 8 and 12 days of culture on the media containing low (1 μM) P or high (1 mM) P concentration. Wild-type seedlings were cultured on media supplemented with varying concentrations of ASC or on ascorbate—free media (control). Seedlings of *vtc1* mutant were cultured on ascorbate—free media. Each determination is based on at least three squash preparations. Mean value and $\pm$SD are indicated. Data represent mean and standard deviation. Asterisk indicates significant differences from the control at $P < 0.05$.

and elongate. It was found that both the wild-type seedlings and *vtc1* mutants were characterized with a similar value of root hair density. This parameter was also not significantly affected by P availability (Figure 5(a)).

Root hair formation was affected by ASC supplementation to the medium. In the presence of high P concentration, ASC decreased the number of root hairs in a dose-dependent manner. When plants grown under P deficiency were analyzed, both, stimulatory and inhibitory effects of the ASC concentration applied were observed. The lowest ASC concentration (100 μM) was found to, slightly but significantly, stimulate the density of root hairs. Higher concentrations did not significantly affect root hair formation. However, a significant the inhibitory effect was observed when the highest concentration (500 μM) was applied (Figure 5(a)).

Root hair growth was strongly affected by P availability. Seedlings grown under limited P availability formed long root hairs in the apical part of the root, while those formed by plants grown on P-sufficient medium were approximately 4 times shorter (Figures 5(b), 6(a), and 6(f)). Root hairs of *vtc1* mutant seedlings were comparable in length with wild-type,

when cultured under low P amount. In contrast to wild-type plants, root hairs of *vtc1* mutants were significantly longer than those of the wild-type when plants were grown under high P concentration (Figures 5(b), 6(f), and 6(j)). Wild-type plants grown in the presence of 100 μM ASC formed longer root hairs when compared to untreated control under both, low and high P, however, only under high P concentration the difference was significant (Figure 5(b)). ASC concentrations of 200 and 300 μM did not affect the root hair length, however, it was strongly reduced when 400 or 500 μM ASC was added to the medium (Figures 5(b) and 6).

Root hair development and elongation is regulated by reactive oxygen species (ROS) production in a root-hair-forming zone [23]. Therefore, we asked whether changing ASC concentration in roots affected H_2O_2 concentration in the rooting zone. In order to assess the H_2O_2 concentrations in the root-hair-forming zone, the roots were stained with DCFH which is oxidized by H_2O_2 to the highly fluorescent 2′,7′-dichlorofluorescein (DCF). Within the root-forming zone, trichoblast cells, were characterized by the highest DCF fluorescence (Figure 7). Irrespective of P concentration

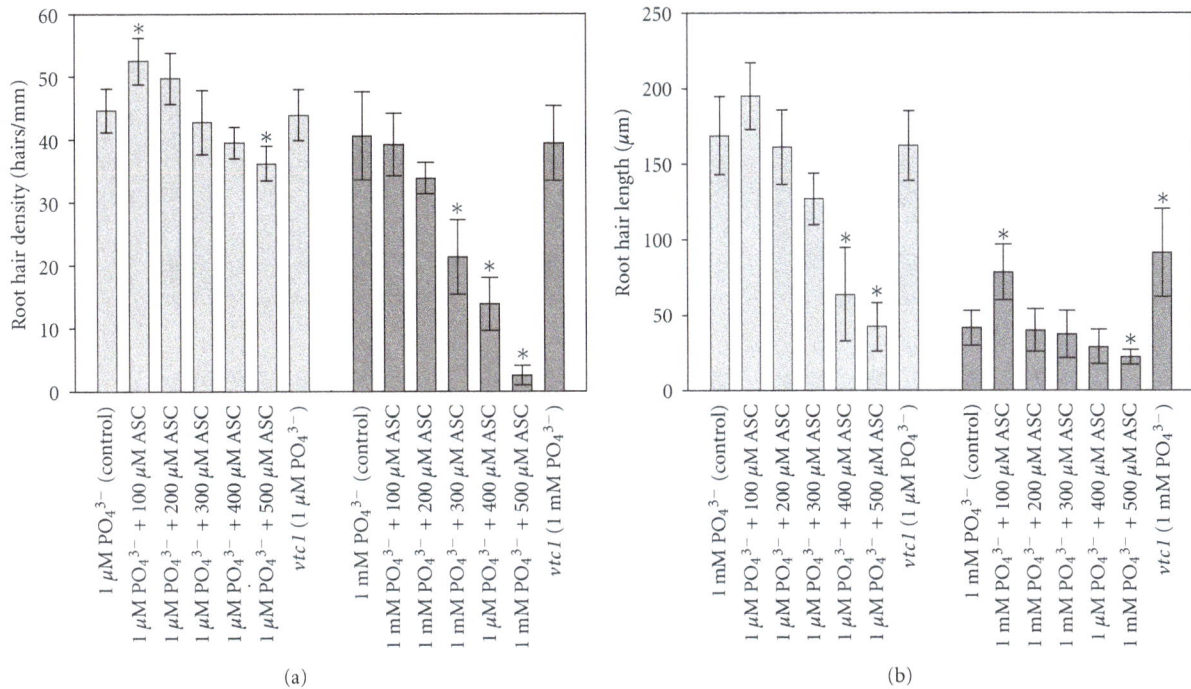

(a)

(b)

FIGURE 5: Effect of phosphate availability and ascorbate concentration on the root hair length and density of *Arabidopsis* wild-type (Col 0) and *vtc1* mutant seedlings. Wild-type seedlings were cultured on media supplemented with varying concentrations of ASC or on ascorbate—free media (control). Seedlings of *vtc1* mutant were cultured on ascorbate—free media. The length is the mean ($\pm$SD) of 50 root hairs, root hair density (mean $\pm$SD) was determined on 25 roots. Asterisk indicates significant differences from the control at $P < 0.05$.

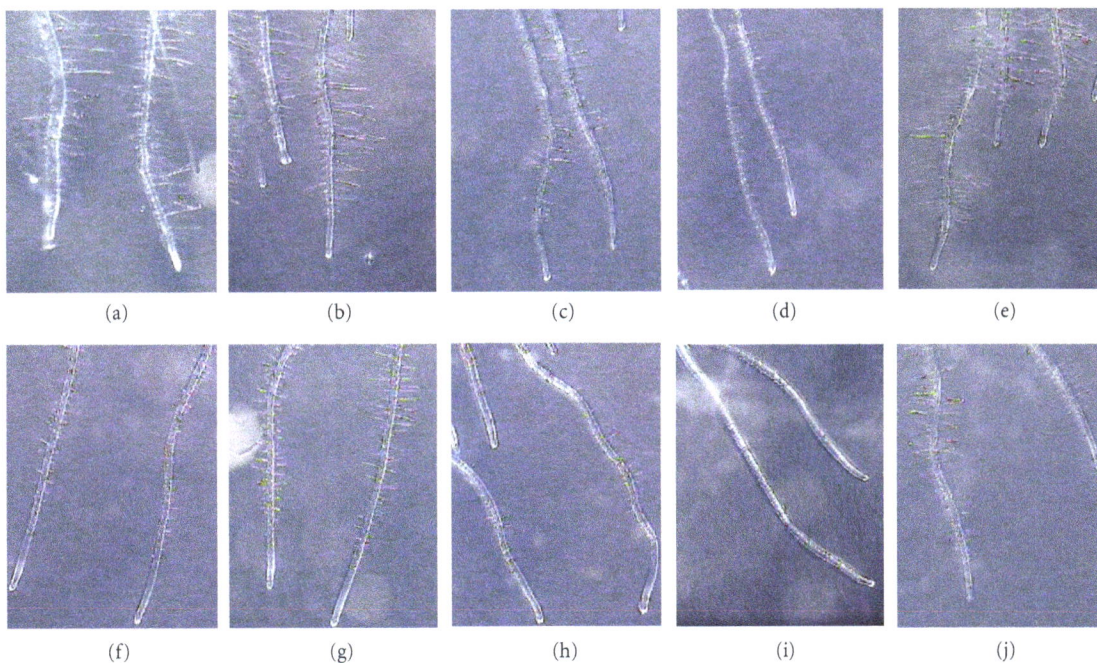

(a)　　(b)　　(c)　　(d)　　(e)

(f)　　(g)　　(h)　　(i)　　(j)

FIGURE 6: Effect of phosphate availability and ascorbate concentration on the development of root hairs in the apical part of primary roots of *Arabidopsis* wild-type (Col 0) and *vtc1* seedlings. The pictures show a representative view of a root tips and root hair zone from seedlings grown for 5 days on the media containing low (1 μM) P or high (1 mM) P concentration. Wild-type (Col 0) seedlings were grown in the presence of the low (1 μM) P concentration (a) or on the same medium supplemented with 100 μM ASC (b), 300 μM ASC (c) or 500 μM ASC (d). Another sets of wild-type seedlings were cultured on high (1 mM) P medium (f) or on the same medium supplemented with 100 μM ASC (g), 300 μM ASC (h) or 500 μM ASC (i). Seedlings of *vtc1* mutants were grown on ascorbate—free media supplemented with low (1 μM) P (e) or high (1 mM) P (j) concentration.

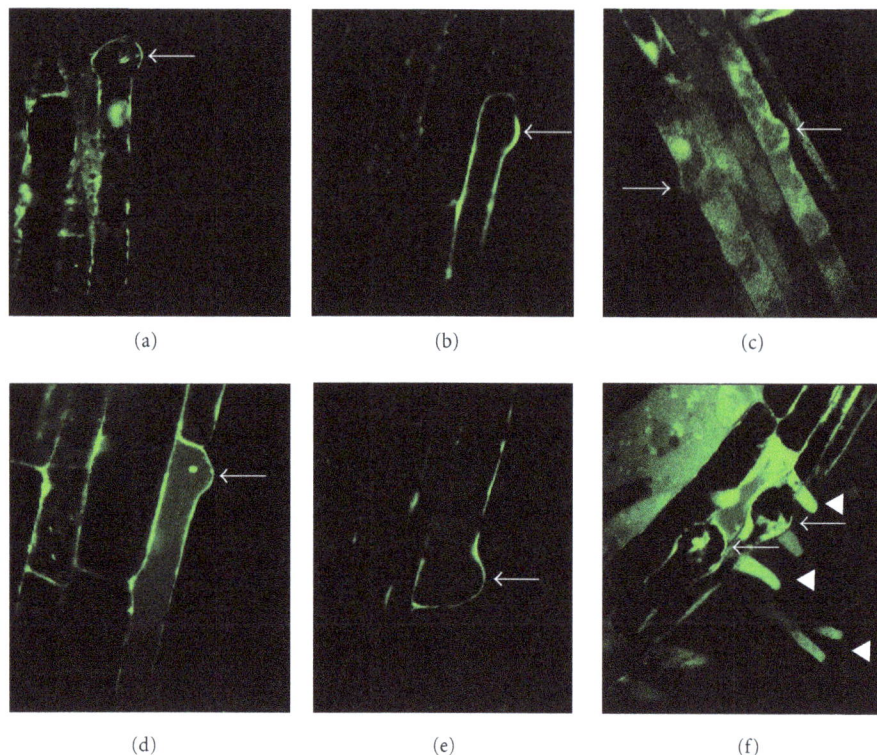

FIGURE 7: The sites of H_2O_2 production in the root-hair-forming zone of the roots of *Arabidopsis* wild type (Col 0) and *vtc1* seedlings. H_2O_2 was visualized by DCF fluorescence. H_2O_2 localization was determined in the differentiation zones of roots of wild-type seedlings grown on the medium containing 1 μM P (a), on the 1 μM P medium supplemented with 500 μM ASC (b), on the medium containing 1 mM P (d), or on the 1 mM P medium supplemented with 500 μM ASC (e). Seedlings of *vtc1* mutants were grown on ascorbate—free media supplemented with low (1 μM) P (c) or high (1 mM) P (f) concentration.

in the culture medium, we observed less DCF fluorescence (indicating reduced H_2O_2 concentration) in the root-hair-forming zone when plants were treated with 500 μM ASC (Figures 7(b) and 7(e)), compared to untreated control (Figures 7(a) and 7(d)). *vtc1* mutants were characterized by more fluorescent cells in the root-hair-forming zone of when compared to wild-type plants (Figures 7(c) and 7(f)).

4. Discussion

It has been established that under P deficiency, the elongation of the *Arabidopsis* primary root is inhibited [2, 4, 5, 9]. We demonstrate that the differences in the length of primary root observed at low and high P are accompanied by changes in ascorbate content and redox status. Root tips of P-starved plants are characterized by significantly lower ASC content than the organs of those grown under high P conditions. Under low P conditions, almost the entire ASC pool is reduced, while under high P concentration, an oxidized form of ascorbate constitutes about 40% of total pool of this antioxidant in the apical part of roots (Figure 3, Table 2).

Under low P, increase in the ASC content stimulated the growth of primary root. It was also observed that the seedlings of *vtc1* mutant, defective in ASC synthesis, produced shorter primary roots than wild-type seedlings (Figures 1 and 2(b)). These data suggest that the increase

in endogenous ascorbate may partly reverse the inhibitory effect of low P availability on primary root elongation. Therefore, the lowering of ASC concentrations in roots of seedlings grown under P deficiency may participate in the mechanism of primary root growth inhibition. However, it should be noted that stimulatory effect of ASC was limited to a relatively narrow range of concentrations and growth inhibition was observed when plants were treated with higher concentrations of ASC.

In plants grown under high P concentrations, the length of the primary root was decreased upon addition of ASC to the medium. However, under high P availability, primary root elongation was also decreased in *vtc1* mutants (Figure 1). This observation is consistent with data reported by Olmos et al. [24] who also found the reduction in primary root length in *vtc1* plants when compared to wild-type plants. These results suggest that an optimal ASC concentration in roots is required to maintain a high rate of primary root elongation in plants grown under high P amounts. However, primary root length of *vtc1* is significantly longer on high P medium compared to low P medium (Figures 1, 2(d), and 2(i)), whereas roots of *vtc1* mutant accumulate similar ascorbate concentrations if grown under both phosphate conditions (Figure 3). These findings suggest, that besides ASC, other factors are responsible for high root-elongation rate under high P availability.

Lateral root formation was inhibited by ASC, if low P concentration was present in the medium, whereas this parameter was not significantly affected by ASC-treatment under high P concentration. Moreover, irrespective of P availability, lateral root formation was increased in *vtc1* mutant (Table 1(b)). This suggests that decreased ASC concentration promotes lateral root proliferation. Contrary to our results, Olmos et al. [24] did not report a difference in the lateral root number between *vtc1* and wild-type plants. However, they reported an increase in lateral root formation by *vtc2*—another ASC-deficient mutant [24]. An increased number of lateral roots in *vtc* mutants is consistent with the stimulatory role of ROS in lateral root formation and growth. ROS, required for lateral root formation, are produced by activation of the AtrbohC NADPH oxidase [23] and specifically localize in the lateral root primordia [25]. In its role as an antioxidant, ASC removes ROS which follows that low ASC availability in roots will favor lateral root formation [24].

Changes in the cell proliferation play an essential role in the onset of plant responses to P deficiency. A significant part of inorganic phosphate is used for DNA synthesis in dividing cells [4]. Under low P conditions, meristematic activity in the main root is blocked or slowed down and relocated to the sites of lateral root formation. Reduction in growth of primary root and lateral roots is due to a determinate, low P-induced, root developmental programme that inhibits cell division in the primary root meristem and promotes differentiation within the root tip [4, 26]. Consistently with aforementioned results, in our experiment, cytokinesis frequency in root tips of plants grown under P deficiency was significantly lower, compared to plants cultured at high P concentration (Figure 4).

Since ASC was demonstrated to stimulate G1-S transition, cell division is one of the possible ASC targets in the regulation of primary root growth [13, 14, 27, 28]. Therefore, we checked the effect of ASC concentration on the frequency of cell divisions in root tips of plants grown under low or high P amounts. Mitotic activity in the root tips of *vtc1* mutants grown under P-deficient medium was lower than the wild-type plants treated with the same P concentration. The number of dividing cells increased in plants treated with ASC when grown under low P. The strongest stimulation occurred when 300 μM ASC was applied. Higher and lower concentrations were less efficient in promoting cell divisions (Figure 4). ASC in 300 μM concentration was also the most efficient in alleviating the inhibitory effect of low P on root elongation (Figure 1). This suggests that stimulatory effect of exogenous ASC on root elongation in plants grown under low P may result from the stimulatory effect of this antioxidant on cell divisions in the root meristem. In contrast to plants grown under low P, under high P the mitotic activity in primary root apices was significantly inhibited by an increased ASC contents. Low frequency of cell divisions was also detected in root tips of ASC-deficient *vtc1* mutants grown at high P (Figure 4). Contrasting effects of ASC treatment on mitotic activity in low P and high P-grown plants suggest that maintaining the high rate of cell divisions in apical root meristem requires an optimal

concentration of this antioxidant. This idea is also supported by the finding that at low P, where initial ASC concentrations are low, the stimulatory effect of ASC on cell divisions is the highest at 300 μM ASC but the stimulatory effect decreases in the presence of higher concentrations. In roots of plants grown under high P, where endogenous ASC concentration in roots is high, ascorbate supplementation may result in supraoptimal ASC concentration which seems to inhibit cell divisions.

Recently, it has been proposed that inhibition of apical root meristem activity in low P is a consequence of increased Fe uptake and its subsequent toxicity [29–31]. Fe under biological conditions can generate toxic hydroxyl radicals via the Fenton reaction [32]. As an antioxidant, ASC may directly scavenge harmful radicals produced under iron overload, thus protecting meristematic cells in P-starved roots from oxidative stress. Alternatively, ASC may be involved in Fe homeostasis in the root apex, as it has been shown that ASC regulates Fe sequestration by iron-storage protein—ferritin [27]. Both mechanisms may possibly be involved in the stimulatory effect of exogenous ASC on cell division activity in root tips in plants grown under low P [33], however further experiments are required to test the afore-mentioned idea.

Root hair development is an important adaptation strategy under nutrient deficiency. In order to optimize P uptake, *Arabidopsis* plants growing on P-limiting media form longer root hairs when compared to those cultured under P-sufficient conditions [34]. Because a characteristic feature of roots of P-deficient plants is the formation of long root hairs close to the root tip, in this work we measured the length of root hairs in the distal part of the root. ASC supplementation to the medium inhibited root hair elongation under both low and high P amounts, if the highest; 400 or 500 μM ASC was applied. However, the reduction of ASC concentration in *vtc1* resulted in the stimulation of root hair growth when plants were grown under high P concentration (Figures 5(b) and 6(e)). Because root hair formation and elongation are dependent on ROS production in the trichoblasts [23], an inhibition of root hair length by ASC may result from ROS scavenging by ASC. To test this idea, we visualized ROS production in the root-hair-forming zones of plants grown on ASC-unsupplemented media or media supplemented with 500 μM ASC; a concentration which significantly decreased both root hair density (Figure 5(a)) and length (Figures 5(b), 6(d) and 6(i)). Lower levels of DCF fluorescence, indicating reduced ROS concentrations, were recorded in root-hair-forming zones of ASC-treated plants (Figures 7(b) and 7(e)) when compared to relevant parts of the roots of wild-type plants (Figure 7(a) and 7(d)) or *vtc1* mutants (Figure 7(c) and 7(f)) grown on ASC-free media, which suggests that ROS-scavenging in the root-hair-forming zone may possibly be responsible for reduced root hair length and density in seedlings grown in the presence of 500 μM ASC.

It is intriguing that 100 μM ASC, in contrast to higher ASC concentrations, stimulated root hair growth in plants grown under high P concentration (Figure 5(b)). These findings require further studies to be fully explained, however, they raise an idea that endogenous ASC may be engaged in the regulation of root hair elongation by controlling ROS levels in the root-hair-forming zone.

In conclusion, our data suggest that ascorbate may be engaged in the regulation of primary root elongation by P availability. The results reported here demonstrate that there are clear differences in ASC content and redox state in the apical parts of primary roots of plants grown under different regimes of phosphate availability. The idea of the involvement of ASC in P-dependent root growth response is supported by the finding that raising the endogenous ASC content by feeding plants with ASC partly reverses primary root growth inhibition in seedlings subjected to low P treatment. The effects of ASC on root growth are possibly mediated by its effect on cell division activity in the apical root meristem.

Acknowledgment

This work was financially supported by a Grant of the Rector of Nicolaus Copernicus University (Grant no. 525-B).

References

[1] L. C. Williamson, S. P. C. P. Ribrioux, A. H. Fitter, and H. M. O. Leyser, "Phosphate availability regulates root system architecture in *Arabidopsis*," *Plant Physiology*, vol. 126, no. 2, pp. 875–882, 2001.

[2] J. López-Bucio, E. Hernández-Abreu, L. Sánchez-Calderón, M. F. Nieto-Jacobo, J. Simpson, and L. Herrera-Estrella, "Phosphate availability alters architecture and causes changes in hormone sensitivity in the *Arabidopsis* root system," *Plant Physiology*, vol. 129, no. 1, pp. 244–256, 2002.

[3] J. López-Bucio, E. Hernández-Abreu, L. Sánchez-Calderón et al., "An auxin transport independent pathway is involved in phosphate stress-induced root architectural alterations in *Arabidopsis*. Identification of BIG as a mediator of auxin in pericycle cell activation," *Plant Physiology*, vol. 137, no. 2, pp. 681–691, 2005.

[4] L. Sánchez-Calderón, J. López-Bucio, A. Chacón-López, A. Gutiérrez-Ortega, E. Hernández-Abreu, and L. Herrera-Estrella, "Characterization of low phosphorus insensitive mutants reveals a crosstalk between low phosphorus-induced determinate root development and the activation of genes involved in the adaptation of *Arabidopsis* to phosphorus deficiency," *Plant Physiology*, vol. 140, no. 3, pp. 879–889, 2006.

[5] L. Sánchez-Calderón, J. López-Bucio, A. Chacón-López et al., "Phosphate starvation induces a determinate developmental program in the roots of *Arabidopsis thaliana*," *Plant and Cell Physiology*, vol. 46, no. 1, pp. 174–184, 2005.

[6] S. Svistoonoff, A. Creff, M. Reymond et al., "Root tip contact with low-phosphate media reprograms plant root architecture," *Nature Genetics*, vol. 39, no. 6, pp. 792–796, 2007.

[7] C. A. Ticconi, R. D. Lucero, S. Sakhonwasee et al., "ER-resident proteins PDR2 and LPR1 mediate the developmental response of root meristems to phosphate availability," *Proceedings of the National Academy of Sciences of the United States of America*, vol. 106, no. 33, pp. 14174–14179, 2009.

[8] H. Rouached, A. B. Arpat, and Y. Poirier, "Regulation of phosphate starvation responses in plants: signaling players and cross-talks," *Molecular Plant*, vol. 3, no. 2, pp. 288–299, 2010.

[9] P. Nacry, G. Canivenc, B. Muller et al., "A role for auxin redistribution in the responses of the root system architecture to phosphate starvation in *Arabidopsis*," *Plant Physiology*, vol. 138, no. 4, pp. 2061–2074, 2005.

[10] R. Shin and D. P. Schachtman, "Hydrogen peroxide mediates plant root cell response to nutrient deprivation," *Proceedings of the National Academy of Sciences of the United States of America*, vol. 101, no. 23, pp. 8827–8832, 2004.

[11] R. Shin, R. H. Berg, and D. P. Schachtman, "Reactive oxygen species and root hairs in *Arabidopsis* root response to nitrogen, phosphorus and potassium deficiency," *Plant and Cell Physiology*, vol. 46, no. 8, pp. 1350–1357, 2005.

[12] R. Liso, G. Calabrese, M. B. Bitonti, and O. Arrigoni, "Relationship between ascorbic acid and cell division," *Experimental Cell Research*, vol. 150, no. 2, pp. 314–320, 1984.

[13] N. M. Kerk and L. J. Feldman, "A biochemical model for the initiation and maintenance of the quiescent center: implications for organization of root meristems," *Development*, vol. 121, no. 9, pp. 2825–2833, 1995.

[14] G. Potters, N. Horemans, R. J. Caubergs, and H. Asard, "Ascorbate and dehydroascorbate influence cell cycle progression in a tobacco cell suspension," *Plant Physiology*, vol. 124, no. 1, pp. 17–20, 2000.

[15] R. Liso, A. M. Innocenti, M. B. Bitonti, and O. Arrigoni, "Ascorbic acid induced progression of quiescent centre cells from G1 to S phase," *New Phytologist*, vol. 110, no. 4, pp. 469–471, 1988.

[16] L. S. Lin and J. E. Varner, "Expression of ascorbic acid oxidase in zucchini squash (*Cucurbita pepo* L.)," *Plant Physiology*, vol. 96, no. 1, pp. 159–165, 1991.

[17] M. A. Green and S. C. Fry, "Apoplastic degradation of ascorbate: novel enzymes and metabolites permeating the plant cell wall," *Plant Biosystems*, vol. 139, no. 1, pp. 2–7, 2005.

[18] P. L. Conklin, S. A. Saracco, S. R. Norris, and R. L. Last, "Identification of ascorbic acid-deficient *Arabidopsis thaliana* mutants," *Genetics*, vol. 154, no. 2, pp. 847–856, 2000.

[19] T. Murashige and F. Skoog, "A revised medium for rapid growth and bioassays with tobacco tissue culture," *Physiologia Plantarum*, vol. 15, no. 3, pp. 437–497, 1962.

[20] K. Kampfenkel, M. Van Montagu, and D. Inzé, "Extraction and determination of ascorbate and dehydroascorbate from plant tissue," *Analytical Biochemistry*, vol. 225, no. 1, pp. 165–167, 1995.

[21] P. Schopfer, C. Plachy, and G. Frahry, "Release of reactive oxygen intermediates (superoxide radicals, hydrogen peroxide, and hydroxyl radicals) and peroxidase in germinating radish seeds controlled by light, gibberellin, and abscisic acid," *Plant Physiology*, vol. 125, no. 4, pp. 1591–1602, 2001.

[22] X. Zhang, L. Zhang, F. Dong, J. Gao, D. W. Galbraith, and C. P. Song, "Hydrogen peroxide is involved in abscisic acid-induced stomatal closure in *Vicia faba*," *Plant Physiology*, vol. 126, no. 4, pp. 1438–1448, 2001.

[23] J. Foreman, V. Demidchik, J. H. F. Bothwell et al., "Reactive oxygen species produced by NADPH oxidase regulate plant cell growth," *Nature*, vol. 422, no. 6930, pp. 442–446, 2003.

[24] E. Olmos, G. Kiddle, T. K. Pellny, S. Kumar, and C. H. Foyer, "Modulation of plant morphology, root architecture, and cell structure by low vitamin C in *Arabidopsis thaliana*," *Journal of Experimental Botany*, vol. 57, no. 8, pp. 1645–1655, 2006.

[25] J. Tyburski, K. Dunajska, and A. Tretyn, "Reactive oxygen species localization in roots of *Arabidopsis thaliana* seedlings grown under phosphate deficiency," *Plant Growth Regulation*, vol. 59, no. 1, pp. 27–36, 2009.

[26] F. Lai, J. Thacker, Y. Li, and P. Doerner, "Cell division activity determines the magnitude of phosphate starvation responses in *Arabidopsis*," *Plant Journal*, vol. 50, no. 3, pp. 545–556, 2007.

[27] G. Potters, L. De Gara, H. Asard, and N. Horemans, "Ascorbate and glutathione: guardians of the cell cycle, partners in crime?" *Plant Physiology and Biochemistry*, vol. 40, no. 6-8, pp. 537–548, 2002.

[28] K. Jiang, Y. L. Meng, and L. J. Feldman, "Quiescent center formation in maize roots is associated with an auxin-regulated oxidizing environment," *Development*, vol. 130, no. 7, pp. 1429–1438, 2003.

[29] J. T. Ward, B. Lahner, E. Yakubova, D. E. Salt, and K. G. Raghothama, "The effect of iron on the primary root elongation of *Arabidopsis* during phosphate deficiency," *Plant Physiology*, vol. 147, no. 3, pp. 1181–1191, 2008.

[30] L. Zheng, F. Huang, R. Narsai et al., "Physiological and transcriptome analysis of iron and phosphorus interaction in rice seedlings," *Plant Physiology*, vol. 151, no. 1, pp. 262–274, 2009.

[31] S. Abel, "Phosphate sensing in root development," *Current Opinion in Plant Biology*, vol. 14, no. 3, pp. 303–309, 2011.

[32] J. Jeong and M. L. Guerinot, "Homing in on iron homeostasis in plants," *Trends in Plant Science*, vol. 14, no. 5, pp. 280–285, 2009.

[33] J. Tyburski, K. Dunajska, and A. Tretyn, "A role for redox factors in shaping root architecture under phosphorus deficiency," *Plant Signaling and Behavior*, vol. 5, no. 1, pp. 64–66, 2010.

[34] T. R. Bates and J. P. Lynch, "Stimulation of root hair elongation in *Arabidopsis thaliana* by low phosphorus availability," *Plant, Cell and Environment*, vol. 19, no. 5, pp. 529–538, 1996.

19

Review of the Genus *Andropogon* (Poaceae: Andropogoneae) in America Based on Cytogenetic Studies

Nicolás Nagahama[1] and Guillermo A. Norrmann[2]

[1] *Instituto Multidisciplinario de Biología Vegetal (IMBIV-CONICET), C.C. 495, 5000 Córdoba, Argentina*
[2] *Facultad de Ciencias Agrarias (FCA),UNNE y Instituto de Botánica del Nordeste (IBONE-CONICET), 3400 Corrientes, Argentina*

Correspondence should be addressed to Guillermo A. Norrmann, gnorrmann@hotmail.com

Academic Editor: Jaume Pellicer

Andropogon is a pantropical grass genus comprising 100–120 species and found mainly in the grasslands of Africa and the Americas. In the new world the genus is represented by approximately sixty (diploids or hexaploids) species grouped in three sections. The hexaploid condition occurs only in the Americas and the full process of this origin is still uncertain, although cytogenetic analysis coupled with taxonomic evidence have provided strong support for new hypothesis. Stebbins proposed the first hypothesis suggesting that the origin of polyploidy in species of *Andropogon* in North America resulted from duplication of the genome of some diploid species, and then by intergeneric crosses with species of a related genus. Since then, numerous studies were performed to clarify the evolutionary history of the genus in America. In this paper, we present a review of cytogenetic studies in the American *Andropogon* species during the last four decades.

1. Introduction

Andropogon L. is a pantropical genus of grasses estimated to contain 100 [1] to 120 [2] species, distributed mainly in the grasslands of Africa and the Americas. *Andropogon* is one of the traditional genera of grasses. Over the course of its circumscription, the genus has included more than 400 species [3] which were subsequently split into several genera [4]. Even considering *Andropogon sensu stricto*, that is, excluding allied genera such as *Bothriochloa* Kuntze, *Dichanthium* Willem., and *Schizachyrium* Nees, the genus remains somewhat heterogeneous [5], especially in the Americas [6, 7]. In the new world the genus is represented by approximately 60 species (see Table 1). The basic chromosome number of the genus is $x = 10$ [6, 8–11], with only a few exceptions [10]. Most African species are diploids or tetraploids ($2n = 2x = 20; 4x = 40$) [9, 11] and American *Andropogon* species are usually diploid or hexaploid ($2n = 2x = 20$ or $6x = 60$) [2, 6, 7, 12, 13], also with only a few exceptions (see [14]).

Stapf [4] proposed four sections in the genus for African's species: (1) *Andropogon* Stapf, (2) *Leptopogon* Stapf, (3) *Notosolen* Stapf, and (4) *Piestium* Stapf. Gould [6] suggested the incorporation of American species into the first three taxonomic sections mentioned above; these are currently recognized by Clayton and Renvoize [1]. The Americas are exceptionally rich in *Leptopogon* Section members, but poor and with hazy boundaries in the other two sections.

Andropogon has diversified into a larger number of species in America and Africa (see Tables 1 and 2) than in Asia or Europe [1, 11]. Genetic differences between American and African (and within) species are poorly understood. Chromosomal evolution, such as polyploidy, appears to be more extensive in America, as hexaploids are almost entirely restricted to this continent (and especially to South America).

In 1985, based on chromosome counts and morphological issues, Norrmann [7] suggested a difference among species from Argentina that could be applied to other South American species. The number of chromosomes allowed

TABLE 1: Species of *Andropogon* distributed in America.

Taxa	Distribution	2n	Section
A. aequatoriensis Hitchc.	South America	Probably 60	*Leptopogon*
A. arctatus Chapm.	Northern America	20	*Leptopogon* (*A. virginicus* complex)
A. arenarius Hack.	South America	60	*Leptopogon* (*A. lateralis* complex)
A. barretoi Norrmann and Quarin	South America	60	*Notosolen*
A. bicornis L.	Americas	60	*Leptopogon* (*A. lateralis* complex)
A. bourgaei Hack.	Northern America	?	*Leptopogon*
A. brachystachyus Chapm.	Northern America	20	*Leptopogon*
A. brasiliensis A. Zanin and Longhi-Wagner	South America	?	?
A. cabanisii Hack.	Northern America	?	*Leptopogon*
A. campestris Trin.	South America	Probably 60	*Leptopogon*
A. campii Swalen	South America	?	*Leptopogon*
A. canaliglumis Norrmann, Swenson and Caponio	Central America	Probably 60	*Leptopogon*
A. carinatus Nees	South America	Probably 20	*Leptopogon*
A. cordatus Swallen	South America	Probably 60	*Leptopogon*
A. crassus Sohns	South America	Probably 60	*Notosolen*
A. crispifolius Guala and Filgueiras	South America	Probably 60	*Notosolen*
A. cubensis Hack.	Central America	Probably 20	?
A. diuturnus Sohns	South America	Probably 20	*Leptopogon*
A. durifolius Renvoize	South America	Probably 60	*Notosolen*
A. elliotii Chapm.	Northern America	20	*Leptopogon*
A. ekmanii Norrmann, Swenson and Caponio	Central America	Probably 60	*Leptopogon* (*A. lateralis* complex)
A. exaratus Hack.	South America	60, 60+2B	*Notosolen*
A. flavescens J. Presl	South America	Probably 60	*Andropogon*
A. floridanus Scribn.	Northern America	20	*Leptopogon* (*A. virginicus* complex)
A. gerardii Vitman	Northern America	60, 70, 80, 90	*Andropogon*
A. glaucescens Kunth	South America	Probably 60	*Andropogon*
A. glaucophyllus Roseng., B.R. Arrill. and Izag.	South America	60	*Notosolen*
A. glaziovii Hack.	South America	60	*Leptopogon* (*A. lateralis* complex)
A. glomeratus (Walter) Britton, Sterns and Poggenb.	Northern America	20	*Leptopogon* (*A. virginicus* complex)
A. glomeratus var. *glomeratus* (Walter) Britton, Sterns and Poggenb.	Northern America	20	*Leptopogon* (*A. virginicus* complex)
A. glomeratus var. *hirsutior* (Hack.) C. Mohr	Northern America	20	*Leptopogon* (*A. virginicus* complex)
A. glomeratus var. *pumilus* (Vasey) L. H. Dewey	Northern America	20	*Leptopogon* (*A. virginicus* complex)
A. gyrans var. *gyrans* Ashe	Northern America	20	*Leptopogon* (*A. virginicus* complex)
A. gyrans var. *stenophyllus* (Hack.) C. S. Campb.	Northern America	20	*Leptopogon* (*A. virginicus* complex)
A. hallii Hack.	Northern America	60, 70, 100	*Andropogon*
A. herzogii Hack.	South America	Probably 60	*Leptopogon*
A. hondurensis (R.W. Pohl) Wipff	Central America	80	*Andropogon*
A. hypogynus Hack.	South and Central America	60	*Leptopogon* (*A. lateralis* complex)

TABLE 1: Continued.

Taxa	Distribution	$2n$	Section
A. indetonsus Sohns	South America	Probably 60	Leptopogon
A. lateralis Nees	South and Central America	60, 60+2B	Leptopogon (A. lateralis complex)
A. leucostachyus Kunth	South and Central America	20	Leptopogon
A. liebmannii Hack.	Northern America	20	Leptopogon
A. liebmannii Hack. var. pungensis (Ashe) C. S. Campb.	Northern America	20	Leptopogon
A. lindmanii Hack.	South America	60	Leptopogon (A. lateralis complex)
A. longiberbis Hack.	Northern America	20	Leptopogon (A. virginicus complex)
A. longiramosus Sohns	South America	?	Leptopogon
A. macrothrix Trin.	South America	20	Leptopogon
A. monocladus A. Zanin and Longhi-Wagner	South America	?	?
A. nashianus Hitchc.	Central America	Probably 20	Leptopogon
A. palustris Pilg.	South America	Probably 20	?
A. perdignus Sohns	South America	Probably 60	Andropogon
A. pohlianus Hack.	South America	Probably 60	Notosolen
A. pringlei Scribn. and Merr.	Northern America	?	probably Leptopogon
A. reedii Hitchc. and Ekman	Central America	Probably 20	Leptopogon
A. reinoldii León	Central America	Probably 20	Leptopogon
A. sanlorenzousa Killeen	South America	Probably 20	Leptopogon
A. scabriglumis Swallen	South America	Probably 60	Leptopogon
A. selloanus (Hack.) Hack.	South and Central America	20	Leptopogon
A. sincoranus Renvoize	South America	?	Leptopogon
A. spadiceus Swallen	Northern America	?	?
A. ternarius Michx.	Northern America	40, 60	Leptopogon
A. ternarius var cabanisii (Hack) Fernald and Griscom	Northern America	?	Leptopogon
A. ternatus (Spreng.) Nees	South America	30	Leptopogon
A. tracyi Nash	Northern America	20	Leptopogon (A. virginicus complex)
A. urbanianus Hitchc.	Central America	80	Leptopogon
A. virginicus L.	Northern America	20	Leptopogon (A. virginicus complex)
A. virginicus var. glaucus Hack.	Northern America	20	Leptopogon (A. virginicus complex)
A. virginicus var. virginicus L.	Northern America	20	Leptopogon (A. virginicus complex)
A. vetus Sohns	South America	Probably 60	?

discriminating the genus *Andropogon* into two main groups: a few species are diploid ($2n = 2x = 20$), small in size with colonizing habit; other ten species are hexaploid ($2n = 6x = 60$), larger in size and with varied habitat. South American species consistently display one ploidy level, with hexaploids being considered of alloploid origin [7]. In contrast, there are a few exceptions in North America. For example, two species show intraspecific variation: *Andropogon ternarius* with $2n = 40$ and 60 (which needs reconfirmation) and *Andropogon gerardii* with $2n = 60$ and 90.

To resolve genomic relationships between *Andropogon* species in the new world, many studies were performed until the present. In this paper we present a review of the published results and a general discussion of them.

2. Andropogon Section

In the Americas the *Andropogon* section is well represented by two species *A. gerardii* Vitman and *A. hallii* Hack., distributed mainly in the Northern Hemisphere [14]. The two species are predominantly hexaploid ($2n = 60$) plants [15–20], however there are populations with high frequency of enneaploids ($2n = 90$) [19, 21–23]. Both species cross in habitat hybridizing zones (e.g., Nebraska sand hills) and hybrid swarms are formed (see [14, 24]). Hybridization in the *A. gerardii-A. hallii* complex was recorded as early as 1891, when an individual was collected in Kansas and described as *A. chrysocomus* Nash [24]. Although hybrids in this combination are fertile, they disappear outside the hybridization habitat, indicating that the species are ecologically distinct [14]. Other members of the section worth mentioning are *A. glaucescens* in South America (no chromosomes count) and *A. hondurensis* (R.W. Pohl) Wipff, with chromosome counts of $2n = 80$.

Stebbins [8] suggested that the polyploid origin of *A. gerardii* in North America could be caused by polyploidization of some diploid of the "Cotton Belt" region, resulting in the constitution of the tetraploid *A. ternarius* (included into the *Leptopogon* section), and then by intergeneric crosses of this tetraploid with species of *Bothriochloa*, which at that time were still regarded as members of *Andropogon*, sect. *Amphilopis*. *Bothriochloa* includes several species that are adapted to the more arid portions of western North America, and therefore such an origin would be compatible with the more xeric nature of *A. gerardii*. Since at the present time no diploid or tetraploid *Bothriochloa* species exists in North America, Stebbins [8] assumed that the ancestor of these higher polyploids is now extinct.

Norrmann et al. [19] described the meiotic and reproductive behavior in $6x$ and $9x$ cytotypes of *A. gerardii*, and the viability of their hybrids. The meiosis in *A. gerardii* was regular in the hexaploids but irregular in the enneaploids. The hexaploid cytotypes ($2n = 6x = 60$) are fully fertile and produce gametes that uniformly contain 30 chromosomes. Minimal embryo sac abortion and good seed production follow. In the enneaploids, "heptaploids," "octoploids," and aneuploids with $2n = 68 - 78$, gametes frequently abort.

Under controlled pollination, the two common cytotypes can be crossed, producing progeny with a range of chromosome numbers with less fertility [25].

In some natural populations of *A. gerardii* high frequencies of hexaploids and enneaploids, also plants with an intermediate chromosome numbers occur [19, 21, 23]. Populations dominated by or composed of only enneaploids would be much less fertile than mixed populations [19], and indeed such populations are rare to nonexistent [23]. Norrmann and Keeler [25] suggested that the predominance of the hexaploids is related to the higher level of fitness and this could eliminate other cytotypes. In addition the authors suggest that the enneaploids are produced from a hexaploid's unreduced gamete combining with a reduced gamete ($2n = 60 + 30 = 90$).

3. Notosolen Section

Only three South American ($2n = 6x = 60$) species [13] included in this section have yielded chromosome counts: *A. barretoi* Norrmann and Quarin, *A. exaratus* Hack. and

A. glaucophyllus Roseng., B. R. Arrill. and Izag. The section was considered the most primitive in South America, because of its closeness with species from West Africa [1, 26].

The geographic distribution of these species is relatively restricted and they are not sympatric anymore, even though they live no more than 1000 mi. from each other. However, the hybrid combinations in artificial crossings between the more geographically distant species (*A. exaratus* × *A. glaucophyllus*) are possible and they are fully fertile [11]. Because of the fertility of interspecific hybrids, Norrmann [11] proposed that they have a highly related genomic composition and a probably common origin. The hybrid combinations between *A. barretoi*, *A. exaratus*, *A. glaucophyllus* and the trihybrid (*A. exaratus* × *A. glaucophyllus*) × *A. barretoi* are under analysis.

4. Leptopogon Section

Leptopogon is considered the most advanced section within the genus [1, 2, 11] and is characterized by the presence of a concave nerveless first glume of the sessile spikelet [26]. In the Americas, the section is mainly represented by two complexes: *A. virginicus* L., distributed in North America [9] and the *A. lateralis* Ness complex covering South and Central America.

Documented American diploids of this section are represented by twelve species. Nine of them belong to the *A. virginicus* complex (Table 1), and the other three species are distributed in South and Central America: *A. leucostachyus* Kunth, *A. macrothrix* Trin., and *A. selloanus* (Hack.) Hack. Another South American species, *A. ternatus* (Spreng.) Nees. maintains permanent triploidy ($2n = 3x = 30$) by transmitting one genome through the egg cell and two genomes through the sperm nucleus [27]. This species may be best regarded as a diploid with an additional accessory chromosomes set [7, 27].

The hexaploid species are all included in the *A. lateralis* complex and are represented by 10 species restricted to South and Central America, except for *A. bicornis* which has the widest geographical distribution in the group and is also present in North America.

Other uncommon ploidy levels are represented in *A. Notosolen* Michx. ($2n = 40$) and *A. urbanianus* Hitchc. ($2n = 80$).

4.1. Andropogon Virginicus Complex. In North America the *A. virginicus* complex is a closely interrelated group of nine diploid species [9, 28]. These species frequently grow together but rarely produce apparent hybrids [9]. They are effectively reproductively isolated from one another without being separated by large morphological gaps.

Norrmann et al. [29] by genomic in situ hybridization (GISH) studies observed that the South American diploids *A. selloanus* and *A. macrothrix*, and the North American diploid *A. gyrans* Ashe (*A. virginicus* complex member), share the basic S genome (Figures 1(a)–1(d)). This was previously proposed based on classical hybridization and meiotic chromosome behavior studies by Galdeano and Norrmann [12] for the first two species and reveals that the S

TABLE 2: Species of *Andropogon* distributed in Africa.

Taxa	Distribution	2n	Section
A. *abyssinicus* R. Br. ex Fresen.	East Africa	32	*Andropogon*
A. *africanus* Franch.	Africa	40	*Leptopogon*
A. *amethystinus* Steud.	Africa	20, c.30	*Andropogon*
A. *amplectens* Nees	Southern Africa	40	*Piestium*
A. *appendiculatus* Nees	Southern Africa	20, 40, 60	*Notosolen*
A. *ascinodis* C.B. Clarke	Africa and India	40	probably *Notosolen*
A. *auriculatus* Stapf	West Africa	?	?
A. *brachyatherus* Hochst.	Southern Africa	20	?
A. *brazzae* Franch.	Southern Africa	20, 40	?
A. *canaliculatus* Schumach.	Africa	20	*Piestium*
A. *chevalieri* Reznik	West Africa	?	?
A. *chinensis* (Nees) Merrill	Africa and Asia	?	*Piestium*
A. *chrysostachyus* Steud.	East Africa	?	?
A. *curvifolius* Clayton	West Africa	20	*Leptopogon*
A. *distachyos* L.	Africa/Europe	36, 40	*Andropogon*
A. *dummeri* Stapf	West Africa	20	?
A. eucomus Nees	Africa	20, 40	?
A. *filifolius* (Nees) Steud	Southern Africa	?	*Piestium*
A. *gabonensis* Stapf	West Africa	20, 21	?
A. *gayanus* Kunth	Africa and Asia	20, 35, 40, 42, 43, 44	*Notosolen*
A. *gayanus* Kunth var *bisquamulatus* (Hochst.) Hack.	Africa and Asia	40	*Notosolen*
A. *gayanus* Kunth var *gayanus*	Africa and Asia	40	*Notosolen*
A. *gayanus* Kunth var *squamulatus* (Hochst) Stapf.	Africa and Asia	40	*Notosolen*
A. *guianensis* Kunth ex Steud.	Africa	40	?
A. *heterantherus* Stapf	East Africa	?	*Piestium*
A. *huillensis* Rendle	Southern Africa	20, 60, 100	*Leptopogon*
A. *ivorensis* Adjan. and Clayton	West Africa	40	?
A. *kilimandscharicus* Pilger	Africa	20	*Andropogon*
A. *laxatus* Stapf	Africa	?	*Leptopogon*
A. *lima* (Hack.) Stapf	Africa	?	*Andropogon*
A. *macrophyllus* Stapf	West Africa	40	?
A. *mannii* Hook. f.	Africa	14	*Andropogon*
A. *patris* Robyns	Africa	20	?
A. *perligulatus* Stapf	Africa	20	?
A. *pinguipes* Stapf	West Africa	?	?
A. *pratensis* Hochst.	West Africa	?	*Andropogon*
A. *pseudapricus* Stapf	Africa	20, 40	*Piestium*
A. *pusillus* Hook. f.	West Africa	?	?
A. *schinzii* Hack.	Africa	20, 40	*Piestium*
A. *schirensis* Hochst. ex A. Rich.	Africa	20, 40	*Piestium*
A. *tectorum* Schumach. and Thonn.	West Africa	20, 23, 30, 40	*Notosolen*
A. *textilis* Rendle	East Africa	?	*Piestium*

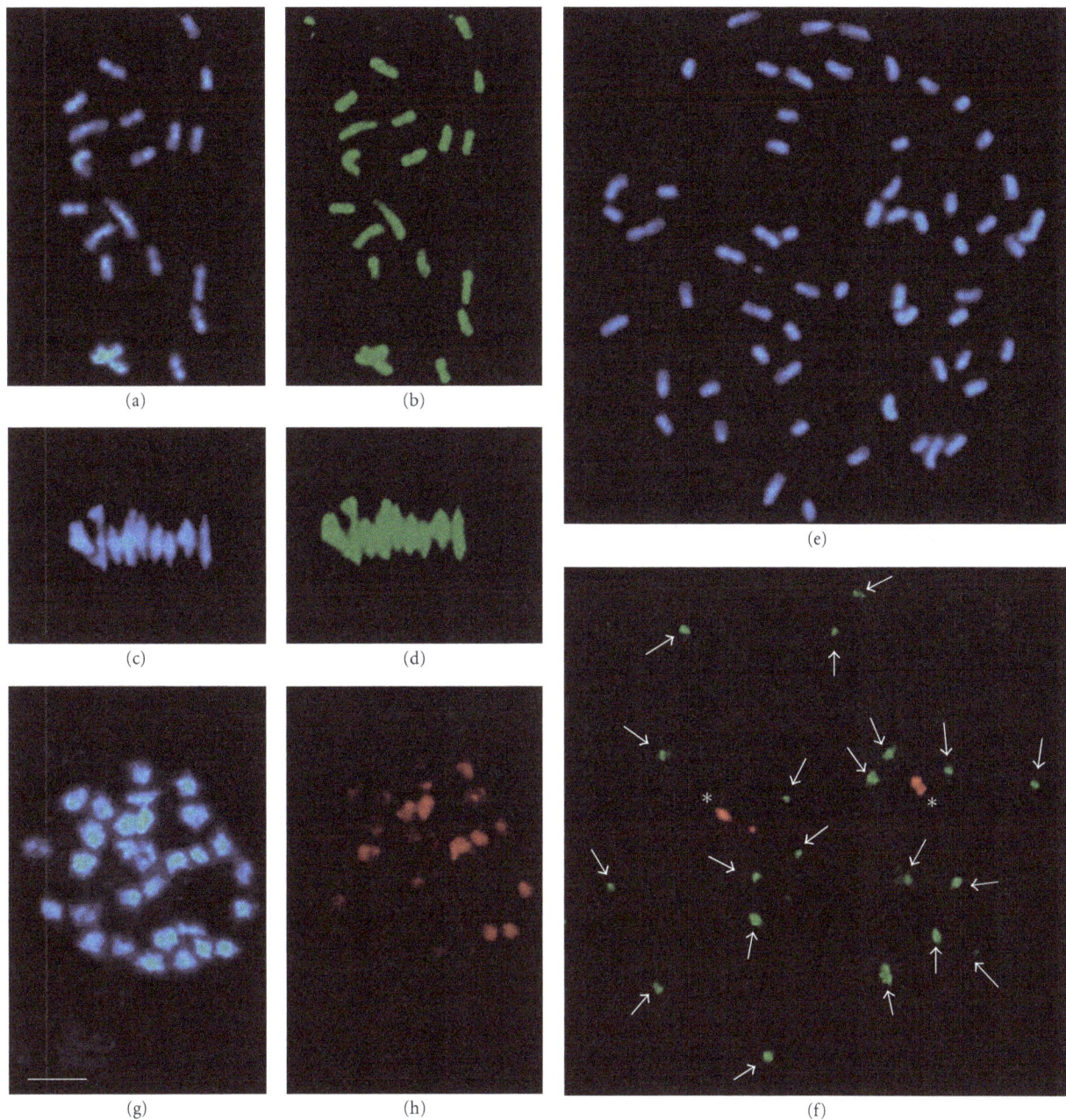

FIGURE 1: GISH on mitotic and meiotic metaphase chromosomes of (a–d) diploid ($2n = 2x = 20$) and (e–h) hexaploid ($2n = 6x = 60$) *Andropogon* species. (a) and (b) Mitotic chromosomes of SA diploid *A. selloanus* probed with genomic DNA from SA diploid *A. macrothrix* and detected with green fluorescence. (a) DAPI-stained chromosomes; (b) GISH showing all 20 chromosomes fluorescing green. (c) and (d) Meiotic chromosomes of SA diploid *A. selloanus* probed with genomic DNA from the NA diploid *A. gyrans* and detected with green fluorescence. (c) DAPI-stained chromosomes; (d) GISH showing all 10 bivalents fluorescing green. (e) and (f) Mitotic chromosomes of SA hexaploid *A. lateralis* probed simultaneously with genomic DNA from the SA diploid *A. selloanus* (green fluorescence) and pTa71 (red fluorescence). (e) DAPI-stained chromosomes; (f) GISH showing 20 chromosomes fluorescing green (*S* genome chromosomes, arrows) and two sites of red hybridization (asterisks) corresponding to the location of the 18S.25S rDNA on two chromosomes that do not fluoresce green and thus do not originate from the S genome. (g) and (h) Meiotic chromosomes of SA hexaploid *A. lateralis* probed with genomic DNA from the NA diploid *A. gyrans* and detected with red fluorescence. (g) DAPI-stained chromosomes; (h) GISH showing approximately 10 bivalents fluorescing red. Scale bar = 2 μm. Figure extracted from [29].

genome, originally defined for the South American diploids, is also shared by the North American diploid *A. gyrans*. Since *A. gyrans* is a member of the *Andropogon virginicus* group, whose monophyly was demonstrated by classical taxonomy [9], it is likely that the remaining members of the *A. virginicus* group also contain the S genome [29].

4.2. Andropogon lateralis Complex. This section is geographically distributed in South and Central America and is constituted entirely by hexaploid species: *A. arenarius* Hack., *A. bicornis* L., *A. glaziovii* Hack., *A. hypogynus* Hack., *A. lateralis* Nees, *A.* × *subtilior* (Hack.) Norrmann (pro. spp.), *A.* × *lindmanii* Hack. (pro. spp.), and *A.* × *coloratus* Hack. (pro. spp.), among others, which present the anther size and the number of pollen grains in fertile sessile spikelets strongly reduced compared with those of pedicellate spikelets. This synapomorphy of dimorphic anthers defines this complex [2, 9]. Within this complex, natural interspecific hybrids have been reported, where populations of different species live in sympatry. Three combinations were reported by Campbell and Windish [2] and two more by Norrmann [11]. Of the ten taxa that comprise the complex in the southern area of South America, five are legitimate species and the others are interspecific hybrids [30].

Norrmann et al. [29] performed GISH studies on two hexaploid species of the *A. lateralis* complex: *A. lateralis* and *A. bicornis*. Hybridization of genomic DNA from the South American diploid *A. selloanus* onto mitotic chromosomes of the South American hexaploid *A. lateralis* resulted in only 20 out of the 60 chromosomes showing strong green fluorescence (Figures 1(e) and 1(f)). These results indicate that *A. lateralis* is an allohexaploid in which the S genome comprises only one of the other genomes. Interestingly, however, the 20 S genome chromosomes were not uniformly labeled along their entire length (as in the diploids, see Figures 1(a)–1(d)); instead, the labeling was mainly concentrated in the pericentromeric regions. These results suggest that there has been some divergence of the repetitive sequences in the distal regions of the S genome chromosomes since the allopolyploid was formed so that they no longer hybridize to the S genome probe.

When meiotic chromosomes of *A. lateralis* were probed with genomic DNA from the North American diploid *A. gyrans*, the overall results were similar to those using the South American diploid *A. selloanus* as a probe, although slight differences in labeling intensity were sometimes observed, suggesting once again that there has been some divergence of the repetitive DNA sequences between the S genome in the North American diploid and hexaploid species as suggested above for *A. lateralis* (Figures 1(g)–1(h)).

Norrmann [30] analyzed the chromosomes and meiotic behavior between interspecific hybrids into the *A. lateralis* complex and observed that all studied hybrids showed $2n = 60$ chromosomes which pair to form up to 30 bivalents per pollen mother cell. The high frequency of bivalents observed in all crosses (30 observed, of 30 maximum) points to the existence of ancient chromosomal homology or homoeology in all species treated, with small differences among the "three" basic genomes (see [29]).

5. Intersectional Analysis

Norrmann [11] crossed *A. barretoi* (*Notosolen*) with *A. gerardii* (section *Andropogon*) and observed a high chromosome pairing in these hybrids. But this pairing does not result from true homology, according to genomic in situ (GISH) experiments carried out recently (Norrmann and Leitch, unpubl. data), which evidence very low homologies among chromosomes from each parental species. The formation of multivalent as a divergence phenomenon is strongly suggested by the odd meiosis, with irregular segregation and formation of multiple nuclei.

Intersectional hybrids among *A. lateralis* (*Leptopogon*) and *A. exaratus* (*Notosolen*) occur in nature and can be experimentally produced [11]. GISH experiments carried on this hybrid revealed very low homologies among these species (Norrmann and Leitch, in prep). On the other hand, upon direct labeling of *A. gerardii* (*Andropogon*) onto *A. bicornis* (*Leptopogon*), much more homologies appear (Norrmann and Leitch, in prep). All these results suggest section *Notosolen* has no close relationships to *Andropogon* or to *Leptopogon* sections.

Finally, preliminary results on *A. gerardii* chromosomes hybridizing to probes from *A. gyrans* suggest a genomic formula SS S^1S^1XX for *A. gerardii* with one genome close to *A. gyrans* (S), another less related (S^1), and a third unrelated (X) (Nagahama and Norrmann, unpubl. data).

6. Discussion

To resolve genomic relationships between *Andropogon* species, previous studies have successfully made interspecific hybrids among diploids [12], among diploids and triploids [27], among hexaploids [11, 19, 30, 31], between hexaploids with enneaploids and inner aneuploids [19, 25] and among diploid and hexaploid species [29]. While the later study suggests that the diploid South American species *A. selloanus* and *A. macrothrix* and the North American diploid *A. gyrans* share a common genome, relationships between the North and South American species are still unclear. This was due in part to the failure to make diploid × hexaploid crosses in several combinations [12].

Stebbins [8] suggested that the North American hexaploids (*A. gerardii*) probably were originated in the new world through processes of polyploidization of diploid species of *Andropogon*, followed by hybridization with species of *Bothriochloa*. Several events have happened since then. First, this hypothesis was proposed by the time hexaploids in South America were not known, as the first chromosome counts were published between 1985 [7] and 1986 [2]. Second, GISH experiences suggest that *Bothriochloa* and *Andropogon* have stronger chromosomal divergences than thought before (Norrmann, unpubl. data). Finally, the similarity suggested by Stebbins [8] among species of the genera *Andropogon* and *Bothriochloa* actually are recognized as evolutionary convergence [1]. Moreover, Stebbins' hypothesis suggests that *A. gerardii* and other polyploid complexes were distributed on the plains of central North America about 5 million years ago. However, due to the last glaciations, it is known that the colonization of the North American prairies by this species was recent, and this happened not earlier than 10,000 years ago. Nowadays, there is consensus in the origin of *A. gerardii* in Central America or Northern South America, and after the retreat of the ice, it would have colonized the North American plains. This hypothesis is also supported by the octoploid *A. hondurensis*, distributed in Central America, due to that this species is related with *A. gerardii*, being considered in the past as subspecies of *A. gerardii* (see [32]).

Stebbins' hypothesis has two parts: first, the conformation of a tetraploid from diploids of the cotton belt and second, the hybridization with *Bothriochloa* species. As we have explained before, second part needs modification but part 1, that is, the generation of hexaploids as an American evolutionary process stands still as the more solid hypothesis, at least for *A. gerardii* and the *A. lateralis* complex, and would also have occurred in genera related to *Andropogon*, for example in *Bothriochloa* [8, 33, 34]. This hypothesis is sustained by GISH experiments pointing to the S genome as being part of *A. gerardii* and the *A. lateralis* genome. Different forms of the S genome are present in American diploid species, as *A. selloanus* or the *A. virginicus* complex and none of these species lives in Africa. Therefore, the origin of *A. gerardii* and the *A. lateralis* complexes could be American, with the providers of the other/s genomes still not found.

On the other hand, it is worth mentioning other hypothesis for the origin of the hexaploids. Norrmann [11] suggested that one or more ancestral hexaploids might have been established both in America and Africa at least in the Cretaceous (60 million years ago). The lack of hexaploid species in Africa could be due to these polyploids proliferating adaptively in America, and not in Africa due to selective pressure, because the continent underwent dramatic changes and rigorous conditions after the separation from the Americas [35, 36]. Another possibility is that the hexaploids do exist, but they have not been found. There are records of an African hexaploid (*A. huillensis* Rendle) [37], but in this species also are recorded chromosome counts with $2n = 20$ [38] and $2n = 100$ [39], suggesting that this polyploidization was because of genome duplication (autopolyploid), and not as in the American hexaploids which are allopolyploids.

Our view of the *Andropogons* in America is much complete nowadays than it was in 1975. Great advances have been made in major issues, as the understanding of the *A. virginicus*, the *A. lateralis*, and the *A. gerardii-A.hallii* complexes and their cytogenetics.

GISH technique has proved useful and overcomes the difficulty in making $2x$–$6x$ hybrids and studying intersectional hybrids. Also, preliminary results based on molecular marker analysis suggest that in the *A. lateralis* complex there are at least two clearly different genomes. On the other hand, the three genomes of sect. *Notosolen* appear to be related (Nagahama and Norrmann, unpubl. data).

We need to be cautious about the comprehension of the whole genus since we are still based on chromosome counts made for only a portion of the species. No cytogenetic information is available from northern Brazilian species, or

from Venezuela, Colombia, Equator, Peru (*A. glaucescens* and *A. flavescens* from the *Andropogon* section) and other species from Central America and West Indies. If diploids could be differentiated from hexaploids by its size, as Norrmann proposed [7], then many South and Central American species could be candidates to look for other genome sources different from S (see Table 1).

Finally, old grasslands hexaploid species as *A. gerardii* and members of the *A. lateralis* complex are under the anthropogenic pressure. The days of the North American plains covered with *A. gerardii* feeding bison, or Venezuelan "llanos" and "campos" of southern South America (NE Argentina, S Brazil, Paraguay, Uruguay) dominated by *A. lateralis* and *A. hypogynus* are not the actual picture, but the species are still there. Concern should perhaps be put on the few American members of the *Notosolen* section: *A. exaratus* survives well because it lives on the marshes, but *A. glaucophyllus* (in dunes of southern Brazil and Uruguay) is loosing presence. The worse situation we are aware is that of *A. barretoi*, which can only be found alongside the road Santa María to Porto Alegre, as related to one of us by Professor Ismar Barreto, the real discoverer of the species back in 1982. Today, the habitat stretches no more than 10 mi. along the roadside.

References

[1] W. D. Clayton and S. A. Renvoize, *Genera Graminum. Grasses of the World*, Kew Bulletin Additional Series 13, 1986.

[2] C. S. Campbell and P. G. Windisch, "Chromosome numbers and their taxonomic implications for eight Brazilian *Andropogons* (Poaceae: Andropogoneae)," *Brittonia*, vol. 38, no. 4, pp. 411–414, 1986.

[3] E. G. Steudel, *Synopsis Plantarum Glumacearum*, vol. 1 of *Synopsis Plantarum Graminearum*, J.B. Metzler, Stuttgart, Germany, 1855.

[4] O. Stapf, "Gramineae," in *Flora of Tropical Africa*, D. Prain, Ed., vol. 9, pp. 208–265, Reeve, London, UK, 1919.

[5] W. D. Clayton, "*Andropogon pteropholis*," *Hooker's Icones Plantarum*, vol. 37: t. 3644, 1967.

[6] F. W. Gould, "The grass genus *Andropogon* in the United States," *Brittonia*, vol. 19, no. 1, pp. 70–76, 1967.

[7] G. A. Norrmann, "Estudios citogenéticos en especies argentinas de *Andropogon* (Gramineae)," *Boletin de la Sociedad Argentina de Botanica*, vol. 24, pp. 137–149, 1985.

[8] L. G. Stebbins, "The role of polyploid complexes in the evolution of North American grasslands," *Taxon*, vol. 24, pp. 91–106, 1975.

[9] C. S. Campbell, "Systematics of the *Andropogon virginicus* complex (Gramineae)," *Journal of Arnold Arboretum of Harvard University*, vol. 64, pp. 171–254, 1983.

[10] G. Davidse, T. Hoshino, and B. K. Simon, "Chromosome counts of Zimbabwean grasses (Poaceae) and an analysis of polyploidy in the grass flora of Zimbabwe," *South African Journal of Botany*, vol. 52, pp. 521–528, 1986.

[11] G. A. Norrmann, *Biosistemática y relaciones filogenéticas en especies hexaploides sudamericanas de* Andropogon *(Gramineae)*, Ph.D. thesis, Facultad de Ciencias Exactas Físicas y Naturales, Universidad Nacional de Córdoba, 1999.

[12] F. Galdeano and G. Norrmann, "Natural hybridization between two South American diploid species of *Andropogon*

(Gramineae)," *Journal of the Torrey Botanical Society*, vol. 127, no. 2, pp. 101–106, 2000.

[13] G. A. Norrmann and F. Scarel, "Biología reproductiva de cuatro especies sudamericanas hexaploides de *Andropogon* L. (Gramineae, Andropogoneae)," *Kurtziana*, vol. 28, pp. 173–180, 2000.

[14] A. Boe, K. Keeler, G. A. Norrmann, and S. Hatch, "The indigenous bluestems (*Bothriochloa, Andropogon* and *Schizachyrium*) of the western Hemisphere and gamba grass (*Andropogon gayanus*)," in *Warm Season Grasses*, L. Moser, B. Byron, and L. Sollenberger, Eds., Agronomy Series: 45, pp. 873–908, Madison, Wis, USA, American Society of Agronomy, Crop Science Society of America, Soil Science Society of America, 2004.

[15] E. L. Nielsen, "Grass studies. III. Additional somatic chromosome compliments," *American Journal of Botany*, vol. 26, pp. 366–372, 1939.

[16] F. W. Gould, "Cromosome counts and cytotaxonomic notes on grasses of the tribe Andropogoneae," *American Journal of Botany*, vol. 43, pp. 395–404, 1956.

[17] G. W. Dewald and S. M. Jalal, "Meiotic behavior and fertility interrelationship in *Andropogon scoparius* and *A. gerardii*," *Cytologia*, vol. 39, pp. 215–223, 1974.

[18] R. D. Riley and K. P. Vogel, "Chromosome numbers of released cultivars of switchgrass, Indian-grass, big bluestem, and sand bluestem," *Crop Science*, vol. 22, pp. 1081–1083, 1982.

[19] G. A. Norrmann, C. L. Quarín, and K. H. Keeler, "Evolutionary implications of meiotic chromosome behavior, reproductive biology, and hybridization in 6x and 9x cytotypes of *Andropogon gerardii* (Poaceae)," *American Journal of Botany*, vol. 84, no. 2, pp. 201–207, 1997.

[20] K. P. Vogel, "Improving warm-season forage grasses using selection, breading, and biotechnology," in *Native Warm-Season Grasses: Research Trends and Issues*, K. J. Moore and B. E. Anderson, Eds., pp. 83–106, Crop Science Society of America Special Publication 30, American Society of Agronomy, Madison, Wis, USA, 2000.

[21] K. H. Keeler and G. A. Davis, "Comparison of common cytotypes of *Andropogon gerardii* (Andropogoneae, Poaceae)," *American Journal of Botany*, vol. 86, no. 7, pp. 974–979, 1999.

[22] K. H. Keeler, "Distribution of polyploid variation in big bluestem (*Andropogon gerardii*, Poaccea) across the tallgrass prairie region," *Genome*, vol. 33, no. 1, pp. 95–100, 1990.

[23] K. H. Keeler, "Local polyploid variation in the native prairie grass *A. gerardii*," *American Journal of Botany*, vol. 79, pp. 1229–1232, 1992.

[24] J. K. Wipff, "Nomenclatural combinations in the *Andropogon gerardii* complex (Poaceae: Andropogoneae)," *Phytologia*, vol. 80, pp. 343–347, 1996.

[25] G. A. Norrmann and K. H. Keeler, "Cytotypes of *Andropogon gerardii* Vitman (Poaceae): fertility and reproduction of aneuploids," *Botanical Journal of the Linnean Society*, vol. 141, no. 1, pp. 95–103, 2003.

[26] W. D. Clayton, "Gramineae," in *Flora of West Tropical Africa*, F. N. Hepper, Ed., vol. 3, part 2, pp. 349–512, Crown Agents for Overseas Governments and Administrations, London, UK, 2nd edition, 1972.

[27] G. A. Norrmann and C. L. Quarin, "Permanent odd polyploidy in a grass (*Andropogon ternatus*)," *Genome*, vol. 29, pp. 340–344, 1987.

[28] C. S. Campbell, "Hybridization between *Andropogon glomeratus* var. *pumilus* and *A. longiberbis* (Gramineae) in Central Florida," *Brittonia*, vol. 34, pp. 146–150, 1982.

[29] G. Norrmann, L. Hanson, S. Renvoize, and I. J. Leitch, "Genomic relationships among diploid and hexaploid species of *Andropogon* (Poaceae)," *Genome*, vol. 47, no. 6, pp. 1220–1224, 2004.

[30] G. A. Norrmann, "Natural hybridization in the *Andropogon lateralis* complex (Andropogoneae, Poaceae) and its impact on taxonomic literature," *Botanical Journal of the Linnean Society*, vol. 159, no. 1, pp. 136–154, 2009.

[31] L. C. Peters and L. V. Newell, "Hybridization between divergent types of big bluestem, *Andropogon gerardii* Vitman, and sand bluestem, *Andropogon hallii* E. Hackel," *Crop Science*, vol. 1, pp. 359–363, 1961.

[32] G. Davidse and R. W. Pohl, "Chromosome numbers, meiotic behavior, and notes on some grasses from Central America and The West Indies," *Canadian Journal of Botany*, vol. 50, pp. 1441–1452, 1972.

[33] F. W. Gould., "A citotaxonomic study in the genus *Andropogon*," *American Journal of Botany*, vol. 40, pp. 297–306, 1953.

[34] L. R. Scrivanti, I. Caponio, A. M. Anton, and G. A. Norrmann, "Chromosome number in South American species of *Bothriochloa* (Poaceae: Andropogoneae) and evolutionary history of the genus," *Plant Biology*, vol. 12, no. 6, pp. 910–916, 2010.

[35] P. W. Hattersley and L. Watson, "Diversification of photosynthesis," in *Grass Evolution and Domestication*, G. P. Chapman, Ed., pp. 38–116, Cambridge University Press, Cambridge, UK, 1992.

[36] P. H. Raven and D. Axelrod, "Angiosperm biogeography and past continental movements," *Annals of the Missouri Botanical Garden*, vol. 61, pp. 539–673, 1974.

[37] J. M. J. De Wet, "Chromosome numbers and some morphological attributes of various South African grasses," *American Journal of Botany*, vol. 47, pp. 44–49, 1960.

[38] M. Dujardin, "Chromosome numbers of some tropical African grasses from western Zaire," *Canadian Journal of Botany*, vol. 57, pp. 864–876, 1978.

[39] J. J. Spies and H. Du Plessis, "Chromosome studies on African plants. 3," *Bothalia*, vol. 17, pp. 257–259, 1987.

20

Seed and Embryo Germination in *Ardisia crenata*

Takahiro Tezuka, Hisa Yokoyama, Hideyuki Tanaka, Shuji Shiozaki, and Masayuki Oda

Graduate School of Life and Environmental Sciences, Osaka Prefecture University, 1-1 Gakuen-cho, Nakaku, Sakai, Osaka 599-8531, Japan

Correspondence should be addressed to Masayuki Oda, moda@plant.osakafu-u.ac.jp

Academic Editor: Philip J. White

Ardisia crenata is an evergreen shrub with attractive bright red berries. Although this species is usually propagated by seed, the seeds take a long time to germinate with conventional sowing methods. We investigated the germination capacity of seeds and embryos collected in different months and the effects of seed storage conditions, germination temperature, water permeability of the seed coat, and the endosperm on seed germination. Seeds and embryos collected in late September or later showed good germination rates. Seeds germinated more rapidly after longer periods of storage at low temperature (approximately 5°C), and those stored in dry conditions showed lower emergence frequency than those stored in wet conditions. Seeds germinated at 15–30°C, but not at 5–10°C. Removal of the seed coat enhanced water uptake and seed germination. Seeds with various proportions of the removed seed coat were sown on a medium supplemented with sucrose. The germination frequency increased as the size of the remaining endosperm decreased. However, the opposite results were obtained when seeds were sown on a medium without sucrose. We concluded that the optimal temperature of 25°C is the most critical factor for seed germination in *A. crenata*.

1. Introduction

The genus *Ardisia* (Myrsinaceae) contains approximately 400–500 species of evergreen shrubs, which are distributed in the subtropical and tropical regions around the world [1]. Several *Ardisia* species have been used for medicinal purposes, because they contain a wide array of biologically active phytochemical constituents such as bergenin and ardisin [2].

Ardisia crenata is one of the most widely grown species in the genus and is distributed throughout East Asia, including Japan. It grows to approximately 1 m high and has glossy, dark green leaves, and attractive bright red berries. It is grown as an ornamental plant in gardens and as a houseplant. Although vegetative propagation methods such as cuttings are sometimes used to propagate *A. crenata* plants [3], the use of seeds has advantages for mass propagation. However, it can take more than 13 weeks to achieve 80% of seed germination [3]. This long period for seed germination hinders and delays the breeding of this shrub.

Seed germination is controlled by several environmental factors, such as seed moisture content, temperature, and light. Seed condition also affects germination; for example, the seed coat may be water impermeable, or the mature seed may contain an underdeveloped embryo that only grows to full size after imbibition [4, 5]. In Japan, fully ripened fruits of *A. crenata* are usually harvested in winter. The seeds are washed with water to remove the fruit pulp and are then buried in wet sand for storage under natural low-temperature conditions. These seeds are sown in the following spring and germinate in early summer. Although this is the traditional method of cultivation, the seeds require a long germination period and show considerable variations in the timing of germination. Furthermore, it is unclear whether high-soil moisture and low temperature during seed storage are essential for seed germination and what temperature is optimal for seed germination.

In the present study, we investigated the germination capacity of seeds and embryos collected in different months. We investigated seed responses to low temperature and soil moisture levels during seed storage. In addition, the effects of germination temperature, water permeability of the seed coat, and the effects of the seed coat and endosperm on germination were investigated.

2. Materials and Methods

2.1. Plant Materials. Fruits were harvested from *A. crenata* plants cultivated in a field at Osaka Prefecture University. Seeds were washed with water to remove the fruit pulp and used in the experiments.

2.2. Germination or Emergence Test of Seeds and Embryos. Seeds collected on 18 January 2005 were sown according to the conventional method as follows. Seeds were buried in vermiculite-filled plastic boxes (6 cm × 6 cm × 9 cm height) and were stored outdoors under ambient low-temperature conditions at approximately 5°C with protection from the wind and rain. The vermiculite was constantly moistened by spraying with water. The seeds were sown in a 288-cell tray (10 mL/cell) filled with wet vermiculite at 12 weeks after the initiation of seed storage (13 April). The seeds were placed outdoors, and observations of seedling emergence were recorded for 126 days after sowing (DAS).

Embryo culture was carried out as follows. Seeds collected on 9 January 2007 were sterilized with 70% ethanol for 30 s and then with 5% sodium hypochlorite for 15 min. Embryos were excised from the sterilized seeds, placed in flat-bottomed test tubes (25 mm diameter, 100 mm length) containing MS medium [6] supplemented with 3% sucrose and 0.85% agar (pH 5.8), and then incubated at 25°C in the dark. Observations of embryo germination were recorded for 126 days after the initiation of incubation (DAI).

Seeds collected at monthly intervals (23 August, 27 September, 25 October, and 22 November, 2007) were used to investigate whether seed and embryo germination capacity was related to seed developmental stage. Some nonsterilized seeds were sown in Petri dishes containing wet filter paper and then incubated at 25°C in the dark. Other seeds were used for embryo culture. For seeds collected on 27 September, 25 October, and 22 November, the embryos were removed and cultured *in vitro* as described above. It was impossible to isolate embryos from seeds collected on 23 August because the embryos were immature and too small to excise. Therefore, the seeds harvested on 23 August were aseptically sown in flat-bottomed test tubes containing MS medium supplemented with 3% sucrose and 0.85% agar (pH 5.8) and then cultured at 25°C in the dark. Observations of germination of embryos and seeds were recorded for 126 DAS or DAI.

2.3. Effects of Seed Storage Conditions. To determine whether *A. crenata* seeds show seed dormancy, which is broken by low temperature, we investigated the ability of seeds stored at low temperature (approximately 5°C) for different periods to germinate. Seeds collected on 18 January 2005 were buried in vermiculite-filled plastic boxes and were stored outdoors under ambient low-temperature conditions with protection

from the wind and rain. The vermiculite was constantly moistened by spraying with water. The seeds were transferred to an incubator at 20°C at 4-week intervals (18 January, 15 February, and 15 March). The seeds were temporarily removed from the plastic boxes every 7 days and their germination evaluated. Observations were recorded for 84 days after the initiation of storage.

Seed dormancy is also related to seed moisture content. Therefore, we evaluated seedling emergence from seeds stored under wet or dry conditions. Seeds collected on 18 January 2005 were buried in vermiculite-filled plastic boxes and stored outdoors under ambient low-temperature conditions with protection from the wind and rain. The vermiculite was constantly moistened by spraying with water (wet conditions) or was not moistened (dry conditions). The seeds were sown in a 288-cell tray filled with wet vermiculite at 12 weeks after the initiation of seed storage (13 April) and placed outdoors. Observations of seed emergence were recorded for 126 DAS.

2.4. Effects of Temperature on Germination. To determine the optimal temperature for seed germination, seeds were sown in plastic boxes filled with wet vermiculite immediately after harvest on 18 January 2005 and incubated at 5, 10, 15, 20, 25, or 30°C in the dark. The seeds were temporarily removed from the plastic boxes every 7 days and their germination evaluated. Observations were recorded for 84 DAS.

2.5. Effects of Scarification and Seed Coat Removal. The effect of the seed coat on seed germination was investigated. Seeds were collected on 13 June 2006. The seed coats of some seeds were scarified or completely removed. The seeds (intact seeds, scarified seeds, and seeds without the seed coat) were sown in Petri dishes containing wet filter paper and incubated at 25°C in the dark. Observations of seed germination were recorded for 22 DAS.

Methylene blue has been used to determine the water permeability of seeds [7]. To examine the water permeability of the seed coat, intact seeds and seeds without seed coats were prepared using seeds collected on 25 October 2007. These seeds were immersed in 1% methylene blue for 24 h, washed with water, cut, and observed under a stereomicroscope (Olympus SZ60, Tokyo, Japan).

2.6. Effects of Endosperm Removal. Isolated embryos germinated immediately, whereas seeds took several weeks to germinate (see below; Table 1). Therefore, it was hypothesized that inhibitors present in the endosperm interfered with seed germination. To test this hypothesis, the seed coat and some or all of the endosperm were removed from embryos (Figure 1), and the growth of the seeds/embryos on MS medium with or without sucrose was evaluated. Seeds were collected on 23 November 2006 and sterilized with 70% ethanol for 30 s and then with 5% sodium hypochlorite for 15 min. Five types of seeds were prepared: intact seeds (Figure 1(a)), seeds without seed coat (Figure 1(b)), seeds without seed coat and micropylar endosperm (Figure 1(c)), seeds without seed coat and half

TABLE 1: Seedling emergence and germination of isolated *A. crenata* embryos.

Method	Number of seeds or embryos		Emergence or germination (%)	Number of days to emergence or germination[a]
	Sown or cultured	Emerged or germinated		
Seed sowing	30	27	90	71.8 ± 2.4
Embryo culture	10	10	100	11.9 ± 1.1

Observations were continued for 126 days after sowing or incubation.
[a]Data are expressed as mean ± SE.

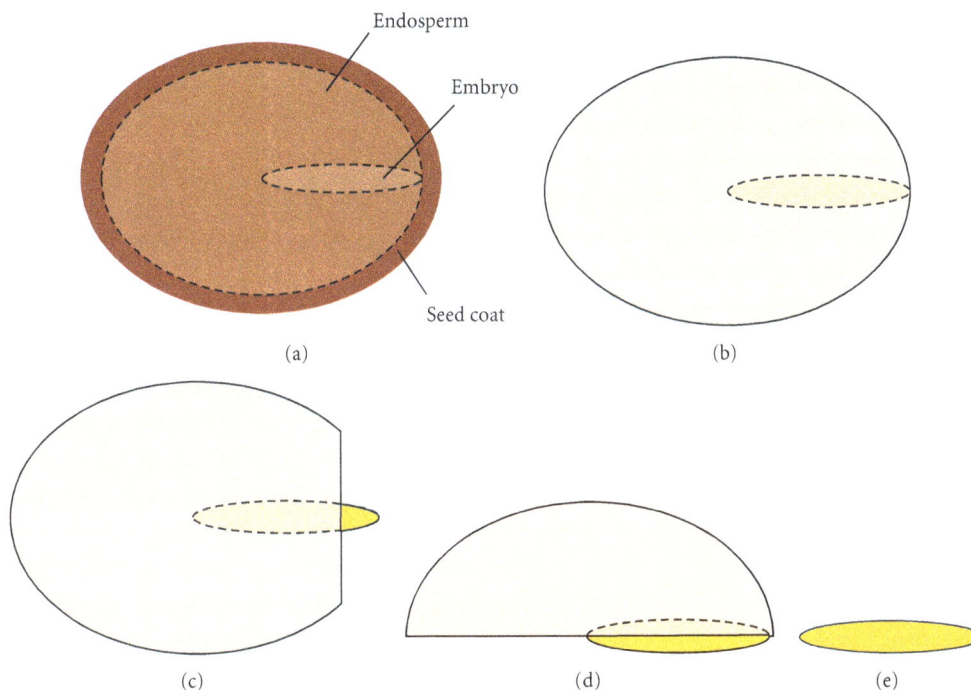

FIGURE 1: Schematic illustration of *A. crenata* seeds with seed coat and some or all of the endosperm removed. These seeds were used to investigate the effects of the endosperm/seed coat on seed germination. (a) Intact seed with seed coat; (b) seed without seed coat; (c) seed without seed coat and micropylar endosperm; (d) seed without seed coat and half of the endosperm; (e) seed without seed coat and endosperm (isolated embryo).

of the endosperm (Figure 1(d)), and seeds without both seed coat and endosperm (only embryos; Figure 1(e)). Four types of these seeds (excluding seeds without the seed coat) were placed in flat-bottomed test tubes that contained MS medium supplemented with 3% sucrose and 0.85% agar (pH 5.8) and then cultured at 25°C in the dark. Observations of seed germination were recorded for 42 DAI.

Seeds collected on 28 December 2006 were also treated as mentioned above. To examine the effects of sucrose on germination, five types of seeds listed above were placed in flat-bottomed test tubes that contained MS medium supplemented with 0.85% agar (pH 5.8) and then cultured at 25°C in the dark. Observations of seed germination were recorded for 42 DAI.

3. Results

3.1. Germination or Emergence Test of Seeds and Embryos. Out of 30 seeds sown using the conventional method, 27 seedlings emerged at 71.8 ± 2.4 DAS. Embryos cultured at

25°C germinated earlier; all 10 embryos germinated at 11.9 ± 1.1 DAI (Table 1).

To determine the effects of harvest time on embryo and seed germination, seeds and embryos collected at monthly intervals (23 August, 27 September, 25 October, and 22 November) were sown or cultured at 25°C (Table 2). The length of immature embryos in seeds collected on 23 August could not be determined, but seeds collected on 27 September contained embryos 2.37 ± 0.05 mm long. The length of the embryos did not increase significantly in seeds collected on 25 October and 22 November. When seeds were sown *in vivo*, those collected on 23 August did not germinate. However, seeds collected in the other three months germinated. On the other hand, 20% of the seeds collected on 23 August germinated when sown *in vitro*. All embryos in seeds collected on 27 September or later germinated.

3.2. Effects of Seed Storage Conditions. There was no change in embryo length during low-temperature storage (approximately 5°C) (data not shown). Germination was not

TABLE 2: Germination of *A. crenata* seeds and embryos collected in different months in 2007.

Date of seed collection	Length of embryo (mm)[a]	Germination (%)	
		Seed[b]	Embryo[b]
23 August	—	0	20[c]
27 September	2.37 ± 0.05	100	100
25 October	2.27 ± 0.06	100	100
22 November	2.32 ± 0.06	100	100

Observations were continued for 126 days after sowing or incubation.
[a]Data are expressed as mean ± SE ($n = 20$).
[b]$n = 20$.
[c]Intact seeds harvested on 23 August were sown *in vitro* because the embryos were immature.

TABLE 3: Germination of *A. crenata* seeds stored at low temperature.

Number of days		Number of seeds		Germination (%)	Number of days to germination after transfer to 20°C[a]
5°C	20°C	Sown	Germinated		
0	84	50	50	100	40.6 ± 0.7a
28	56	50	50	100	31.5 ± 1.4b
56	28	50	48	96	23.1 ± 0.7c
84	0	50	0	0	—

Observations were continued for 84 days after storage.
[a]Data are expressed as mean ± SE. Means followed by a different letter are significantly different ($P < 0.05$, Tukey's multiple comparison test).

observed in seeds stored at 5°C for 84 days (Table 3). Conversely, most seeds transferred from 5°C to 20°C germinated. The number of days to germination was significantly different depending on the timing of transfer to 20°C. Seeds stored at 5°C for 0, 28, and 56 days after storage required 40.6 ± 0.7, 31.5 ± 1.4, and 23.1 ± 0.7 days to germinate, respectively.

Seeds stored under wet conditions showed 90% emergence at 71.8 ± 2.4 DAS (Table 4). Those stored under dry conditions showed 73% emergence after a longer germination period (81.1 ± 2.2 DAS).

3.3. Effects of Temperature on Germination. All seeds germinated at 20°C–30°C and 58% of seeds germinated at 15°C. No seeds germinated at 5°C–10°C (Table 5). The most rapid seed germination was at 25°C (30.1 ± 0.7 DAS). The number of days to germinate at 30, 20, and 15°C was 32.2 ± 0.7, 40.6 ± 0.7, and 77.0 ± 1.4, respectively.

3.4. Effects of Scarification and Seed Coat Removal. Germination of seeds lacking the seed coat was observed from 7 DAS at 25°C, and the germination frequency reached 100% at 18 DAS (Figure 2(a)). In contrast, germination was delayed in intact seeds (with the seed coat) and scarified seeds. These seeds started to germinate at 10 DAS, and the germination frequency of intact and scarified seeds reached 100% at 20 and 21 DAS, respectively (Figure 2(a)).

Intact seeds and seeds without the seed coat were immersed in methylene blue, and the water permeability of the seeds was investigated. In intact seeds, only the seed coat was stained, which indicated that the seed coat prevented

imbibition (Figure 2(b)). Conversely, the endosperm was stained in seeds without the seed coat (Figure 2(c)).

3.5. Effects of Endosperm Removal. Among the four types of seeds (Figure 1) cultured on MS medium supplemented with sucrose, isolated embryos showed the highest and 100% of embryos had germinated at 14 DAI (Figure 3(a)). Seeds without the seed coat and with half of the endosperm showed the next highest germination frequency, followed by seeds without the seed coat and micropylar endosperm. No germination was observed in intact seeds during 42 DAI.

When five types of seeds (Figure 1) were cultured on MS medium without sucrose, isolated embryos and seeds without the seed coat and half of the endosperm did not germinate, whereas the other three types of seeds germinated (Figure 3(b)). On MS medium without sucrose, seeds lacking the seed coat showed the best germinability (100% germination by 35 DAI).

4. Discussion

Although *A. crenata* seeds germinated with the conventional method, they required a long period for seed germination (Table 1). To facilitate the seedling propagation and to accelerate the generation turnover in breeding programs, development of a rapid seed germination method is desirable. For these purposes, embryo culture was the most suitable method (Table 1). Alternatively, the time required for seed germination could be reduced by optimizing storage conditions and germination temperature, and by removing the seed coat.

TABLE 4: Emergence of *A. crenata* seeds stored under wet or dry conditions.

Storage condition	Number of seeds		Emergence (%)	Number of days to emergence[a]
	Sown	Emerged		
Wet	30	27	90	71.8 ± 2.4a
Dry	30	22	73	81.1 ± 2.2b

Observations were continued for 126 days after sowing.
[a]Data are expressed as mean ± SE. Means followed by a different letter are significantly different ($P < 0.05$, Tukey's test).

TABLE 5: Germination of *A. crenata* seeds at different temperatures.

Temperature (°C)	Number of seeds		Germination (%)	Number of days to germination[a]
	Sown	Germinated		
5	50	0	0	—
10	50	0	0	—
15	50	29	58	77.0 ± 1.4a
20	50	50	100	40.6 ± 0.7b
25	50	50	100	30.1 ± 0.7c
30	50	50	100	32.2 ± 0.7c

Observations were continued for 84 days after sowing.
[a]Data are expressed as mean ± SE. Means followed by a different letter are significantly different ($P < 0.05$, Tukey's multiple comparison test).

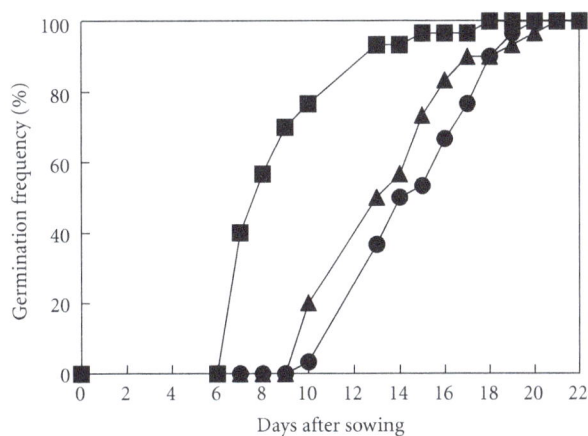

(a)

(b)   (c)

FIGURE 2: Effects of scarification and seed coat removal on seed germination in *A. crenata*. (a) Germination frequency of intact seeds (circles), scarified seeds (triangles), and seeds without the seed coat (squares). (b) and (c) Water permeability of seeds. Intact seeds (b) and seeds without the seed coat (c) were immersed in 1% methylene blue, and cut surfaces were photographed.

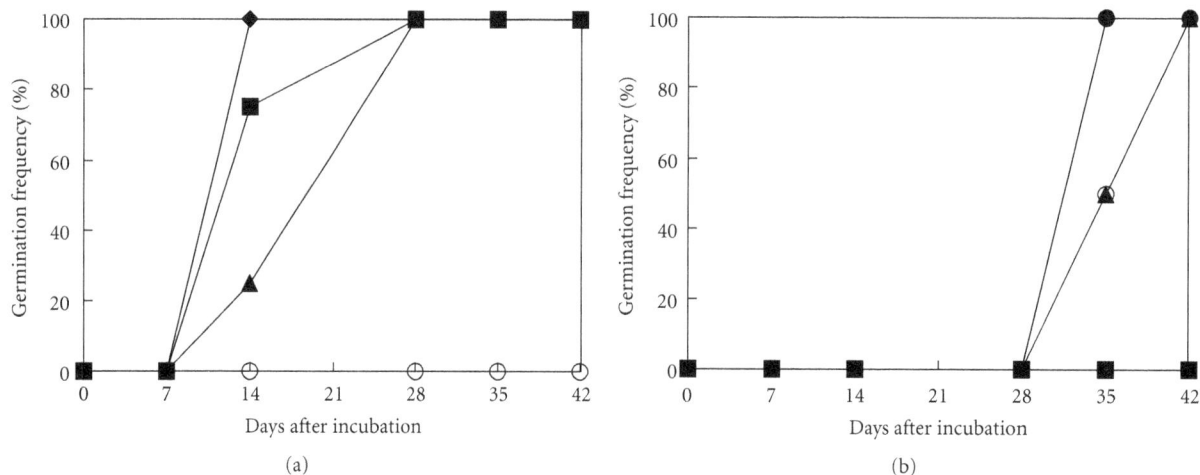

FIGURE 3: Germination frequencies of *A. crenata* seeds with seed coat and some or all of the endosperm removed (see Figure 1). Seeds were cultured on MS medium supplemented with sucrose (a) or without sucrose (b). Open circle: intact seeds with seed coat; filled circle: seeds without seed coat; triangles: seeds without seed coat and micropylar endosperm; squares: seeds without seed coat and half of the endosperm; diamonds: seeds without seed coat and endosperm (isolated embryos). Note that open circles and triangles, and squares and diamonds overlapped in (b).

In Japan, *A. crenata* usually flowers in July and subsequently produces berries. Seeds harvested in late August contained immature embryos. Consequently the germination frequencies of these seeds sown *in vivo* and embryos cultured *in vitro* at 25°C were 0% and 20%, respectively (Table 2). By late September, embryos were more developed and 100% of the embryos and seeds germinated. A decrease in the germinability or induction of further dormancy in seeds harvested in late October or later was not observed. Therefore, seeds of *A. crenata* were considered to reach maximum germinability during September.

Seeds of *A. crenata* germinated at 20°C, regardless of storage at low temperature. However, the seeds germinated more rapidly after longer periods of storage at low temperature (Table 3). Because there was no change in embryo length during storage, it is likely that the earlier germination was related to breaking of dormancy, rather than the stage of embryo development.

Seeds stored in dry conditions showed lower germination frequency and slower emergence than those stored in wet conditions (Table 4). Yang et al. [8] also reported that seeds stored under dry conditions for 90 days did not germinate in *A. crenata* var. *bicolor*. Moisture conditions during seed storage often influence seed germination. Partial drying of dormant seeds of *Zizania palustris* increased the germination response [9]. Conversely, in *Cucumis sativus* var. *hardwickii*, seed germination was improved by storing seeds at higher humidity [10]. Drying of seeds induced seed dormancy (secondary dormancy) in *Sorghum vulgare* [11] and *Panicum virgatum* [12]. Further research is required to determine whether secondary dormancy is induced if seed viability is lost during dry storage of *A. crenata* seeds.

Among the germination temperatures tested, *A. crenata* seeds germinated at 15–30°C but not at 5°C–10°C (Table 5).

In the traditional method of seed germination in Japan, *A. crenata* seeds are sown in spring and germinate in early summer. This long period required for seed germination is because the suitable temperature for seed germination is relatively high (20°C–30°C).

Seed coat removal enhanced seed germination in *A. crenata* (Figure 2(a)). A water permeability test with methylene blue solution showed that seed coat removal enhanced water uptake by seeds (Figures 2(b) and 2(c)). Stabell et al. [13] reported that the seed coat regulates dormancy partly by restricting diffusion of O_2 to the embryo in *Cynoglossum officinale*. Therefore, the seed coat in *A. crenata* might inhibit water uptake as well as O_2 uptake during seed germination.

In *A. crenata*, isolated embryos germinated more rapidly than seeds (Table 1). Therefore, we considered that certain inhibitors such as abscisic acid (ABA) contained in the endosperm might interfere with seed germination. In *Arabidopsis*, ABA synthesized in the endosperm induces seed dormancy [14, 15]. In the present study, when several types of *A. crenata* seeds were sown in medium containing sucrose, the germination frequency increased with a decreasing size of the remaining endosperm (Figure 3(a)). However, in sucrose-free medium, isolated embryos, and seeds without the seed coat and half of the endosperm did not germinate, whereas seeds retaining a greater proportion of the endosperm germinated (Figure 3(b)). Carbohydrates such as sucrose play an important role in physiological processes in plants. During seed germination, carbohydrate reserves serve as an energy source [16]. With this point in mind, it was hypothesized that sucrose in the medium was more suitable as an energy source for germination than reserve materials in the endosperm in *A. crenata*. Therefore, the promotion of the germination by endosperm removal may be because of absorption of sucrose, which is readily available as an energy source, and not the result of the removal of inhibitors contained in the endosperm.

In conclusion, the optimal temperature of 25°C was the most critical factor for seed germination in *A. crenata*. For accelerated emergence of *A. crenata* seedlings, the seeds should be harvested in late September and immediately sown at 25°C. Seed germination is further enhanced by removing the seed coat. If seeds must be stored before sowing, they should be stored at a low temperature (approximately 5°C) under moist conditions.

References

[1] C. Chen and J. J. Pipoly, "Myrsinaceae," in *Flora of China*, Z. Wu and P. H. Raven, Eds., vol. 15, pp. 1–38, Missouri Botanical Garden Press, St. Louis, Mo, USA, 1996.

[2] H. Kobayashi and E. de Mejía, "The genus *Ardisia*: a novel source of health-promoting compounds and phytopharmaceuticals," *Journal of Ethnopharmacology*, vol. 96, no. 3, pp. 347–354, 2005.

[3] M. S. Roh, A. K. Lee, and J. K. Suh, "Production of high quality *Ardisia* plants by stem tip cuttings," *Scientia Horticulturae*, vol. 104, no. 3, pp. 293–303, 2005.

[4] R. L. Geneve, "Impact of temperature on seed dormancy," *HortScience*, vol. 38, no. 3, pp. 336–341, 2003.

[5] J. M. Baskin and C. C. Baskin, "A classification system for seed dormancy," *Seed Science Research*, vol. 14, no. 1, pp. 1–16, 2004.

[6] T. Murashige and F. Skoog, "A revised medium for rapid growth and bio assays with tobacco tissue cultures," *Physiologia Plantarum*, vol. 15, pp. 473–497, 1962.

[7] A. Orozco-Segovia, J. Márquez-Guzmán, M. E. Sánchez-Coronado, A. Gamboa de Buen, J. M. Baskin, and C. C. Baskin, "Seed anatomy and water uptake in relation to seed dormancy in *Opuntia tomentosa* (Cactaceae, Opuntioideae)," *Annals of Botany*, vol. 99, no. 4, pp. 581–592, 2007.

[8] Q. H. Yang, W. H. Ye, Z. M. Wang, and X. J. Yin, "Seed germination physiology of *Ardisia crenata* var. *bicolor*," *Seed Science and Technology*, vol. 37, no. 2, pp. 291–302, 2009.

[9] C. D. Aldridge and R. J. Probert, "Effects of partial drying on seed germination in the aquatic grasses *Zizania palustris* L. and *Porteresia coarctata* (Roxb.) Tateoka," *Seed Science Research*, vol. 2, pp. 199–205, 1992.

[10] L. A. Weston, R. L. Geneve, and J. E. Staub, "Seed dormancy in *Cucumis sativus* var. hardwickii (Royle) Alef.," *Scientia Horticulturae*, vol. 50, no. 1-2, pp. 35–46, 1992.

[11] G. E. Nutile and L. W. Woodstock, "The influence of dormancy-inducing desiccation treatments on the respiration and germination of sorghum," *Physiologia Plantarum*, vol. 20, pp. 554–561, 1967.

[12] Z. X. Shen, D. J. Parrish, D. D. Wolf, and G. E. Welbaum, "Stratification in switchgrass seeds is reversed and hastened by drying," *Crop Science*, vol. 41, no. 5, pp. 1546–1551, 2001.

[13] E. Stabell, M. K. Upadhyaya, and B. E. Ellis, "Role of seed coat in regulation of seed dormancy in houndstongue (*Cynoglossum officinale*)," *Weed Science*, vol. 46, no. 3, pp. 344–350, 1998.

[14] V. Lefebvre, H. North, A. Frey et al., "Functional analysis of Arabidopsis NCED6 and NCED9 genes indicates that ABA synthesized in the endosperm is involved in the induction of seed dormancy," *The Plant Journal*, vol. 45, no. 3, pp. 309–319, 2006.

[15] K. P. Lee, U. Piskurewicz, V. Turečková, M. Strnad, and L. Lopez-Molina, "A seed coat bedding assay shows that RGL2-dependent release of abscisic acid by the endosperm controls embryo growth in *Arabidopsis* dormant seeds," *Proceedings of the National Academy of Sciences of the United States of America*, vol. 107, no. 44, pp. 19108–19113, 2010.

[16] T. R. Johnson, M. E. Kane, and H. E. Pérez, "Examining the interaction of light, nutrients and carbohydrates on seed germination and early seedling development of *Bletia purpurea* (Orchidaceae)," *Plant Growth Regulation*, vol. 63, no. 1, pp. 89–99, 2011.

21

Phenology of Some Phanerogams (Trees and Shrubs) of Northwestern Punjab, India

Gurveen Kaur, Bhupinder Pal Singh, and Avinash Kaur Nagpal

Department of Botanical and Environmental Sciences, Guru Nanak Dev University, Amritsar 143005, India

Correspondence should be addressed to Gurveen Kaur; bajwagurveen@gmail.com and
Avinash Kaur Nagpal; avnagpal.dobes@gndu.ac.in

Academic Editor: William K. Smith

Plants perform various vegetative and reproductive functions throughout the year in order to persist in their habitats. The study of these events including their timing and how the environment influences the timing of these events is known as phenology. This study of the timing of seasonal biological activities of plants is very important to know about plant's survival and its reproductive success. The variation in the phenological activities is due to change in different abiotic conditions. This paper deals with the study of phenological activities like bud formation, flowering time, fruiting time, and seed formation for some leguminous plants of Amritsar, Punjab (a state in the northwest of India) for three consecutive years from 2009 till 2011.

1. Introduction

The timing of various phenological activities such as germination, bud break, flowering, fruit dehiscence, and leaf drop is important for survival and reproductive success of many plant species. Abiotic environmental conditions such as rain, change in temperature, presence/absence of pollinators, competitors, and herbivores have been shown to play a significant role in timing of various phenological events [1–6]. Natural selection has also been considered to play some role in determining the phenological patterns of plant species [7]. Phenological studies are also important in understanding species interrelations and their interaction with the environment. Variations in phenophases among individuals of the same species or different species have been linked to environmental perturbations [8].

Considerable amount of phenological data is available on different plant species from different parts of the world including tropical savanna and semideciduous forest of Venezuelan Ilanos (South America) [9], dry tropical forest in Ghana [10], NE Spain [11–13], Panama [14], Mexico [15], tropical rain forest in Malaya [16], semiarid grassland in the Rock mountain, USA [17], and tibetan plateau [18]. A number of studies on phenology of different plant species from different parts of India have also been undertaken which include those from a subtropical humid forest in North-Eastern India [19, 20], Kumaun Himalayan forests [21, 22], deciduous forest of Bandipur in peninsular India [23], Shervaroys, Southern India [24], tropical moist forest of Western Ghats in Karnataka [25], Hathinala Forest in Uttar Pradesh [26], alpine expanse of North-West Himalaya [27], Orissa coast [28], tropical montane forests in the Nilgiris [8], Kolhapur region (Maharashtra) [29], and Katerniaghat wildlife sanctuary situated in the Himalayan Terai region in Uttar Pradesh [30], Northeastern India [31]. However, no such information is available on plants of Punjab plains, hence we initiated phenological studies in trees and shrubs of this region of India. To begin with, we have compiled information on fifteen leguminous species from Amritsar.

The family leguminosae is a large and economically important family of flowering plants and has traditionally been divided into three subfamilies Papilionoideae (Faboideae), Caesalpinioideae, and Mimosoideae. These have been recognized as independent families Fabaceae (Papilionaceae), Caesalpiniaceae, and Mimosaceae in several recent systems of classification.

The present paper deals with phenological observations of fifteen leguminous plants growing at four different sites in Amritsar over a period of three years from 2009 till 2011 as these sites represent most of the plant diversity of this region.

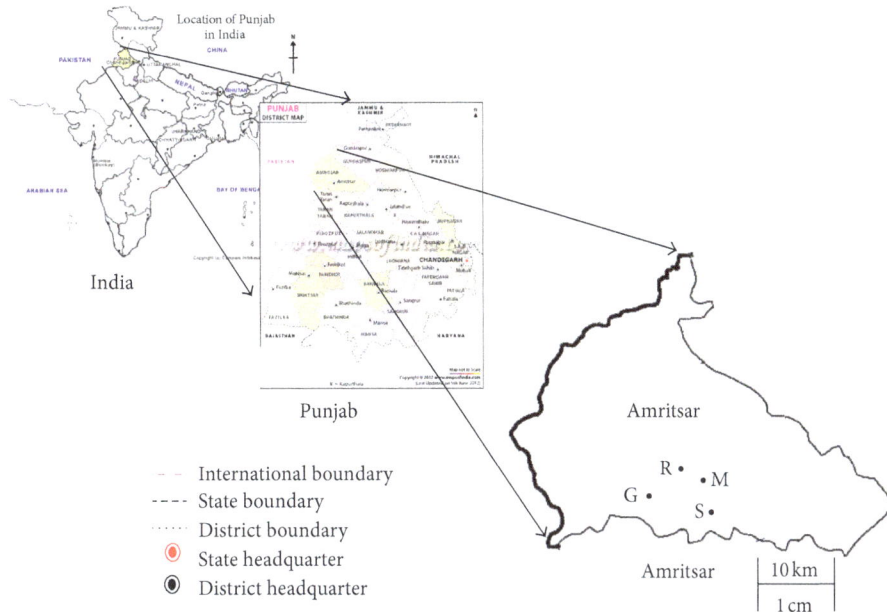

FIGURE 1: Amritsar, the Northwestern part of India. G: Guru Nanak Dev University; S: Sakatri Bagh; M: Maharaja Ranjit Singh Park; R: Ranjit Avenue.

2. Materials and Methods

2.1. Study Area. The Amritsar district is located in northwestern part of the Punjab state and lies between $31°28'30''$ to $32°03'15''$ north latitude and $74°29'30''$ to $75°24'15''$ east longitude. Total area of the district is $5056\,km^2$ with tropical dry deciduous type of vegetation [32]. Natural vegetation is fragmented and is at present available only in narrow strips and patterns.

Amritsar has a semiarid climate, typical of northwestern India and experiences four seasons primarily:

(i) winter season (December to February) with temperature ranges from 4°C (39°F) to about 19°C (66°F);

(ii) summer season (April to June) where temperatures can reach 45°C (113°F);

(iii) monsoon season (July to September);

(iv) postmonsoon season (September to October).

There is a transitional period between winter and summer in March and early April (called as spring), as well as a transitional season between postmonsoon season and winter in October and November (called as autumn). Annual rainfall in Amritsar is about 680 millimetres (ground water information booklet, Amritsar district, Punjab by Central Ground Water Board, Ministry of Water Resources, Government of India, north western region, Chandigarh).

The whole area of Amritsar was surveyed and four different sampling sites, that is, Guru Nanak Dev University (G), Sakatri Bagh (S), Maharaja Ranjit Singh Park (M), and Ranjit Avenue (R) were selected for data collection. These sites were chosen as they represent most of the plant diversity of Amritsar. Figure 1 shows the location of four sites on the map of Amritsar. The maps were downloaded from Maps of India and Google images, and the climatic data of Amritsar was obtained from the wunderground website (http://www.wunderground.com/). This weather station (VIAR) is located at Raja Sansi International Airpot, Amritsar at an elevation of 234 m. Table 1 shows the minimum, mean, and maximum temperature values from January to December for three consecutive years from 2009 to 2011 of Amritsar.

Table 2 lists fifteen plant species selected for the present study along with their family, sampling site, growth form, and leaf habit. The data for each species was collected from a minimum of two sampling sites mentioned in Table 2. At each site, one mature and healthy plant from the population of that species was fixed as reference for the study. The observations were made on the marked plant and 10–15 of its closest neighbours (minimum 6 plants if more plants were not available for a particular plant species at a particular site) for a period of 3 years starting from January 2009 up to December, 2011. This selection regime was based on a study by Pilar and Gabriel [12]. The areas with the plants under study were visited daily to record presence/absence of different phenological events or phenophases: flower bud formation (B), flowering (FL), fruit setting (FR), and seed set and dispersal (S). The information for the initial date and the last date when these various phenophases were observed was recorded. A phenophase was considered to be active in the population just when it was observed in at least 5% of the crown in a minimum of 20% of the studied plants [12]. Phenophase calendars for each species were prepared and were studied for the interpretation of the overall results. For the whole set of studied species, the frequency of occurrence of different phenophases in each month was calculated.

3. Results

The results on percentage of 15 plant species taken all together showing timing of different phenophases (Buds, Flowers,

TABLE 1: Monthly minimum, mean, and maximum temperature in Amritsar for the years 2009, 2010, and 2011.

| | 2009 | | | 2010 | | | 2011 | | |
	Min.	Mean	Max.	Min.	Mean	Max.	Min.	Mean	Max.
Jan.	6	12	18	3	8	14	5	10	16
Feb.	9	16	23	10	16	22	9	14	20
Mar.	12	20	27	16	23	30	13	20	27
Apr.	18	26	34	21	30	38	17	25	33
May	23	31	39	25	32	40	25	32	40
Jun.	25	33	41	26	32	39	26	31	37
Jul.	26	31	36	27	31	34	27	30	34
Aug.	27	30	35	26	29	33	26	29	32
Sep.	23	28	33	24	28	32	25	28	32
Oct.	21	26	32	20	25	31	18	25	32
Nov.	11	18	24	12	19	27	13	20	27
Dec.	5	13	21	5	12	20	6	13	21

TABLE 2: List of plant species studied with their abbreviation, family, sampling site, growth form (T: tree; S: shrub; C: climber) and leaf habit (E: evergreen and D: deciduous).

S. No.	Species	Abbreviation	Family	Sampling site*	Growth form	Leaf habit
1	*Bauhinia variegata* Linn.	Bva	Caesalpiniaceae	G, S, M, R	T	D
2	*Caesalpinia pulcherrima* (Linn.) Sw.	Cpu	Caesalpiniaceae	G, M, R	S	E
3	*Cassia fistula* Linn.	Cfi	Caesalpiniaceae	G, S, M, R	T	D
4	*Cassia glauca* Lam.	Cgl	Caesalpiniaceae	G, S, R	S	E
5	*Cassia siamea* Lamk.	Csi	Caesalpiniaceae	G, R	T	E
6	*Delonix regia* Rafin.	Dre	Caesalpiniaceae	G, S, M	T	D
7	*Parkinsonia aculeata* Linn.	Pac	Caesalpiniaceae	G, R	T	D
8	*Acacia auriculiformis* A. Cunn.	Aau	Mimosaceae	G, M, R	T	E
9	*Acacia nilotica* Delile	Ani	Mimosaceae	G, R	T	E
10	*Albizia lebbeck* Benth.	Ale	Mimosaceae	G, S	T	D
11	*Calliandra tweedii* Benth.	Ctw	Mimosaceae	G, R	S	E
12	*Prosopis juliflora* DC.	Pju	Mimosaceae	G, S, M	T	E
13	*Abrus precatorius* Linn.	Ape	Papilionaceae	G, S, R	C	D
14	*Butea monosperma* (Lam.) Kuntze	Bmo	Papilionaceae	G, R	T	D
15	*Dalbergia sissoo* Roxb.	Dsi	Papilionaceae	G, S, M, R	T	D

*Sampling sites shown in Figure 1.

Fruits, and Seeds) during the study period are shown in Figure 2. This figure reveals that percentage of plant species exhibiting each phenophase, that is, buds, flowers, fruits, and seeds showed almost a similar trend for three consecutive years within each phenophase. Bud formation showed a peak in the beginning of summer (60%) and another small peak during late monsoon and was almost evenly distributed through the rest of the year for all the three years. Similar trend was observed for flowering which also showed a peak (66.67%) during beginning of summer and another small peak (53.33%) during postmonsoon period for all the three years. Maximum percentage of plant species showing fruiting was observed in the months of March, April, and May (60%) which resulted in maximum seed set in the months of June, July, and August during the study period. Comparison of results on observations of flowering in different plant species during the three consecutive years 2009, 2010, and 2011, reveals that maximum percentage of plant species (66.67%) showing flowering was observed in the month of April during 2011.

Figure 3 shows phenological patterns of individual plant species studied. For each plant species, the whole period of appearance of different phenophases-like bud formation, flowering, fruiting, and seed set has been indicated for three consecutive years. The study reveals high phonological diversity for different phenophases studied among fifteen plant species. However, for individuals of the same species, no differences in the appearance of different phenophases were observed.

During the study period, it was observed that some species like *Bauhinia variegata* Linn., *Caesalpinia pulcherrima* (Linn.) Sw., *Cassia glauca* Lam., *Cassia siamea* Lamk., and so forth showed conspicuous buds (2–4 weeks) before flowering, but in case of *Dalbergia sissoo* Roxb bud formation was observed just few days (less than a week) before prominent flowering was recorded. Most of the species flowered from spring to late summer (e.g., *Parkinsonia aculeata* Linn. and *Acacia auriculiformis* A. Cunn.). *Butea monosperma* (Lam.) Kuntze exhibited flowering during the monsoon and postmonsoon period only. Some of the species that showed

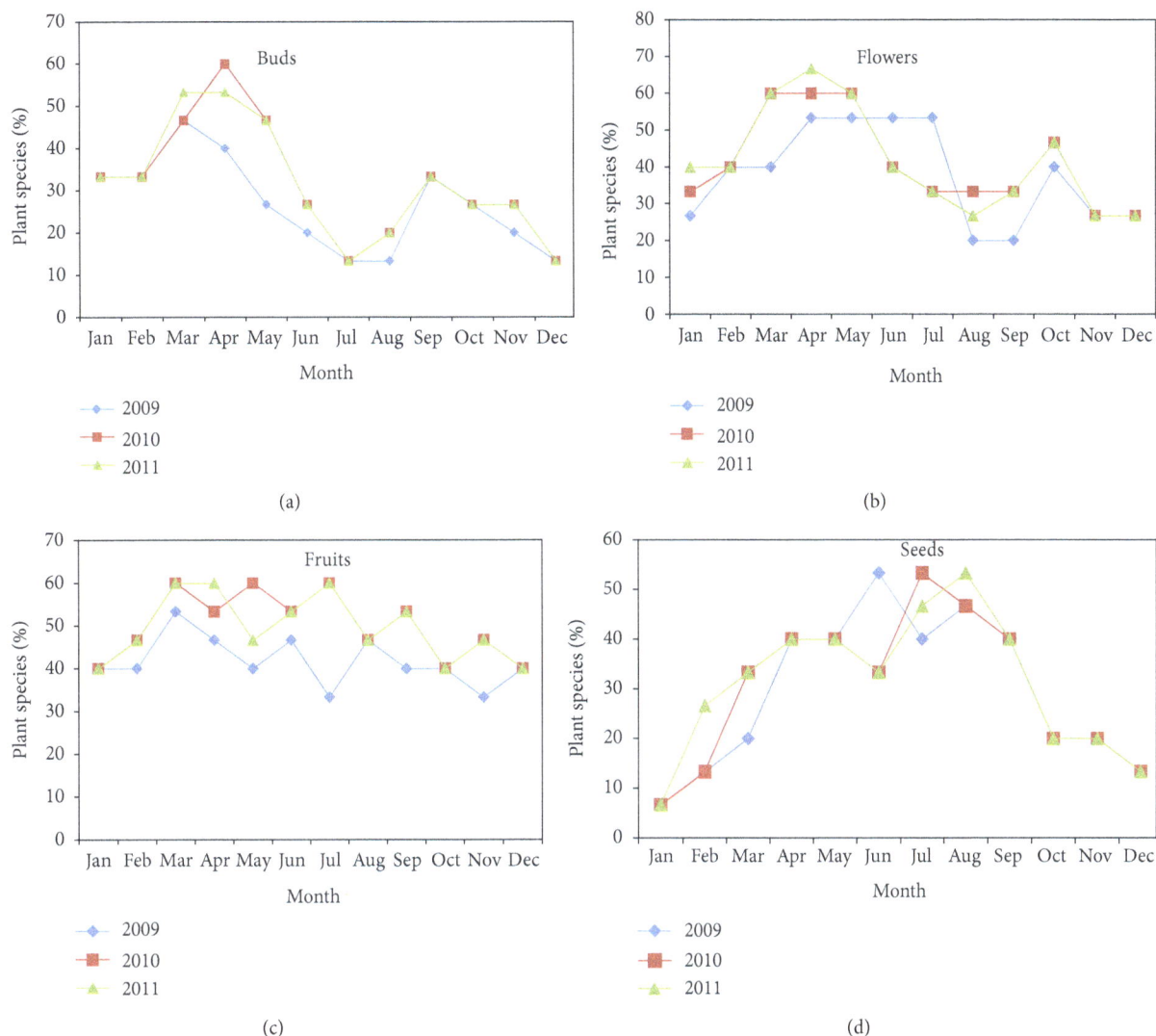

FIGURE 2: Percentage of plant species under study showing each phenophase: buds (a), flowers (b), fruits (c), and seeds (d) during 2009 to 2011.

flowering during winter months extending to early summer included *Caesalpinia pulcherrima* (Linn.) Sw., *Cassia siamea* Lamk., and *Delonix regia* Rafin. Variation in the flowering time period was observed for different species ranging from 2 months (*Abrus precatorius* Linn.) to 7 months (*Cassia glauca* Lam.). Similarly the variation in fruiting period was observed from 3 months as in the case of *Cassia siamea* Lamk., and so forth, to have fruiting almost throughout the year as for *Dalbergia sissoo* Roxb. By the end of summer, the seeds of most of the plant species were mature and ready for dispersal. There were two time periods in the year when leaf shedding was observed, that is, in summer and in the autumn depending on the type of leaf habit of that plant species.

4. Discussion

Our study reveals high phenological diversity for the four phenological patterns (buds, flowers, fruits, and seeds) among fifteen leguminous plant species growing in Amritsar

under similar environmental conditions for the three consecutive years, that is, 2009, 2010, and 2011. Changes in plant phenological patterns have been associated with the species specific plant structural architecture, availability and transfer of nutrients [33], plant growth rates [34], temperature [35], and water [36].

A comparison of observations on flowering time for the three years under study reveals an advancement of flowering time by about 2–4 weeks in the year 2010 for some species like *Acacia nilotica* Delile, *Abrus precatorius* Linn., *Bauhinia variegata* Linn., *Dalbergia sissoo* Roxb., *Delonix regia* Rafin., and *Parkinsonia aculeata* Linn. as compared to that observed in 2009 (Figure 3). This change in phenological shift can be attributed to increase in temperature in 2010 as compared to that reported for 2009 (Table 1). The total length of flowering period was extended by 2–4 weeks in 2010-11 as compared to 2009 for species like *Bauhinia variegata* Linn., *Caesalpinia pulcherrima* (Linn.) Sw., *Cassia fistula* Linn., *Parkinsonia aculeata* Linn., *Acacia nilotica* Delile, *Calliandra tweedii* Benth., and *Prosopis juliflora* DC.

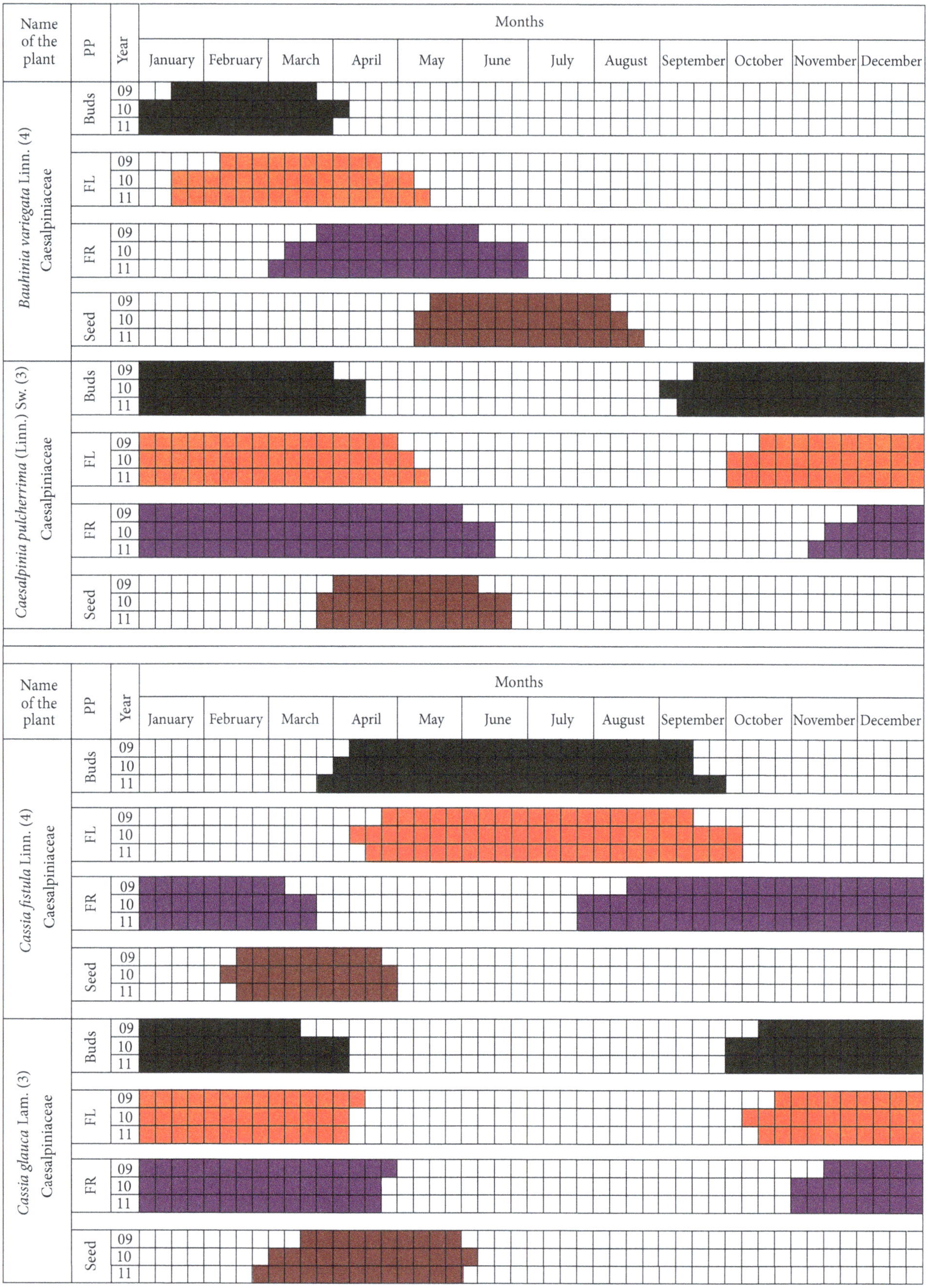

FIGURE 3: Continued.

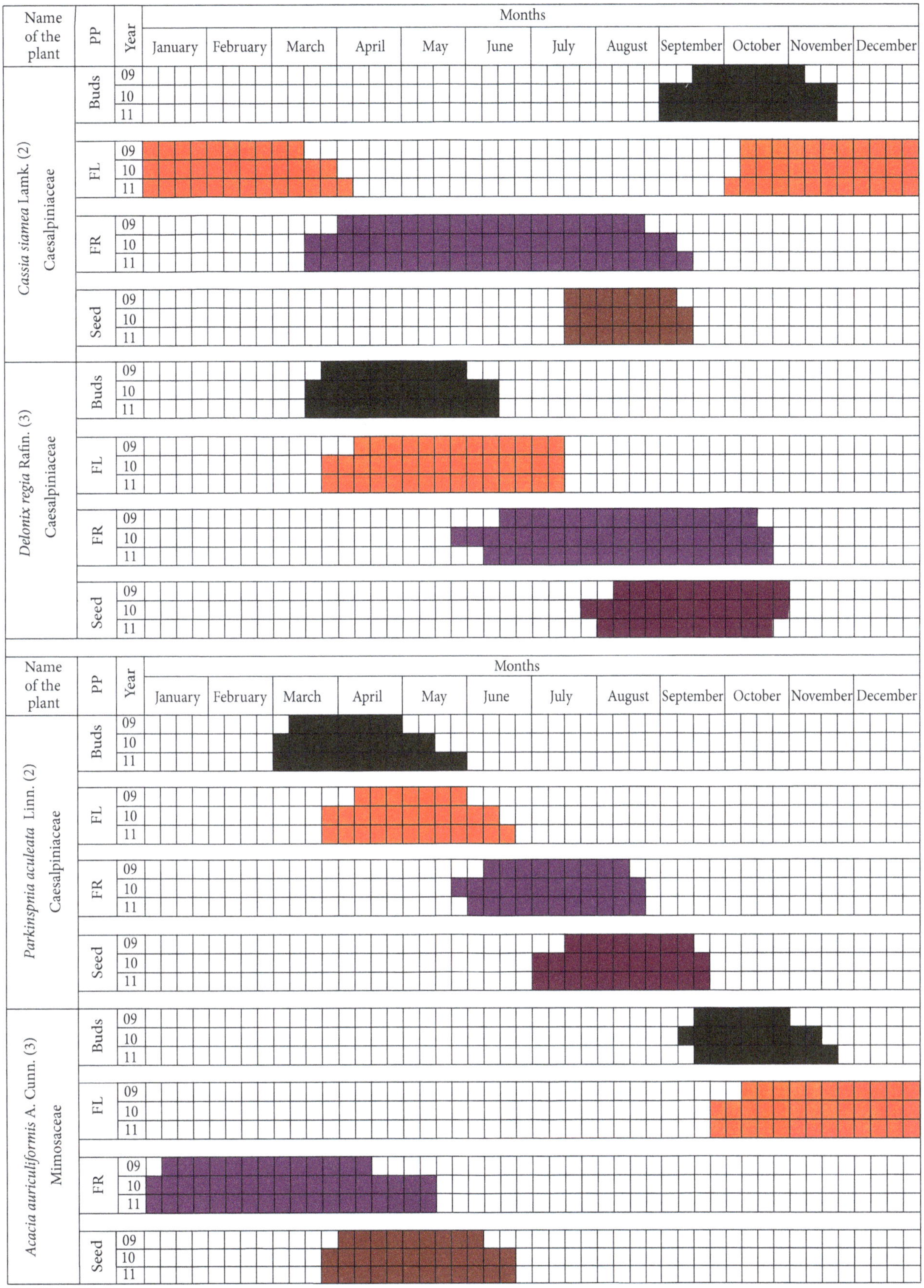

FIGURE 3: Continued.

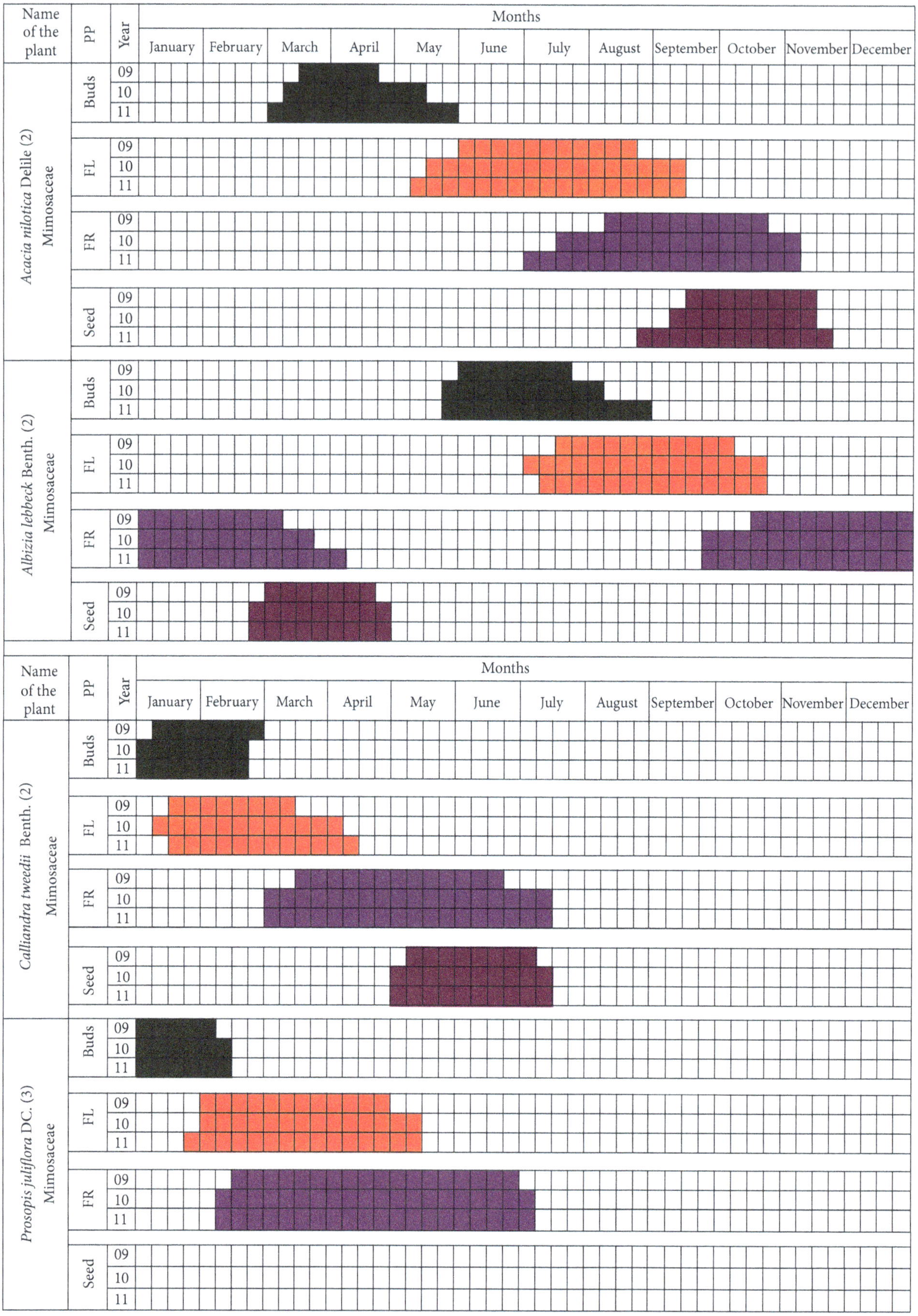

FIGURE 3: Continued.

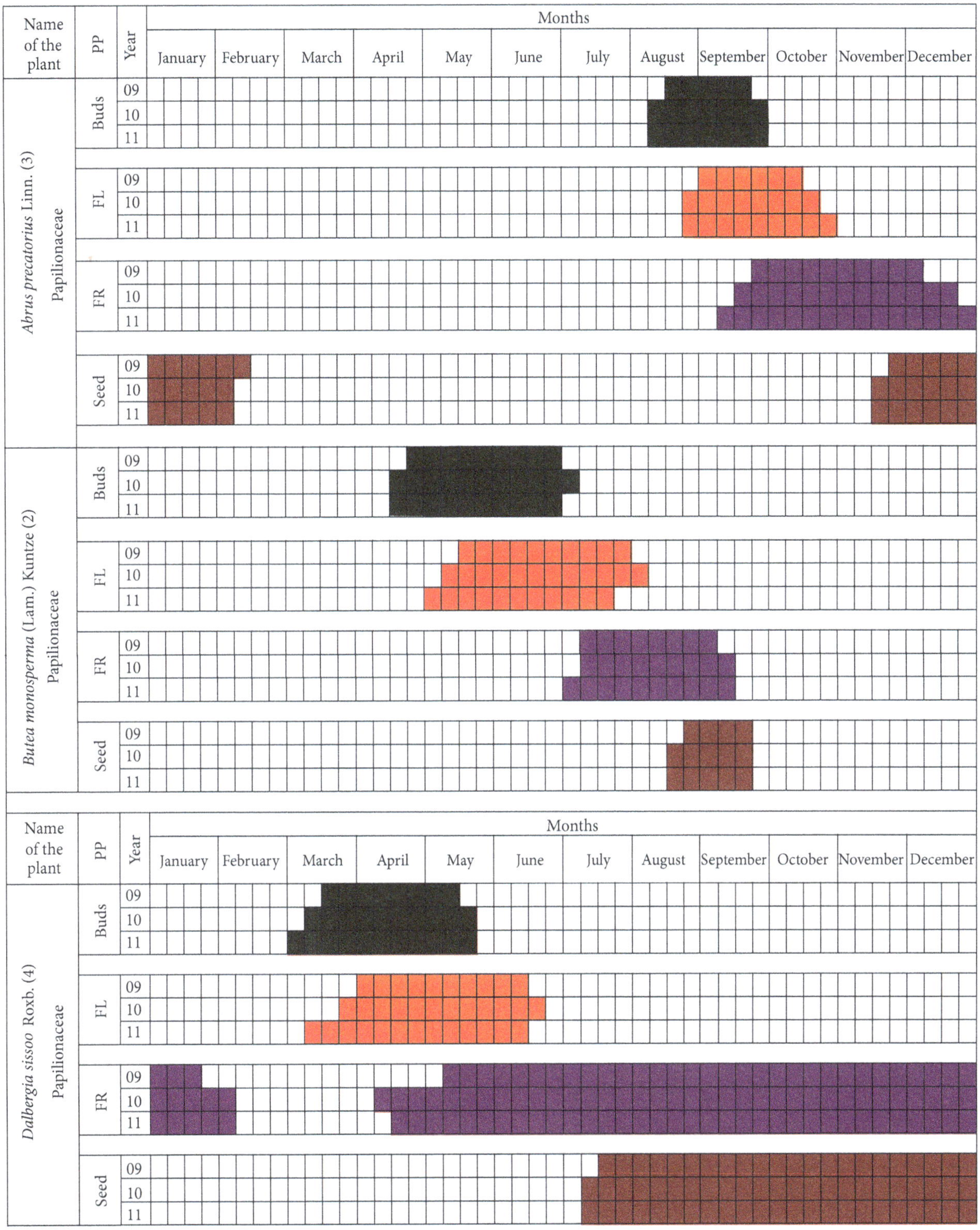

FIGURE 3: Phenological diagrams of fifteen plant species belonging to families caesalpiniaceae, mimosaceae, and papilionaceae. The dark colored bars indicate the whole period of appearance of different phenophases including buds: bud formation; FL: flowering; FR: fruiting, and seed: seed formation for three consecutive years. Numbers in parenthesis beside the species names are the number of considered populations. (PP: phenophase).

Some other studies have also demonstrated an association of an advancement in flowering date with climate change. One study has reported an increase in the mean terrestrial surface temperature by more than 0.2°C per decade ever since 1970 [37]. A number of other studies have shown an advancement of phenological events as a result of increase in temperature [38–41]. With the increase in the temperature, most of the species showed earlier flowering time period [39] and even increase in their time period of flowering (e.g., *Bauhinia variegate* Linn., *Prosopis juliflora* DC.). Similarly early flowering was observed for 24 out of the 32 plant species studied in a semiarid grassland by Lesica and Kittelson [17].

The period of maximum activity of bud formation in the months of spring and the flowering period (started in spring and extended to summer and the beginning of autumn) coincided with the observations of Pilar and Gabriel [12]. Flower development dates have been shown to be synchronous among individuals of the same plant species studied which seems to be important in increasing the chances of outcross pollinations as suggested by Ollerton and Lack [42].

These changes in the phenological patterns can result in adverse effects on insect pollinators as well as herbivores (if the plant species is present in barren or wild areas) that depend on those species for food [43, 44]. The changes in phenological patterns of plant's response to different temperature and rainfall availability have been shown to be species specific [45, 46]. Hence changes in vegetation community species composition might also be responsible for changes in the phenology of the studied plants. The present study reveals that different plants of the same family (Leguminosae) flower at different times during the same year growing at same location under similar environmental conditions which can be attributed to species-specific plant structural architecture.

5. Conclusion

The present study reveals high phenological diversity for four different phenophases among fifteen leguminous plant species growing in Amritsar with tropical dry deciduous type of vegetation. This study would be of great help in knowing the timing of different phenophases of the studied plants which can be of interest to people of this region (or where similar climatic conditions prevail) who wish to plan their gardens and wish to have flowers in their gardens round the year. So, they can select plants which flower during different parts of the year. This type of study can provide important insights into the biology of the plants concerned and reveal phenological pattern of surveyed species. This study would also be of great help for comparison over long duration of time. For example, to see if there is any change in the phenological patterns of the same plant species in next 10 or so years. Such comparative study could not be possible at this time since no relevant literature is available for this region.

Acknowledgment

The authors thank Department of Science and Technology, Government of India, for INSPIRE Fellowship to Ms. Gurveen Kaur.

References

[1] E. R. Heithaus, "The role of plant-pollinator interactions in determining community structure," *Annals of the Missouri Botanical Garden*, vol. 61, pp. 675–691, 1974.

[2] G. W. Frankie, "Tropical forest phenology and pollinator plant coevolution," in *Coevolution of Animals and Plants*, L. E. Gilbert and P. H. Raven, Eds., pp. 192–209, University of Texas Press, Austin, Tex, USA, 1975.

[3] H. R. Pulliam and M. R. Brand, "The production and utilization of seeds in plains grassland of southeastern Arizona," *Ecology*, vol. 56, pp. 1158–1166, 1975.

[4] P. A. Opler, G. W. Frankie, and H. G. Baker, "Rainfall as a factor in the release, timing and synchronization of anthesis by tropical trees and shrubs," *Journal of Biogeography*, vol. 3, pp. 231–236, 1976.

[5] J. N. Thompson and M. F. Willson, "Evolution of temperature fruit/bird interactions: phenological strategies," *Evolution*, vol. 33, pp. 973–982, 1979.

[6] E. W. Stiles, "Patterns of fruit presentation and seed dispersal in bird disseminated woody plants in the eastern deciduous forest," *The American Naturalist*, vol. 116, pp. 670–688, 1980.

[7] G. F. Estabrook, J. A. Winsor, A. G. Stephenson, and H. F. Howe, "When are two phenological patterns different?" *Botanical Gazette*, vol. 143, no. 3, pp. 374–378, 1982.

[8] H. S. Suresh and R. Sukumar, "Vegetative phenology of tropical montane forests in the Nilgiris, South India," *Journal of the National Science Foundation of Sri Lanka*, vol. 39, no. 4, pp. 333–343, 2011.

[9] M. Monasterio and G. Sarmiento, "Phenological strategies of plant species in the tropical savanna and the semideciduous forest of the Venezuelan Ilanos," *Journal of Biogeography*, vol. 3, pp. 352–356, 1976.

[10] D. Lieberman, "Seasonality and phenology in a dry tropical forest in Ghana," *Journal of Ecology*, vol. 70, no. 3, pp. 791–806, 1982.

[11] R. Milla, P. Castro-Díez, and G. Montserrat-Martí, "Phenology of Mediterranean woody plants from NE Spain: synchrony, seasonality, and relationships among phenophases," *Flora*, vol. 205, no. 3, pp. 190–199, 2010.

[12] C.-D. Pilar and M.-M. Gabriel, "Phenological pattern of fifteen Mediterranean phanaerophytes from *Quercus ilex* communities of NE-Spain," *Plant Ecology*, vol. 139, no. 1, pp. 103–112, 1998.

[13] G. Montserrat-Martí and C. Pérez-Rontomé, "Fruit growth dynamics and their effects on the phenological pattern of native *Pistacia* populations in NE Spain," *Flora*, vol. 197, no. 3, pp. 161–174, 2002.

[14] D. De Steven, D. M. Windsor, and F. E. Putz, "Vegetative and reproductive phonologies of a palm assemblage in Panama," *Biotropica*, vol. 19, pp. 342–356, 1987.

[15] S. H. Bullock and J. A. Solis-Magallanes, "Phenology of canopy trees of a tropical deciduous forest in Mexico," *Biotropica*, vol. 22, pp. 22–35, 1990.

[16] L. Medway, "Phenology of a tropical rain forest in Malaya," *Biological Journal of the Linnean Society*, vol. 4, no. 2, pp. 117–146, 1972.

[17] P. Lesica and P. M. Kittelson, "Precipitation and temperature are associated with advanced flowering phenology in a semi-arid grassland," *Journal of Arid Environments*, vol. 74, pp. 1013–1017, 2010.

[18] M. Shen, Y. Tang, J. Chen, X. Zhu, and Y. Zheng, "Influences of temperature and precipitation before the growing season on spring phenology in grasslands of the central and eastern Qinghai-Tibetan Plateau," *Agricultural and Forest Meteorology*, vol. 151, no. 12, pp. 1711–1722, 2011.

[19] R. P. Shukla and P. S. Ramakrishnan, "Phenology of trees in a sub-tropical humid forest in north-eastern India," *Vegetatio*, vol. 49, no. 2, pp. 103–109, 1982.

[20] R. P. Shukla and P. S. Ramakrishnan, "Leaf dynamics of tropical trees related to successional status," *New Phytologist*, vol. 97, no. 4, pp. 697–706, 1984.

[21] P. K. Ralhan, R. K. Khanna, S. P. Singh, and J. S. Singh, "Phenological characteristics of the tree layer of Kumaun Himalayan forests," *Vegetatio*, vol. 60, no. 2, pp. 91–101, 1985.

[22] Y. P. S. Pangtey, R. S. Rawal, N. S. Bankoti, and S. S. Samant, "Phenology of high-altitude plants of Kumaun in Central Himalaya, India," *International Journal of Biometeorology*, vol. 34, no. 2, pp. 122–127, 1990.

[23] S. N. Prasad and M. Hegde, "Phenology and seasonality in the tropical deciduous forest of Bandipur, South India," *Proceedings: Plant Sciences*, vol. 96, no. 2, pp. 121–133, 1986.

[24] N. Sivaraj and K. V. Krishnamurthy, "Flowering phenology in the vegetation of Shervaroys, South India," *Vegetatio*, vol. 79, no. 1-2, pp. 85–88, 1988.

[25] D. M. Bhat and K. S. Murali, "Phenology of understorey species of tropical moist forest of Western Ghats region of Uttara Kannada district in South India," *Current Science*, vol. 81, no. 7, pp. 799–805, 2001.

[26] K. P. Singh and C. P. Kushwaha, "Diversity of flowering and fruiting phenology of trees in a tropical deciduous forest in India," *Annals of Botany*, vol. 97, no. 2, pp. 265–276, 2006.

[27] R. K. Vashisthe, N. Rawat, A. K. Chaturvedi, B. P. Nautiyal, P. Prasad, and M. C. Nautiyal, "An exploration on the phenology of different growth forms of an alpine expanse of north-west Himalaya, India," *New York Science Journal*, vol. 2, pp. 29–42, 2009.

[28] V. P. Upadhyay and P. K. Mishra, "Phenology of mangroves tree species on Orissa coast, India," *Tropical Ecology*, vol. 51, no. 2, pp. 289–295, 2010.

[29] A. R. Kasarkar and D. K. Kulkarni, "Phenological studies of family zingiberaceae with special reference to *Alpinia* and *Zingiber* from Kolhapur region(MS) India," *Bioscience Discovery*, vol. 2, pp. 322–327, 2011.

[30] O. Bajpai, A. Kumar, A. K. Mishra, N. Sahu, S. K. Behera, and L. B. Chaudhary, "Phenological study of two dominant tree species in tropical moist deciduous forest from the Northern India," *International Journal of Botany*, vol. 8, pp. 66–72, 2012.

[31] A. Lokho and Y. Kumar, "Reproductive phenology and morphology analysis of Indian *Dendrobium* Sw. (Orchidaceae) from the northeast region," *International Journal of Scientific and Research Publications*, vol. 2, pp. 1–14, 2012.

[32] S. H. G. Champion and S. K. Seth, *A Revised Survey of the Forest Types of India*, The manager of Publications, Delhi, India, 1968.

[33] R. E. Sosebee and H. H. Wiebe, "Effect of phenological development on radiophosphorus translocation from leaves in crested wheatgrass," *Oecologia*, vol. 13, no. 2, pp. 103–112, 1973.

[34] F. G. Taylor Jr., *Phenodynamics of Production in a Mesic Deciduous Forest. US/IBP Eastern Deciduous Forest Biome*, Oak Ridges National Laboratory, Oak Ridge, Tenn, USA, 1972.

[35] M. Y. Nuttonson, *Wheat-Climate Relationships and the Use of Phenology in Ascertaining the Thermal and Photo-Thermal Requirements of Wheat; Based on Data of North America and Some Thermally Analogous Areas of North America, in the Soviet Union and in Finland*, American Institute for Crop Ecology, Washington, DC, USA, 1955.

[36] J. P. Blaisdell, "Seasonal development and yield of native plants on the upper Snake River plains and their relation to certain climatic factors," *US Department of Agriculture Technical Bulletin*, vol. 1190, pp. 1–68, 1958.

[37] J. Hansen, M. Sato, R. Ruedy, K. Lo, D. W. Lea, and M. Medina-Elizade, "Global temperature change," *Proceedings of the National Academy of Sciences of the United States of America*, vol. 103, no. 39, pp. 14288–14293, 2006.

[38] Y. Julien and J. A. Sobrino, "Global land surface phenology trends from GIMMS database," *International Journal of Remote Sensing*, vol. 30, no. 13, pp. 3495–3513, 2009.

[39] A. Menzel, T. H. Sparks, N. Estrella et al., "European phenological response to climate change matches the warming pattern," *Global Change Biology*, vol. 12, no. 10, pp. 1969–1976, 2006.

[40] C. Parmesan and G. Yohe, "A globally coherent fingerprint of climate change impacts across natural systems," *Nature*, vol. 421, no. 6918, pp. 37–42, 2003.

[41] S. Piao, J. Fang, L. Zhou, P. Ciais, and B. Zhu, "Variations in satellite-derived phenology in China's temperate vegetation," *Global Change Biology*, vol. 12, no. 4, pp. 672–685, 2006.

[42] J. Ollerton and A. J. Lack, "Flowering phenology: an example of relaxation of natural selection?" *Trends in Ecology and Evolution*, vol. 7, no. 8, pp. 274–276, 1992.

[43] F. Saavedra, D. W. Inouye, M. V. Price, and J. Harte, "Changes in flowering and abundance of *Delphinium nuttallianum* (Ranunculaceae) in response to a subalpine climate warming experiment," *Global Change Biology*, vol. 9, no. 6, pp. 885–894, 2003.

[44] N. C. Stenseth and A. Mysterud, "Climate, changing phenology, and other life history traits: nonlinearity and match-mismatch to the environment," *Proceedings of the National Academy of Sciences of the United States of America*, vol. 99, no. 21, pp. 13379–13381, 2002.

[45] I. Ibáñez, R. B. Primack, A. J. Miller-Rushing et al., "Forecasting phenology under global warming," *Philosophical Transactions of the Royal Society B*, vol. 365, no. 1555, pp. 3247–3260, 2010.

[46] C. Parmesan, "Influences of species, latitudes and methodologies on estimates of phenological response to global warming," *Global Change Biology*, vol. 13, no. 9, pp. 1860–1872, 2007.

22

Adaptation to High Temperature and Water Deficit in the Common Bean (*Phaseolus vulgaris* L.) during the Reproductive Period

Hide Omae,[1] Ashok Kumar,[2] and Mariko Shono[1]

[1] Tropical Agriculture Research Front, Japan International Research Center for Agricultural Sciences (JIRCAS), 1091-1, Maezato, Ishigaki, Okinawa 907-0002, Japan
[2] Department of Agronomy, CCS Haryana Agricultural University, Hisar 125004, India

Correspondence should be addressed to Hide Omae, homae@affrc.go.jp

Academic Editor: William K. Smith

This paper reviews the adaption to heat and drought stresses in *Phaseolus vulgaris*, a grain and vegetable crop widely grown in both the Old and New World. Substantial genotypic differences are found in morphophysiological characteristics such as phenology, partitioning, plant-water relations, photosynthetic parameters, and shoot growth, which are related to reproductive responses. The associations between (a) days to podding and leaf water content and (b) the number of pods per plant and seed yield are consistent across different environments and experiments. Leaf water content is maintained by reductions in leaf water potential and shoot extension in response to heat and drought stress. Heat-tolerant cultivars have higher biomass allocation to pods and higher pod set in branches. These traits can be used as a marker to screen germplasm for heat and drought tolerance. In this paper, we briefly review the results of our studies carried out on heat and drought tolerance in the common bean at the Tropical Agriculture Research Front, Ishigaki, Japan.

1. Introduction

Transitory or constantly high temperatures cause an array of morphoanatomical, physiological, and biochemical changes in plants, which affect plant growth and development and may lead to a drastic reduction in economic yield. The adverse effects of heat stress can be mitigated by developing crop plants with improved thermotolerance using various genetic approaches [1]. However, achieving this requires a thorough understanding of the physiological responses of plants to high temperature, the mechanisms of heat tolerance, and potential strategies for improving crop thermotolerance.

The common bean (*Phaseoluls vulgaris* L.) is originally a crop of the New World [2], but it is now grown extensively in all major continental areas [3]. Its production spans from 52°N to 32°S latitude [4] and from near sea level in the continental US and Europe to elevations of more than 3000 m in Andean South America. The common bean has two major gene pools [5], the Andean and the Mesoamerican, based on their centers of origin in South and Central America, respectively [6]. Within these gene pools are a total of six races, including three Mesoamerican (Mesoamerica, Durango, and Jalisco) and three Andean (Peru, Nueva Granada, and Chile) [7, 8]. An additional Mesoamerican race has been designated Guatemala, which includes certain climbing beans from Central America [9].

After domestication, the common bean spread across Mesoamerica and South America and, after the European discovery of the Americas, to Europe and Africa, where it was cultivated in diverse environments and agricultural conditions [10]. As much as 60% of bean production in the developing world occurs under conditions of significant drought stress [11]. This includes large areas in Mexico and Africa where the growing season is short and the rainfall unreliable; regions of Central America where beans are planted after maize and may be subjected to the abrupt cessation of the rains; areas of Brazil where overall rainfall may be adequate but the growing period is interrupted by significant periods without precipitation. In the highlands of Mexico, beans are subjected to extended periods of intermittent drought. The only traits that have proven to be valuable in tolerating both terminal (end-of-season) and intermittent drought are earliness and partitioning toward reproductive structures, resulting in a greater harvest index [12, 13]. Bean breeders in Mexico have developed bean cultivars with indeterminate prostrate growth habits similar to pinto bean landraces in the semiarid highlands [14]. Cultivars such as Pinto Villa use phenotypic plasticity to respond to intermittent drought [15]. Interracial and intergene-pool crosses have been made in Mexico to combine different drought tolerance traits [16].

In lowland environments, terminal drought stress can be aggravated by high temperatures [11]. In Central America and the Caribbean, breeders have focused on heat as a constraint to expanding bean production in the lowland tropics [11, 17]. They have made significant progress in developing bean cultivars with improved levels of heat tolerance [18, 19]. In the subtropical island of Okinawa, Japan, vegetable production in the summer season is very difficult due to high temperatures and intense solar radiation, along with associated effects such as drought and infestation by insects and other pests [20]. High temperature in the summer is causing drastic reductions in common bean yield [21–24]. The heat-tolerant cultivar Haibushi was developed by the Okinawa Subtropical Station (now the Tropical Agriculture Research Front), JIRCAS, Okinawa, by screening the germplasm collected from Southeast Asian countries [25].

Development during the reproductive growth stage in the common bean is sensitive to temperature. High temperatures during this stage result in a reduction in pod and seed set due to enhanced abscission of flower buds, flowers, and pods [26–28]. Pollen-stigma interaction, pollen germination, pollen tube growth, and fertilization are all negatively affected by high temperature [29–32], with the lowest pod set observed in plants exposed to high temperature 1–6 days prior to anthesis [11]. Exposure to 35/20°C or 35°C reduced pollen viability (evaluated by pollen staining) [31]. Lower pod and seed set caused by high temperature at anthesis (32/21°C [29] and 35/20°C [28], resp.) were related to pollen injury, as assessed by pollen stainability and reciprocal pollinations. Continued exposure to 35/20°C did not affect embryo sac structure, but fertilization failed and it degenerated after anthesis [33]. Lower pod and seed set after the exposure of common bean plants to high temperature (32/27°C) are the combined result of both lower pollen viability (evaluated by pod and seed set resulting from reciprocal hand pollination) and impaired female performance in a large proportion of the flowers [32]. High temperature (33/25°C) affects the endoplasmic reticulum structure and blocks its function in the tapetum and then induces earlier-than-usual degeneration of the tapetum. Pollen sterility is associated with tapetal degeneration [34]. Weaver et al. [35] reported a close relationship between pollen stainability and tolerance to high-temperature stress among bean selections. Pollen staining by acetocarmine has been used widely for the rapid determination of pollen sterility occurring under

environmental stresses [36, 37]. A highly positive correlation was observed between pod set and pollen stainability in flowers that were affected by high temperature (32/28°C for 24 h) 8 to 11 days before anthesis [23], which corresponds to the early microspore stage in the common bean [38].

It is recognized that high temperature affects many physiological processes, including photosynthesis and the translocation of photosynthetic production, across a wide range of crops [39–41]. For example, in studies on birch trees, river birch was found to maintain the high net photosynthetic rates (P_n) at high temperature, ranging from 25 to 40°C, while the P_n of paper birch was reduced the most. Inhibition of P_n at higher temperatures was due largely to nonstomatal limitations in both taxa [41]. At high temperature (40°C), Norchip, the most heat-tolerant cultivar of potato, synthesized small heat shock proteins for a longer time period than the other cultivars. The levels of an 18 kDa small heat shock protein increased up to 24 h in Norchip and Desiree, which are heat-tolerant cultivars, whereas the levels started to decrease after 4 h in Russet Burbank and after 12 h in Atlantic, which are heat-sensitive cultivars [39]. Suzuki et al. [42] examined the effect of succinic acid 2,2-dimethylhydrazide (SADH) on the drought tolerance of bean plants. In SADH-applied plants, leaf water potential below which photosynthetic rate decreased was lower than that in control plants. Phenological adjustment and shoot biomass distribution on seed yield of drought-stressed common bean were assessed in two locations in Mexico [43]. Days to flowering and days to physiological maturity showed a negative and significant relationship with seed yield. Under drought stress, a significant reduction in the harvest index was observed in susceptible cultivars. Genotypic variation was detected in all partitioning indices, chiefly harvest index and relative sink strength by drought stress [44]. The crop faces water deficit due to excessive transpiration caused by high temperature (31/27°C) [45]. Even short diurnal fluctuations in the plant's water status [46] at the time of anthesis could adversely affect the development and function of its reproductive organs [24].

Phenological adjustment, plant-water relations, photosynthetic parameters, and shoot growth are all related to reproductive responses and thus may play an important role in heat and drought tolerance in the common bean. In this paper, we reviewed the results of our own studies on the above factors, but focused on photosynthesis in relation to leaf water status, genotypic differences in water status in relation to reproductive responses, genotypic differences in drought tolerance in relation to vegetative growth, and the seasonal performance of cultivars to elucidate the way in which heat tolerance and water deficit are related to reproductive responses in the common bean.

2. Photosynthesis and Leaf Water Status

Under field conditions in a hot summer season, the heat-tolerant cultivars differ markedly in leaf water status, leaf conductance, and intercellular CO_2 concentration, while there are no consistent differences in photosynthesis and transpiration rates, which vary within a narrow range [47]. This indicates that the effect of high temperature on the biochemical factors controlling intercellular CO_2 assimilation is similar in all the cultivars. The midday leaf water potential decreases with increasing air temperature, but the decline is greater in heat-tolerant cultivar Haibushi and strain Ishigaki-2 than in the remaining cultivars/strains. A steeper water potential gradient from soil to plant may enhance the ability of plants to absorb water at a faster rate [48]. This would reduce the development of severe internal water deficit in the reproductive organs and increase their survival and growth. Sinclair and Ludlow [49] support our assumption that photosynthesis, protein synthesis, NO_3 reduction, and leaf senescence are better correlated with changes in tissue water content than with leaf water potential. It is worth noting that the heat-tolerant cultivar Haibushi and strain Ishigaki-2 display an association between (a) photosynthesis and leaf conductance and (b) leaf water potential, while this is absent in the heat-sensitive cultivars [47]. This indicates that the heat-tolerant cultivars possess better stomatal control over CO_2 and H_2O exchange in leaves in response to high temperature. This is evidenced by the fact that the sensitive cultivar Kentucky Wonder and strain 92783 show greater water loss [50].

3. Genotypic Differences in Water Status in relation to Reproductive Responses

Haibushi, a heat-tolerant cultivar, displays better leaf water status than Kentucky Wonder, a heat-sensitive cultivar, which exhausted soil water quickly, resulting in a greater deterioration in water status [51]. The reduction in leaf water content with water potential occurred faster with the increase in high temperature and is larger in the heat-sensitive than in the heat-tolerant cultivar [52]. Under field conditions, strains 86884 and 92783, collected from Southeast Asia countries [25] and cultivar Kentucky Wonder failed to show any relationship between leaf water potential and water content and produced very few pods despite the higher pollen fertility. In contrast, in strains 45817, Ishigaki-2, and 3028520 and cultivars Kurodane Kinugasa and Haibushi, relatively higher leaf water content was maintained with declining water potential and a larger number of pods were set [50]. Osmotic adjustment and cell wall elasticity enable the plants to maintain higher water content, turgor, and other turgor-related processes during water deficit [53, 54]. This allows plant organs to survive longer in tolerant than in sensitive types. The cultivars with a smaller midday drop in leaf water content showed a higher pod-setting ratio and consequently had higher yield than the plants with a larger midday drop in leaf water content [55].

4. Genotypic Differences in Drought Tolerance in relation to Vegetative Growth

The common bean cultivars display distinct responses to prolonged drought stress under field conditions. The responses of photosynthetic parameters and shoot extension to leaf water status are related to soil water content. A decrease in soil water causes a decline in leaf water status. The high-yielding cultivars display a smaller reduction in leaf water content but a larger reduction in leaf water potential than the poor yielders. Such differences in leaf water content and leaf water potential may arise due to differences in osmotic adjustment [48, 56, 57] and cell wall elasticity [53]. Coyne et al. [58] argue that a steeper leaf water potential gradient from soil to plant may enhance the ability of the plants to extract soil water at low soil water content. The reduction in leaf water potential due to water stress is linearly correlated with reductions in shoot extension rate and leaf water content.

A discriminant analysis revealed that the five cultivars display two distinct types of responses [59]. One group includes cultivars Haibushi, Kurodane-Kinugasa, and strain Ishigaki-2, which showed a large reduction of about 16–20% in both shoot extension and water potential, and they also produced a higher number of pods per plant and seed yield than cultivar Kentucky Wonder and strain 92783. Kentucky Wonder and 92783, which form a separate group, displayed a comparatively smaller reduction (4–8%) in both water potential and shoot growth. In contrast, the former group displayed a smaller reduction in leaf water content, while the latter group showed a larger reduction in leaf water content. This suggests that tissue water content is kept high by restricting excessive vegetative growth and a large reduction in water potential. The reduction in shoot growth due to stress contributes to a build-up of water-economizing traits, such as specific leaf weight and succulence index.

5. Seasonal Performance of Cultivars

The performance of common bean cultivars Haibushi, Kentucky Wonder, and Kurodane Kinugasa and the strains Ishigaki-2, 45817, 92783, 86884, and 3028520 was evaluated between 2003 and 2005 in many field and controlled-environment experiments during the winter and summer seasons. Across the seasons, days to pod formation was positively associated with the number of pods per plant, seeds per pod, seed weight, and yield ($r > 0.97$). On the contrary, among the cultivars/strains, shorter duration to podding or flowering resulted in a higher number of pods per plant ($r = 0.93$) and number of seeds per pod ($r = 0.82$). Haibushi and Ishigaki-2 consistently produced a higher number of pods per plant and seed yield across the seasons and environments than the remaining cultivars. The number of pods per plant is the most important yield attribute and is precisely determined by thermal units and the duration between emergence and flowering. Porfirio and James [44] report that a high partitioning index (chiefly harvest index) shows high heritability, contributing to drought stress in the common bean. Thus, we can evaluate this character as genetic variation for adaptation to high temperature and drought.

6. Morphological Characters and Partitioning for Adaptation to High Temperature

The partitioning of dry matter (the ratio of dry weight of individual parts to that of total dry matter) was analyzed in the common bean at four temperature regimes (24/20, 27/23, 30/26, and 33/29°C) [60, 61]. Haibushi, a heat-tolerant cultivar, has a higher pod weight per plant, number of pods per plant, average pod weight, pod set ratio, number of branches, and rate of biomass allocation to pods, but lower rates of biomass allocation to leaves, stems, and roots, than Kentucky Wonder, a heat-sensitive cultivar, across all temperature regimes [61]. A sharp decline in dry matter partitioning to pods is observed at 33/29°C [60]. In the temperature range of 24/20 to 30/26°C, Haibushi showed higher partitioning to pods than Kentucky Wonder, independent of temperature. On the contrary, Kentucky Wonder showed higher partitioning to pods at 27/23°C than at 24/20°C. These results show that higher biomass allocation to pods and higher pod set in branches, which vary with the cultivar and temperature, play an important role in achieving a higher harvest index in the heat-tolerant compared to the heat-sensitive cultivars. Konsens et al. [27] recognize that high night temperature promotes branching in the common bean. Drought stresses induce genotypic variation of shoot biomass accumulation, pod and seed number, and biomass partitioning index [43, 44].

7. Concluding Remarks

Our results reveal that leaf water content is involved in heat and drought tolerance in the common bean, but the supporting system for maintaining high water content is unclear. Leaf water content is better correlated with leaf vapor pressure deficit, internal CO_2 concentration, and leaf conductance than with water potential. Therefore, plant water status can be explained better in terms of leaf water content in the common bean. Evaluation of the association between (a) number of pods per plant and seed yield and (b) midday drop of leaf water content provides clear evidence that leaf water content is responsible for the genotypic variations in heat and drought tolerance. A small reduction in leaf water content is displayed by the tolerant cultivars, which show larger reductions in shoot extension and leaf water potential than the sensitive cultivars. Therefore, we can conclude that leaf water content is an important physiological trait for improved productivity and that it can be used as a screening tool for heat and drought tolerance in the common bean.

References

[1] A. Wahid, S. Gelani, M. Ashraf, and M. R. Foolad, "Heat tolerance in plants: an overview," *Environmental and Experimental Botany*, vol. 61, no. 3, pp. 199–223, 2007.

[2] H. S. Gentry, "Origin of the common bean, *Phaseolus vulgaris*," *Economic Botany*, vol. 23, no. 1, pp. 55–69, 1969.

[3] J. Šustar-Vozlič, M. Maras, B. Javornik, and V. Meglič, "Genetic diversity and origin of slovene common bean (*Phaseolus vulgaris* L.) germplasm as revealed by AFLP markers and phaseolin analysis," *Journal of the American Society for Horticultural Science*, vol. 131, no. 2, pp. 242–249, 2006.

[4] A. V. Schoonhoven and O. Voysest, *Common Beans: Research for Crop Improvement*, CIAT, Cali, Colombia, 1991.

[5] M. W. Blair, M. C. Giraldo, H. F. Buendía, E. Tovar, M. C. Duque, and S. E. Beebe, "Microsatellite marker diversity in common bean (*Phaseolus vulgaris* L.)," *Theoretical and Applied Genetics*, vol. 113, no. 1, pp. 100–109, 2006.

[6] P. Gepts and D. Debouck, "Origin, domestication, and evolution of the common bean (*Phaseolus vulgaris* L.)," in *Common Beans: Research for Crop Improvement*, A. V. Schoonhoven, Ed., pp. 7–53, CIAT, Calif, USA, 1991.

[7] S. P. Singh, P. Gepts, and D. G. Debouck, "Races of common bean (*Phaseolus vulgaris*, Fabaceae)," *Economic Botany*, vol. 45, no. 3, pp. 379–396, 1991.

[8] S. P. Singh, R. Nodari, and P. Gepts, "Genetic diversity in cultivated common bean—I. Allozymes," *Crop Science*, vol. 31, pp. 19–23, 1991.

[9] S. Beebe, P. W. Skroch, J. Tohme, M. C. Duque, F. Pedraza, and J. Nienhuis, "Structure of genetic diversity among common bean landraces of Middle American origin based on correspondence analysis of RAPD," *Crop Science*, vol. 40, no. 1, pp. 264–273, 2000.

[10] M. I. Chacón S, B. Pickersgill, and D. G. Debouck, "Domestication patterns in common bean (*Phaseolus vulgaris* L.) and the origin of the Mesoamerican and Andean cultivated races," *Theoretical and Applied Genetics*, vol. 110, no. 3, pp. 432–444, 2005.

[11] P. H. Graham and P. Ranalli, "Common bean (*Phaseolus vulgaris* L.)," *Field Crops Research*, vol. 53, no. 1–3, pp. 131–146, 1997.

[12] J. A. Acosta-Gallegos and M. W. Adams, "Plant traits and yield stability of dry bean (*Phaseolus vulgaris* L.) cultivars under drought stress," *Journal of Agricultural Science*, vol. 117, pp. 213–219, 1991.

[13] E. F. Foster, A. Pajarito, and J. Acosta-Gallegos, "Moisture stress impact on N partitioning, N remobilization and N-use efficiency in beans (*Phaseolus vulgaris* L.)," *Journal of Agricultural Science*, vol. 124, no. 1, pp. 27–37, 1995.

[14] J. A. Acosta Gallegos and J. Kohashi Shibata, "Effect of water stress on growth and yield of indeterminate dry-bean (*Phaseolus vulgaris*) cultivars," *Field Crops Research*, vol. 20, no. 2, pp. 81–93, 1989.

[15] J. A. Acosta-Gallegos, R. Ochoa-Marquez, M. P. Arrieta-Montiel et al., "Registration of Pinto Villa common bean," *Crop Science*, vol. 35, p. 1211, 1989.

[16] K. A. Schneider, R. Rosales-Serna, F. Ibarra-Perez et al., "Improving common bean performance under drought stress," *Crop Science*, vol. 37, no. 1, pp. 43–50, 1997.

[17] J. S. Beaver, J. C. Rosas, J. Myers et al., "Contributions of the Bean/Cowpea CRSP to cultivar and germplasm development in common bean," *Field Crops Research*, vol. 82, no. 2-3, pp. 87–102, 2003.

[18] J. C. Rosas, A. Castro, J. S. Beaver, C. A. Perez, A. Morales, and R. Lepiz, "Mejoramiento genetico para tolerancia a altas temperaturas y Resistencia a mosaico dorado en frijol comun," *Agronomia Mesoamericana*, vol. 11, no. 1, pp. 1–10, 2000.

[19] J. C. Rosas, J. C. Hernandez, and J. A. Castro, "Registration of "Bribri" small red bean(race Mesoamerica)," *Crop Science*, vol. 43, pp. 430–431, 2003.

[20] A. Kumar, H. Omae, Y. Egawa, K. Kashiwaba, and M. Shono, "Adaptation to heat and drought stresses in snap bean (*Phaseolus vulgaris*) during the reproductive stage of development,"

Japan Agricultural Research Quarterly, vol. 40, no. 3, pp. 213–216, 2006.

[21] H. Nakano, M. Kobayashi, and T. Terauchi, "Sensitive stages to heat stress in pod setting of common bean (*Phaseolus vulgaris* L.)," *Japanese Journal of Tropical Agriculture*, vol. 42, pp. 78–84, 1998.

[22] H. Nakano, M. Kobayashi, and T. Terauchi, "Heat acclimation and de-acclimation for pod setting in heat-tolerant varieties of common bean (*Phaseolus vulgaris* L.)," *Japanese Journal of Tropical Agriculture*, vol. 44, pp. 123–129, 2000.

[23] K. Suzuki, T. Tsukaguchi, H. Takeda, and Y. Egawa, "Decrease of pollen stainability of green bean at high temperatures and relationship to heat tolerance," *Journal of the American Society for Horticultural Science*, vol. 126, no. 5, pp. 571–574, 2001.

[24] T. Tsukaguchi, Y. Kawamitsu, H. Takeda, K. Suzuki, and Y. Egawa, "Water status of flower buds and leaves as affected by high temperature in heat-tolerant and heat-sensitive cultivars of snap bean (*Phaseolus vulgaris* L.)," *Plant Production Science*, vol. 6, no. 1, pp. 24–27, 2003.

[25] H. Nakano, T. Momonoki, T. Miyashige et al., "Haibushi, a new variety of snap bean tolerant to heat stress," *JIRCAS Journal*, vol. 5, pp. 1–12, 1997.

[26] F. E. Ahmed, A. E. Hall, and D. A. Demason, "Heat injury during floral development in cowpea (*Vigna unguiculata*, Favaceae)," *American Journal of Botany*, vol. 79, pp. 784–791, 1992.

[27] I. Konsens, M. Ofir, and J. Kigel, "The effect of temperature on the production and abscission of flowers and pods in snap bean (*Phaseolus vulgaris* L.)," *Annals of Botany*, vol. 67, no. 5, pp. 391–399, 1991.

[28] V. A. Monterroso and H. C. Wien, "Flower and pod abscission due to heat stress in beans," *Journal of American society for Horticultural Science*, vol. 115, pp. 631–634, 1990.

[29] M. L. Weaver and H. Timm, "Influence of temperature and plant water status on pollen fertility in beans," *Journal of American Society for Horticultural Science*, vol. 113, pp. 31–35, 1988.

[30] J. H. Anthony, D. C. Carl, and D. T. Iwan, "Influence of high temperature on pollen grain viability and pollen tube growth in the styles of *Phaseolus vulgaris* L," *Journal of American Society for Horticultural Science*, vol. 105, pp. 12–14, 1980.

[31] A. J. Halterlein, C. D. Clayberg, and I. D. Teare, "Influence of high temperature on pollen grain viability and pollen tube growth in the styles of *Phaseolus vulgaris* L," *American Society for Horticultural Science*, vol. 105, pp. 12–14, 1980.

[32] Y. Gross and J. Kigel, "Differential sensitivity to high temperature of stages in the reproductive development of common bean (*Phaseolus vulgaris* L.)," *Field Crops Research*, vol. 36, no. 3, pp. 201–212, 1994.

[33] D. P. Ormrod, C. J. Wooley, G. W. Eaton, and E. H. Stobbe, "Effect of temperature on embryo sac development in *Phaseolus vulgaris* L," *Canadian Journal of Botany*, vol. 44, pp. 948–950, 1967.

[34] K. Suzuki, H. Takeda, T. Tsukaguchi, and Y. Egawa, "Ultrastructural study on degeneration of tapetum in anther of snap bean (*Phaseolus vulgaris* L.) under heat stress," *Sexual Plant Reproduction*, vol. 13, no. 6, pp. 293–299, 2000.

[35] M. L. Weaver, H. Timm, M. J. Silbernagel, and D. W. Burke, "Pollen staining and high-temperature tolerance of bean," *Journal of American Society for Horticultural Science*, vol. 110, pp. 797–799, 1985.

[36] G. E. Marks, "An aceto-carmine glycerol jelly for use in pollen-fertility counts," *Stain Technology*, vol. 29, no. 5, p. 277, 1954.

[37] M. Takagaki, M. Kakinuma, S. You, and T. Ito, "Effect of temperature on pollen fertility and pollen germination of three

[38] H. Watanabe, "Studies on the unfruitfulness of the beans—II. Influence of the temperature on the flower bud differentiation and blooming," *Journal of Japanese Journal for Horticultural Science* , vol. 22, pp. 36–42, 1953 (Japanese).

[39] Y. J. Ahn, K. Claussen, and J. L. Zimmerman, "Genotypic differences in the heat-shock response and thermotolerance in four potato cultivars," *Plant Science*, vol. 166, no. 4, pp. 901–911, 2004.

[40] A. Bar-Tsur, J. Rudich, and B. Bravdo, "High temperature effects on CO_2 gas exchange in heat-tolerant and sensitive tomatoes," *Journal of American Society for Horticultural Science*, vol. 110, pp. 582–586, 1985.

[41] T. G. Ranney and M. M. Peet, "Heat tolerance of five taxa of birch (Betula): physiological responses to supraoptimal leaf temperatures," *Journal of the American Society for Horticultural Science*, vol. 119, no. 2, pp. 243–248, 1994.

[42] S. Suzuki, K. Yamada, and T. Takano, "Effect of succinic acid 2, 2-dimethylhydrazide on drought tolerance of bean plant," *Science Report*, vol. 23, pp. 15–22, 1987.

[43] R. Rosales-Serna, J. Kohashi-Shibata, J. A. Acosta-Gallegos, C. Trejo-López, J. Ortiz-Cereceres, and J. D. Kelly, "Biomass distribution, maturity acceleration and yield in drought-stressed common bean cultivars," *Field Crops Research*, vol. 85, no. 2-3, pp. 203–211, 2004.

[44] R. V. Porfirio and D. K. James, "Traits related to drought resistance in common bean," *Euphytica*, vol. 99, no. 2, pp. 127–136, 1998.

[45] H. Omae, A. Kumar, Y. Egawa, K. Kashiwaba, and M. Shono, "Water consumption in different heat tolerant cultivars of snap bean (*Phaseolus vulgaris* L.)," in *Proceedings of the 4th International Crop Science Congress*, vol. 1.3.4, 2004.

[46] T. Tsukaguchi, H. Fukamachi, K. Ozawa, H. Takeda, K. Suzuki, and Y. Egawa, "Diurnal change in water balance of heat-tolerant snap bean (*Phaseolus vulgaris*) cultivar and its association with growth under high temperature," *Plant Production Science*, vol. 8, no. 4, pp. 375–382, 2005.

[47] A. Kumar, H. Omae, Y. Egawa, K. Kashiwaba, and M. Shono, "Some physiological responses of snap bean (*Phseolus vulgaris* L.) to water stress during reproductive period," in *Proceedings of the International Conference on Sustainable Crop Production in Stress Environment: Management and Genetic Option*, pp. 226–227, JNKVV, Jabarpur, India, 2005.

[48] A. Kumar, P. Singh, D. P. Singh, H. Singh, and H. C. Sharma, "Differences in osmoregulation in *Brassica* species," *Annals of Botany*, vol. 54, no. 4, pp. 537–542, 1984.

[49] T. R. Sinclair and M. M. Ludlow, "Who taught plants thermodynamics? The unfulfilled potential of plant water potential," *Australian Journal of Plant Physiology*, vol. 12, pp. 213–217, 1985.

[50] H. Omae, A. Kumar, K. Kashiwaba, and M. Shono, "Genotypic differences in plant water status and relationship with reproductive responses in snap bean (*Phaseolus vulgaris* L.) during water stress," *Japanese Journal of Tropical Agriculture*, vol. 49, pp. 1–7, 2005.

[51] A. Kumar, H. Omae, Y. Egawa, K. Kashiwaba, and M. Shono, "Influence of water and high temperature stresses on leaf water status of high temperature tolerant and sensitive cultivars of snap bean (*Phaseolus vulgaris* L.)," *Japanese Journal of Tropical Agriculture*, vol. 49, pp. 109–118, 2005.

[52] H. Omae, A. Kumar, K. Kashiwaba, and M. Shono, "Influence of level and duration of high temperature treatments on plant

pepper (*Capsicum annunum* L.) varieties," *Japanese Journal of Tropical Agriculture*, vol. 39, pp. 247–249, 1995 (Japanese).

water status in snap bean (*Phaseolus vulgaris* L.)," *Japanese Journal of Tropical Agriculture*, vol. 49, pp. 238–242, 2005.

[53] A. Kumar and J. Elston, "Genotypic differences in leaf water relations between *Brassica juncea* and B. napus," *Annals of Botany*, vol. 70, no. 1, pp. 3–9, 1992.

[54] A. Kumar and D. P. Singh, "Use of physiological indices as a screening technique for drought tolerance in oilseed *Brassica* species," *Annals of Botany*, vol. 81, no. 3, pp. 413–420, 1998.

[55] H. Omae, A. Kumar, Y. Egawa, K. Kashiwaba, and M. Shono, "Midday drop of leaf water content related to drought tolerance in snap bean (*Phaseolus vulgaris* L.)," *Plant Production Science*, vol. 8, no. 4, pp. 465–467, 2005.

[56] J. M. Morgan, R. A. Hare, and R. J. Fletcher, "Genetic variation in osmoregulation in bread and durum wheats and its relationship to grain yield in a range of field environments," *Australian Journal of Agricultural Research*, vol. 37, pp. 449–457, 1986.

[57] S. W. Ritchie, H. T. Nguyen, and A. S. Holaday, "Leaf water content and gas-exchange parameters of two wheat genotypes differing in drought resistance," *Crop Science*, vol. 30, pp. 105–111, 1990.

[58] P. I. Coyne, J. A. Bradford, and C. I. Dewald, "Leaf water relations and gas exchange in relation to forage production in four Asiatic blue stems," *Crop Science*, vol. 22, pp. 1036–1040, 1982.

[59] H. Omae, A. Kumar, K. Kashiwaba, and M. Shono, "Assessing drought tolerance of snap bean (*Phaseolus vulgaris*) from genotypic differences in leaf water relations, shoot growth and photosynthetic parameters," *Plant Production Science*, vol. 10, no. 1, pp. 28–35, 2007.

[60] H. Omae, A. Kumar, K. Kashiwaba, and M. Shono, "Influence of temperature shift after flowering on dry matter partitioning in two cultivars of snap bean (*Phaseolus vulgaris*) that differ in heat tolerance," *Plant Production Science*, vol. 10, no. 1, pp. 14–19, 2007.

[61] H. Omae, A. Kumar, K. Kashiwaba, and M. Shono, "Influence of high temperature on morphological characters, biomass allocation, and yield components in snap bean (*Phaseolus vulgaris* L.)," *Plant Production Science*, vol. 9, no. 3, pp. 200–205, 2006.

Permissions

The contributors of this book come from diverse backgrounds, making this book a truly international effort. This book will bring forth new frontiers with its revolutionizing research information and detailed analysis of the nascent developments around the world.

We would like to thank all the contributing authors for lending their expertise to make the book truly unique. They have played a crucial role in the development of this book. Without their invaluable contributions this book wouldn't have been possible. They have made vital efforts to compile up to date information on the varied aspects of this subject to make this book a valuable addition to the collection of many professionals and students.

This book was conceptualized with the vision of imparting up-to-date information and advanced data in this field. To ensure the same, a matchless editorial board was set up. Every individual on the board went through rigorous rounds of assessment to prove their worth. After which they invested a large part of their time researching and compiling the most relevant data for our readers. Conferences and sessions were held from time to time between the editorial board and the contributing authors to present the data in the most comprehensible form. The editorial team has worked tirelessly to provide valuable and valid information to help people across the globe.

Every chapter published in this book has been scrutinized by our experts. Their significance has been extensively debated. The topics covered herein carry significant findings which will fuel the growth of the discipline. They may even be implemented as practical applications or may be referred to as a beginning point for another development. Chapters in this book were first published by Hindawi Publishing Corporation; hereby published with permission under the Creative Commons Attribution License or equivalent.

The editorial board has been involved in producing this book since its inception. They have spent rigorous hours researching and exploring the diverse topics which have resulted in the successful publishing of this book. They have passed on their knowledge of decades through this book. To expedite this challenging task, the publisher supported the team at every step. A small team of assistant editors was also appointed to further simplify the editing procedure and attain best results for the readers.

Our editorial team has been hand-picked from every corner of the world. Their multi-ethnicity adds dynamic inputs to the discussions which result in innovative outcomes. These outcomes are then further discussed with the researchers and contributors who give their valuable feedback and opinion regarding the same. The feedback is then collaborated with the researches and they are edited in a comprehensive manner to aid the understanding of the subject.

Apart from the editorial board, the designing team has also invested a significant amount of their time in understanding the subject and creating the most relevant covers. They scrutinized every image to scout for the most suitable representation of the subject and create an appropriate cover for the book.

The publishing team has been involved in this book since its early stages. They were actively engaged in every process, be it collecting the data, connecting with the contributors or procuring relevant information. The team has been an ardent support to the editorial, designing and production team. Their endless efforts to recruit the best for this project, has resulted in the accomplishment of this book. They are a veteran in the field of academics and their pool of knowledge is as vast as their experience in printing. Their expertise and guidance has proved useful at every step. Their uncompromising quality standards have made this book an exceptional effort. Their encouragement from time to time has been an inspiration for everyone.

The publisher and the editorial board hope that this book will prove to be a valuable piece of knowledge for researchers, students, practitioners and scholars across the globe.

List of Contributors

Nava-Gutiérrez Yolanda
Laboratorio de Fisiolog´ıa Vegetal Molecular, Centro de Investigaci´on Cient´ıfica de Yucat´an, Calle 43 No. 130, Colonia Chuburn´a de Hidalgo, M´erida, YUC, M´exico 97200, Mexico
Laboratorio de Micorrizas, Centro de Investigaci´on en Ciencias Biol´ogicas (CICB), Universidad Aut´onoma de Tlaxcala, Ixtacuixtla, TLAX, M´exico 90000, Mexico

Jorge M. Santamar1a
Laboratorio de Fisiolog´ıa Vegetal Molecular, Centro de Investigaci´on Cient´ıfica de Yucat´an, Calle 43 No. 130, Colonia Chuburn´a de Hidalgo, M´erida, YUC, M´exico 97200, Mexico

Ronald Ferrera-Cerrato
Area de Microbiolog´ıa, Especialidad de Edafolog´ıa, Instituto de Recursos Naturales, Colegio de Postgraduados Campus, Montecillo, MEX, M´exico 56230, Mexico

Parissa Taheri and Saeed Tarighi
Department of Crop Protection, Faculty of Agriculture, Ferdowsi University of Mashhad, P.O. Box 1163, Mashhad 9177948978, Iran

Alla B. Kholina and Nina M. Voronkova
Institute of Biology and Soil Science, Far Eastern Branch of Russian Academy of Sciences, 159 Prospect 100-letiya Vladivostoka, Vladivostok 690022, Russia

Pallavi Sharma and Rama Shanker Dubey
Department of Biochemistry, Faculty of Science, Banaras Hindu University, Varanasi 221005, India

Ambuj Bhushan Jha
Crop Development Centre, Department of Plant Sciences, College of Agriculture and Bioresources, University of Saskatchewan, 51 Campus Drive, Saskatoon SK, Canada SK S7N 5A8

Mohammad Pessarakli
School of Plant Sciences, The University of Arizona, Forbes Building, Room 303, P.O. Box 210036, Tucson, AZ 85721-0036, USA

Sonia Silva
CESAM and Department of Biology, University of Aveiro, 3810-193 Aveiro, Portugal

J. Gordon Burleigh and W. Brad Barbazuk
Department of Biology, University of Florida, Gainesville, FL 32611, USA

John M. Davis and Alison M. Morse
School of Forest Resources and Conservation, University of Florida, Gainesville, FL 32611, USA

Pamela S. Soltis
Florida Museum of Natural History, University of Florida, Gainesville, FL 32611, USA

Bruno Ladeiro
MAP-Bio Plant, Porto, Portugal

Sonia Pinho
Biology Department, University of Aveiro, 3810-193 Aveiro, Portugal

Bruno Ladeiro
MAP-Bio Plant, Biology Department, Faculty of Science, University of Porto, 4169-007 Porto, Portugal

Dhananjay K. Pandey, Arvind K. Singh and Bhupendra Chaudhary
School of Biotechnology, Gautam Buddha University, Greater Noida 201 308, India

Helena Oliveira
Department of Biology, CESAM, University of Aveiro, 3810-193 Aveiro, Portugal

D. K. Letourneau and J. A. Hagen
Department of Environmental Studies, University of California, Santa Cruz, CA 95064, USA

Magdy Hussein Abd El-Twab
Department of Botany and Microbiology, Faculty of Science, Minia University, El-Minia 61519, Egypt

Katsuhiko Kondo
Laboratory of Plant Genetics and Breeding Research, Department of Agriculture, Faculty of Agriculture, Tokyo University of Agriculture, Funako, Kanagawa, Atsugi 1737, Japan

Renuka Diwan, Amit Shinde and Nutan Malpathak
Department of Botany, University of Pune, Pune Maharashtra 411007, India

Stephen Blackmore Frieda Christie, Fiona Inches, Neil Watherston and Alexandra H. Wortley
Royal Botanic Garden Edinburgh, 20a Inverleith Row, Edinburgh EH3 5LR, UK

See-Chung Chin
Singapore Botanic Gardens, 1 Cluny Road, Singapore 259569, Singapore

Lindsay Chong Seng
Plant Coversation Action Group, P.O. Box 392, Victoria, Mah´e, Seychelles

Putri Winda Utami
Center for Plant Conservation-Bogor Botanical Gardens, Kebun Raya Indonesia, Indonesian Institute of Sciences, Jl. Ir. H. Juanda No. 13 P.O. BOX 309, Bogor 16003, Indonesia

Andreas Kempe, Martin Sommer and Christoph Neinhuis
Department of Biology, Institute of Botany, Faculty of Science, Technische Universit¨at Dresden, 01062 Dresden, Germany

Jie He, Hazelman Norhafis and Lin Qin
Natural Sciences and Science Education Academic Group, National Institute of Education, Nanyang Technological University, 1 NanyangWalk, Singapore 637 616

Hong-Mei Liu
Key Laboratory of Southern Subtropical Plant Diversity, Fairylake Botanical Garden, Shenzhen & Chinese Academy of Sciences, Shenzhen 518004, China

Robert J. Dyer
Department of Life Sciences, Natural History Museum, London SW7 5BD, UK
Research Group in Biodiversity Genomics and Environmental Sciences, Imperial College London, Silwood Park Campus, Ascot, Berkshire SL5 7PY, UK

Zhi-You Guo
Department of Biological Sciences, Qiannan Normal College for Nationalities, Duyun 558000, China

ZhenMeng and Jian-Hui Li
Computer Network Information Center, Chinese Academy of Sciences, Beijing 100190, China

Harald Schneider
Department of Life Sciences, Natural History Museum, London SW7 5BD, UK
State Key Laboratory of Systematic and Evolutionary Botany, Institute of Botany, Chinese Academy of Sciences, Beijing 100093, China

Jarosław Tyburski, Kamila Dunajska-Ordak, Monika Skorupa and Andrzej Tretyn
Chair of Plant Physiology and Biotechnology, Nicolaus Copernicus University, Gagarina 9, 87-100 Toru´n, Poland

Nicolás Nagahama
Instituto Multidisciplinario de Biolog´ıa Vegetal (IMBIV-CONICET), C.C. 495, 5000 C´ordoba, Argentina

Guillermo A. Norrmann
Facultad de Ciencias Agrarias (FCA), UNNE y Instituto de Bot´anica del Nordeste (IBONE-CONICET), 3400 Corrientes, Argentina

Takahiro Tezuka, Hisa Yokoyama, Hideyuki Tanaka, Shuji Shiozak and Masayuki Oda
Graduate School of Life and Environmental Sciences, Osaka Prefecture University, 1-1 Gakuen-cho, Nakaku, Sakai, Osaka 599-8531, Japan

Gurveen Kaur, Bhupinder Pal Singh and Avinash Kaur Nagpal
Department of Botanical and Environmental Sciences, Guru Nanak Dev University, Amritsar 143005, India

Hide Omae and Mariko Shono
Tropical Agriculture Research Front, Japan International Research Center for Agricultural Sciences (JIRCAS), 1091-1, Maezato, Ishigaki, Okinawa 907-0002, Japan

Ashok Kumar
Department of Agronomy, CCS Haryana Agricultural University, Hisar 125004, India